An Introduction to Biomechanics

Solids and Fluids, Analysis and Design

Springer
New York
Berlin
Heidelberg
Hong Kong
London
Milan
Paris
Tokyo

Jay D. Humphrey   Sherry L. Delange

An Introduction to Biomechanics

Solids and Fluids, Analysis and Design

With 303 Illustrations

Springer

Jay D. Humphrey and Sherry L. Delange
Texas A&M University
Department of Biomedical Engineering
and M.E. DeBakey Institute
233 Zachry Engineering Center
College Station, TX 77843-3120 USA
jhumphrey@tamu.edu

Library of Congress Cataloging-in-Publication Data
Humphrey, Jay Dowell, 1959–
    An introduction to biomechanics : solids and fluids, analysis and design /
Jay D. Humphrey, Sherry L. Delange.
      p.   cm.
    Includes bibliographical references and index.
    ISBN 0-387-40249-7 (hbk. : alk. paper)
    1. Biomechanics.   I. Delange, Sherry L.   II. Title.
QH513.H86 2003
610¢.28—dc21   2003050660

ISBN 0-387-40249-7   Printed on acid-free paper.

Printed in the United States of America.

9 8 7 6 5 4 3 2 1   SPIN 10934562

www.springer-ny.com

Springer-Verlag   New York Berlin Heidelberg
A member of BertelsmannSpringer Science+Business Media GmbH

◆

To my sisters,
Linda Hill and Bonnie Zinn,
for their love, support,
and example of what it means to serve others on a daily basis.
J.D. Humphrey

◆

To my loving husband Christopher,
parents, Stephen and Mary O'Rourke,
and sister, Shelly O'Rourke,
for giving meaning and purpose to my life.
S.L. Delange

◆

Preface

Biomechanics aims to explain the mechanics of life and living. From molecules to organisms, everything must obey the laws of mechanics.

—Y.C. Fung (1990)

It is purported that Leonardo da Vinci once said, "by means of this [mechanics], all animated bodies that have movement perform all their actions." Although this assertion is obviously overstated, it serves to remind us that scientists have long thought that mechanics plays an extremely important role in governing biological as well as physical actions. Indeed, perhaps one of the most exciting recent discoveries in cell biology is that of mechanotransduction. It is now known that many cell types express different genes (i.e., perform different functions) in response to even small changes in their mechanical environment. Because cells are the fundamental structural and functional units of all living things, the importance of mechanics in biology and medicine is thereby far reaching! One goal of this book is to serve as an introduction to a few of the many, many applications of biomechanics—one of the cornerstones of biomedical engineering. Before proceeding, however, a few words on the scope and philosophy of approach.

There are four general areas of mechanics: continuum, statistical, quantum, and relativistic. Each is important, but this text focuses on biomechanics from a continuum perspective, which we will see embraces many aspects of biomedical engineering at various length and time scales.

Introductory textbooks on mechanics sometimes give the wrong impression that the subject is primarily a collection of solutions to individual problems—nothing could be further from the truth. As a branch of classical physics, continuum mechanics is a deductive science founded upon a few basic postulates and concepts through which all problems must be formulated and then solved. Mechanics should be recognized, therefore, as a consistent, focused approach to the solution of classes of problems rather than as a collection of special results. Another goal of this textbook is to introduce the student to biosolid and biofluid mechanics such that it is the underlying, consistent approach that is learned and reinforced throughout. Indeed, the ultimate goal here is to enable the reader to formulate and solve real-life problems, many of which have yet to be identified. In other words, the primary goal of a student should not be to learn how to solve the specific problems (illustrative examples and exercises) in this text; we, as a community, already know their solution. Textbook problems should be used simply as a means to practice the underlying *approach of mechanics*, to gain confidence in formulating and solving problems, and to develop intuition.

Although this philosophy of learning the fundamentals is as old as mechanics itself, it has at no time in our history been more important. With continued advances in computer technology and engineering software, the biomechanicist will have increasingly remarkable experimental, computational, and design tools at his/her disposal to address the incredibly complex real-life problems of biomechanics. The only way to ensure that these tools are used well, rather than misused, is to understand the underlying general approach as well as the specific assumptions (with associated limitations) within a given formulation. For example, a finite element program should not be treated as a black box capable of finding any solution of interest; rather, it should be used cautiously as a tool only by one who understands how the program actually works. Toward this end, note one caveat. It has been appropriately said that undergraduates should be told the truth, nothing but the truth, but *not* the whole truth. Why not the whole truth? From a purist perspective, we do not know the whole truth, scientific knowledge being relative to current advances. From a practical perspective, however, continuum biomechanics has tremendous breadth and depth and it is impossible in an introductory course to scratch the surface of the whole truth. Therefore the interested student is strongly encouraged to pursue intermediate and advanced study in biomechanics, which will successively reveal more and more of the beauty and, indeed, the power of biomechanics. Biomechanics is a lifelong pursuit, one with many rewards.

Whereas graduate courses on biomechanics are often best taught using a problem-based paradigm, we suggest that that an introductory course on biomechanics should be taught using a traditional discipline-based paradigm; that is, graduate courses are often best taught by focusing on a particular tissue, organ, or system, or, alternatively, on a specific disease or treatment modality, and then by bringing to bear all tools (experimental, computational, theoretical, biological) that aid in the solution of that class of problems. An introduction to biomechanics should be different, however. To see the overall approach used in mechanics, it is best to introduce all of the general tools (e.g., concept of stress, strain, and equilibrium) and then to illustrate their use via multiple similar problems that build in complexity but continually reinforce the same approach. Hence, this book is divided according to approach (e.g., via chapters on beam theory and Navier–Stokes solutions), not according to areas of research such as cardiovascular, musculoskeletal, pulmonary, or cell mechanics. Therefore, we employ illustrative problems from various fields of study, often within single chapters.

As a first course in biomechanics, the primary prerequisites are the sequence of courses on calculus for engineers (including vectors, which are reviewed in Chapter 7) and some basic biology (reviewed in Chapter 1); many students will likely have had a course in engineering statics (briefly reviewed in Chapter 1), which will help but it is not necessary. Although a course on differential equations would also be helpful (briefly reviewed in Chapter 8), related methods needed herein will be reviewed at the appropriate time. Given the availability of personal computers and useful software packages, the student will be asked to obtain numerical solutions to many exercises. For example, additional solutions and exercises will be provided at http://biomed.tamu.edu to allow an interactive exploration of the effects of various parameters and boundary conditions on particular solutions. Again, however, the requisite mathematical tools will be discussed as needed.

I would like to close with a quote from the 1998 Bioengineering Consortium (BECON) Report of the National Institutes of Health:

The success of reductionist and molecular approaches in modern medical science has led to an explosion of information, but progress in integrating information has lagged . . . Mathematical models provide a rational approach for integrating this ocean of data, as well as providing deep insight into biological processes.

Biomechanics provides us with a means to model mathematically many biological behaviors and processes, thus biomechanics will continue to play a central role in both basic and applied research. The key, therefore, is to learn well the basic approaches.

J.D. Humphrey
College Station, TX
January 2003

Comments from a Student to a Student

Although one tends to teach the way he or she was taught, this textbook is designed to be different. For example, rather than introduce biomechanics through a sequence of increasingly more involved and detailed problems, with each illustrating new foundational concepts, we choose to introduce the basic concepts first and then to illustrate and reinforce the use of these concepts through the consideration of increasing more complex problems. In addition, rather than have two professors coauthor the book from their two perspectives, we chose to have a professor and student coauthor the book from their two perspectives—teacher and learner. Ms. Delange completed a sequence of three biomechanics courses at Texas A&M University (BMEN 302 Biosolid Mechanics, BMEN 421 Biofluid Mechanics, and BMEN 689 Cardiovascular Mechanics) offered by Professor Humphrey. The goal of this joint effort, therefore, is to present the material in a way that a professor feels is most beneficial and yet in a way that a student feels is most easily assimilated. Here, therefore, consider comments from a student to a student.

When taking my first course from Professor Humphrey "Biosolid Mechanics", he explained mechanics in such a way that made sense to me. He introduced the idea of a continuum and that classes of problems that fall within the realm of continuum mechanics, whether it be solid or fluid mechanics, are governed by the same fundamental relations. The backbone of mechanics was revealed and the basic/fundamental equations were derived from a single perspective. This introductory text introduces these basic concepts, which are essential to all problems in biomechanics. It presents a unified approach that helps the student to understand and learn the basic concepts and allows one to build upon these concepts to formulate and solve problems of increasing difficulty.

The concepts introduced in Part II of this text are reinforced in Part III, as we apply the same governing equations to different classes of problems, again deriving necessary equations as we move along. As a student, I found the course in "Biofluid Mechanics" to be easier than "Biosolid Mechanics" partly because I had already learned the general approach to formulating and solving problems. In these ways, I feel that this book reflects the positive aspects from my learning experiences at Texas A&M.

All in all, as a student, I appreciated having a unified problem-solving process presented to me and reinforced throughout each course as well as knowing from where the basic/fundamental equations were derived. I also appreciated reference to real-life problems for motivation, and derivations that skipped very few steps. This book reflects these ideas, upon which one can build.

Sherry L. Delange
College Station, TX
January 2003

Acknowledgments

We especially thank two Texas A&M University undergraduate students, Anne Price and Elizabeth Wang, who worked diligently for many months to help type much of the manuscript, compose the figures, and proofread the text. Some of the figures are reprinted from an earlier book, *Cardiovascular Solid Mechanics*, and were drawn by another undergraduate student, William Rogers. Current graduate students (G. David, C. Farley, R. Gleason, M. Heistand, J. Hu, S. Na, and P. Wells) proofread portions of the text, which is gratefully acknowledged, and Professor Larry A. Taber from Washington University read portions of the text and provided valuable input. JDH would also like to thank daughters Kaitlyn and Sarah, and wife Rita, for continued understanding and support. We also thank Dr. John H. Linehan and The Whitaker Foundation for providing partial financial support to complete this project through their Teaching Materials Program. The American Heart Association, Army Research Office, National Institutes of Health, National Science Foundation, Texas Advanced Research Program, and The Whitaker Foundation funded numerous research projects from which some of the material was taken. Finally, we thank Paula M. Callaghan, Senior Life Sciences Editor at Springer-Verlag New York, for wonderful encouragement, suggestions, and support throughout.

Contents

Preface vii
Comments from a Student to a Student xi
Acknowledgments xiii

Part I   Background 1

1. Introduction 3
   1.1 Point of Departure 3
   1.2 Health Care Applications 5
   1.3 What Is Continuum Mechanics? 8
   1.4 A Brief on Cell Biology 11
   1.5 The Extracellular Matrix 16
   1.6 Mechanotransduction in Cells 22
   1.7 General Method of Approach 24
   Appendix 1: Engineering Statics 26
   Exercises 39

Part II   Biosolid Mechanics 43

2. Stress, Strain, and Constitutive Relations 45
   2.1 Introduction 45
   2.2 Concept of Stress 46
   2.3 Stress Transformations 56
   2.4 Principal Stresses and Maximum Shear 62
   2.5 Concept of Strain 67
   2.6 Constitutive Behavior 81
   2.7 Mechanical Properties of Bone 94
   Appendix 2: Material Properties 96
   Exercises 97

3. Equilibrium, Universal Solutions, and Inflation   104
   3.1   General Equilibrium Equations   104
   3.2   Navier–Space Equilibrium Equations   108
   3.3   Axially Loaded Rods   110
   3.4   Pressurization and Extension of a Thin-Walled
        Tube   118
   3.5   Pressurization of a Thin Spherical Structure   130
   3.6   Thick-Walled Cylinders   135
   Appendix 3: First Moments of Area   142
   Exercises   147

4. Extension and Torsion   154
   4.1   Deformations due to Extension   155
   4.2   Shear Stress due to Torsion   169
   4.3   Principal Stresses and Strains in Torsion   176
   4.4   Angle of Twist due to Torque   181
   4.5   Experimental Design: Bone Properties   188
   4.6   Experimental Design: Papillary Muscles   189
   4.7   Inflation, Extension and Twist   193
   Appendix 4: Second Moments of Area   194
   Exercises   197

5. Beam Bending and Column Buckling   202
   5.1   Shear Forces and Bending Moments   203
   5.2   Stresses in Beams   211
   5.3   Deformation in Beams   227
   5.4   Transducer Design: The AFM   235
   5.5   Principle of Superposition   239
   5.6   Column Buckling   245
   Appendix 5: Parallel Axis Theorem and Composite
   Sections   254
   Exercises   260

6. Some Nonlinear Problems   271
   6.1   Kinematics   271
   6.2   Pseudoelastic Constitutive Relations   277
   6.3   Design of Biaxial Tests on Planar Membranes   287
   6.4   Stability of Elastomeric Balloons   293
   6.5   Residual Stress and Arteries   303
   6.6   A Role of Vascular Smooth Muscle   315
   Appendix 6: Matrices   321
   Exercises   324

Part III   Biofluid Mechanics   329

7. **Stress, Motion, and Constitutive Relations**   331
   7.1   Introduction   331
   7.2   Stress and Pressure   332
   7.3   Kinematics: The Study of Motion   333
   7.4   Constitutive Behavior   351
   7.5   Blood Characteristics   359
   7.6   Cone-and-Plate Viscometry   364
   Appendix 7: Vector Calculus Review   368
   Exercises   371

8. **Fundamental Balance Relations**   379
   8.1   Balance of Mass   380
   8.2   Balance of Linear Momentum   383
   8.3   Navier–Stokes Equations   386
   8.4   Euler Equation   392
   8.5   The Bernoulli Equation   396
   8.6   Measurement of Pressure and Flow   411
   8.7   Navier–Stokes Worksheets   415
   Appendix 8: Differential Equations   419
   Exercises   421

9. **Some Exact Solutions**   426
   9.1   Flow Between Parallel Flat Plates   427
   9.2   Steady Flow in Circular Tubes   444
   9.3   Circumferential Flow Between Concentric Cylinders . .   452
   9.4   Steady Flow in an Elliptical Cross Section   462
   9.5   Pulsatile Flow   465
   9.6   Non-Newtonian Flow in a Circular Tube   473
   Appendix 9: Biological Parameters   477
   Exercises   479

10. **Control Volume and Semi-empirical Methods**   491
   10.1   Fundamental Equations   491
   10.2   Control Volume Analyses in Rigid Conduits   500
   10.3   Control Volume Analyses in Deforming Containers . . .   509
   10.4   Murray's Law and Optimal Design   513
   10.5   Buckingham Pi and Experimental Design   518
   10.6   Pipe Flow   528
   10.7   Conclusion   542
   Appendix 10: Thermodynamics   543
   Exercises   545

Part IV   Closure 555

11.   Coupled Solid–Fluid Problems 557
    11.1   Vein Mechanobiology 557
    11.2   Diffusion Through a Membrane 559
    11.3   Dynamics of a Saccular Aneurysm 571
    11.4   Viscoelasticity: QLV and Beyond 582
    11.5   Lubrication of Articular Joints 597
    11.6   Thermomechanics, Electromechanics, and
          Chemomechanics 603
    Exercises 606

12.   Epilogue 612
    12.1   Future Needs in Biomechanics 612
    12.2   Need for Lifelong Learning 616
    12.3   Conclusion 618

References 619
Index 625
About the Authors 631

Part I
Background

1
Introduction

1.1 Point of Departure

Biology is the study of living things; mechanics is the study of motions and the applied loads that cause them. Biomechanics can be defined, therefore, as the study of the motions experienced by living things in response to applied loads. Herein, however, we consider that *biomechanics is the development, extension, and application of mechanics for the purposes of understanding better the influence of mechanical loads on the structure, properties, and function of living things*. Thus, the domain of biomechanics is very broad. It includes, among many other things, studying the effects of wind loads or gravity on the growth of plants, the mechanical properties of foodstuffs, the flight of birds, the drag-reducing properties of the skin of dolphins, and human athletic performance. Additionally, biomechanics addresses many issues of health as well as disease, injury, and their treatment in both humans and animals. This shall be our primary motivation herein; thus, it is easy to see that biomechanics is fundamental to the rapidly growing field of biomedical engineering.

It is not possible to identify a true "father of biomechanics," but many point to either Leonardo da Vinci (1452–1519) or Galileo Galilei (1564–1642). Among many other things, da Vinci was interested in a means by which man could fly, and to this end, he studied the mechanics of the flight of birds. Mankind's attempt to base the *design* of engineering systems on nature's way of doing something (e.g., the honeycomb structure within a beehive, a bat's radar system, etc.) is called bionics, which remains a very important area within biomechanics. In contrast to da Vinci, Galileo was interested in the intrinsic strength of bones and, in particular, its relation to the structural design of bones. Based on a preliminary *analysis*, he suggested that bones are hollow, for this improves the strength-to-weight ratio. Clearly, then, biomechanics focuses on both design and analysis, each of which is fundamental to engineering.

Jumping forward to the late nineteenth century, Wilhelm Roux put forth the idea of a "quantitative self-regulating mechanism" that results in functional adaptation by tissues, organs, and organisms, an idea that was consistent with the concept of a stress-mediated organization of the microstructure of bone that was put forth by Julius Wolff in 1884. Briefly, Wolff suggested that the fine structure within bones (i.e., oriented trabeculae) is governed by lines of tension that result from the applied loads. Although his analysis was not correct, the basic idea was extremely important. For more on "Wolff's law of bone remodeling," see Chapter 4 as well as Roesler (1987). Indeed, we will return many times to this observation that mechanical loads control tissue structure and function, which has given rise to the very important area of research called *mechanobiology*.

Many other savants were interested in biomechanical applications. They include Hooke (1635–1703), Euler (1707–1783), Young (1773–1829), Poiseuille (1799–1869), and von Helmholtz (1821–1894). Despite the caliber of scientists who have sought answers in biomechanics over the centuries, our field did not truly come into its own until the mid-1960s. Although historians will likely argue over the reasons for this, it is suggested here that five nearly concurrent developments provided both increased motivation and increased capabilities in biomechanics. Recall that the 1960s was the decade of mankind's pursuit of the Moon. When faced with the question, "How will man respond to the altered loads associated with space travel, including a reduced gravitational load on the Moon?," clinical medicine could not provide the answers, for it is based largely on observations. There was a need, therefore, for a predictive science, one focusing on how the body responds to mechanical loads. In addition, note that much of biomechanics deals with the response of soft tissues (i.e., tissues other than bones and teeth). It has long been known that soft tissues exhibit complex nonlinear behaviors that could not be described by the classical mechanics of continua developed in the eighteenth and nineteenth centuries. Rather, biomechanics had to await the post-World War II renaissance in continuum mechanics (~1948–1965) through which the nonlinear theories achieved a more complete and rational foundation. During this same period, 1950s to 1960s, technological developments gave rise to the digital computer. Computers are essential in biomechanics for solving many important but complex boundary and initial value problems, for controlling complicated experiments, and for performing nonlinear analyses of the data. Paralleling the development of computers was the advancing of powerful numerical methods of analysis, including the finite element method, which was introduced in 1956 and has become a standard tool in the biomechanicist's arsenal for attacking basic and applied problems. Finally, it is not coincidental that biomechanics emerged at the time that modern biology was born, which was due in large part to the identification in the 1950s of the basic structure of proteins (by Pauling) and DNA (by Watson and Crick). In summary then, *the 1950s and 1960s provided important new motivations as well as theoretical, experimental, and technological advances that allowed the emergence of biomechanics.* This is, of course, only a synopsis of some of the essential historical developments. The interested reader is encouraged to investigate further the history of our field.

Although biomechanics encompasses a broad range of topics, the purposes of this book are twofold: first, to introduce fundamental concepts and results from solid and fluid mechanics that can be applied to many different problems of importance in biology and medicine and, second, to illustrate some of the many possible applications by focusing on the mechanics of human health, disease, and injury. Hence, to motivate our study further, let us briefly review some of the many cases wherein biomechanics can and must contribute to the advancement of health care. Once we have sufficient motivation, we shall then briefly review results from Cell and Matrix Biology, results on which we shall build in Chapters 2–11.

1.2 Health Care Applications

There are many obvious examples wherein biomechanics plays a central role in the delivery of health care, roles that literally span all levels from the molecule to the person. Beginning with the latter, a simple example of an important biomechanical contribution is the design of efficient wheelchairs. By efficient, of course, we mean having sufficient strength with minimal weight, but also ease of maneuverability, ease of transport in a car or van, flexibility in the positioning of the patient, and even affordability. One does not realize the importance of what may seem to be such a simple device until a family member is incapacitated and in need. Selection of materials, design, experimentation, and stress analysis each play important roles in the engineering of an efficient wheelchair. Another common example at the level of the whole person is the design of transportation systems that improve occupant safety. Again, one only needs to see the devastation wrought on a family when someone is injured severely in a vehicular accident to appreciate the need for biomechanical solutions to improve safety in transportation.

Intracranial saccular aneurysms are balloonlike dilatations of the arterial wall that tend to form in or near bifurcations in the circle of Willis (Fig. 1.1), the primary network of arteries that supply blood to the brain. Although the natural history of saccular aneurysms is not well understood, it is generally accepted that mechanical factors play important roles. Hemodynamic forces may contribute to the initial local weakening of the wall, intramural forces that balance the distending blood pressure may contribute to the enlargement of the lesion from a small bulge to a sac over 25 mm in diameter (*note*: the parent artery is often less than 4 mm in diameter), and it is thought that rupture occurs when the intramural forces exceed the strength of the wall. Ruptured saccular aneurysms are the primary cause of spontaneous subarachnoid hemorrhage (i.e., bleeding within the brain due to nontraumatic cause) and thus are responsible for

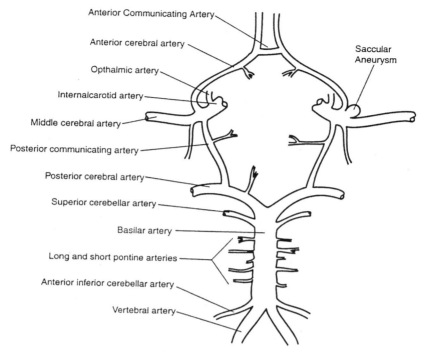

FIGURE 1.1 Schema of the circle of Willis, the primary network of arteries that supplies blood to the brain. Note the intracranial saccular aneurysm, which is a focal dilatation of the arterial wall on the left middle cerebral artery (with the circle viewed from the base of the brain). Such lesions tend to be thin-walled and susceptible to rupture. From Humphrey and Canham (2000), with permission from Kluwer Academic Publishers.

significant morbidity and mortality. Understanding the biomechanics of aneurysms at the tissue level is thus potentially very important in neurosurgery.

On yet another scale, it was discovered around 1974 that endothelial cells, which line all blood vessels, are very sensitive and responsive to the forces imparted on them by the flowing blood. In particular, these cells express different genes, and thus produce different molecules, depending on the magnitude and direction of the blood-flow-induced forces (Fig. 1.2). Many different situations alter the flow of blood and thus the forces felt by the endothelium: exercise or the lack thereof, diseases such as atherosclerosis and aneurysms, a microgravity environment on the space shuttle, the implantation of medical devices including artificial arteries or left ventricular assist devices, and even the surgical creation of arterio-venous fistulas for kidney dialysis. To understand and ultimately to control endothelial

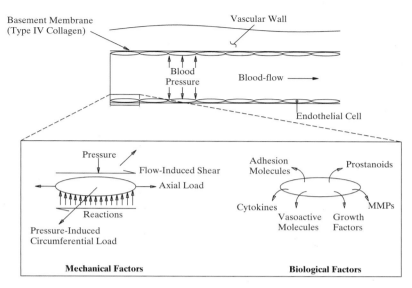

FIGURE 1.2 Schema of the monolayer of endothelial cells that lines the inner surface of a blood vessel, with a free-body diagram showing various mechanical loads that act on a single cell: flow-induced shear forces; radial forces due to the blood pressure; circumferential forces due to cell–cell contacts and the distension due to the pressure; and axial forces due to cell–cell contacts and the prestretch that appears to arise during development. Also shown are classes of molecules that are produced by endothelial cells in response to changes in these mechanical loads. MMPs denotes matrix metalloproteinases—molecules that degrade extracellular matrix.

function, we must understand the associated biomechanics—how the fluid-induced forces deform a cell, how the cell senses these forces, and how the transduction of these forces controls gene expression. It is thought, for example, that loads applied to the surface of a cell are transmitted to the proteins within the cell through membrane-bound protein receptors. Hence, from the wheelchair to individual proteins in the cell membrane, and everywhere in between, biomechanics has a vital role to play in analysis and design that seeks to improve health care.

Figure 1.3 is a rendition of the drawing of a man by da Vinci that emphasizes interesting symmetries of the body. Shown, too, are some of the many examples wherein mechanics plays a key role: from understanding why abdominal aortic aneurysms rupture, to identifying the failure strength of the anterior cruciate ligament (ACL) in an elite athlete, which must be protected during training and competition; from designing an artificial heart valve that must open and close over 30 million times per year, to understanding why artificial hip implants loosen over time and cause pain; from understanding what pressure must be applied to an angioplasty balloon to

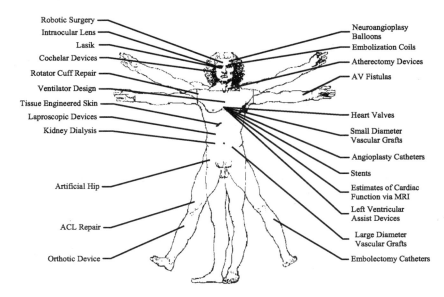

FIGURE 1.3 A schema of da Vinci's man showing a few of the many different aspects of human physiology, pathophysiology, and injury that can be addressed using biomechanics.

open a diseased artery, to understanding how deep and how many incisions should be made to modify the curvature of the cornea to correct for visual problems; from understanding the role of stresses in biological growth for the purpose of engineering tissue replacements, to designing a mechanical ventilator for those in respiratory distress; from using computer-aided modeling to guide robotic-assisted surgery, to designing needles that induce less damage to the arterial or venous wall; from designing an orthotic device for supporting an injured limb, to specifying a rehabilitation schedule that promotes tissue healing. In these and many, many other cases, biomechanics plays vital roles in the research laboratory, biomedical device industry, and hospital on a daily basis.

1.3 What Is Continuum Mechanics?

The functioning of the body and, likewise, the success of many clinical interventions depend on chemical, electrical, mechanical, and thermal processes. Nevertheless, we shall focus herein solely on the mechanics. Recall, therefore, that classical physics is typically thought to consist of a number of related areas of study: acoustics, electromagnetics, mechanics, optics, and thermodynamics. Thus, most of classical physics is concerned with the behavior of matter on a "natural" scale of observation or experience. Although its foundations and applications continue to be vibrant areas of research, the fundamental ideas upon which classical physics rests (due to Gibbs, Huygens, Maxwell, Newton, and others) were identified prior to the twentieth century. In contrast, modern physics is concerned primarily with phenomena at "extreme" scales of observation and thus includes atomic (or nuclear) physics, low-temperature physics, quantum mechanics, and relativity. Clearly, biomedical engineering is supported by, and relies on, both classical and modern physics. Without the latter, important diagnostic tools such as CAT (computerized axial tomography) scans and MRI (magnetic resonance imaging) would not be possible. In this introductory text, however, we shall rely solely upon classical mechanics.

Classical mechanics is typically thought to offer two basic approaches: continuum mechanics and statistical mechanics. Consider, for example, a simple glass of water at room temperature. On the natural scale of observation, we see and can think of the water as a continuous medium. In reality, however, we know that water is a collection of discrete, interacting molecules composed of hydrogen and oxygen atoms, and we know that there are gaps between the H_2O molecules and even gaps between the electrons and nucleus of each of the atoms. In statistical mechanics, we attempt to describe the (statistical) mean behavior of the individual molecules so as to understand gross behaviors on a natural scale of observation. In continuum mechanics, we also consider a volume-averaged mean behavior, but one that is independent of any consideration of the individual molecules. Perhaps a good example that illustrates when the continuum and statistical approaches are each useful is the analysis of drag on the Saturn V rocket that carried the *Apollo* spacecraft into space. When the rocket took off, the drag due to the frictional interaction between the surface of the rocket and the molecules of the air could be studied within a continuum context because there were so many closely spaced molecules that a gross, volume-averaged description of their properties was meaningful. In the upper atmosphere, however, the molecules of the air may be far enough apart that one should consider statistically their individual behaviors. In other words, *the continuum assumption (or hypothesis) tends to be reasonable when δ/λ ≪ 1, where δ is a characteristic length scale of the microstructure and λ is a characteristic length scale of the physical problem of interest*. For the rocket, δ may be the distance between the individual molecules of the air and λ the diameter of the rocket. In this case, the ratio of δ/λ is much less than 1 near the ground but perhaps on the order of 1 in the upper atmosphere. With regard to biomechanics, consider the following. If one is interested, for example, in the forces felt by cells (on average) within the wall of a large artery due to the distending blood pressure, the characteristic length scales would be micrometers (μm) for the microstructure (e.g., size of the cell and diameters of the fibers in the extracellular matrix) and millimeters (mm) for the physical problem (wall thickness). Thus, $\delta/\lambda \sim \mu$m/mm

~ 0.001 which is much less than 1 and the continuum assumption would be expected to be reasonable. Similarly, if one is interested in the velocity of blood at the centerline of a large artery, the characteristic length scales would again be micrometers for the microstructure (diameter of a red blood cell) and millimeters for the physical problem (luminal diameter), and again $\delta/\lambda \ll 1$. The situation would be very different in a capillary, however, wherein $\delta/\lambda \sim 1$ because the diameter of the red blood cell and capillary are both about 5–8 μm. We shall see throughout this text that the continuum assumption tends to be very useful in a wide variety of problems of design and analysis in biomechanics; hence, it is adopted throughout. Nevertheless, we are well advised to remember the following: "Whether the continuum approach is justified, in any particular case, is a matter, not for the philosophy or methodology of science, but for experimental test" (Truesdell and Noll, 1965, p. 5). In other words, the utility of any of our designs or analyses must first be checked in the laboratory.

Recall, too, that matter is typically thought to exist in one of three phases: solid, liquid, or gas. Mechanics tends to be divided along these lines into *solid mechanics* and *fluid mechanics*, where fluid mechanics includes the study of both liquids and gases; that is, one can define a fluid as a substance that assumes (within short times) the shape of the container in which it is placed, whereas a solid tends to resist such shape changes unless so forced. Referring to Figure 1.4, therefore, note that solid and fluid mechanics are generally studied in the order of increasing complexity, which has (artificially) given rise to subfields of study. Although no solid is rigid, the assump-

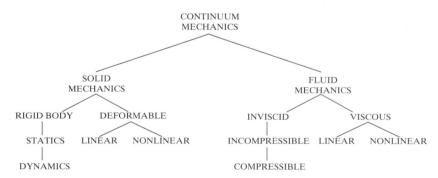

FIGURE 1.4 Flowchart of traditional divisions of study within continuum mechanics. Note that solid mechanics and fluid mechanics focus primarily on solidlike and fluidlike behaviors, not materials in their solid versus fluid/gaseous phases. Note, too, that linear and nonlinear refer to material behaviors, not the governing differential equations of motion. As we shall see in Chapter 11, many materials simultaneously exhibit solidlike (e.g., elastic) and fluidlike (e.g., viscous) behaviors, which gives rise to the study of viscoelasticity and the theory of mixtures, both of which are important areas within continuum biomechanics.

tion of a rigid body can lead to many useful designs and analyses, as, for example, in satellite dynamics. Likewise, all fluids resist the forces that cause them to deform, or flow. Again, however, neglecting this intrinsic resistance to flow (or, viscosity) can lead to many useful engineering solutions, particularly in aerodynamics. Hence, despite being based on unrealistic assumptions, rigid-body solid mechanics and inviscid fluid mechanics are both useful and convenient starting points for study.[1]

Our focus herein is on deformable solids and viscous fluids, for which it is often convenient to study separately the linear and nonlinear behaviors (Fig. 1.4), which give rise to additional subfields of study such as elasticity and plasticity (in solid mechanics) or Newtonian and non-Newtonian fluid mechanics. Although many problems in biomechanics necessitate dealing with the complexities associated with nonlinear behaviors (e.g., the stiffening response of soft tissues to increasing loads or the flow-dependent viscosity of blood), we shall focus primarily on the linear behavior of both solids and fluids. Not only do such problems serve as a natural preparation for the consideration of the more complex problems found in advanced courses, but many solutions to linear problems are fundamental to clinical and industrial applications as well as to basic research. For an introduction to nonlinear cardiovascular solid mechanics, see Humphrey (2002).

1.4 A Brief on Cell Biology[2]

The word "cell" comes from the Latin *cellulea*, meaning "little rooms." This terminology was coined by Hooke (1635–1703) who was perhaps the first to describe a cellular structure, which in his case was remnant cell walls in a thin slice of cork. Today, by the word "cell," we mean "the fundamental, structural, and functional unit of living organisms" (Dorland's Medical Dictionary, 1988). For a detailed discussion of cell biology, see the wonderful work by Alberts et al. (2002) or similar texts; here, we simply offer a brief introduction.

Most cells consist of various organelles (i.e., organized structures having specific functions), the cytosol, the cytoskeleton, and an outer membrane (Fig. 1.5). The most conspicuous organelle is the nucleus, which contains the genetic information, chromosomal DNA. The nucleus consists of its own porous membrane or envelope, which mediates all transport into and out of the nucleus, a nucleoplasm that contains a fibrous scaffold, and a nucleolus that produces the ribosomes that are responsible for translating mRNA data for protein synthesis. The primary functions of the nucleus,

---

[1] It is assumed herein that the student has had an introduction to mechanics, which typically covers rigid body statics and sometimes dynamics. A brief review of statics is found in Appendix 1.

[2] Much of Sections 1.4 and 1.5 are from Humphrey (2002).

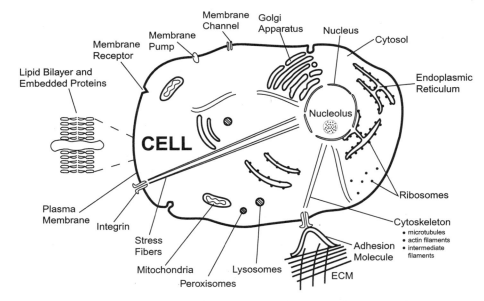

FIGURE 1.5 Schema of a mammalian cell showing its three primary constituents: the cell membrane (with various receptors, pumps, channels, and transmembrane proteins), the cytoplasm (including many different types of organelles, the cytoskeleton, and the cytosol), and the nucleus. From a mechanics perspective, the three primary proteins of the cytoskeleton (actin, intermediate filaments, and microtubules) are of particular importance. [From Humphrey (2002), with permission.]

therefore, are to archive and replicate the genetic code as needed to direct cellular activity. Whereas the cells in a given organism contain the same genetic information (the genotype), each cell does not "express" the same genes. The genes that are expressed define the phenotype; hence, skin cells are different from bone cells and so on. That cells are able to express different genes in response to changing external stimuli, particularly mechanical loads, will prove to be very important in biomechanics and, thus, is discussed separately in Section 1.6. Other organelles within a cell include the mitochondria, endoplasmic reticulum (rough and smooth), and the Golgi apparatus. Mitochondria provide the cell with usable energy by oxidizing foodstuffs (e.g., sugars) to make adenosine triphosphate (ATP). A typical cell may have over 1000 distributed mitochondria, which, together, may constitute up to one-fourth of the total cell volume. The rough endoplasmic reticulum represents an interconnected space that specializes in the synthesis of proteins; it connects to the outer portion of the nuclear membrane and is intimately associated with ribosomes—carriers of the RNA. The smooth endoplasmic reticulum is tubular in structure; although it aids in the packaging of proteins, it specializes in the synthesis of lipids and steroids. The Golgi apparatus plays a key role in the synthesis of polysaccharides as well as in the modification, packaging, and transport of various macromolecules; this transport includes secretion into the extracellular space. In addition to these organelles, which are responsible for the conversion of energy or processing of products, lysosomes and peroxisomes are responsible for the degradation of various substances within the cell. Lysosomes are capable of digesting proteins, carbohydrates, and fats and thereby aid in both the breakdown of foodstuffs and the removal of unnecessary cellular components. With an internal pH of about 5, lysosomes accomplish this degradation via various acidic enzymes, including nucleases, proteases, and lipases. Peroxisomes are capable of generating and degrading hydrogen peroxide, which is cytotoxic, and they assist in the detoxification of other compounds (e.g., formaldehyde). Of course, cells also ingest extracellular substances via a process called phagocytosis, which facilitates a controlled intracellular degradation by the lysosomes and peroxisomes. A controlled degradation of "old" constituents plays an important role in the biomechanics of tissue maintenance, adaptation, and wound healing.

The cytoplasm is defined as that part of the interior of the cell that does not include the nucleus. Thus, it consists of all the other organelles, the cytoskeleton, and the cytosol. The cytosol constitutes up to one-half of the total cell volume and consists primarily of water.[3] The cytoskeleton consists primarily of three classes of filamentous proteins: actin, which is often the most abundant protein in a cell; microtubules, which are formed from tubulin; and intermediate filaments, which include vimentin, lamins, and keratins. These cytoskeletal filaments have diameters of 5–25 nm and they can polymerize to form linear units that span distances between organelles or even over the entire length of a cell. Collectively, these filamentous proteins along with hundreds of different types of accessory proteins endow the cell with much of its internal structure, they aid in cell division, they enable cell mobility, and they maintain cell shape. The cytoskeleton is thus fundamental to cell mechanics. Moreover, much of the water and other proteins within the cytosol are bound to the cytoskeleton, which aids in the selective positioning or movement of components within the cell. The cytoskeleton is a dynamic structure, continually reorganizing to meet the needs of the cells. For example, the intermediate filaments can increase in density in response to increased mechanical stress. Likewise, stress fibers consisting of temporary bundles of actin often form within fibroblasts. They serve to connect the strong network of intermediate filaments that surround the nucleus to the plasma membrane at sites where it is connected to the extracellular matrix via transmembrane linker proteins (e.g., integrins). This arrangement (Fig. 1.6) may allow the stress fibers to transduce the level of tension in the extracellular matrix to the nucleus and thus to control gene expression (i.e., mechanotransduction). Conversely, stress fibers in fibrob-

---

[3] Note: 70% of the total cell volume is due to water.

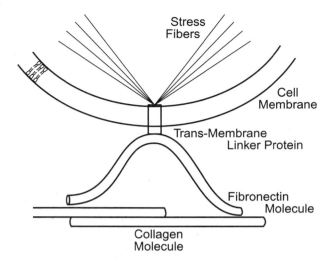

FIGURE 1.6 Schema of some of the constituents that participate in cell–matrix interactions that are important to the mechanobiology. The transmembrane protein that
"links" the cytoskeletal (e.g., stress fibers) and extracellular (e.g., fibronectin and
collagen) proteins is often a member of the family of integrins. [From Humphrey
(2002), with permission.]

lasts also allow them to exert tension on the extracellular matrix, which is
particularly useful during morphogenesis or repair in wound healing.
Understanding the mechanics of growth and remodeling is one of the most
important open problems in biomechanics at this time; this general area is
discussed more in Section 1.6.

Note that striated muscle (e.g., that makes up the myocardium of the
heart wall or skeletal muscle) contains an additional, specialized intracellular constituent—the myofibril. These contractile elements are approximately 1–2 μm in diameter, they span the length of the cell, and they consist
of a chain of shorter (2.2 μm) units, called sarcomeres. According to the
sliding filament model proposed in 1954, sarcomeres consist of overlapping
thin (actin) and thick (myosin) filaments. It is thought that the myosin has
tiny "cross-bridges" that attach, detach, and reattach in a ratcheting fashion
with the actin, which thereby produces movement associated with the
contraction of muscle (Fig. 1.7). Smooth muscle cells similarly rely on
actin–myosin interactions although they do not have a sarcomere structure.
Thus, studying the biomechanics of muscular organs such as the heart,
blood vessels, diaphragm, or uterus as well as studying locomotion at the
organism level all require an understanding of the associated cell biology.

The cell membrane separates the cellular contents from their surround-
ings. It consists primarily of a phospholipid bilayer with embedded proteins
and is on the order of 5 nm thick (cf. Fig. 1.5). Held together by noncova-

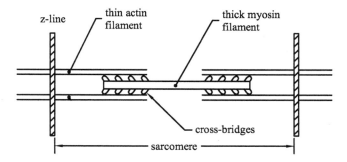

FIGURE 1.7 Schema of the cross-bridge mechanism that is thought to control the contraction and relaxation of muscle. In particular, the cross-bridges allow a ratcheting motion between the thick myosin filaments and the thin actin filaments. Calcium plays a key role in this process.

lent bonds, this membrane is described in biology texts as having "fluidity"; that is, the lipid molecules can exhibit rapid lateral diffusion, which is to say that they can readily exchange places with each other. It appears that this fluidity endows the membrane with a self-sealing capability and it plays a role in some processes of transport across the membrane (e.g., ion transport facilitated by glycolipids). The embedded proteins likewise play many roles: They may participate in the conduction of electrical signals or the transport of various substances across the membrane by serving as selective channels, gates, and pumps. Alternatively, these proteins may serve as enzymes to catalyze specific reactions, they may act as selective receptors that bind extracellular substances to the cell membrane, or they may serve as anchors for the attachment of intracellular cytoskeletal filaments or extracellular proteins to the membrane (Fig. 1.6). The latter is accomplished primarily via a special class of transmembrane proteins, the integrins, which consist of two noncovalently associated glycoproteins referred to as α and β units (there are at least 14 different α units and 9 different β units). Some integrins bind to specific proteins (e.g., laminin or fibronectin), whereas others bind to multiple proteins by recognizing a particular amino acid sequence (e.g., arginine–glycine–aspartic acid, or RGD). Integrins are found in large numbers, but their binding to a particular ligand tends to be weak. This would be advantageous in cell migration, for example, wherein local adhesion would be short-lived. Cells can regulate the activity of their integrins, and, conversely, gene expression can be mediated by the extracellular matrix via the integrins. Finally, note that some of the embedded membrane proteins are decorated with carbohydrates; this glycocalyx, or "sugar coat," appears to protect the cell from mechanical and chemical damage and may participate in certain transient adhesion processes.

Cells can be interconnected via three types of junctions: occluding, or tight, junctions seal cells together; anchoring junctions mechanically attach cells to other cells or extracellular matrix at specific sites; and communicating (e.g., gap) junctions allow cell-to-cell exchange of electrical or chemical signals. At any particular instant in the mature organism, most cells are simply performing their primary function (e.g., muscle cells are contracting and fibroblasts are synthesizing extracellular matrix). Nonetheless, normal tissue maintenance also typically requires a delicate balance between continuous cell replication and cell death; in the adult, for example, millions of cells are produced each minute to replace cells that are damaged, killed, or simply experience a normal cell death (apoptosis). Of course, cells reproduce by duplicating their contents and dividing in two. Although we will not consider the details of the cell cycle (see Alberts et al., 2002), note that it appears that cells require multiple external signals before they will divide. Growth factors, for example, are special proteins that bind to specific receptors on the cell membrane and encourage cell division. According to Gooch et al. (1998),

Growth factors can stimulate or inhibit cell division, differentiation, and migration. They up- or down-regulate cellular processes such as gene expression, DNA and protein synthesis, and autocrine and paracrine factor expression. [They] . . . can interact with one another in an additive, cooperative, synergistic, or antagonistic manner. They may cause dissimilar responses when applied to different cell types or tissues, and their effect on a certain type of cell or tissue may vary according to concentration or time of application.

Among the over 50 different growth factors in humans are the platelet-derived growth factors (PDGFs), fibroblast growth factors (FGFs), and transforming growth factors (TGFs). Mechanical stresses and injuries have both been shown to modulate the secretion of growth factors; hence, tissues that normally have a slow turnover of cells (replication and death) can experience rapid increases in turnover in response to certain mechanical stimuli. Understanding and quantifying these homeostatic control mechanisms is a newly identified, important topic in biomechanics and mechanobiology.

This is but a cursory introduction to the general structure and function of the cell, yet it serves as sufficient motivation for our purposes. Of course, in most cases, we will not be interested in a single cell, but rather large populations of communicating cells. In this regard, the role of the extracellular matrix, in which most cells are embedded, is of utmost importance. Let us now consider this important component in more detail.

1.5 The Extracellular Matrix

It is axiomatic in continuum mechanics that the properties of a material result from its internal constitution, including the *distributions*, *orientations*, and *interconnections* of the constituents. Examination of microstructure is essential, therefore, for quantifying the mechanical behavior and analyzing the internal distribution of forces. In most tissues and organs in the body, the microstructure depends largely on the extracellular matrix (ECM).

The ECM serves multiple functions: It endows a tissue with strength and resilience and thereby maintains its shape; it serves as a biologically active scaffolding on which cells can migrate or adhere; it may regulate the phenotype of the cells; it serves as an anchor for many proteins, including growth factors and enzymes such as proteases and their inhibitors; and it provides an aqueous environment for the diffusion of nutrients, ions, hormones, and metabolites between the cell and the capillary network. In many respects, therefore, it is the ECM that regulates cell shape, orientation, movement, and metabolic activity. It is the cells (e.g., fibroblasts), however, that fashion and maintain the ECM. Hence, the ECM and cells have a strong symbiotic relation.

The ECM consists primarily of proteins (e.g., collagen, elastin, fibronectin, and laminin), glycosaminoglycans (GAGs), and bound and unbound water (Fawcett, 1986; Ayad et al., 1994; Ninomiya et al., 1998; Alberts et al., 2002). The GAGs are often bound covalently to protein cores, thus forming proteoglycans. Although collagen was long regarded to be a single protein, more than 16 distinct forms have been identified. Collectively, the collagens are the most abundant protein in the body (~25–30% of all protein), common forms being types I, II, III, and IV, as well as types V, VI, and VIII. Types I and III form fibers and provide structural support in tension; they are found in tendons, skin, bone, the heart, arteries, and cornea. Type II collagen occurs as fibrils; it is found largely in cartilage, which also contains significant proteoglycans. Type IV collagen forms as a porous network (basement membrane) that acts as a scaffolding for epithelial and endothelial cells (adhesion being aided by fibronectin or laminin); it is found, for example, in the lens capsule of the eye as well as in the inner layer of blood vessels. Types V and VI collagen tend to associate with smooth muscle cells, whereas type VIII tends to associate with endothelial cells. For more on the collagens, see Kucharz (1992).

Synthesized by various cells (Fig. 1.8), the collagen molecule consists of three polypeptide α chains, each containing 1300–1700 amino acid residues. The majority of these residues (~1000–1400) are organized into a central triple-helix motif (Ayad et al., 1994), which is on the order of 285 nm long and 1.4 nm in diameter. The triple helix results from the repetition of a triplet of amino acid residues of the form $(G\text{-}X\text{-}Y)_n$, where G stands for Glycine, the simplest amino acid, and X and Y may be any of the other 19 common amino acids, although often proline or hydroxyproline. The triple-helix structure is stabilized by abundant interchain hydrogen bonds, many via the hydroxyprolines. Intramolecular covalent cross-links in or near the nonhelical ends of the molecule provide further structural stability, often via hydroxylysine. Type IV collagen also has extensive disulfide bonds. Details on the biosynthesis of collagen can be found in Nimni (1992)

COLLAGEN STRUCTURE

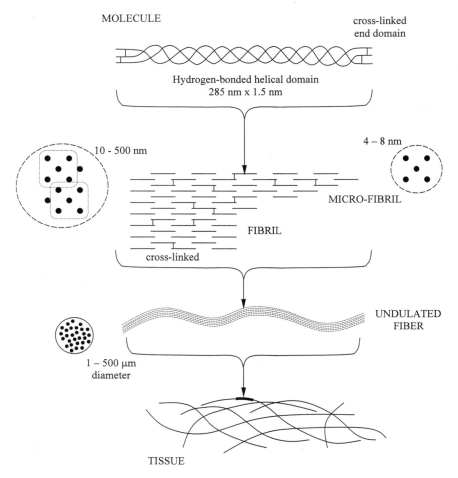

FIGURE 1.8 Schema of collagen at different levels of organization: from the triple-helix molecule consisting of three α-helices of repeating triplets of amimo acids (G-X-Y), where G is glycine and X and Y are often proline or hydroxyproline, to an undulated fiber that could be found in arteries, cartilage, cornea, the heart, lungs, skin, tendons, and many other tissues. [From Humphrey (2002), with permission.]

and Kucharz (1992); details on the chemical structure can be found in Ayad et al. (1994).

Vascular type IV collagen is synthesized, for example, by endothelial cells, whereas types I and III collagen are synthesized by fibroblasts and smooth muscle cells; it takes the cell on the order of 10–60 min to synthesize a complete intracellular collagen precursor, called procollagen (Nimni, 1992). Following secretion by the cell, newly synthesized type I collagen molecules undergo extracellular modifications prior to "self"-assembly (polymerization) into 4–8-nm-diameter microfibrils consisting of repeating quarter-staggered (which gives the characteristic 67-nm periodicity) groups of four to five molecules in cross section. This assembly results from electrostatic and hydrophobic bonding (which liberates previously bound water) between molecules. The specific directional assembly may be aided by narrow extracellular channels within the plasma membrane of oriented cells [e.g., the fibroblast (Birk et al., 1989)]. Note, therefore, that the orientation of cells appears to be governed by the local force field (Carver et al., 1991) and so too for the orientation of the collagen [e.g., in tendons, the collagen tends to be oriented uniaxially, whereas in skin, it is distributed primarily in a two-dimensional (2-D) fashion]. The microfibrils, in turn, are organized into successively larger fibrils (~10–500 nm in diameter) and ultimately fibers (1–500 μm in diameter), the specific diameter of which is also thought to be dictated largely by the mechanical force field in the ECM. The extracellularly organized fibrils and fibers are stabilized by interchain cross-links that occur primarily through the conversion of lysine and hydroxylysine (in the nonhelical portion of the molecule), via the enzyme lysyl oxidase, into peptide-bound aldehydes. Further aldehyde cross-linking of collagen is important industrially with regard to the engineering of bioprosthetic heart valves, which must exhibit sufficient biocompatibility, strength, and efficiency as a valve. The need to understand the microstructure, which governs these characteristics, is thus clear. Finally, note that additional cross-links also form in type III collagen via intermolecular disulfide bonds. Cross-links can be either reducible or nonreducible; reducible cross-links can be broken, for example, during thermal treatment.[4] Overall, the degree of cross-linking tends to increase with age, which results in concomitant stiffening; pathological stiffening, via the addition of glycosylated cross-links, occurs in diabetes. Note, too, that collagen fibers are usually undulated at physiologic loads; thus, they exhibit their true stiffness only when straightened under the action of applied loads. For example, the tensile strength of nearly straight uniaxially oriented type I collagen in tendons can be 100 MPa in mature tissue.

Finally, the half-life of collagen varies tremendously throughout the body: it is only a few days in the periodontal ligament but typically many months in tendons and possibly years in bones.[5] In the cardiovascular system, the half-life of collagen is on the order of 15–90 days. Regardless of the spe-

---

[4] Advances in laser, microwave, and radio-frequency technologies continue to encourage new uses of thermal energy to treat disease and injury (Humphrey, 2003b).

[5] In contrast, many cellular proteins have half-lives of hours or days (Alberts et al., 2002).

cific half-life, maintenance of physiologic levels of collagen depends on a delicate balance between continual synthesis and degradation, the kinetics of which is complex but may be assumed to be of first order (Niedermuller et al., 1977; Gelman et al., 1979). Degradation can be accomplished by blood-plasma-borne serine proteases, the extracellular release of matrix metalloproteinases (MMPs), as, for example, by macrophages or via intracellular lysosomal activity within phagocytotic fibroblasts (Ten Cate and Deporter, 1975). As noted earlier, phagocytosis can be a highly selective mechanism of degradation. It also appears that much of the synthesized collagen is degraded prior to its incorporation into the ECM (McAnulty and Laurent, 1987). The reason for this is not clear, but may simply reflect an internal mechanism for culling imperfectly synthesized molecules (i.e., a cellular quality control). *In response to disease or injury, however, the rates and control of the continual degradation and synthesis can change dramatically*, as needed. Wound healing in skin is a prime example of an accelerated turnover of collagen, which in this case may result in a collagenous scar.

Strictly speaking, elastic fibers in the ECM consist of two components—one microfibrillar (10 nm in diameter) and one amorphous. Whereas the former is not well understood, the amorphous (major) portion is called elastin. It consists of a polypeptide chain of ~786 amino acid residues, the majority of which are glycine, alanine, and proline. Elastin is synthesized in minutes, as the precursor proelastin, via normal pathways—mRNA, endoplasmic reticulum, Golgi apparatus, and so forth. Moreover, it appears that synthesis can be assumed to be a first-order process, one that is completed in less than 1 h (Davidson and Giro, 1986). In the vasculature, this synthesis is accomplished primarily by smooth muscle, but also by specialized fibroblasts and, perhaps, endothelial cells. Once secreted into the extracellular space, the soluble proelastin is cross-linked to form the insoluble (stable) elastin meshwork. Two unique amino acids, desmosine and isodesmosine, are largely responsible for the formation of distributed covalent cross-links between the relatively loose and unstructured chains. It is the loose, amorphous, but highly cross-linked structure of elastin which results in a meshwork that exhibits an elastic (i.e., nondissipative or recoverable) response over large deformations (indeed, elastin appears to be the most elastic protein in the body). Moreover, a high concentration of nonpolar hydrophobic amino acids renders elastin one of the most chemically, thermally, and protease-resistant proteins in the body. Indeed, in contrast to collagen, the turnover of elastin is much slower in the adult, perhaps on the order of years to decades (Lefevre and Rucker, 1980). Much of the production of elastin may thus occur during development. The protease elastase, which can be secreted by macrophages, is capable of degrading elastin, however. Such degradation appears to play a role in the formation of aneurysms in the vasculature. For more details on elastin, see Robert and Hornebeck (1989).

Elastic fibers appear to consist of aggregated 10-nm-diameter microfibrils embedded in the amorphous elastin. These fibers can be from 0.2 to 5.0 μm in diameter, and they tend to branch and form networks or sheets. When straight, elastic fibers can experience uniaxial extensions of 150% without breaking (compared to less than 10% for collagen), and they return to their original configuration when unloaded. Indeed, it has been said that the primary role of elastic fibers is to store and then return mechanical energy.

Other important components of the ECM include the aforementioned fibronectin and laminin, both of which play important roles in cell adhesion (cf. Fig. 1.6). Fibronectin consists of ~2476 amino acid residues; it is a widely distributed glycoprotein—synthesized by fibroblasts, endothelial cells, and smooth muscle cells—that mediates cellular interactions and migration. For example, fibronectin binds fibroblasts to underlying collagen substrates, thereby playing an important role in normal development, growth, remodeling, and wound healing. It may likewise play a role in the aggregation of platelets. The ability of fibronectin to bind to different proteins and cells is due to the presence of different binding sites, which depend in part on the aforementioned RGD sequence. The laminins constitute a family of large glycoproteins (over 3000 amino acid residues) that are associated with the basement membrane; they self-assemble into a feltlike sheet. Laminin, one of the first proteins produced in the embryo, has numerous functional binding domains, as, for example, for heparan sulfate, type IV collagen, and various cells. Hence, like fibronectin, this protein plays an important role in the migration and anchoring of cells.

Proteoglycans represent a relatively small portion of the ECM in most tissues and have no preferred structural organization; they play important roles nonetheless. Proteoglycans consist of a core protein to which is attached multiple glycosaminoglycan (GAG) chains via covalent bonds. GAGs are linear polymers that contain repeating disaccharide units, the principal ones being hyaluronic acid, chondroitin sulfates, dermatan sulfates, keratan sulfates, heparan sulfates, and heparin. Because GAGs tend to occupy large volumes compared to their mass, and because they are highly negatively charged, they tend to imbibe considerable water into the ECM. Water, in turn, enables many of the necessary diffusive processes within the ECM and enables the tissue to withstand compressive loads (this is particularly important in cartilage). Moreover, hyaluronic acid, for example, gives the aqueous portion of the ECM its fluidlike consistency, or viscosity. It is for this reason that the nonfibrous portion of the ECM is often referred to as an amorphous ground substance or gel matrix.

Referring to Section 1.4, note that the core protein of the proteoglycan is made on membrane-bound ribosomes and transported to the endoplasmic reticulum. Upon passage to the Golgi apparatus, GAGs are affixed to the core and possibly modified (Alberts et al., 2002). By associating with the fibrous proteins in the ECM, proteoglycans and individual glycosaminoglycans create a highly complex 3-D structure embodied with chemical reactivity and intercellular signaling pathways. For example, fibroblast growth factor (FGF) binds to heparan sulfate, which may not only localize the FGF, it may also activate it. The ubiquitous transforming growth factor (TGF) likewise binds to numerous proteoglycans. Similarly, proteases and protease inhibitors may bind to proteoglycans, thus localizing activity, inhibiting activity, or providing a storage mechanism for later use.

In addition to the binding of specific cells to fibronectin and laminin, recall from Section 1.4 that cell–matrix interactions are often mediated by the integrins. For example, the integrins that are connected to intracellular actin can "pull" on extracellular proteins to which they are bound. Alternatively, tensions in the ECM may be sensed by the nucleus of a cell via the ECM–integrin–cytoskeletal connections. It is through the integrins, therefore, that cells influence the ECM and the ECM may provide inputs for cell growth.

Finally, when discussing the extracellular matrix in tissue and organs, the role of fibroblasts cannot be overemphasized. Fibroblasts belong to the differentiated cell family known as connective tissue cells [other members in this family include osteocytes, chondrocytes, adipocytes, and smooth muscle cells (Alberts et al., 2002)]. Fibroblasts are the least differentiated member of this family and are found throughout the body. Their primary responsibility is regulation of the collagen-rich ECM. For example, in response to tissue damage, fibroblasts will quickly migrate to the site of injury, proliferate, and then synthesize new collagen. Such activity is regulated in part by growth factors, in particular FGFs and TGF-β. Likewise, macrophages are essential in regulating the ECM: They dispose of dead cells and degrade unneeded matrix material. Macrophages are mononuclear phagocytes that arise from stem cells in the bone marrow, enter the bloodstream as monocytes, and eventually enter tissues wherein they increase in size and phagocytic activity. Macrophages secrete a wide variety of products in addition to proteases, including coagulation factors, prostaglandins, and cytokines.

1.6 Mechanotransduction in Cells

As noted earlier, one of the most exciting and important recent findings in cell biology is that mechanical stimuli have a direct influence on gene expression in many different cell types. Such cells have been classified as *mechanocytes*, which include chondrocytes, endothelial cells, epithelial cells, fibroblasts, macrophages, myocytes, and osteoblasts. Consider, for example, the endothelial cell. Endothelial cells form a contiguous monolayer throughout the vasculature (Fig. 1.2). Because the luminal surface of the endothelial cell is decorated with the glycosaminoglycan heparan sulfate, it was long thought that these cells served primarily as a smooth, nonthrombogenic surface that minimizes blood clots and thus facilitates blood flow. We now know that this is but one of the many functions of the endothelium. In response to local increases in blood flow, endothelial cells increase their production of nitric oxide (NO), a potent vasodilator; conversely, in response to local decreases in blood flow, endothelial cells increase their production of endothelin-1 (ET-1), a potent vasoconstrictor; that is, by altering its production of vasoactive molecules that diffuse into the wall and cause vascular smooth muscle cell relaxation or contraction, the endothelium is able to help control the diameter of the blood vessel in response to changing hemodynamic demands. Of course, sympathetic and parasympathetic signals as well as circulating hormones also contribute to the control of blood vessel diameter.

In addition to its mechanosensitive control of the production of vasoactive molecules (e.g., NO, ET-1), growth regulatory molecules (e.g., PDGF, FGF), cytokines (e.g., IL-1,6), and adhesion molecules (e.g., vasular cell adhesion molecule VCAM-1, monocyte chemotactic protein MCP-1), the endothelium also changes its shape and ultrastructure in response to changing hemodynamic loads. In vivo and in vitro studies both reveal that these cells realign to follow the direction of the blood flow and they realign perpendicular to an applied uniaxial stretching of a substrate on which they are adhered (*Note*: whereas the flow of blood along the axis of an artery causes cells to align in the axial direction, the distending blood pressure stretches the vessel circumferentially, which, being perpendicular to the axial direction, also causes the cells to align in the axial direction; hence, these two effects are complementary.) Additionally, an increased blood flow induces an increase in the density of flow-aligned stress fibers (i.e., specialized actin filaments). For beautiful time-lapse figures of these changes, see Galbraith et al. (1998).

One of the key questions facing biomechanics and mechanobiology, therefore, is how are these many different changes effected? In other words, how does a cell sense a changing mechanical environment and how is this signal transduced to the nucleus wherein different genes are expressed? This question becomes more acute when we realize, for example, that vascular smooth muscle cells independently express different genes in response to the changing hemodynamics even though they are not in direct contact with the pressure-driven blood flow. As noted by Zhu et al. (2000), which is a very readable, nice review of cell mechanics, critical questions are as follows: How do forces applied to a tissue distribute around the surface of a cell? How are these forces balanced within the interior of the cell? How does this internal force field induce a biological response? See, too, the review by Stamenovic and Ingber (2002). Although we do not have answers to these and similarly important questions, competing hypotheses and theories are under consideration. The student is encouraged to read, for example, the provocative paper by Ingber et al. (2000), which contrasts two ideas on how the intracellular forces balance the externally applied loads. One idea is based on tensegrity (tensional integrity), an architectural concept advanced by Buckminster Fuller wherein a stable structure is constructed from self-equilibrating tensional and compressive elements; the other idea focuses on the combined fluidlike and solidlike behaviors exhibited by cells under different conditions. Clearly, there is a pressing need for more data on the mechanical properties of cells; fortunately, new experimental tools such as laser tweezers (see Chapter 3) and the atomic force microscope (Chapter 5) allow increased insight into cell mechanics and, indeed, the various constituents that they are comprised of, including actin filaments, intermediate filaments, microtubules, the plasma membrane, and even the cytosol. The need to understand cellular responses also leads naturally to a focus on molecular biomechanics (i.e., how individual molecules respond to applied loads). Zhu et al. (2000) point out, for example, that in response to applied loads, a molecule may rotate/translate, it may deform, or it may unfold/refold. By changing the conformation of a molecule, one can change its biochemical character, as, for example, the availability of binding sites. In summary then, there is a need for mechanics at all scales in biology—from the molecule to the cell to the organ to the organism. Although much is known, much remains to be discovered.

1.7 General Method of Approach

The biomechanical behavior of biological tissues and organs results from the integrated manifestation of the many components that comprise the structure and their interactions. Although we may not always be directly interested in cellular- or molecular-level phenomena, as, for example, when calculating the forces within the wall of an aneurysm to evaluate its rupture potential or when designing a wheelchair, some knowledge of the associated cell and matrix biology can always provide important insight. In the case of an aneurysm, its fibroblasts regulate the continuous production and removal of intramural collagen in response to changes in the intramural forces; in the case of the wheelchair, the skin may break down at the cellular level (e.g., decubitus ulcer) in response to frictional forces, which must be designed against. Throughout this book, therefore, we will continually refer to the biology that motivates the mechanics.

Whereas Issac Newton (1642–1727) developed a "discrete" mechanics in which his fundamental postulates were assumed to apply to individual mass points (whether the Earth or an apple), Leonard Euler (1707–1783) showed that these same postulates apply to every mathematical point within a body. We submit, therefore, that *every continuum biomechanics problem can be addressed via the fundamental postulates of continuum mechanics by specifying three things*: the geometry (i.e., the domain of interest), the constitutive relations (i.e., how the material responds to applied loads under conditions of interest), and the applied loads (or associated boundary conditions). Moreover, we agree with Fung (1990) and others that the key to success in this approach is often the identification of robust constitutive relations. We discuss specific constitutive relations in Chapters 2, 6, 7, and 11. Here, however, simply note that there are five steps in every constitutive formulation[6]:

Delineation of general characteristic behaviors
Establishment of an appropriate theoretical framework
Identification of specific functional forms of the constitutive relation
Calculation of the values of the material parameters
Evaluation of the predictive capability of the final constitutive relation

Specifically, the first step is to *observe* the particular behaviors of interest and then, by *induction*, to delineate general characteristics of the material's response to the applied loads. In practice, this step is as difficult as it is critical. In many cases, the biomechanicist must distill the results from tens to hundreds of papers in the biological and clinical literature to delineate the underlying mechanism or general characteristics of importance. Once accomplished, one then attempts to formulate a general hypothesis and establish a theoretical framework; robust theories should rely on the axiomatic and *deductive* foundations of mathematics and mechanics. Two theories that we will consider in detail in this book are the theories of the linearly elastic behavior of solids and the linearly viscous behavior of fluids. Next, one must perform *experiments* to test the hypothesis or theory, which includes identification of specific functional relationships between quantities of interest and calculation of the values of the associated material parameters. Because of the unique behaviors exhibited by living tissues, performing theoretically motivated experiments may necessitate the design and construction of a novel experimental system or transducer. Moreover, based on comparisons to experimental data, one will often need to refine the hypothesis or theory and then to perform additional experiments and data analysis. This iterative procedure continues until the associated constitutive relation has predictive capability, which must be verified against additional observations or experimental data. Only then can one begin to answer applied questions of interest, often via numerical simulations (i.e., computations) and then animal and clinical trials. See Figure 1.9 for a summary of this overall approach, which is best appreciated via the examples that are provided in Chapters 2–11. In conclusion, we emphasize that *a constitutive relation is but a mathematical descriptor of particular behaviors exhibited by a material under conditions of interest; it is not a descriptor of a material per se.* Hence, multiple theories will likely be needed to describe the myriad of behaviors exhibited by a given molecule, cell, tissue, or organ under different conditions. Moreover, although we should always seek to understand and quantify the basic mechanisms by which responses to applied loads occur, this is often difficult or impossible; hence, we must

---

[6] A former student suggested that these five important steps in a constitutive formulation are remembered easily via the acrostic DEICE.

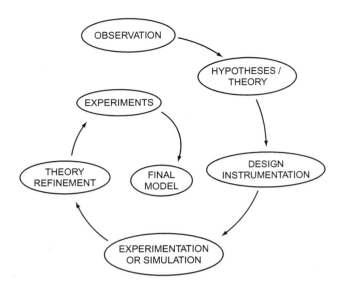

FIGURE 1.9 Illustration of the scientific approach employed by many in biomechanics. In particular, note that observations and experiments are equally important, but very different. The latter must be designed based on a hypothesis or theory for the purpose of testing an idea. Because science is "relative truth," we often need to iterate to improve our models of the physical and biological worlds.

sometimes rely on phenomenological descriptors or empirical correlations. Regardless of approach, the main goal of biomechanics must remain clear—to improve health care delivery via careful and appropriate design and analysis.

Appendix 1: Engineering Statics

From physics, we recall that Newton put forth three "laws" that form the foundation of classical mechanics: (1) a body at rest will remain at rest, or a body in motion at a constant velocity will remain at a constant velocity, unless acted upon by an external force; (2) in an inertial frame of reference, the time rate of change of the linear momentum must balance the applied external forces; (3) for every force that acts on a body there is an equal and opposite force that acts on some other body. Of these, the second law of motion is of particular importance herein; it is a statement of the balance of linear momentum. To this postulate, we can add the balance of mass, balance of angular momentum, balance of energy, and the entropy inequality. These *five basic postulates form the foundation of continuum mechanics* from introductory statics and dynamics to the sophisticated theories addressed in graduate courses.

Here, however, let us only consider the two balance of momentum equations:

$$\sum F = ma, \quad \sum M = I\alpha, \tag{A1.1}$$

where F is an applied force (i.e., a vectorial push or pull), m is the mass of a material particle, a is its acceleration vector, M is the applied moment (i.e., a force acting at a distance), I is the inertia, and α is the angular acceleration vector. Hence, these balance equations state that relative to an inertial frame, the sum of all forces acting on a body must balance the time rate of change of the linear momentum of the body and that the sum of all moments must balance the time rate of change of the angular momentum. (Note: For a particle of constant mass, the time derivative of the linear momentum mv, where v is the velocity, is simply ma, and similarly for the angular momentum.) In addition, recall that a moment M is defined *with respect to* a reference point, say o. M_o is thus defined as[7]

$$M_o = r_A \times F \tag{A1.2}$$

where r_A is a position vector that connects point o to any point A along the line of action of F. The *line of action* is simply a line that coincides with the force vector but extends well beyond it in each direction. It is important to be comfortable with the calculation of moments; hence, consider the following example.

Example A1.1 Prove that $r \times F \equiv r_\perp \times F$ where $r_\perp$ is the shortest distance between o and the line of action of F.

Solution: For simplicity, let us consider the 2-D case. Referring to Figure 1.10 let $r_\perp = d\hat{e}_\perp$ and $r = r_x \hat{i} + r_y \hat{j} = r_x' \hat{i}' + r_y' \hat{j}'$ depending on which coordinate system is found to be most convenient. Likewise, the force vector can be written as $F = F_x \hat{i} + F_y \hat{j} = F_x' \hat{i}' + F_y' \hat{j}'$. Because $F_x' = 0$, it is easiest to consider

$$M_o = r \times F = \left(r_x' \hat{i}' + r_y' \hat{j}' \right) \times \left(F_y' \hat{j}' \right)$$

or

$$M_o = (r_x' F_y') \hat{k}' + 0.$$

Note, however, that $d \equiv r_x'$ and $\hat{e}_\perp \equiv \hat{i}'$, thus proving $r \times F = r_\perp \times F$.

---

[7] Vector operations are reviewed in Appendix 7, if needed.

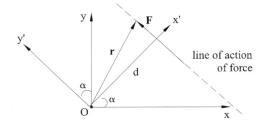

FIGURE 1.10 Schema of a force F acting in the x-y plane relative to an origin O. Although forces are vectors and thus defined independent of a coordinate system (i.e., an origin and basis), they must be resolved into components relative to convenient coordinate systems to permit computations.

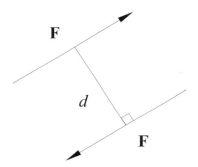

FIGURE 1.11 Schema of couple C (or, pure moment), which is constructed by equal and opposite forces separated by a distance d.

Whereas moments should be computed with respect to a fixed reference point, there exists a special moment whose value is independent of that reference point. Such a moment is called a *couple*. It is constructed via two equal and opposite forces F separated by a distance d, and it has a magnitude of $|F|d$. Figure 1.11 illustrates, in two dimensions, the couple $C = -Fd\hat{k}$. In general, however, we will typically seek to compute $M = r \times F$; hence, let us review a simple method to determine r. Any position vector r from o to A can be determined by subtracting the coordinate locations of A and o with respect to a common coordinate system. *A coordinate system is defined by an origin and a basis*, of course, where a basis is a linearly independent set of (unit) vectors, as, for example, the triplet $(\hat{i}, \hat{j}, \hat{k})$ in a 3-D Cartesian space. For example, if o is located at (x_o, y_o, z_o) and A is at (x_A, y_A, z_A), then the position vector r_A between o and A is

$$r_A = (x_A - x_o)\hat{i} + (y_A - y_o)\hat{j} + (z_A - z_o)\hat{k}. \tag{A1.3}$$

In Engineering Statics, both accelerations are zero and Eqs. (A1.1) reduce to

$$\sum F = 0, \qquad \sum M = 0, \qquad (A1.4)$$

which are our "two" basic equations of mechanical equilibrium. We say two equations, but because each is vectorial, we actually have three total (scalar) equilibrium equations in 2-D problems and six total equations in 3-D problems. Regardless, an important observation is that *if a body is in equilibrium, then each of its parts are in equilibrium.* This observation allows us to make fictitious cuts to isolate parts of a body in order to quantify internal forces and moments that are necessary to maintain equilibrium. For example, consider Figure 1.12: The upper part of the figure shows that the body is maintained in equilibrium by two externally applied forces, f_1 and f_2, which must be equal and opposite; the lower part of the figure shows that these forces are balanced by equal and opposite internal forces f that act on the cut surface, plus possibly equal and opposite moments, which we neglect at first. For example, *equilibrium of the whole* requires

$$f_1 + f_2 = 0 \rightarrow f_1 = -f_2, \qquad (A1.5)$$

whereas *equilibrium of the parts* requires

$$f_2 - f = 0 \rightarrow f = f_2 \quad (\text{to the } \leftarrow) \qquad (A1.6)$$

and

$$f - f_2 = 0 \rightarrow f = f_2 \quad (\text{to the } \rightarrow). \qquad (A1.7)$$

We will see that in solid mechanics, it is often useful to draw such free-body diagrams to analyze the distribution of internal forces and moments. *By a free-body diagram, we mean a drawing of the body of interest, free from all external structures, that depicts all externally applied forces, including those due to interactions between the body and its environment.* Because each part

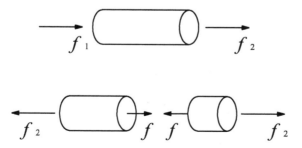

FIGURE 1.12 Free-body diagram of a generic body B subjected to applied end loads f_1 and f_2; a fictitious cut isolates the internal forces f, which are equal and opposite on opposing faces, consistent with Newton's third law.

of a body in equilibrium is also in equilibrium, one often considers multiple free-body diagrams for the same body; that is, multiple judicious, fictitious cuts are often needed to expose and determine all of the internal forces and moments of interest. It is upon these basic ideas of mechanics that we will build; thus, we return to this issue in Chapter 2 and again in Chapter 7. Here, however, let us introduce a few additional topics of statics and consider a few examples. Indeed, because statics is embodied in the two balance relations in Eqs. (A1.4), much of statics simply entails illustrations of the use of these relations in diverse applications.

Example A1.2 Consider the structure in Figure 1.13, a rigid strut fixed at its base and loaded in three dimensions via a cable. Given the applied force and the dimensions and assuming the strut is rigid, find the reactions (forces and moments) at the base of the strut.

Solution: In statics, a *cable* is typically defined as an inextensible structure of negligible mass that only supports a tensile (axial) load. A fixed support is one that completely prevents displacements and rotations of a member at the support. In two dimensions, this means that two displacements (e.g., in x and y) and one rotation (about the z axis) are prevented by horizontal and vertical reaction forces and one reaction moment; in three dimensions, all three displacements and rotations are prevented at a fixed support by three reaction forces and three reaction moments. Letting the reactions at the fixed support be denoted by R_x, R_y, and R_z for the forces and M_x, M_y,

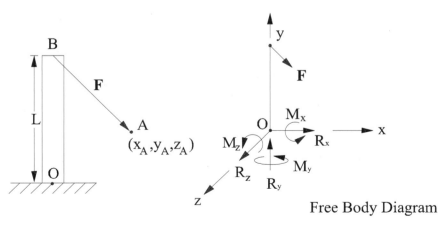

FIGURE 1.13 A rigid, vertical strut is loaded in three dimensions via a rigid cable. A fixed end can, in three dimensions, supply three reaction forces, which resist displacements, and three reaction moments, which resist rotations.

and M_z for the moments, each according to a positive sign convention, and letting the force applied by the cable be $\boldsymbol{F}$, we have from force balance

$$\sum \boldsymbol{F} = \boldsymbol{0} \rightarrow R_x \hat{\boldsymbol{i}} + R_y \hat{\boldsymbol{j}} + R_z \hat{\boldsymbol{k}} + F_x \hat{\boldsymbol{i}} + F_y \hat{\boldsymbol{j}} + F_z \hat{\boldsymbol{k}} = \boldsymbol{0}$$

or

$$(R_x + F_x)\hat{\boldsymbol{i}} + (R_y + F_y)\hat{\boldsymbol{j}} + (R_z + F_z)\hat{\boldsymbol{k}} = \boldsymbol{0} \rightarrow$$
$$R_x = -F_x, \qquad R_y = -F_y, \qquad R_z = -F_z.$$

If we know the magnitude of the force, say $T \equiv |\boldsymbol{F}|$, then

$$\boldsymbol{F} = |\boldsymbol{F}|\hat{\boldsymbol{e}} = T\left(\frac{\boldsymbol{r}_{BA}}{|\boldsymbol{r}_{BA}|}\right) = T\left(\frac{(x_A - x_B)\hat{\boldsymbol{i}} + (y_A - y_B)\hat{\boldsymbol{j}} + (z_A - z_B)\hat{\boldsymbol{k}}}{\sqrt{(x_A - x_B)^2 + (y_A - y_B)^2 + (z_A - z_B)^2}}\right),$$

which thereby yields the components of $\boldsymbol{F}$ in terms of (x, y, z). If, consistent with Figure 1.13, $x_A > x_B$ and $z_A > z_B$, then the values of F_x and F_z will be positive; if $y_B > y_A$, then the value of F_y will be negative, which is to say that the assumed direction of R_y is correct, whereas those for R_x and R_z are not; that is, *a negative value tells us to switch the assumed direction of a particular component of a force (or moment).* It is usually best to assume that all reactions are positive and to let equilibrium determine the actual directions. Likewise, for moment balance, we have

$$\sum \boldsymbol{M}_o = \boldsymbol{0} \rightarrow M_x \hat{\boldsymbol{i}} + M_y \hat{\boldsymbol{j}} + M_z \hat{\boldsymbol{k}} + \boldsymbol{r} \times \boldsymbol{F} = \boldsymbol{0},$$

where $\boldsymbol{r}$ is a position vector from the origin o to any point along the line of action of the force, say point A. Thus,

$$\boldsymbol{r} = (x_A - 0)\hat{\boldsymbol{i}} + (y_A - 0)\hat{\boldsymbol{j}} + (z_A - 0)\hat{\boldsymbol{k}},$$

whereby, remembering that $\hat{\boldsymbol{i}} \cdot \hat{\boldsymbol{i}} = 1$, $\hat{\boldsymbol{i}} \cdot \hat{\boldsymbol{j}} = 0, \dots$, but $\hat{\boldsymbol{i}} \times \hat{\boldsymbol{i}} = \boldsymbol{0}$, $\hat{\boldsymbol{i}} \times \hat{\boldsymbol{j}} = \hat{\boldsymbol{k}}$, $\dots$,

$$\boldsymbol{r} \times \boldsymbol{F} = (y_A F_z - z_A F_y)\hat{\boldsymbol{i}} + (z_A F_x - x_A F_z)\hat{\boldsymbol{j}} + (x_A F_y - y_A F_x)\hat{\boldsymbol{k}}.$$

Consequently,

$$M_x + (y_A F_z - z_A F_y) = 0, \qquad M_y + (z_A F_x - x_A F_z) = 0,$$
$$M_z + (x_A F_y - y_A F_x) = 0,$$

from which the reaction moments are computed easily. Although it is critical to rely on the mathematics to solve 3-D problems, it is also important to consider simple special cases, for they provide important checks and they help to develop our intuition. For example, with $(x_B, y_B, z_B) = (0, L, 0)$, consider a special case where $(x_A, y_A, z_A) = (L, L, 0)$ and, thus, $(F_x, F_y, F_z) = (T, 0, 0)$, which is to say a simple horizontal load at the top of the strut. In this case, we find, as expected,

$$R_x = -F_x = -T, \qquad R_y = 0, \qquad R_z = 0,$$
$$M_x = 0, \qquad M_y = 0, \qquad M_z = TL.$$

Conversely, if we have $(x_A, y_A, z_A) = (0, 2L, 0)$ and $(F_x, F_y, F_z) = (0, T, 0)$ (i.e., a simple tensile end load acting on the strut), then

$$R_x = 0, \qquad R_y = -F_y = -T, \qquad R_z = 0,$$
$$M_x = 0, \qquad M_y = 0, \qquad M_z = 0,$$

which, again, is expected if the line of action of F goes through point o.

Example A1.3 Find the tensions $T_1 = |F_1|$ and $T_2 = |F_2|$ in the two cables in Figure 1.14 that support the weight W.

Solution: A free-body diagram of the (whole) weight reveals that the lines of action of the three forces go through a common point o. Thus, $\Sigma M)_o \equiv 0$ is satisfied identically. Hence, we have two remaining equilibrium equations to find our two unknowns T_1 and T_2. In vector form, $F_1 + F_2 + W = 0$, where $W = W(-\hat{j})$, the component equations of which are

$$\sum F_x = 0 \rightarrow F_{x_1} + F_{x_2} + 0 = 0,$$
$$\sum F_y = 0 \rightarrow F_{y_1} + F_{y_2} - W = 0.$$

Written this way, however, it appears that we have four unknowns (the x, and y components of two force vectors) and just two equations. Clearly, we need more information. Note, therefore, that

$$F_1 = F_{x_1}\hat{i} + F_{y_2}\hat{j} = T_1\hat{e}_1 = T_1(\cos\phi_1\hat{i} + \sin\phi_1\hat{j}),$$
$$F_2 = F_{x_2}\hat{i} + F_{y_2}\hat{j} = T_2\hat{e}_2 = T_2[\cos\phi_2(-\hat{i}) + \sin\phi_2\hat{j}],$$

where ϕ_1 and ϕ_2 are assumed to be known. Hence, the x and y components of the two yet unknown forces actually represent but two unknowns, which can be found (do it) to be

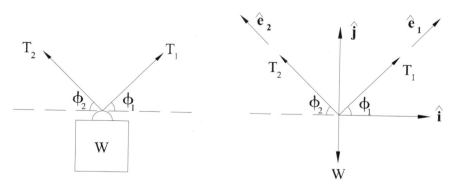

FIGURE 1.14 A weight W is supported by two cables, each having individual tensions $T_1 = |F_1|$ and $T_2 = |F_2|$. The free-body diagram is shown to the right.

$$T_1 = \frac{W}{(\cos\phi_1/\cos\phi_2)(\sin\phi_2) + \sin\phi_1},$$

$$T_2 = \frac{W}{\sin\phi_2 + \sin\phi_1(\cos\phi_2/\cos\phi_1)}.$$

Note, therefore, that if $\phi_1 = \phi_2 = \phi$, then $T_1 = T_2 = W/(2\sin\phi)$, which would go to infinity if $\phi = 0$; that is, a cable cannot support a transverse load without a change in shape.

Pulleys are often used in the biomechanics laboratory (e.g., to calibrate load cells) and in the clinical setting (e.g., to apply traction to a broken limb). An *ideal pulley* is one in which there is no friction (i.e., no resistance to rotation). Consequently, a cable (e.g., a suture, a thin string, or a true metal cable) has the same tension "going onto" and "coming off of" a frictionless pulley.

Example A1.4 Prove that the tension in a cable is the same on each side of an ideal pulley.

Solution: This observation is proved easily considering Figure 1.15. A free-body diagram of the whole structure reveals that the reactions (in two dimensions) at the fixed support are given by

$$\sum F_x' = 0 \rightarrow R_x' - T_1\cos\alpha - T_2\sin\alpha = 0,$$

$$\sum F_y' = 0 \rightarrow R_y' - T_1\sin\alpha - T_2\cos\alpha = 0,$$

$$\sum M_z')_A = 0 \rightarrow M_z\hat{k} + \left[(L\cos\alpha)\hat{i} + (L\sin\alpha + a)\hat{j}\right] \times (-T_1\hat{i})$$

$$+ \left[(L\cos\alpha + a)\hat{i} + (L\sin\alpha)\hat{j}\right] \times (-T_2\hat{j}) = 0\hat{k}$$

in terms of the known quantities, T_1, T_2, L, a, and α. Moreover, a free-body diagram of the pulley alone reveals that

$$\sum M_z')_B = 0 \rightarrow T_1a - T_2a = 0 \rightarrow T_1 = T_2.$$

Not all cables run over a frictionless surface, however. *Friction* is defined as a force of resistance that acts on a body to prevent or retard its slipping with respect to a second body with which it acts. The friction force f thus occurs in a plane tangent to the two contacting bodies and has been found experimentally to be proportional to the normal force N between the con-

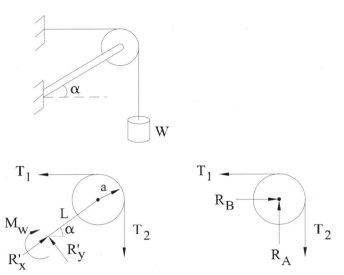

FIGURE 1.15 An ideal (frictionless) pulley of radius a is mounted from a wall at angle α and length L. Free-body diagrams of the whole pulley and its rotating part alone isolate reaction and internal forces and moments of interest.

tacting bodies; the constant of proportionality is the so-called coefficient of friction μ. Thus,

$$f = \mu N. \tag{A1.8}$$

The maximum value of f (i.e., the value just prior to slipping) is given via the coefficient of static friction μ_s; thus $f_{max} = \mu_s N$.

Example A1.5 Find the relationship between the force in a cable as it goes onto and off of a surface with friction.

Solution: Figure 1.16 shows a cable (or belt) that is pulled over a rough surface. In this case, the force (or tension) on the side corresponding to the direction of motion is larger than that on the "feed direction"; that is, $T_L > T_S$. A free-body diagram of a small part of the belt, wherein T_L is only slightly greater than T_S, reveals the pointwise equilibrium result (governing differential equation). Note that the frictional force is denoted by Δf, which we know from physics is related to the normal force ΔN via $\Delta f = \mu_s \Delta N$, where μ_s is the coefficient of static friction. Hence, equilibrium requires

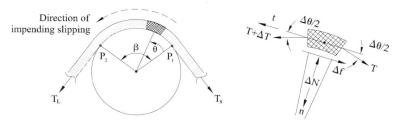

FIGURE 1.16 Free-body diagram of a belt being "pulled" over a rough, cylindrical surface. A differential element (free-body diagram of a part) allows derivation of the governing differential equation.

$$^+\!\uparrow \sum F_{\text{normal}} = 0 \rightarrow -T\sin\left(\frac{\Delta\theta}{2}\right) - (T+\Delta T)\sin\left(\frac{\Delta\theta}{2}\right) + \Delta N = 0,$$

$$\underset{+}{\leftarrow} \sum F_{\text{tangent}} = 0 \rightarrow (T+\Delta T)\cos\left(\frac{\Delta\theta}{2}\right) - T\cos\left(\frac{\Delta\theta}{2}\right) - \Delta f = 0.$$

If we assume $\Delta\theta \ll 1$, then $\sin(\Delta\theta/2) \approx \Delta\theta/2$, $\cos(\Delta\theta/2) \approx 1$, and $\Delta T \approx \Delta f = \mu_s \Delta N$. Hence, the normal force equation requires that

$$-2T\left(\frac{\Delta\theta}{2}\right) - \Delta T\left(\frac{\Delta\theta}{2}\right) = -\Delta N = -\frac{\Delta T}{\mu_s}.$$

If we ignore the higher-order term $\Delta T \Delta\theta/2$ with respect to the other terms, we have

$$T\Delta\theta = \frac{\Delta T}{\mu_s} \rightarrow \frac{\Delta T}{\Delta\theta} = \mu_s T,$$

or in the limit as $\Delta\theta \rightarrow 0$,

$$\frac{dT}{d\theta} = \mu_s T.$$

This first-order differential equation, with a constant coefficient, admits a solution via integration, namely

$$\int \frac{1}{T}\frac{dT}{d\theta}d\theta = \int \mu_s\, d\theta \rightarrow \ln T = \mu_s\theta + c_1.$$

Now, if $T = T_s$ when $\theta = 0$ (i.e., we establish our coordinate system where the belt first contacts the surface), then

$$\ln T_s = 0 + c_1 \rightarrow \ln T - \ln T_s = \mu_s\theta,$$

or

$$\ln\!\left(\frac{T}{T_s}\right) = \mu_s\theta \to T = T_s e^{\mu_s\theta}.$$

The maximum tension is thus $T_L = T_s e^{\mu_s\beta}$, where $\beta = \theta_L - \theta_s$ in Figure 1.16. We see, therefore, that in contrast to the frictionless pulley wherein $T_1 = T_2$, here the ratio of the two tensions is exponentially dependent on the angle of contact and the coefficient of friction.

Finally, let us consider a structure referred to as a *truss*. Such a structure is composed of elements that may support tension or compression along their long axis and are joined together at their ends by pins or welds. It is further assumed that each member is rigid, which is to say, inextensible. Again, an illustrative example serves well to introduce the associated analysis.

Example A1.6 Consider the structure in Figure 1.17a, a 2-D truss fixed at A by a simple pin and at B by a simple roller. Given the applied force at C and the dimensions and assuming the truss is constructed of rigid members, find the reactions (boundary conditions) at A and B as well as all internal forces.

Solution: Let the reactions at the pin be denoted as A_x and A_y (note: a 2-D pin cannot resist a rotation and thus cannot supply a reaction moment) and similarly let the reaction at the roller be denoted as B_y (note: a roller cannot resist a horizontal motion or a rotation), each according to a positive sign convention (Fig. 1.17b). In such problems, it is best to first solve equilibrium of the whole and then equilibrium of individual parts as needed to find all of the values of interest. For overall equilibrium, we have from force balance,

$$\sum F = 0 \to A_x\hat{i} + A_y\hat{j} + B_y\hat{j} + F\hat{i} = 0$$

or in components,

$$A_x + F = 0 \to A_x = -F, \qquad A_y + B_y = 0 \to A_y = -B_y.$$

Now, for moment balance, let us take moments about point A because the lines of action of two of the four forces go through A. Hence,

$$\sum M)_A = 0 \to 2l\hat{i} \times B_y\hat{j} + \left(l\hat{i} + l\hat{j}\right) \times F\hat{i} = 0 \to 2lB_y\hat{k} - lF\hat{k} = 0.$$

From moment balance and the second force balance equations respectively, we thus have

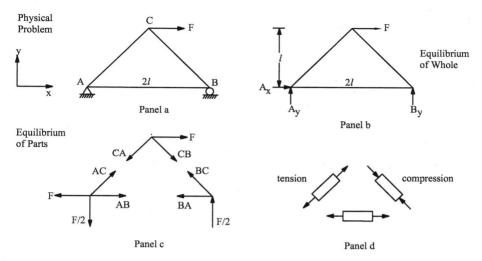

FIGURE 1.17 Shown is a simple 2-D truss and associated free-body diagrams of the whole structure (panel b) and individual parts (panel c). Panel d shows results of the final analysis with regard to which members support tension versus compression.

$$B_y = \frac{F}{2}, \quad A_y = -\frac{F}{2}.$$

It is reemphasized that in two dimensions, we have three scalar equations (summation of force in x and y as well as summation of moments about the z axis at any one point) to find three unknowns; if there are more unknown reactions in the overall problem, it is said to be statically indeterminate, which is to say that we need additional equations to find the reactions. The focus of Chapters 2–5 is the development and use of such additional equations. Here, however, the reactions are now known; thus, we can consider equilibrium of the parts. Toward this end, there are two commonly used approaches: the *method of pins* and the *method of sections*. In the former, one isolates, via a free-body diagram (FBD), the pins at each joint and then enforces equilibrium at each; in the latter, one similarly isolates, via FBDs, sections of the truss and likewise enforces equilibrium. Because of the 2-D nature of each of these subproblems, one can only determine two unknowns in each. We review the method of sections here.

Figure 1.17c shows the truss with three fictitious cuts; note that each cut crosses only two structural elements, thereby isolating only two unknown internal forces each. (Because the lines of action in each "part" go through a point, one can only enforce force balance in the x and y directions and thus solve for only two unknowns in each FBD—Why is this the case?) For the part including pin A, we have,

$$\sum F_x = 0 \rightarrow AC \cos\theta + AB - F = 0,$$

$$\sum F_y = 0 \rightarrow AC \sin\theta - \frac{F}{2} = 0 \rightarrow AC = \frac{F}{2}\frac{2}{\sqrt{2}} = \frac{F}{\sqrt{2}},$$

where $\sin\theta = \sqrt{2}/2$ and $\cos\theta = \sqrt{2}/2$ given the geometry. From the x-direction equation, therefore, we find that $AB = F/2$. Similarly, for the part containing pin B, we have

$$\sum F_x = 0 \rightarrow -BC \cos\theta - BA = 0,$$

$$\sum F_y = 0 \rightarrow BC \sin\theta + \frac{F}{2} = 0 \rightarrow BC = -\frac{F}{\sqrt{2}}$$

and thus from the x-direction equation, $BA = F/2$. Although the solution is now complete (i.e., all reactions and internal forces have been found in terms of F and l), we shall consider the part containing pin C as a consistency check. We have

$$\sum F_x = 0 \rightarrow -CA \sin\theta + CB \sin\theta + F = 0,$$

$$\sum F_y = 0 \rightarrow -CA \cos\theta - CB \cos\theta = 0 \rightarrow CA = -CB,$$

and thus from the x-direction equation, $CB = -F/\sqrt{2}$, and from the y-direction equation, $CA = F/\sqrt{2}$. The correctness of the solutions is thus verified by the consistency check. Finally, note that we assumed a positive sign convention (i.e., tensile load in each member) when denoting each unknown symbolically. When doing so, a positive sign for the solution of an unknown reveals that the direction was assumed correctly, whereas a negative sign reveals that the direction is actually opposite that which was assumed. Hence, as shown in Figure 1.17d, members AB and AC are in tension and BC is in compression.

Although this truss problem is very simple, it reveals most of the important methods of approach in statics: the use of FBDs, equilibrium of the whole followed by equilibrium of individual parts, the use of vectors to sum forces and moments, the need to take moments with respect to specific points, the need to match the number of unknowns and equations, the need to know boundary conditions, and the importance of checking for internal consistency. Indeed, it is for this reason that appreciating the approach to solving simple, illustrative problems serves us well when we approach new, more difficult problems. Although Civil Engineers are typically the ones who study trusses, examination of the cytoskeleton of a cell reveals a truss-like structure and similar approaches have been applied to studying cell mechanics. Of course, the primary assumption that needs to be relaxed when moving from steel structures to cytoskeleton components is that of rigid members. We shall consider in Chapters 2–6 how to begin to address structures and structural members that are not rigid (i.e., solids that deform under the action of applied loads). In such cases, however, we do not "forget" the statics, we merely add new considerations.

Exercises

1.1   Write a four-page (double-spaced, 12-point font) summary of the biomechanical interests of either Leonardo da Vinci or Galileo Galilei, including biographical information. Ensure that references are cited amply and correctly (e.g., see the citation format in current journals such as the *Annals of Biomedical Engineering* or *Biomechanics and Modeling in Mechanobiology*).

1.2   Write a four-page summary of the impact/role of biomechanics in health care research. Illustrate via one or two specific examples.

1.3   Pick a particular biomedical "device" (e.g., a heart valve, an orthotic device, a balloon catheter for angioplasty, an artificial hip, an intraocular device, a tissue engineered skin graft) and review the process of design and analysis that was employed in its development. Submit a four-page summary.

1.4   Identify the top-10 employers of biomedical engineers, with expertise in biomechanics, in your region and discuss their products or service in no more than 4 pages.

1.5   Write a five-to-seven-page summary of Engineering Statics, based upon your prior course work. Pretend that you are charged with giving a review of statics for the Engineering Fundamentals Examination, the first step toward becoming licensed as a Professional Engineer (P.E.), and thus ensure that you review all of the salient features of the subject.

1.6   Write a five-page summary of mechanotransduction in cells. Select a cell type of interest, as, for example, osteoblasts, fibroblasts, or smooth muscle cells.

1.7   Draw a free-body diagram of an epithelial cell that lines the bronchioles and discuss the types of loads that may act on it. Likewise, show a schema of the cytoskeletal architecture in both no-flow and high-flow environments. What morphological or histological changes occur in these airways at the time of birth (i.e., the beginning of breathing)?

1.8   Draw a free-body diagram of an artificial hip implant and discuss the types of load that may act on it. Discuss how different methods of fixation affect the boundary conditions.

1.9   Tissue maintenance depends on a delicate balance between the production and removal of constituents. If the synthesis and degradation of collagen each follow first-order kinetics, namely

$$\frac{d[C]}{dt} = -k[C],$$

where k is a specific reaction rate, find the change in concentration $[C]$ as a function of time.

1.10 The explosion of discoveries in molecular and cellular biology have given rise to new areas of research in bioengineering, including tissue engineering and genetic engineering. Write a four-page review of the state of the art in functional tissue engineering emphasizing the role that biomechanics must play.

1.11 Write a four-page summary of the work by Wolff in the late nineteenth century on remodeling in bone and contrast it with current trends in research in bone mechanics.

1.12 Write a three-page discussion of the differences between *induction* and *deduction*. In particular, consider the roles of Bacon (1561–1626) and Descartes (1596–1650). Because biomechanics combines biology and mechanics, which tend to employ induction and deduction, respectively, discuss how these two different philosophical approaches should be synthesized in modern biomechanical research.

1.13 Write a three-page essay on the difference between *observation* and *experimentation* in the overall scientific method. You may want to consider the commentary of the nineteenth century scientist C. Bernard whose book is entitled *An Introduction to the Study of Experimental Medicine* (reprinted in 1957 by Dover Books, New York).

1.14 Write a four-page summary of the scientific method, including its origins. In particular, define and discuss the role of *hypothesis* in biomechanical research. Illustrate your position by reviewing three to five hypotheses of importance in recent scientific papers.

1.15 Visit the NIH webpage (*www.nih.gov*) and search for information on bioengineering and biomedical engineering. Write a four-page summary of current trends and directions for research.

1.16 Write a four-page summary of new experimental tools that promise to provide new insight into the response of cells and intracellular proteins to applied loads. Consider, for example, the atomic force microscope, laser tweezers, confocal microscopy, magnetic bead cytometry.

1.17 Prove that the value of a couple Fd, where $F = |\mathbf{F}|$ is the magnitude of equal and opposite parallel forces separated by the distance d, is the same regardless of the point about which moments are computed.

1.18 Repeat Example A1.2 with $(x_A, y_A, z_A) = (0, L, L)$ and $|\mathbf{F}| = T$; that is, find the reaction forces and moments.

1.19 In two dimensions, the so-called direction cosines ($\cos \alpha$ and $\cos \beta$ here) are determined easily:

$$\cos \alpha = \frac{A_x}{A}, \qquad \cos \beta = \frac{A_y}{A},$$

where $A = A_x\hat{i} + A_y\hat{j} = A\hat{e}$ with $\hat{e} = \cos\alpha\hat{i} + \sin\alpha\hat{j}$ or $\hat{e} = \cos\alpha\hat{i} + \cos\beta\hat{j}$. Repeat this for three dimensions given angles α, β, and γ, and $A = A_x\hat{j} + A_y\hat{j} + A_z\hat{k}$ with a clear diagram showing all quantities. Because $\hat{e}$ is a unit vector, $|\hat{e}| = 1$, note that α, β, and γ are not independent. Find their interrelationship. Does such an interrelationship make sense in two dimensions?

1.20 Use equilibrium restrictions and vectorial representations for r_1 and r_2 as well as forces F_1 and F_2, with magnitudes $T_1 = |F_1|$ and $T_2 = |F_2|$, respectively, to relate T_1, T_2, d_1, and d_2 in the following figure.

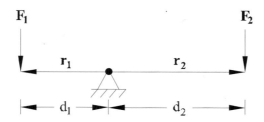

FIGURE 1.18

1.21 The following figure illustrates a simple setup for calibrating a (tension) load cell. Determine the load "felt" by the load cell for each applied weight W. Also, find the reaction supports for the (ideal) pulley.

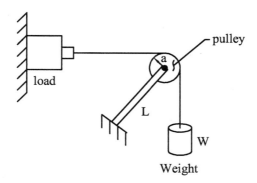

FIGURE 1.19

1.22 Use the following simple setup to design an experiment to determine the coefficient of static friction μ_s between materials A and B. In particular, show that $\mu_s = \tan\theta_s$ where θ_s is the angle at which the relative slippage begins.

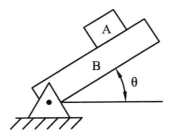

FIGURE 1.20

1.23 Find the internal forces in the truss (see figure) and note the members
that are in compression versus tension.

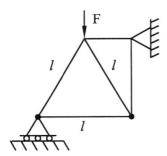

FIGURE 1.21

Part II
Biosolid Mechanics

2
Stress, Strain, and Constitutive Relations

2.1 Introduction

Consider the two structural members in Figure 2.1, each acted upon by an applied weight W that is much larger than the individual weights mg, which we therefore neglect. From statics, we know that if these two members are in equilibrium, then $\Sigma F = 0$ and $\Sigma M = 0$. Free-body diagrams of the whole structure and the individual parts reveal that the reaction and internal forces are the same: $R_y = f_y = W$; that is, from the perspective of statics alone, these two problems are equivalent. Nevertheless, intuition tells us that the behavior of member A need not be the same as that of member B. One may fail before the other. An important question to be answered by mechanics, therefore, may be the following: Which member will likely fail first given increasing weights W? At first glance, we may be inclined to say that A will fail before B, for A is "thinner," and indeed this may well be. Yet, our information is incomplete: We have not specified what A and B are made of; A could be made of a much stronger material than B. Thinking back to statics, we realize that we never specified the properties of the materials or structures that we studied, we simply assumed that they were always rigid (i.e., infinitely stiff). In this book, however, we will see that *the individual properties of materials are central in biomechanics.* For example, we often seek to match the properties of man-made or tissue-engineered replacements to those of the native tissue or organ. Indeed, one of the continuing challenges in biomechanics is accurate characterization, or quantification, of the material behavior of both living tissues and biomaterials.

Returning to Figure 2.1 and the question of whether structure A or B will fail first, we first need to define what is meant by failure. In mechanics, *failure simply implies an inability to perform the intended mechanical function.* Structures A and B could thus fail by the following:

- Material failure, including fracture, tearing or rupture, as, for example, in the tearing of an anterior cruciate ligament

45

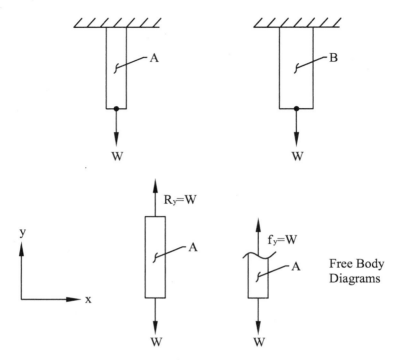

FIGURE 2.1 Contrast the potentially different responses of two simple structural members, A and B, which have the same type of fixed support at the top, the same initial length, and the same axial loading W at the otherwise free-end. Free body diagrams of the whole and a part reveal the reactions at the fixed support and the internal force.

- Deforming excessively, which may or may not include a permanent deformation such as a severely bent (e.g., plastically deformed) surgical instrument, which does not return to its functional shape

Determination of *failure criteria* for materials is thus an important responsibility of the biomedical engineer. Recalling our intuition earlier that structure *A* may fail before *B* because *A* is thinner (Fig. 2.1) suggests that failure criteria cannot be written in terms of the applied loads alone; one must also consider the geometry. This brings us to the concept of stress.

2.2 Concept of Stress

In 1678, Robert Hooke published the anagram (in Latin) *ceiiinossssttuv*, which can be deciphered as, *ut tensio sic vis*, and translated, *as the force, so the extension*; that is, by studying the response of linear metallic springs to

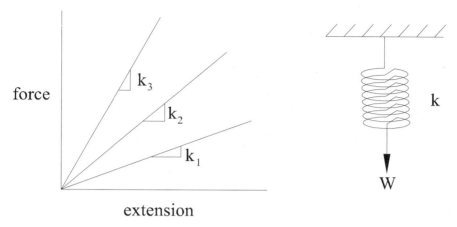

extension

FIGURE 2.2 Force–extension behavior of three different metallic springs, which exhibit linear behaviors and thereby can be quantified by individual spring constants k (or stiffnesses). Although many springs exhibit a linear behavior, nonlinear springs exist as well.

the application of various weights, Hooke realized that there is a one-to-one relationship for many materials between the applied load (force) and the motion (extension). Figure 2.2 shows force–extension curves for three similar but different linear springs, each described by the general formula $f = k(x - x_o)$ where x is the current length, x_o is the original length, and k is the so-called spring constant or stiffness. The results in Figure 2.2 for three different springs suggest that each is characterized by an individual spring constant k_1, k_2, or k_3 (or material property). If we apply the same idea of plotting force versus extension for cylindrical specimens of various materials (e.g., aluminum or stainless steel), we quickly discover that such tests do not characterize the material. If the same loads are applied to the same material in two different labs, which use two different diameter specimens, we find different slopes in the force–extension data. Indeed, the thicker sample, albeit composed of the same material, will appear "stiffer" because it will extend less in response to the same force.[1] Hence, in contrast to Hooke's original idea, there is more to it than just "as the force, so the extension."

In 1757, Leonard Euler realized that a better measure for analysis is a "force intensity" or *stress*. Simply put, Euler defined this intensity as a force acting normal to an area divided by the value of that area (i.e., a pressure-like quantity that we now call a normal stress). During the period 1823–1827, Augustin-Louis Cauchy formalized the concept of stress. Defined as a *force acting over an oriented area* at any point in a body, it is

---

[1] Differences between structural stiffness, which depends in part on geometry, and true material stiffness are important in clinical measurements, as discussed later.

clear that there can be different "stresses" at the same point depending on the orientation of the applied force and the orientation of the area of interest, which implicitly says depending on the choice of a coordinate system (i.e., an origin and basis)—that is, stress is a mathematical construct; its "value" is not unique.

For example, consider a force having only an x component, say Δf_x, which acts over an area ΔA in the current (deformed) configuration of the body (Fig. 2.3). Intuitively, the effect of the same force Δf_x on the same area ΔA will have different effects depending on the orientation of ΔA, which is denoted by the outward unit normal vector n (i.e., $|n| = 1$). For example, if n is in the direction of Δf_x, we call the force a normal force and its intensity (per unit area) a *normal stress*; if n is perpendicular to the direction of Δf_x, we call the force a shearing force and its intensity a *shear stress*. Note, therefore, that although a force could act on an area at any angle, it is generally convenient to resolve the force vector into components that are normal and parallel to the surface. Specifically, then, if we let ΔA_x denote that ΔA has an outward normal $n = \hat{e}_x$ and take the limit as ΔA tends to zero, then we obtain the normal stress:

$$\lim_{\Delta A \to 0} \frac{\Delta f_x}{\Delta A_x} = \frac{df_x}{dA_x} = \sigma_{xx}; \qquad (2.1)$$

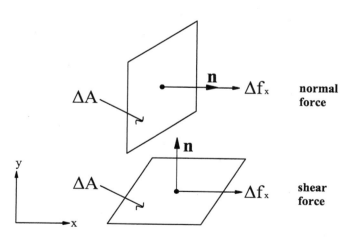

FIGURE 2.3 Schema of the x component of a differential force Δf_x (actually, the mean value of a uniformly applied force) that acts on an area ΔA. Clearly, the effect of this single component of force on the underlying material will depend on the orientation of the area over which it acts: If the area is oriented in the same direction as the force, we expect a tension or compression, whereas if the area is oriented orthogonal to the force, we expect a shearing action. The directions of both the force **f** and the area (given by its outward unit normal vector **n**) are equally important.

that is, we denote stress (i.e., a force acting over an oriented area in the current deformed configuration) with the Greek lowercase sigma, with the first and second subscripts (or indices) associated with the oriented area (i.e., face) on which the force acts and the direction of the applied force, respectively. Hence, with

$$\sigma_{(\text{face})(\text{direction})},\tag{2.2}$$

then

$$\lim_{\Delta A \to 0} \frac{\Delta f_x}{\Delta A_y} = \frac{df_x}{dA_y} = \sigma_{yx}\tag{2.3}$$

for a shear stress in the x-y plane.

Although stresses act in the direction of that component of the force that acts at the point of interest, they are not vectors. Rather, because stress is a force acting over an oriented area, it is associated with two directions, one each for the direction of the force and the outward unit normal $\boldsymbol{n}$. Mathematically, such quantities are called tensors, but we will not exploit this character. It is useful nonetheless to represent the components of stress by arrows that act on the appropriate faces of a body in the appropriate directions. See, for example, Figure 2.4, which shows the so-called *positive sign convention* for a 2-D state of stress relative to a Cartesian coordinate system. In particular, we shall assume that normal stresses are positive

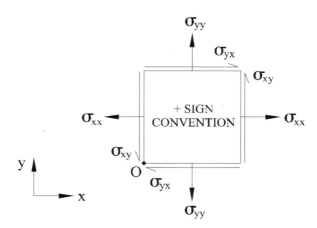

FIGURE 2.4 Positive sign convention for a 2-D state of stress, relative to Cartesian coordinates, that exists at point p but is shown over a square domain for illustrative purposes. The direction of each component is positive on a positive face (i.e., a face with an outward unit normal in the positive coordinate direction) but negative on a negative face. This convention is consistent with normal stresses being positive when tensile and it is consistent with equilibrium at a point (i.e., the balancing of equal and opposite pushes and pulls).

when tensile; this requires that σ_{xx} be directed in a positive direction on a positive face (i.e., one having an outward unit normal in a positive coordinate direction) and conversely that σ_{xx} be directed in a negative direction on a negative face (i.e., one having an outward unit normal in a negative coordinate direction). Indeed, for consistency, we assume the same for the shear stresses, as seen for σ_{yx} and σ_{xy} in Fig. 2.4. (Note: An easy way to remember this positive sign convention is that a positive times a positive is positive and a negative times a negative is a positive; hence, the positive sign convention requires a negative direction stress on a negative face.) As in statics, it is best to use the directions associated with the sign convention; if the computed value turns out to be negative, it simply tells us to switch the assumed direction of that component of stress.

Recall that *if a body is in equilibrium, then each of its parts must also be in equilibrium*—this holds true for any material point p. Note, therefore, that because of our sign convention (Fig. 2.4), the "two" normal stresses σ_{xx} in the figure balance and so too for the x-direction action of the "two" shear stresses σ_{yx}. If the mathematical point p represents a material particle even of infinitesimal dimension, however, we see that equilibrium is not necessarily satisfied; that is, the two σ_{yx}'s would tend to create a couple (i.e., a pure moment, or force acting at a distance) that would tend to rotate the differential region centered at particle p. Equilibrium could be ensured by the addition of an opposing pair σ_{xy}, as seen in Figure 2.4, wherein we have preserved both the positive sign convention (e.g., positive direction on a positive face) and the notation sigma subscript (face, direction). Consequently, $\sigma_{xy} \equiv \sigma_{yx}$ numerically at every point, which can be proven rigorously via the balance of angular momentum (i.e., $\Sigma M = 0$ in this case) as shown below. Committing the sign convention represented in Figure 2.4 to memory serves one well throughout mechanics.

In general, however, we note that each point p could be thought of as an infinitesimal cube that is reduced in size in a limiting process. As such, each point can be thought to have six faces relative to each Cartesian coordinate system. For (x, y, z) coordinates, this implies positive and negative ΔA_x, ΔA_y, and ΔA_z faces. Moreover, given that each point can be acted upon by a force Δf, which has a component representation relative to (x, y, z) as

$$\Delta f = \Delta f_x \hat{i} + \Delta f_y \hat{j} + \Delta f_z \hat{k} \equiv \Delta f_x \hat{e}_x + \Delta f_y \hat{e}_y + \Delta f_z \hat{e}_z, \tag{2.4}$$

where $\hat{i} \equiv \hat{e}_x$ and so forth, there are nine possible measures (i.e., components) of stress at each point p relative to (x, y, z). They are

$$\sigma_{(\text{face})(\text{direction})} = \begin{bmatrix} \dfrac{df_x}{dA_x} & \dfrac{df_y}{dA_x} & \dfrac{df_z}{dA_x} \\[2ex] \dfrac{df_x}{dA_y} & \dfrac{df_y}{dA_y} & \dfrac{df_z}{dA_y} \\[2ex] \dfrac{df_x}{dA_z} & \dfrac{df_y}{dA_z} & \dfrac{df_z}{dA_z} \end{bmatrix} \equiv \begin{bmatrix} \sigma_{xx} & \sigma_{xy} & \sigma_{xz} \\ \sigma_{yx} & \sigma_{yy} & \sigma_{yz} \\ \sigma_{zx} & \sigma_{zy} & \sigma_{zz} \end{bmatrix}, \tag{2.5}$$

which we have written in matrix form for convenience (matrices are reviewed in Appendix 6). The stresses σ_{xx}, σ_{yy}, and σ_{zz} are *normal stresses;* they can cause extension or compression. The stresses σ_{xy}, σ_{xz}, σ_{yx}, σ_{yz}, σ_{zx}, and σ_{zy} are called *shear stresses;* they can cause a body to distort, which is to say to experience changes in internal angles. Consistent with the above, this matrix is symmetric (i.e., $\sigma_{xy} = \sigma_{yx}$, $\sigma_{xz} = \sigma_{zx}$, and $\sigma_{yz} = \sigma_{zy}$). This can be shown formally by letting the dimensions of an infinitesimal element be Δx, Δy, and Δz. Because the components of stress have units of force/area, to sum the moments about an axis such as the z axis in Figure 2.4, we must first multiply the respective component of stress by the area over which it acts and then multiply by the associated moment arm about any point, say o (because two of the four stress components have lines of action that go through o, this point is convenient for computing the moments). Hence,

$$\curvearrowleft + \sum M_z)_o = 0 \rightarrow -\sigma_{yx}(\Delta x \Delta z)\Delta y + \sigma_{xy}(\Delta y \Delta z)\Delta x = 0. \qquad (2.6)$$

Simplifying, therefore, we have the result:

$$\sigma_{xy} = \sigma_{yx}. \qquad (2.7)$$

Similarly, show that $\sigma_{xz} = \sigma_{zx}$ and $\sigma_{yz} = \sigma_{zy}$.

Example 2.1 Referring to Figure 2.5, what are the values of σ_{xx}, σ_{xy}, σ_{yx}, and σ_{yy} in this 2-D state of stress.

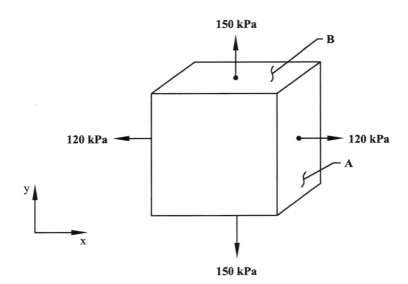

FIGURE 2.5 An illustrative 2-D state of stress acting at a point, components of which act over oriented areas that are expanded for ease of visualization. Although the magnitudes of the components can be considered arbitrary, these values are consistent with in-plane values of stress within a large artery.

Solution: Noting that the right face is an x face (with outward unit normal $\hat{e}_x$) and that the top face is a y face, we have $\sigma_{xx} = 120\,\text{kPa}$, $\sigma_{xy} = 0\,\text{kPa}$, $\sigma_{yy} = 150\,\text{kPa}$, and $\sigma_{yx} = 0\,\text{kPa}$. Being able to identify components of stress $\sigma_{(\text{face})(\text{direction})}$ is an important step in understanding the mechanics.

It cannot be overemphasized that stress is a mathematical concept; it is defined as a measure of a force acting over an oriented area at a point. Mathematically, stress is a tensor, which is defined independent of a coordinate system. Yet, *to solve practical problems, one must always compute components of stress relative to a particular coordinate system.* Because coordinate systems (which are defined by an origin and a set of base vectors) are not unique but can be defined in many different ways, many different sets of components of stress exist at the same point in a body that is subjected to a single set of applied loads. For example, for the three Cartesian coordinate systems shown in Fig. 2.6—defined by $(o;\, \hat{e}_x,\, \hat{e}_y,\, \hat{e}_z)$, $(o;\, \hat{e}'_x,\, \hat{e}'_y,\, \hat{e}'_z)$, and $(o;\, \hat{e}''_x,\, \hat{e}''_y,\, \hat{e}''_z)$—the point p admits three different sets of components of the same stress $\sigma_{(\text{face})(\text{direction})}$:

$$\begin{bmatrix} \sigma_{xx} & \sigma_{xy} & \sigma_{xz} \\ \sigma_{yx} & \sigma_{yy} & \sigma_{yz} \\ \sigma_{zx} & \sigma_{zy} & \sigma_{zz} \end{bmatrix}, \tag{2.8}$$

$$\begin{bmatrix} \sigma'_{xx} & \sigma'_{xy} & \sigma'_{xz} \\ \sigma'_{yx} & \sigma'_{yy} & \sigma'_{yz} \\ \sigma'_{zx} & \sigma'_{zy} & \sigma'_{zz} \end{bmatrix}, \tag{2.9}$$

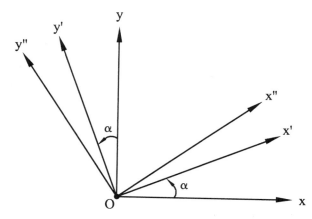

FIGURE 2.6 Interrelations, via single angles, between different Cartesian coordinate systems that share a common origin and a z-axis. The angle α is taken to increase in the direction given by the right-hand rule: counterclockwise, with the z-axis coming out of the paper.

and

$$\begin{bmatrix} \sigma''_{xx} & \sigma''_{xy} & \sigma''_{xz} \\ \sigma''_{yx} & \sigma''_{yy} & \sigma''_{yz} \\ \sigma''_{zx} & \sigma''_{zy} & \sigma''_{zz} \end{bmatrix}. \tag{2.10}$$

[Note: The $(\ldots)'$ and $(\ldots)''$ notation here simply denotes different coordinate systems; it does not imply differentiation as used in many courses on differential equations.]

In some cases, it may be more natural to compute one set of components, say $\sigma_{xx}, \sigma_{xy}, \ldots, \sigma_{zz}$, whereas in other cases, it may be more useful to compute another set of components, say $\sigma'_{xx}, \sigma'_{xy}, \ldots, \sigma'_{zz}$. A good example of this need is the case of a rectangular structure that consists of two members that are glued together on a 45° angle (Fig. 2.7). Because glue is stronger in shear than in extension (empirically compare removing a Post-it® note by applying a normal versus a shear force), it is useful to know how much of the applied force f results in shear versus normal stresses at the glued interface; that is, we would like to know the values of σ'_{xx} and σ'_{xy}, which are computed relative to $(o; \hat{e}'_x, \hat{e}'_y, \hat{e}'_z)$. Yet, from Figure 2.7, it is clearly easier to enforce equilibrium relative to $(o; \hat{e}_x, \hat{e}_y, \hat{e}_z)$; that is, assuming the force is applied uniformly over the surface area on which it acts, it is easy to show (see Section 3.3.2) that $\Sigma F = 0$ yields $\sigma_{xx} = f/A$ and $\sigma_{xy} = 0$ on a cross section with an outward unit normal $n = \hat{e}_x$ that cuts through the glued

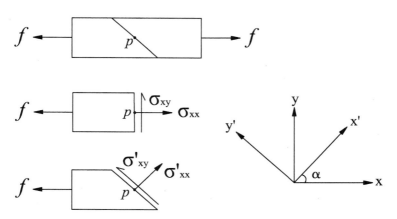

FIGURE 2.7 Free-body diagrams of the same structure cut along two cross sections: one with an outward unit normal in the direction of the applied force (which is natural for solving the equilibrium problem) and one with an outward unit normal to the glued surface (which exposes stresses that act thereon and are important with regard to possible debonding). In each case, the components are identified as $\sigma_{(\text{face})(\text{direction})}$. Note: although we could denote stresses with respect to an x' face and x' direction as $\sigma_{x'x'}$, we prefer to denote them as σ'_{xx} for convenience.

region. It is clear, therefore, that multiple coordinate systems can be useful even in the same problem. Fortunately, we shall discover in the Section 2.3 that the desired values σ'_{xx} and σ'_{xy} can be determined directly from the more easily computed values $\sigma_{xx} = f/A$ and $\sigma_{xy} = 0$; that is, we will not need to solve the equilibrium problem for each coordinate system of interest.

Inasmuch as coordinate systems are introduced for convenience, many different coordinate systems prove useful in the wide variety of problems that fall within the domain of biomechanics. For problems in the circulatory and pulmonary systems, for example, the nearly circular nature of the arteries, capillaries, veins, and bronchioles render cylindrical-polar coordinate systems very useful. For problems involving certain cells, saccular aneurysms, the urinary bladder, and so forth, spherical coordinates are very useful. For problems in cardiac mechanics, particularly for the left ventricle, prolate spheroidal coordinates are useful. For problems in developmental cardiology, toroidal coordinates are convenient. Indeed, the list goes on and on, including more complex coordinate systems. Fortunately, regardless of the coordinate system, our notation $\sigma_{(\text{face})(\text{direction})}$ will hold; that is, we seek measures that describe the intensity of the force relative to both the oriented area on which the force acts and the direction of the applied force. In cylindrical coordinates (r, θ, z), we have (Fig. 2.8)

$$[\sigma] = \begin{bmatrix} \sigma_{rr} & \sigma_{r\theta} & \sigma_{rz} \\ \sigma_{\theta r} & \sigma_{\theta\theta} & \sigma_{\theta z} \\ \sigma_{zr} & \sigma_{z\theta} & \sigma_{zz} \end{bmatrix}, \tag{2.11}$$

and likewise for spherical coordinates (r, θ, ϕ), we have (Fig. 2.9)

$$[\sigma] = \begin{bmatrix} \sigma_{rr} & \sigma_{r\theta} & \sigma_{r\phi} \\ \sigma_{\theta r} & \sigma_{\theta\theta} & \sigma_{\theta\phi} \\ \sigma_{\phi r} & \sigma_{\phi\theta} & \sigma_{\phi\phi} \end{bmatrix}, \tag{2.12}$$

each at every point p. It is important to review and understand that which is represented in these figures.

Independent of the specific coordinate system, a *1-D state of stress* is one in which only one component of stress (e.g., σ_{xx}) is nonzero relative to the prescribed coordinate system; a *2-D state of stress* is one in which four components of stress (e.g., σ_{xx}, σ_{yy}, σ_{xy}, and σ_{yx}, three of which are independent because $\sigma_{xy} = \sigma_{yx}$) may be nonzero relative to the chosen coordinate system; a *3-D state of stress* is one in which all nine components (six of which are independent) may be nonzero in general.

In summary, the concept of stress is a mathematical one. Stress may be computed at each point in a continuum body; when resolved with respect to a coordinate system, there are nine components at each point, although

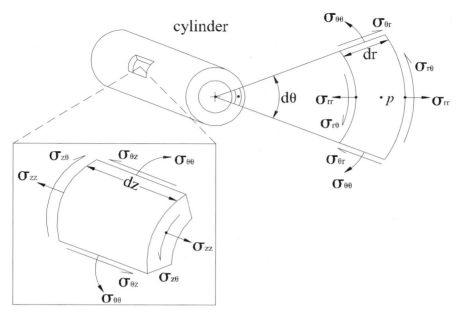

FIGURE 2.8 Components of stress relative to a cylindrical coordinate system, again using the standard notation $\sigma_{(\text{face})(\text{direction})}$.

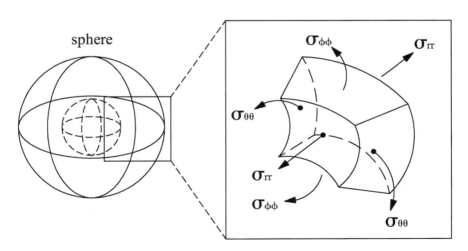

FIGURE 2.9 Normal components of stress relative to a spherical coordinate system, again denoting the components as $\sigma_{(\text{face})(\text{direction})}$. As an exercise, add the shearing components.

only six components are independent relative to each 3-D coordinate system—three normals and three shears. Because coordinate systems can be related via transformation relations, the various components of stress can be related through transformation relations. Let us now derive these useful relations for Cartesian components.

2.3 Stress Transformations

Consider a 2-D state of stress relative to either $(o; \hat{e}_x, \hat{e}_y)$ or $(o; \hat{e}'_x, \hat{e}'_y)$ as shown in Figure 2.10. Because these figures merely represent the stresses that act at point p, we can cut either square part in order to represent components relative to both coordinate systems in a single figure. Anticipating the need to sum forces and moments to enforce equilibrium (of the parts), let the diagonally cut part be of uniform width Δz and length Δy along the vertical cut edge. Hence, the three exposed areas of interest are computed easily, as shown in the Figure 2.11.

From geometry, we have $\sin \alpha = $ opp/hyp and $\cos \alpha = $ adj/hyp, where opp $\equiv \Delta x$ and adj $\equiv \Delta y$. Hence, hyp $= \Delta y/\cos \alpha = \Delta y \sec \alpha$. Now, if we multiply

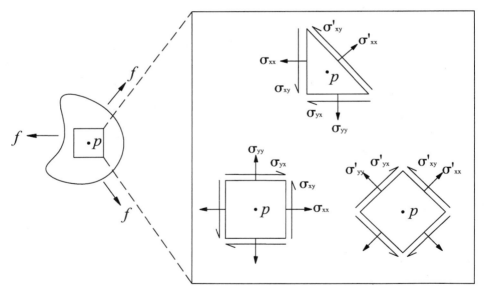

FIGURE 2.10 General 2-D state of stress at the point p emphasizing again that the components are defined with respect to the orientation of the area over which they act (i.e., the face) and the direction of the applied force (i.e., the direction). Hence, different sets of components coexist at the same point. This allows us to make fictitious cuts that expose, on the same element, components relative to different coordinate systems.

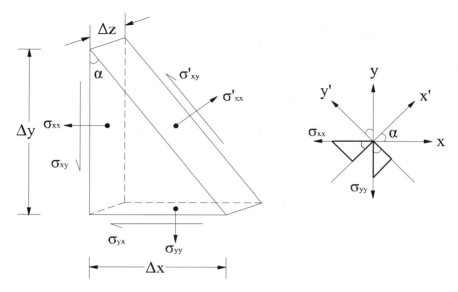

FIGURE 2.11 Detailed diagram of the fictitious element from Figure 2.10 with the 2-D components of stress isolated relative to two different Cartesian coordinate systems. Remembering that if a body is in equilibrium, then each of its parts are in equilibrium, we can therefore use a force balance to relate the components of stress for the two coordinate systems. Alpha is an arbitrary cutting angle.

through by the width Δz, then we have the result that the area that σ'_{xx} and σ'_{xy} act over is given by Δz (hyp) $= \Delta A \sec \alpha$, with $\Delta y \Delta z = \Delta A$ being the area over which σ_{xx} and σ_{xy} act. Similarly, σ_{yy} and σ_{yx} act over $\Delta z(\text{opp}) = \Delta z(\text{hyp})$ $\sin \alpha = \Delta A \sec \alpha \sin \alpha = \Delta A \tan \alpha$. Now, we are ready to sum forces. Balancing forces (i.e., stresses multiplied by the areas over which they act) in the x' direction requires that we find the components in the x' direction. Clearly, the x-directed forces must be multiplied by $\cos \alpha$, whereas the y-directed forces must be multiplied by $\sin \alpha$ to get the x' direction components. Hence, equilibrium yields

$$\sum F_{x'} = 0 = \sigma'_{xx} \Delta A \sec \alpha - (\sigma_{xx} \Delta A) \cos \alpha - (\sigma_{yx} \Delta A \tan \alpha) \cos \alpha$$
$$- (\sigma_{yy} \Delta A \tan \alpha) \sin \alpha - (\sigma_{xy} \Delta A) \sin \alpha$$

or

$$\sigma'_{xx} = \sigma_{xx} \cos^2 \alpha + 2\sigma_{xy} \sin \alpha \cos \alpha + \sigma_{yy} \sin^2 \alpha, \qquad (2.13)$$

wherein we let $\sigma_{xy} = \sigma_{yx}$ from above, and we see that the ΔA cancels throughout, thereby rendering the equation valid for arbitrarily chosen (small) dimensions about point p. It is important to realize, therefore, that the continuum concept of stress actually represents an average force intensity within a small region (neighborhood) centered about the point of interest.

Recalling the trigonometric identities

$$\cos^2 \alpha = \frac{1 + \cos 2\alpha}{2}, \qquad \sin^2 \alpha = \frac{1 - \cos 2\alpha}{2}, \qquad \sin 2\alpha = 2 \sin \alpha \cos \alpha \quad (2.14)$$

Eq. (2.13) can be rewritten as

$$\sigma'_{xx} = \sigma_{xx} \frac{1 + \cos 2\alpha}{2} + \sigma'_{xy} \sin 2\alpha + \sigma_{yy} \frac{1 - \cos 2\alpha}{2}, \qquad (2.15)$$

or

$$\sigma'_{xx} = \frac{\sigma_{xx} + \sigma_{yy}}{2} + \frac{\sigma_{xx} - \sigma_{yy}}{2} \cos 2\alpha + \sigma_{xy} \sin 2\alpha. \qquad (2.16)$$

Given that Eq. (2.13) is a perfectly acceptable way to compute σ'_{xx} from values of stress relative to (o; $\hat{e}_x$, $\hat{e}_y$) for any α, one might ask: Why use the trigonometric identities to obtain the alternate form [Eq. (2.16)]? This is actually a good question, the answer to which comes from *hindsight*. Throughout this text, we must remember that even what may appear to be simple or obvious may have taken great thinkers many years to realize (e.g., nearly 150 years passed between Hooke's ideas on force to Cauchy's on stress). We will see below that Eq. (2.16) is extremely convenient in one particular application. It is also important to remember that we, as students, benefit from the many hours, days, indeed weeks or even years of thought by many which resulted in simplifications we have today.

Forces in the y' direction (Fig. 2.11) can similarly be balanced, namely

$$\sum F_{y'} = 0 = \sigma'_{xy} \Delta A \sec \alpha + (\sigma_{xx} \Delta A) \sin \alpha - (\sigma_{xy} \Delta A) \cos \alpha$$
$$- (\sigma_{yy} \Delta A \tan \alpha) \cos \alpha + (\sigma_{yx} \Delta A \tan \alpha) \sin \alpha,$$

or

$$\sigma'_{xy} = 2 \sin \alpha \cos \alpha \left(\frac{\sigma_{yy} - \sigma_{xx}}{2} \right) + (\cos^2 \alpha - \sin^2 \alpha) \sigma_{xy}. \qquad (2.17)$$

Again, using trigonometric identities, we can rewrite this equation as

$$\sigma'_{xy} = \frac{\sigma_{yy} - \sigma_{xx}}{2} \sin 2\alpha + \sigma_{xy} \cos 2\alpha. \qquad (2.18)$$

Remembering that if a body is in equilibrium, then all of its parts are in equilibrium, we often fictitiously cut a body into multiple different parts to expose, on cut surfaces, specific components of stress of interest. Because the selection of the oblique cutting plane in Figure 2.10 did not isolate a y' surface (i.e., an area with outward unit normal $\hat{e}'_y$), we must consider another free-body diagram that isolates σ'_{yy} and σ'_{yx} (Fig. 2.12). Doing so, we can again balance forces in x' and y'. This is left as an exercise; thus, show that given such a cut,

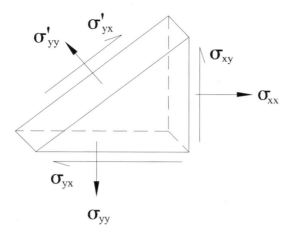

FIGURE 2.12 Alternate fictitious cut (cf. Fig. 2.11) to expose y'-face components of stress σ.

$$\sum F_{x'} = 0 \rightarrow \sigma'_{yx}\Delta A \sec\alpha = -(\sigma_{xx}\Delta A \tan\alpha)\cos\alpha - (\sigma_{xy}\Delta A \tan\alpha)\sin\alpha$$
$$+ (\sigma_{yy}\Delta A)\sin\alpha + (\sigma_{yx}\Delta A)\cos\alpha,$$

or

$$\sigma'_{yx} = 2\sin\alpha\cos\alpha\left(\frac{\sigma_{yy} - \sigma_{xx}}{2}\right) + (\cos^2\alpha - \sin^2\alpha)\sigma_{xy}. \tag{2.19}$$

Again, using trigonometric identities, we have the alternate form

$$\sigma'_{yx} = \frac{\sigma_{yy} - \sigma_{xx}}{2}\sin 2\alpha + \sigma_{xy}\cos 2\alpha, \tag{2.20}$$

which is the same as Eq. (2.18), as it should be (i.e., σ'_{xy} must equal σ'_{yx} to satisfy the balance of angular momentum for a rectangular body cut parallel to x' and y'). Finally, show that

$$\sum F_{y'} = 0 \rightarrow \sigma'_{yy}\Delta A \sec\alpha = (\sigma_{xx}\Delta A \tan\alpha)\sin\alpha - (\sigma_{xy}\Delta A \tan\alpha)\cos\alpha$$
$$+ (\sigma_{yy}\Delta A)\cos\alpha - (\sigma_{yx}\Delta A)\sin\alpha,$$

or

$$\sigma'_{yy} = \sigma_{xx}\sin^2\alpha - 2\sigma_{xy}\sin\alpha\cos\alpha + \sigma_{yy}\cos^2\alpha. \tag{2.21}$$

This equation can then be written as

$$\sigma'_{yy} = \frac{\sigma_{xx} + \sigma_{yy}}{2} + \frac{\sigma_{yy} - \sigma_{xx}}{2}\cos 2\alpha - \sigma_{xy}\sin 2\alpha. \tag{2.22}$$

Together, Eqs. (2.16), (2.18), (2.20), and (2.22) show that the components of a 2-D state of stress relative to one Cartesian coordinate system can be related to those of any other Cartesian system sharing a common origin. All that is needed is the angle α that relates the two coordinate systems; indeed, as a check, we see that at $\alpha = 0$, the $(o; x, y)$ and $(o; x', y')$ coordinate systems coincide, and our transformations yield $\sigma'_{xx} = \sigma_{xx}$ at $\alpha = 0$, and so on, as they should. Although it can be shown that similar transformation relations hold for 3-D states of stress and also for other coordinate systems, we will not go into the details here. Rather, the most important things to realize are that the concept of stress is defined at every point in a continuum body and that the components of the stress (tensor) are not unique; they are determined by the coordinate system of interest. Fortunately, one does not have to solve the equations of equilibrium to determine the value of each component of stress relative to each coordinate system. Rather, *one only needs to solve equilibrium once (in terms of the coordinate system that is most convenient) and then to compute any related component of interest through the transformation relations.* Because these derivations did not require us to specify the material, these relations are good for any solid or fluid as long as the continuum assumption is valid. We will thus use these transformations throughout this book.

Example 2.2 Consider the 2-D state of stress in Figure 2.5. Find the values of stress σ'_{xx}, σ'_{yy}, and σ'_{xy} for $\alpha = 45°$.

Solution: From Eqs. (2.13), (2.21), and (2.17), we have

$$\sigma'_{xx} = 120(\cos 45°)^2 + 2(0)\cos 45° \sin 45° + 150(\sin 45°)^2$$

$$= 120\left(\frac{\sqrt{2}}{2}\right)^2 + 150\left(\frac{\sqrt{2}}{2}\right)^2 = 135\,\text{kPa},$$

$$\sigma'_{yy} = 120\left(\frac{\sqrt{2}}{2}\right)^2 + 150\left(\frac{\sqrt{2}}{2}\right)^2 = 135\,\text{kPa},$$

$$\sigma'_{xy} = 2\left(\frac{\sqrt{2}}{2}\right)\left(\frac{\sqrt{2}}{2}\right)\left(\frac{150 - 120}{2}\right) + 0 = 15\,\text{kPa}.$$

Hence, the state of stress at point p can also be represented as in Figure 2.13. We see, therefore, that a "shearless" state of stress with respect to one coordinate system need not be shearless in general. Indeed, it can be seen from Eq. (2.17) that if $\sigma_{xy} = 0$, then $\sigma'_{xy} = 0$ only if $\sigma_{xx} = \sigma_{yy}$ or $\alpha = 0$ or $\alpha = 90°$.

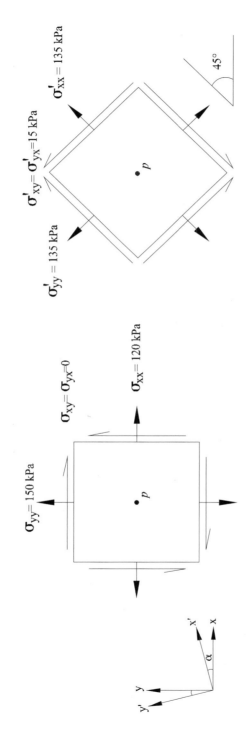

FIGURE 2.13 Two-dimensional state of stress from Figure 2.5 with components computed relative to both the original x-y coordinate system and an x'-y' coordinate system with α = 45°.

2.4 Principal Stresses and Maximum Shear

Given that different values of normal stresses and shear stresses can be computed at the same point in a body depending on the choice of coordinate system (e.g., different Cartesian coordinate systems related via the arbitrary angle α), it is natural to ask if a particular coordinate system exists relative to which the normal or shear stresses are maximum or minimum. The answer, of course, is yes, which will prove very important. For example, if we plot σ'_{xx} as a function of α according to Eq. (2.13) (e.g., for values from Example 2.1 of $\sigma_{xx} = 120 \,\text{kPa}$, $\sigma_{yy} = 150 \,\text{kPa}$, $\sigma_{xy} = 0 \,\text{kPa}$), we obtain the result shown in Figure 2.14, with σ'_{xx} minimum at $\alpha = 0°$ and maximum at $\alpha = 90°$ in this case. Recall from calculus, therefore, that general max/min problems require us to compute a first derivative with respect to the quantity of interest. Hence, to find a maximum or minimum normal stress in two dimensions, relative to Cartesian coordinates, differentiate Eq. (2.16) with respect to α and set the result equal to zero; that is,

$$\frac{d\sigma'_{xx}}{d\alpha} = \frac{\sigma_{xx} - \sigma_{yy}}{2}(-\sin 2\alpha)(2) + \sigma_{xy}(\cos 2\alpha)(2) = 0, \qquad (2.23)$$

or

$$\frac{\sin 2\alpha_p}{\cos 2\alpha_p} = \frac{\sigma_{xy}}{(\sigma_{xx} - \sigma_{yy})/2} = \tan 2\alpha_p. \qquad (2.24)$$

Hence, the maximum or minimum normal stresses σ'_{xx} occur when α is given by

$$\alpha_p = \frac{1}{2}\tan^{-1}\left(\frac{2\sigma_{xy}}{(\sigma_{xx} - \sigma_{yy})}\right). \qquad (2.25)$$

We denote this value of α as α_p because *the maximum/minimum normal stresses are called principal values.* Note that whenever $\sigma_{xy} = 0$, then $\alpha_p = 0$, which is to say, σ_{xx} and σ_{yy} are the max/min values of the normal stress. This was the case in Example 2.2 and thus Figure 2.14. Conversely, if $\sigma_{xx} = \sigma_{yy}$, then $\tan 2\alpha_p = \infty$. Recall that the tangent function goes to infinity at $\pi/2$ radians; hence in this case, $2\alpha_p = \pi/2$ radians, which is to say, $\alpha_p = \pi/4$ radians whenever $\sigma_{xx} = \sigma_{yy}$ regardless of the value of σ_{xy}. All other values of α_p are computed easily.

  Now, if we substitute the value of $\alpha = \alpha_p$ into Eqs. (2.16), (2.18), and (2.22) for σ'_{xx}, σ'_{yy}, and σ'_{xy}, we will find $\sigma_1 \equiv \sigma'_{xx})_{\text{max/min}}$, $\sigma_2 \equiv \sigma'_{yy})_{\text{max/min}}$, and the value of shear associated with these so-called *principal values of stress* σ_1 and σ_2. This is easily done numerically, but it proves useful to note the following. The tangent of an angle equals the opposite over the adjacent. Hence, we can think of a triangle with an angle $2\alpha_p$ and sides as shown in Figure 2.15 (this is hindsight for which we introduced the above trigonometric identities). Hence, we have

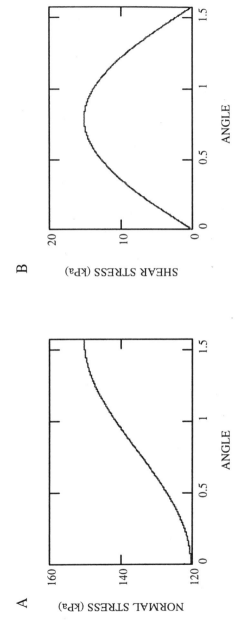

FIGURE 2.14 Plot of the normal and shear stresses σ'_{xx} and σ'_{xy} as a function of $\alpha \in [0, \pi/2]$ radians (i.e., 90°). Note that the local extrema for the normal stress occur at $\alpha = 0$ and 90°, whereas the local maximum for the shear stress is at $\alpha = 45°$.

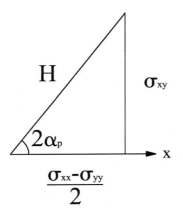

FIGURE 2.15 Trigonometric interpretation of the angle α_p, which is associated with the principal values of stress in two dimensions.

$$\sin 2\alpha_p = \frac{\sigma_{xy}}{H}, \tag{2.26}$$

where

$$H = \sqrt{\left(\frac{\sigma_{xx} - \sigma_{yy}}{2}\right)^2 + \sigma_{xy}^2} \tag{2.27}$$

and, similarly,

$$\cos 2\alpha_p = \frac{(\sigma_{xx} - \sigma_{yy})/2}{H}. \tag{2.28}$$

Using these relations for $\cos 2\alpha_p$ and $\sin 2\alpha_p$ in Eq. (2.16), we have

$$\begin{aligned}
\sigma_{xx}')_{max/min} = \sigma_{xx}'(\alpha = \alpha_p) &\equiv \frac{\sigma_{xx} + \sigma_{yy}}{2} \\
&+ \frac{\sigma_{xx} - \sigma_{yy}}{2}\left(\frac{(\sigma_{xx} - \sigma_{yy})/2}{H}\right) + \sigma_{xy}\left(\frac{\sigma_{xy}}{H}\right),
\end{aligned} \tag{2.29}$$

wherein the second and third terms have a common denominator and can be combined. Multiplying this combined term by unity (i.e., H/H), we have

$$\sigma_{xx}')_{max/min} = \frac{\sigma_{xx} + \sigma_{yy}}{2} + \frac{((\sigma_{xx} - \sigma_{yy})/2)^2 + \sigma_{xy}^2}{H}\frac{H}{H}, \tag{2.30}$$

or, finally,

$$\sigma_1 \equiv \sigma_{xx}')_{max/min} = \frac{\sigma_{xx} + \sigma_{yy}}{2} \pm \sqrt{\left(\frac{\sigma_{xx} - \sigma_{yy}}{2}\right)^2 + \sigma_{xy}^2}. \tag{2.31}$$

Hence, we see that it is easy to compute one of the principal stresses. Show that $\sigma_{yy}')_{max/min}$ yields the same result; that is, the two principal values of stress are given by the same equation with the plus/minus signs preceding the radical delineating the two.

Next, note that if we compute σ'_{xy} at $\alpha = \alpha_p$, we obtain from Eq. (2.18)

$$\sigma'_{xy}(\alpha = \alpha_p) = \left(\frac{\sigma_{yy} - \sigma_{xx}}{2}\right)\left(\frac{\sigma_{xy}}{H}\right) + \sigma_{xy}\left(\frac{(\sigma_{xx} - \sigma_{yy})/2}{H}\right) = 0; \quad (2.32)$$

that is, the shear stress associated with the max/min normal (or principal) stresses is always zero. In other words, *a principal state of stress simply imposes extension or compression, not shear*, relative to the principal directions (defined by α_p).

Finally, one can ask similarly: At what value of α is the shear maximum or minimum? In this case, we differentiate Eq. (2.18) with respect to α and set the result equal to zero. Doing so, we obtain

$$\frac{d\sigma'_{xy}}{d\alpha} = \frac{\sigma_{yy} - \sigma_{xx}}{2}(\cos 2\alpha)(2) + \sigma_{xy}(-\sin 2\alpha)(2) = 0, \quad (2.33)$$

or

$$\frac{\sin 2\alpha_s}{\cos 2\alpha_s} = \frac{(\sigma_{yy} - \sigma_{xx})/2}{\sigma_{xy}} = \tan 2\alpha_s. \quad (2.34)$$

Denoting the value of α at which the shear is max/min as α_s, we thus have

$$2\alpha_s = \tan^{-1}\left(\frac{\sigma_{yy} - \sigma_{xx}}{2\sigma_{xy}}\right). \quad (2.35)$$

Here, we see that if $\sigma_{yy} = \sigma_{xx}$, then $\alpha_s = 0$ and the associated σ_{xy} is an extremum; conversely, if $\sigma_{xy} = 0$, then $2\alpha_s = \pi/2$ or $\alpha_s = \pi/4$. Recalling that the shear stress is zero when the state of stress is principal, this reveals that α_s and α_p differ by $\pi/4$ or $45°$. Substituting the value of α_s into Eq. (2.18) and using ideas similar to those in Figure 2.15, we find that

$$\sigma'_{xy})_{\text{max/min}} = \sigma'_{xy}(\alpha = \alpha_s) = \frac{\sigma_{yy} - \sigma_{xx}}{2}\left(\frac{\sigma_{yy} - \sigma_{xx}}{2H}\right) + \sigma_{xy}\left(\frac{\sigma_{xy}}{H}\right), \quad (2.36)$$

which can be written as

$$\sigma'_{xy})_{\text{max/min}} = \pm\sqrt{\left(\frac{\sigma_{yy} - \sigma_{xx}}{2}\right)^2 + \sigma_{xy}^2} \quad (2.37)$$

or because of the squared term, it is often written (which is the same as H above)

$$\tau_m \equiv \sigma'_{xy})_{\text{max/min}} = \pm\sqrt{\left(\frac{\sigma_{xx} - \sigma_{yy}}{2}\right)^2 + \sigma_{xy}^2}. \quad (2.38)$$

Here, note two things. First, the normal stresses at $\alpha = \alpha_s$ are nonzero in general (this is different from the vanishing shear at $\alpha = \alpha_p$), but computed easily. Second, if the principal stresses occur at $\alpha = 0$, then σ_{xx} and σ_{yy} are principal, whereas $\sigma_{xy} = 0$. In this case, $\sigma'_{xy})_{max/min}$ is simply one-half the difference between the principal values [cf. Eq. (2.38)]. Indeed, it can be shown (do it) that this is the case in general:

$$\tau_m = \pm \frac{\sigma_1 - \sigma_2}{2}, \tag{2.39}$$

where σ_1 and σ_2 are the principal values, usually ordered $\sigma_1 > \sigma_2$.

Example 2.3 For the state of stress in Example 2.1 ($\sigma_{xx} = 120\,$kPa, $\sigma_{yy} = 150$ kPa, $\sigma_{xy} = 0\,$kPa), find α_p and α_s and discuss.

Solution: From Eq. (2.25), we have

$$\alpha_p = \frac{1}{2} \tan^{-1} \left(\frac{\sigma_{xy}}{(\sigma_{xx} - \sigma_{yy})/2} \right) = \frac{1}{2} \tan^{-1} \left(-\frac{0}{15} \right) = 0$$

and, therefore, the $(o; \hat{e}_x, \hat{e}_y)$ coordinate system is principal; that is, the values of σ'_{xx} and σ'_{yy} are max/min at $\alpha = 0$, which is consistent with Figure 2.14 and the finding in Example 2.2 that σ'_{xx} ($\alpha = 45°$) $= \sigma'_{yy}$ ($\alpha = 45°$) $= 135$ kPa, which is an intermediate value between 120 and 150 kPa.
    From Eq. (2.35), we have

$$\alpha_s = \frac{1}{2} \tan^{-1} \left(\frac{150 - 120}{2(0)} \right) = \frac{1}{2} \tan^{-1} (\infty) = \frac{1}{2} \left(\frac{\pi}{2} \right) = \frac{\pi}{4}$$

or 45°. Hence, the value of $\sigma'_{xy})_{max/min} = 15\,$kPa, as computed in Example 2.2.

Example 2.4 Given the 2-D state of stress $\sigma_{xx} = -p$, $\sigma_{yy} = -p$, and $\sigma_{xy} = 0$, show that such a "hydrostatic state of stress" exists relative to all coordinate systems.

Solution: Recall from Eq. (2.13) that

$$\sigma'_{xx} = \sigma_{xx} \cos^2 \alpha + 2\sigma_{xy} \sin\alpha \cos\alpha + \sigma_{yy} \sin^2 \alpha;$$

hence for our state of stress,

$$\sigma'_{xx} = -p(\cos^2 \alpha + \sin^2 \alpha) = -p \quad \forall \alpha$$

and similarly for σ'_{yy}. Likewise, recall from Eq. (2.17) that

$$\sigma'_{xy} = \sin\alpha \cos\alpha (\sigma_{yy} - \sigma_{xx}) + (\cos^2 \alpha - \sin^2 \alpha)\sigma_{xy},$$

and thus for our state of stress,

$$\sigma'_{xy} = \sin\alpha \cos\alpha (-p + p) + 0 = 0 \quad \forall \alpha.$$

A similar finding can be shown in three dimensions. Thus, *a hydrostatic state of stress* (in two dimensions, $\sigma_{xx} = \sigma_{yy} = -p$ and $\sigma_{xy} = 0$, or in three dimensions, $\sigma_{xx} = \sigma_{yy} = \sigma_{zz} = -p$ and $\sigma_{xy} = \sigma_{yz} = \sigma_{zx} = 0$) *is principal relative to all coordinate systems*. This is a very special case.

Finally, the student should be aware that Otto Mohr showed in 1895 that the simple trigonometric structure of these relations [Eqs. (2.26)–(2.28)] for max/min components of stress can be represented easily in a 2-D diagram called Mohr's circle. The interested reader is encouraged to explore this representation via any standard textbook entitled *Strength of Materials* or *Mechanics of Materials*. We shall not discuss Mohr's circles herein because the computer (or calculator) has rendered these computations so easy (compared to the slide rule) that Mohr's circle is no longer needed even though some still use it because of its visual appeal. Rather, we refer the reader to Table 2.1, which reviews the methods discussed herein.

2.5 Concept of Strain

Mechanics is, of course, the study of forces and the associated motions. In dynamics, we tend to study the motion (i.e., kinematics) in terms of quantities like the velocity vector v or the acceleration vector a. These will likewise prove central to our discussion of biofluid mechanics in Chapters 7–10. In biosolid mechanics, however, our primary interest is usually the displacement vector u. Basically, a *displacement vector* quantifies the difference between where we (a point) are, denoted by a position vector x, and where we were originally, denoted by a position vector X. Thus, $u = x - X$ (Fig. 2.16). Because each point in a body can displace separately (provided certain compatibilities are maintained between neighboring points, except in cases of fracture, of course), the displacement vector can vary with position and time, namely

$$u(X, t) = x(X, t) - X, \tag{2.40}$$

where the position vector x also depends on which point (i.e., originally located by X) is being tracked. Because u is a vector, it has components relative to the selected coordinate system. With respect to Cartesian coordinates, we may write

$$u = u_x \hat{i} + u_y \hat{j} + u_z \hat{k} \equiv u_x \hat{e}_x + u_y \hat{e}_y + u_z \hat{e}_z, \tag{2.41}$$

where u_x, u_y, and u_z are the components relative to the chosen Cartesian coordinate system. As a simple example, consider a slender structural member that is fixed at its upper end and loaded by a uniformly distributed force at its other end (Fig. 2.17). From Section 2.2, we can show that each cross section (e.g., that obtained via the cutting plane D-D) has a stress

TABLE 2.1. Flowchart showing our approach for determining max/min values of stress (or strain) relative to preferred coordinate axes.

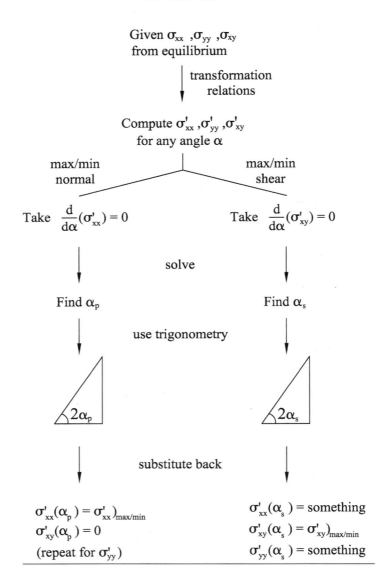

MAXIMUM COMPONENTS OF STRESS

Given σ_{xx} , σ_{yy} , σ_{xy}
from equilibrium

| transformation relations

Compute σ'_{xx} , σ'_{yy} , σ'_{xy}
for any angle α

max/min
normal

max/min
shear

Take $\dfrac{d}{d\alpha}(\sigma'_{xx}) = 0$

Take $\dfrac{d}{d\alpha}(\sigma'_{xy}) = 0$

solve

Find α_p

Find α_s

use trigonometry

$2\alpha_p$

$2\alpha_s$

substitute back

$\sigma'_{xx}(\alpha_p) = \sigma'_{xx}{}_{\text{max/min}}$
$\sigma'_{xy}(\alpha_p) = 0$
(repeat for σ'_{yy})

$\sigma'_{xx}(\alpha_s) = \text{something}$
$\sigma'_{xy}(\alpha_s) = \sigma'_{xy}{}_{\text{max/min}}$
$\sigma'_{yy}(\alpha_s) = \text{something}$

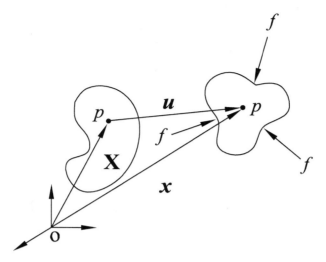

FIGURE 2.16 Schema of the displacement vector u of a generic point p from its location X in an undeformed reference configuration to its location x in a deformed configuration.

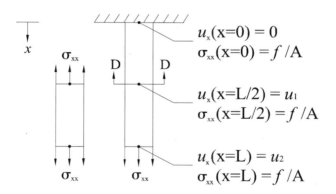

$$u_x(x=0) = 0$$
$$\sigma_{xx}(x=0) = f/A$$

$$u_x(x=L/2) = u_1$$
$$\sigma_{xx}(x=L/2) = f/A$$

$$u_x(x=L) = u_2$$
$$\sigma_{xx}(x=L) = f/A$$

FIGURE 2.17 Displacements at various locations in a uniformly loaded, vertically suspended structural member. Note, in particular, that the value of the displacement varies from point to point (i.e., it is nonuniform), whereas the value of the xx component of stress does not vary.

$\sigma_{xx} = f/A$, where f is the total axial force and A is the cross-sectional area. If the member is of homogeneous composition, we would expect this same stress at each point to cause the same response. An obvious question then is whether the displacement u can serve as a good measure of this response: Do we expect a one-to-one relation between the stress and the displacement at a given point? A quick examination of the problem reveals that the answer is no. Whereas the value of σ_{xx} is the same at all points in this member, the displacement clearly differs from point to point. Because of the fixed support at $X = 0$, the u_x displacement there is zero. Conversely, the u_x displacement at the end of the member is a maximum.[2] As it turns out, the displacement gradient $\partial u_x/\partial X$, like stress, is the same at each point in this simple problem. Without going into details, Cauchy showed during the period 1827–1841 that certain combinations of displacement gradients (called strains) are convenient for relating to the stress. Indeed, because stress and strain are both mathematical concepts, or definitions, various nineteenth and early twentieth-century investigators (Almansi, Green, Kirchhoff, and others) showed that different definitions of stress and strain can be equally useful in different situations. One of the commonly used definitions of strain in biomechanics is that due to George Green in 1841. In terms of Cartesian components, it can be computed via

$$E_{XX} = \frac{\partial u_X}{\partial X} + \frac{1}{2}\left[\left(\frac{\partial u_X}{\partial X}\right)^2 + \left(\frac{\partial u_Y}{\partial X}\right)^2 + \left(\frac{\partial u_Z}{\partial X}\right)^2\right],$$

$$E_{YY} = \frac{\partial u_Y}{\partial Y} + \frac{1}{2}\left[\left(\frac{\partial u_X}{\partial Y}\right)^2 + \left(\frac{\partial u_Y}{\partial Y}\right)^2 + \left(\frac{\partial u_Z}{\partial Y}\right)^2\right],$$

$$E_{ZZ} = \frac{\partial u_Z}{\partial Z} + \frac{1}{2}\left[\left(\frac{\partial u_X}{\partial Z}\right)^2 + \left(\frac{\partial u_Y}{\partial Z}\right)^2 + \left(\frac{\partial u_Z}{\partial Z}\right)^2\right],$$

$$E_{XY} = \frac{1}{2}\left(\frac{\partial u_X}{\partial Y} + \frac{\partial u_Y}{\partial X} + \frac{\partial u_X}{\partial X}\frac{\partial u_X}{\partial Y} + \frac{\partial u_Y}{\partial X}\frac{\partial u_Y}{\partial Y} + \frac{\partial u_Z}{\partial X}\frac{\partial u_Z}{\partial Y}\right) = E_{YX},$$

$$E_{YZ} = \frac{1}{2}\left(\frac{\partial u_Y}{\partial Z} + \frac{\partial u_Z}{\partial Y} + \frac{\partial u_X}{\partial Y}\frac{\partial u_X}{\partial Z} + \frac{\partial u_Y}{\partial Y}\frac{\partial u_Y}{\partial Z} + \frac{\partial u_Z}{\partial Y}\frac{\partial u_Z}{\partial Z}\right) = E_{ZY},$$

$$E_{ZX} = \frac{1}{2}\left(\frac{\partial u_Z}{\partial X} + \frac{\partial u_X}{\partial Z} + \frac{\partial u_X}{\partial Z}\frac{\partial u_X}{\partial X} + \frac{\partial u_Y}{\partial Z}\frac{\partial u_Y}{\partial X} + \frac{\partial u_Z}{\partial Z}\frac{\partial u_Z}{\partial X}\right) = E_{XZ}. \quad (2.42)$$

Relations are similar, but more complex, for other coordinate systems such as cylindrical and spherical. For a complete derivation and interpretation of these relations, see Humphrey (2002). Suffice it to say, however, that one of the reasons that these relations are so useful is that they are insensitive to rigid-body translations or rotations; that is, the components of the *Green strain* measure only the deformation part of a total motion, where we note that it is the deformation (changes in length or internal angle due to applied loads) that we wish to relate to the stress. Clearly, however,

---

[2] If the overall deformation is homogeneous, careful experimental measurements show that $u_x = \Lambda X - X = (\Lambda - 1)X$, where Λ is just a number, a so-called stretch ratio. Stretch ratios are used extensively in Chapter 6.

Green's definition of strain is nonlinear (quadratic) in terms of the displacement gradients. Even for the simple (idealized) example in Figure 2.17, which consists only of an axial extension and associated lateral thinning, we have

$$E_{XX} = \frac{\partial u_X}{\partial X} + \frac{1}{2}\left(\frac{\partial u_X}{\partial X}\right)^2, \qquad E_{YY} = \frac{\partial u_Y}{\partial Y} + \frac{1}{2}\left(\frac{\partial u_Y}{\partial Y}\right)^2,$$

$$E_{ZZ} = \frac{\partial u_Z}{\partial Z} + \frac{1}{2}\left(\frac{\partial u_Z}{\partial Z}\right)^2. \qquad (2.43)$$

As it turns out, the nonlinear terms can introduce considerable complexity into the solution of the full boundary value problem. We will consider such problems in Chapters 6 and 11.

Here, let us consider a tremendous simplification. IF the displacement is small, then $x \sim X$ from $u_X = x - X$ and similarly for $y \sim Y$ and $z \sim Z$; IF the displacement gradients are small, then the nonlinear terms can be neglected in comparison to the linear terms (e.g., if $\partial u_X/\partial X \sim 0.001$, then $\frac{1}{2}(\partial u_X/\partial X)^2$ ~ 0.0000005 is small in comparison); and IF the rigid-body rotations are small (see below), then the Green strains can be *approximated* as

$$[\varepsilon] = \begin{bmatrix} \varepsilon_{xx} & \varepsilon_{xy} & \varepsilon_{xz} \\ \varepsilon_{yx} & \varepsilon_{yy} & \varepsilon_{yz} \\ \varepsilon_{zx} & \varepsilon_{zy} & \varepsilon_{zz} \end{bmatrix}, \qquad (2.44)$$

where

$$\varepsilon_{xx} = \frac{\partial u_x}{\partial x}, \qquad \varepsilon_{xy} = \frac{1}{2}\left(\frac{\partial u_x}{\partial y} + \frac{\partial u_y}{\partial x}\right) = \varepsilon_{yx},$$

$$\varepsilon_{yy} = \frac{\partial u_y}{\partial y}, \qquad \varepsilon_{yz} = \frac{1}{2}\left(\frac{\partial u_z}{\partial y} + \frac{\partial u_y}{\partial z}\right) = \varepsilon_{zy},$$

$$\varepsilon_{zz} = \frac{\partial u_z}{\partial z}, \qquad \varepsilon_{xz} = \frac{1}{2}\left(\frac{\partial u_z}{\partial x} + \frac{\partial u_x}{\partial z}\right) = \varepsilon_{zx}, \qquad (2.45)$$

where ε_{xx}, ε_{yy}, and ε_{zz} are the extensional components and ε_{xy}, ε_{yz}, and ε_{xz} are the shear components of the linearized strain.

Similarly for cylindricals, $\boldsymbol{u} = u_r\hat{\boldsymbol{e}}_r + u_\theta\hat{\boldsymbol{e}}_\theta + u_z\hat{\boldsymbol{e}}_z$ and the linearized (often called small) strains are

$$[\varepsilon] = \begin{bmatrix} \varepsilon_{rr} & \varepsilon_{r\theta} & \varepsilon_{rz} \\ \varepsilon_{\theta r} & \varepsilon_{\theta\theta} & \varepsilon_{\theta z} \\ \varepsilon_{zr} & \varepsilon_{z\theta} & \varepsilon_{zz} \end{bmatrix}, \qquad (2.46)$$

where

$$\varepsilon_{rr} = \frac{\partial u_r}{\partial r}, \qquad \varepsilon_{r\theta} = \frac{1}{2}\left(\frac{1}{r}\frac{\partial u_r}{\partial \theta} + \frac{\partial u_\theta}{\partial r} - \frac{u_\theta}{r}\right) = \varepsilon_{\theta r},$$

$$\varepsilon_{\theta\theta} = \frac{u_r}{r} + \frac{1}{r}\frac{\partial u_\theta}{\partial \theta}, \qquad \varepsilon_{\theta z} = \frac{1}{2}\left(\frac{\partial u_\theta}{\partial z} + \frac{1}{r}\frac{\partial u_z}{\partial \theta}\right) = \varepsilon_{z\theta}, \qquad (2.47)$$

$$\varepsilon_{zz} = \frac{\partial u_z}{\partial z}, \qquad \varepsilon_{rz} = \frac{1}{2}\left(\frac{\partial u_r}{\partial z} + \frac{\partial u_z}{\partial r}\right) = \varepsilon_{zr}.$$

Finally, for sphericals, the linearized strains are

$$[\varepsilon] = \begin{bmatrix} \varepsilon_{rr} & \varepsilon_{r\theta} & \varepsilon_{r\phi} \\ \varepsilon_{\theta r} & \varepsilon_{\theta\theta} & \varepsilon_{\theta\phi} \\ \varepsilon_{\phi r} & \varepsilon_{\phi\theta} & \varepsilon_{\phi\phi} \end{bmatrix}, \qquad (2.48)$$

where

$$\varepsilon_{rr} = \frac{\partial u_r}{\partial r}, \qquad\qquad \varepsilon_{r\theta} = \frac{1}{2}\left(\frac{1}{r}\frac{\partial u_r}{\partial \theta} + \frac{\partial u_\theta}{\partial r} - \frac{u_\theta}{r}\right) = \varepsilon_{\theta r},$$

$$\varepsilon_{\theta\theta} = \frac{u_r}{r} + \frac{1}{r}\frac{\partial u_\theta}{\partial \theta}, \qquad \varepsilon_{\theta\phi} = \frac{1}{2}\left(\frac{1}{r\sin\theta}\frac{\partial u_\theta}{\partial \phi} + \frac{1}{r}\frac{\partial u_\phi}{\partial \theta} - \frac{u_\phi}{r}\cot\phi\right)$$

$$= \varepsilon_{\phi\theta},$$

$$\varepsilon_{\phi\phi} = \frac{1}{r\sin\theta}\frac{\partial u_\phi}{\partial \phi} + \frac{u_\theta}{r}\cot\theta + \frac{u_r}{r}, \quad \varepsilon_{r\phi} = \frac{1}{2}\left(\frac{1}{r\sin\theta}\frac{\partial u_r}{\partial \phi} + \frac{\partial u_\phi}{\partial r} - \frac{u_\phi}{r}\right) = \varepsilon_{\phi\theta}.$$

$$(2.49)$$

Of course, the exact (nonlinear) components can likewise be represented as 3 × 3 matrices because they too consist of nine components (six independent) relative to a particular coordinate system. It is also very important to note that we have not derived the exact (nonlinear) or the approximate (linear) relations for strain; we have merely listed the results. In many introductory textbooks, the linearized relations are often derived poorly, primarily in an attempt to make the derivation "accessible" to the beginning reader. We prefer to adhere to the adage stated in the preface: To tell the truth, nothing but the truth, but not the whole truth until the student is ready to appreciate the whole truth. Hence, rather than derive these relations poorly, let us merely consider a few 1-D or 2-D examples to illustrate their meaning and usage. First, consider a motion described by the displacement vector $\boldsymbol{u} = u_x\hat{\boldsymbol{e}}_x + u_y\hat{\boldsymbol{e}}_y$, with components

$$u_x = (\Lambda - 1)X, \qquad u_y = 0, \qquad (2.50)$$

where Λ is a number close to unity; that is, as the body deforms, none of its material particles displace vertically, whereas particles may displace in the X direction differently: at $X = 0$, there is no displacement, which is to say that the left edge is fixed, whereas the right edge displaces the most. Hence, current positions of points originally at (X, Y) are given by $x = \Lambda X$ and $y = Y$, which allows us to map material points from original to current places (cf. Fig. 2.16). For example, point $(X, Y) = (0, 0)$ stays put, whereas point $(X, Y) = (1, 1)$ goes to $(x, y) = (\Lambda, 1)$ for any value of Λ. Finally, for Λ near unity (which satisfies the above requirement that the displacement and displacement gradients are both small for this problem with zero rigid-body rotation), the linearized strain is

$$\varepsilon_{xx} = \Lambda - 1, \qquad \varepsilon_{yy} = 0, \qquad \varepsilon_{xy} = 0. \tag{2.51}$$

As can be seen in Figure 2.18a, this motion represents a 1-D extension only. That the lineaized values of strain differ from the exact (nonlinear) values is seen easily given that

$$E_{xx} = (\Lambda - 1) + \frac{1}{2}(\Lambda - 1)^2 = \frac{1}{2}(\Lambda^2 - 1) \equiv \frac{1}{2}(\Lambda + 1)(\Lambda - 1),$$

$$E_{xy} = 0, \qquad E_{yy} = 0.$$

Second, consider a motion described by

$$u_x = \kappa Y, \qquad u_y = 0, \tag{2.52}$$

where κ is a number close to zero. Hence, $x = X + \kappa Y$ and $y = Y$ allows us to map points from the undeformed to the deformed configuration. For example, point $(X, Y) = (0, 0)$ stays put again, whereas point $(X, Y) = (0, 1)$ goes to $(x, y) = (\kappa, 1)$. Moreover,

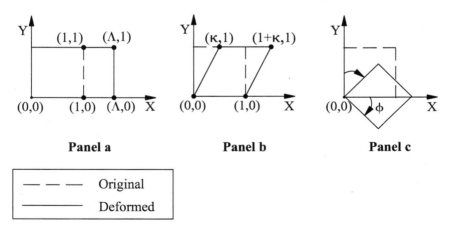

Panel a   **Panel b**   **Panel c**

– – – –	Original
————	Deformed

FIGURE 2.18 Schema of three simple motions: (a) a 1-D extension by the amount Λ, (b) a simple shear via the measure κ, and (c) a rigid-body rotation given by the angle ϕ. In each case, note that we assume that the motion is homogeneous; that is, although the displacements differ from point to point, their spatial gradients do not. In other words, each point experiences the same strain in a homogeneous motion, although each point need not experience the same displacement.

$$\varepsilon_{xx} = 0, \qquad \varepsilon_{yy} = 0, \qquad \varepsilon_{xy} = \frac{1}{2}\kappa, \qquad (2.53)$$

and as can be seen from Figure 2.18b, this motion is one of *simple shear* (if $x = X + \kappa Y$ and $y = Y + \kappa X$, then we would have a *pure shear*). Third, consider a motion given by

$$x = \cos \phi X + \sin \phi Y,$$
$$y = -\sin \phi X + \cos \phi Y, \qquad (2.54)$$

where ϕ is some fixed angle. Hence, the displacements [differences between where we (a point) are (x, y) and where we were (X, Y)] are

$$u_x = (\cos \phi - 1)X + \sin \phi Y,$$
$$u_y = (-\sin \phi)X + (\cos \phi - 1)Y. \qquad (2.55)$$

Consequently, the linearized strains are

$$\varepsilon_{xx} = \cos \phi - 1, \qquad \varepsilon_{yy} = \cos \phi - 1, \qquad \varepsilon_{xy} = \frac{1}{2}(\sin \phi - \sin \phi) = 0. \quad (2.56)$$

Clearly, ε_{xx} and ε_{yy} equal zero *if and only if* $\phi = 0$. If we use these displacements to map the motions of points demarcating a unit square, we find that this case represents a rigid-body rotation about the z axis (Fig. 2.18c). Although we do not expect strains to arise due to rigid-body motions, increasingly larger values of ϕ wrongly suggest increasing extensional strains. As we stated earlier, therefore, *the approximate (linearized) relations for strain are only good for small deformations and small rotations.* This is extremely important to remember in biomechanics, especially in soft tissue biomechanics wherein the deformations and rigid-body motions are often large (finite). This is the case for the heart, for example, which twists, shortens, shears, and becomes much smaller in diameter upon contraction. There are, nonetheless, many articles in the literature that use the small strain measure to study the heart—this is wrong and the reader must beware.

Example 2.5 Show that, in contrast to the linearized measure ε_{xx}, the exact measure E_{XX} is insensitive to the rigid-body motion in Eq. (2.54).

Solution: From Eq. (2.42), the 2-D strain E_{XX} is

$$E_{XX} = \frac{\partial u_X}{\partial X} + \frac{1}{2}\left[\left(\frac{\partial u_X}{\partial X}\right)^2 + \left(\frac{\partial u_Y}{\partial X}\right)^2\right],$$

where for the rigid-body rotation,

$$\frac{\partial u_X}{\partial X} = \cos\phi - 1, \qquad \frac{\partial u_Y}{\partial X} = -\sin\phi.$$

Thus,

$$E_{XX} = \cos\phi - 1 + \frac{1}{2}\left[(\cos\phi - 1)^2 + (-\sin\phi)^2\right]$$

$$= \cos\phi - 1 + \frac{1}{2}(\cos^2\phi - 2\cos\phi + 1 + \sin^2\phi)$$

$$= \cos\phi - 1 + \frac{1}{2}(-2\cos\phi + 2) = 0 \quad \forall\phi.$$

As an exercise, the reader should confirm that $E_{YY} = 0$ and $E_{XY} = 0$ for this rigid-body motion as well.

Observation 2.1 Although we illustrated a few simple states of strain using examples based on the displacements of four points that define a 2-D rectangular domain, we must realize that strains cannot be computed, in general, by simply knowing the displacements at a few points. Strains are computed from displacement gradients, which requires that we know the displacement *field* {i.e., the displacement as a continuous function of position [e.g., $u = u(x, y, z)$]}. Experimentally, however, we cannot measure the displacement at all points; we can only measure the displacements at a finite, often small, number of points. In practice, therefore, one often introduces *interpolation functions*, which allow one to estimate displacements between measurement points. Because the mathematics of interpolation is well established, knowledge of these functions aids the experimentalist in designing the number and placement of markers for measuring displacements. For example, the minimum number of points to estimate the mean 3-D strain in the wall of the heart is four, which forms a tetrahedron.[3] Conversely, the minimum number of points needed to estimate the mean 2-D strain on the surface of the heart is three, which form a triangle. By using multiple sets of four or three markers, one can begin to map region-to-region differences in strain using interpolation functions. The biomechanicist should thus be familiar with interpolation, even though we leave such study for intermediate and advanced courses. See, for example, Humphrey (2002).

---

[3] Note that the discussion in Chapter 5 in Humphrey (2002) contains an error. It correctly notes the need for a minimum of three line segments, but wrongly suggests that they can be obtained from three coplanar points.

In Chapters 3–6, we will seek to *relate the deformations (strains) to the applied loads (stresses) that act on the body*. To do this, we will see that we must use equilibrium equations to determine the stresses that exist at each point, which, in turn, will be related to the strains at the same point through functions that quantify the material behavior (i.e., through constitutive relations). Consequently, it is very important to note the following. Like stress, strain can have different components at each point (given the same deformation) depending on the coordinate system to which it is referred. Fortunately, similar to Eqs. (2.13)–(2.22) for stress, it can be shown that strain transforms in like fashion:

$$\varepsilon'_{xx} = \varepsilon_{xx} \cos^2 \alpha + 2\varepsilon_{xy} \sin\alpha \cos\alpha + \varepsilon_{yy} \sin^2 \alpha,$$

$$\varepsilon'_{yy} = \varepsilon_{xx} \sin^2 \alpha - 2\varepsilon_{xy} \sin\alpha \cos\alpha + \varepsilon_{yy} \cos^2 \alpha, \qquad (2.57)$$

$$\varepsilon'_{xy} = 2\sin\alpha \cos\alpha \left(\frac{\varepsilon_{yy} - \varepsilon_{xx}}{2}\right) + (\cos^2 \alpha - \sin^2 \alpha)\varepsilon_{xy},$$

where α is again the angle that relates the $(o; x, y, z)$ and $(o; x', y', z')$ Cartesian coordinate systems. Note: If $\alpha = 0$, then the components relative to the two systems are equal, as they should be. Similarly, principal values for strain are determined at $\alpha = \alpha_p$ [cf. Eqs. (2.25)–(2.39)], namely

$$\varepsilon_{1,2} = \varepsilon'_{xx})_{\max/\min} = \varepsilon'_{yy})_{\max/\min} = \frac{\varepsilon_{xx} + \varepsilon_{yy}}{2} \pm \sqrt{\left(\frac{\varepsilon_{xx} - \varepsilon_{yy}}{2}\right)^2 + \varepsilon_{xy}^2}, \quad (2.58)$$

with

$$\alpha_p = \frac{1}{2}\tan^{-1}\left(\frac{\varepsilon_{xy}}{(\varepsilon_{xx} - \varepsilon_{yy})/2}\right). \qquad (2.59)$$

Similarly, the maximum value of the shearing strain is determined at $\alpha = \alpha_s$; that is,

$$\varepsilon'_{xy})_{\max/\min} = \varepsilon'_{xy}(\alpha_s) = \pm\sqrt{\left(\frac{\varepsilon_{xx} - \varepsilon_{yy}}{2}\right)^2 + \varepsilon_{xy}^2}, \qquad (2.60)$$

where

$$\alpha_s = \frac{1}{2}\tan^{-1}\left(\frac{\varepsilon_{yy} - \varepsilon_{xx}}{2\varepsilon_{xy}}\right). \qquad (2.61)$$

Whereas the stress transformation equations were derived via equilibrium considerations and thus force balances, here we must take a different approach. Consider, for example, the linearized extensional strain $\varepsilon'_{xx} = \partial u'_x/\partial x'$, just as $\varepsilon_{xx} = \partial u_x/\partial x$. Recall from calculus that two coordinate systems can be related via a coordinate transformation (Fig. 2.6), specifically

$$x' = x\cos\alpha + y\sin\alpha, \qquad y' = -x\sin\alpha + y\cos\alpha \qquad (2.62)$$

whereby $x' = x$ and $y' = y$ if $\alpha = 0$ (i.e., if the coordinate systems coincide). Because displacement is just a vector (i.e., difference between position vectors; Fig. 2.16), we have similar relations for each component,

$$u'_x = u_x \cos\alpha + u_y \sin\alpha, \qquad u'_y = -u_x \sin\alpha + u_y \cos\alpha, \qquad (2.63)$$

where, of course, the displacement components can each vary from point to point in the body: $u_x = u_x(x, y)$ and $u_y = u_y(x, y)$, and likewise $u'_x = u'_x(x', y')$ and $u'_y = u'_y(x', y')$. Yet, from Eq. (2.62), the primed coordinates are a function of the unprimed coordinates, namely $x' = x'(x, y)$ and $y' = y'(x, y)$, and, consequently,

$$u'_x = u'_x(x'(x, y), y'(x, y)), \qquad u'_y = u'_y(x'(x, y), y'(x, y)), \qquad (2.64)$$

which is to say, u'_x and u'_y also depend on position (x, y).

Hence, we can compute strains relative to (x', y') using the chain rule:

$$\varepsilon'_{xx} = \frac{\partial u'_x}{\partial x'} = \frac{\partial u'_x}{\partial x}\frac{\partial x}{\partial x'} + \frac{\partial u'_x}{\partial y}\frac{\partial y}{\partial x'} \qquad (2.65)$$

and so forth. Toward this end, let us first solve for x and y in terms of x' and y' [from Eq. (2.62), which represents two equations and two unknowns]:

$$x = x' \cos\alpha - y' \sin\alpha, \qquad y = x' \sin\alpha + y' \cos\alpha. \qquad (2.66)$$

Now, we have from Eq. (2.65), using Eqs. (2.63) and (2.66),

$$\varepsilon'_{xx} = \left(\frac{\partial u_x}{\partial x}\cos\alpha + \frac{\partial u_y}{\partial x}\sin\alpha\right)(\cos\alpha) + \left(\frac{\partial u_x}{\partial y}\cos\alpha + \frac{\partial u_y}{\partial y}\sin\alpha\right)(\sin\alpha)$$

$$= \frac{\partial u_x}{\partial x}\cos^2\alpha + \left(\frac{\partial u_y}{\partial x} + \frac{\partial u_x}{\partial y}\right)\sin\alpha\cos\alpha + \frac{\partial u_y}{\partial y}\sin^2\alpha, \qquad (2.67)$$

whereby, from Eq. (2.45), we have the desired result,

$$\varepsilon'_{xx} = \varepsilon_{xx}\cos^2\alpha + 2\varepsilon_{xy}\sin\alpha\cos\alpha + \varepsilon_{yy}\sin^2\alpha, \qquad (2.68)$$

which is similar in form to the relation for σ'_{xx}. It is left as an exercise for the reader to find the transformation equations for ε'_{yy} and ε'_{xy}.

As we saw earlier, one way to infer components of strain based on experimental measurements is to place multiple markers (points) on the specimen and to follow their motions. From these motions, we then construct displacement vectors at each point to identify the displacement field, as, for example, $\boldsymbol{u} = \boldsymbol{u}(X, Y, Z)$, from which one can compute the appropriate displacement gradients and thus strains. Indeed, using noncontacting methods [e.g., video, X-ray, magnetic resonance imaging (MRI), laser Doppler] to track the motions of multiple surface or embedded markers is a common way to "measure" strains in soft tissues and even cells. Such approaches are used in applications ranging from gait analysis to quantifying cardiac motion in health and disease. Figure 2.19 shows, for example, that all six components of the finite Green strain are nonzero and changing through-

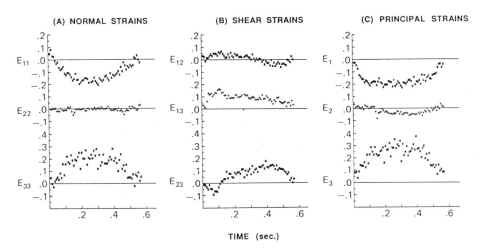

TIME (sec.)

FIGURE 2.19 All six components of the Green strain (extensional and shear) calculated from the motions of small metallic markers that were implanted within the wall of an animal heart. Note that 11, 22, and 33 denote circumferential, axial, and radial components of strain, respectively, with 12, 23, and so forth denoting the associated shears. The principal values, are E_1, E_2, and E_3. Clearly, all six components are nonzero, finite in magnitude, and time varying over the cardiac cycle. [From Waldman et al. (1985), with permission from Lippincot Williams & Wilkins.]

out the cardiac cycle; likewise, they vary from point to point. For more on cardiac motions, see Humphrey (2002).

Here, however, let us consider devices called *strain gauges*, which are useful for inferring surface strains on many engineering structures, from bridges to components on the space shuttle, as well as hard biological tissues. Briefly, in 1856, Lord Kelvin (William Thompson) reported three important observations: The electrical resistance of metallic wires increases with increasing mechanical loads applied along their long axis, different materials have different sensitivities, and the Wheatstone bridge can be used to measure well the changes in resistance. These observations led to the invention of the electrical-resistance strain gauge (Fig. 2.20). These gauges are glued onto the surface of the specimen, which allows them to deform with the underlying specimen. By deforming with the specimen, the electrical resistance changes in the wires, which, in turn, provides (via calibration) the value of the associated extensional strain (i.e., in the predominant direction of the wires). These gauges are very useful when the strains are small, as, for example, in bones, teeth, and biomaterial implants such as an artificial hip. One limitation, however, is that these gauges can only measure extensional components, not shears. To know completely the strain at a point, however, we must know both extensional and shearing

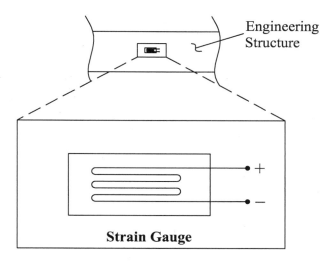

FIGURE 2.20 Schema of an electrical-resistance strain gauge that is glued onto a structure of interest. Such gauges are commonly used in aerospace, civil, and mechanical engineering to measure strains in materials that experience small strains. They are likewise useful in biomechanics for measuring strains in transducers, select biomaterials, and hard tissues such as bone and teeth.

strains in general. Fortunately, theory supports experiment, thus allowing us to make the necessary measurements as shown next.

Example 2.6 Design an experimental set-up using strain gauges whereby one can measure a complete 2-D strain in a small region (i.e., averaged over a small region even though strain is, strictly speaking, defined at a point).

Solution: From Eq. (2.57), we see that an extensional strain relative to a primed coordinate system is related to the 2-D components of strain relative to an original coordinate system. If an extensional strain ε'_{xx} is measurable by a strain gauge, then measuring three extensional strains would provide three equations for the three components ε_{xx}, ε_{yy}, and ε_{xy}; that is, as illustrated in Figure 2.21,

$$\varepsilon'_{xx} = \varepsilon_{xx} \cos^2 \alpha_1 + 2\varepsilon_{xy} \cos\alpha_1 \sin\alpha_1 + \varepsilon_{yy} \sin^2 \alpha_1,$$

$$\varepsilon''_{xx} = \varepsilon_{xx} \cos^2 \alpha_2 + 2\varepsilon_{xy} \cos\alpha_2 \sin\alpha_2 + \varepsilon_{yy} \sin^2 \alpha_2,$$

$$\varepsilon'''_{xx} = \varepsilon_{xx} \cos^2 \alpha_3 + 2\varepsilon_{xy} \cos\alpha_3 \sin\alpha_3 + \varepsilon_{yy} \sin^2 \alpha_3,$$

where the angles α_1, α_2, and α_3 relate the coordinate systems to a baseline x direction. Clearly then, these equations represent three equations in terms

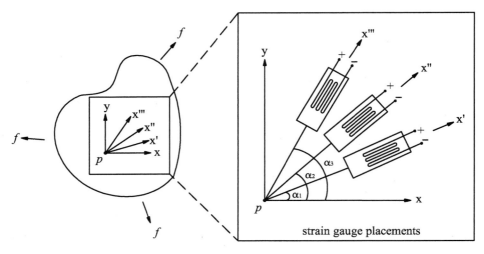

strain gauge placements

FIGURE 2.21 Placement of three strain gauges to form a so-called strain rosette. It is assumed that each gauge is affixed to the surface at a known angle. Although small, strain gauges are obviously of finite, not infinitesimal, size, thus information from rosettes necessarily represent mean values of strain within the region of measurement.

of three unknowns ($\varepsilon_{xx}, \varepsilon_{yy}, \varepsilon_{xy}$) provided that $\varepsilon'_{xx}, \varepsilon''_{xx}, \varepsilon'''_{xx}, \alpha_1, \alpha_2$, and α_3 are measured. (Note: Whereas the strains come from the resistance changes, the angles are known because we are the ones who glue the gauges onto the surface). Although any values of α_1, α_2, and α_3 are fine, certain values are preferred. For example, $\alpha_1 = 0, \alpha_2 = \pi/4$, and $\alpha_3 = \pi/2$ radians or $\alpha_1 = 0$, $\alpha_2 = \pi/3$, and $\alpha_3 = 2\pi/3$ radians are common. For example, let us consider the former case:

$$\varepsilon_{0°} = \varepsilon'_{xx}(\alpha = 0) = \varepsilon_{xx},$$

$$\varepsilon_{45°} = \varepsilon''_{xx}\left(\alpha = \frac{\pi}{4}\right) = \varepsilon_{xx}\left(\frac{\sqrt{2}}{2}\right)^2 + 2\varepsilon_{xy}\left(\frac{\sqrt{2}}{2}\right)\left(\frac{\sqrt{2}}{2}\right) + \varepsilon_{yy}\left(\frac{\sqrt{2}}{2}\right)^2,$$

$$\varepsilon_{90°} = \varepsilon'''_{xx}\left(\alpha = \frac{\pi}{2}\right) = \varepsilon_{yy}.$$

Hence, if $\varepsilon'_{xx}, \varepsilon''_{xx}, \varepsilon'''_{xx}$ are known from the gauges, $\varepsilon_{xx} = \varepsilon_{0°}, \varepsilon_{yy} = \varepsilon_{90°}$, and $\varepsilon_{xy} = \varepsilon_{45°} - \varepsilon_{0°}/2 - \varepsilon_{90°}/2$ are thereby measurable.

We emphasize, therefore, that *theory is indispensable in the design of experiments; it tells us what to measure, why, and to what accuracy.* Moreover, theory reveals the inherent limitations and restrictions. Given that the strain gauge provides information that is averaged along its length and that clusters (rosettes) of gauges further average information over the enclosed region, strain gauges should not be used in areas where large gradients (i.e., point-to-point differences) are expected in the strain field. Again, theory will often reveal the domain of applicability.

In summary, although Hooke's suggestion in the late 17th century, "as the force, so the extension," was profound, we now see that Euler, Cauchy, Green, and others in the eighteenth, nineteenth, and early twentieth centuries showed that the mathematical concepts of stress and strain are often much more useful in continuum mechanics than the physical quantities of force and extension. Being mathematical concepts, however, stress and strain are merely definitions, not physical realities or experimental measurables. Stress and strain can thus be defined in different ways, to suit the particular need, and, fortunately, they can be inferred from experimental "measurables" such as forces, dimensions, and displacements. Because stress and strain are but mathematical concepts, having different components depending on the coordinate system to which they are referred, they cannot be sensed directly by a cell and thus cannot be the stimulus for mechanotransduction (Humphrey, 2001) even though many have suggested otherwise. These quantities can nevertheless be conveniently *correlated* with mechanosensitive responses by cells (e.g., altered gene expression due to a microgravity environment) and thus they can serve as important metrics in phenomenological theories that have predictive capability. More on this later. First, however, let us explore mathematical relationships between stress and strain that serve to quantify the behavior of particular materials.

2.6 Constitutive Behavior

Mathematical relations that describe the response of a material to applied loads under conditions of interest are called *constitutive relations* because this response depends on the internal makeup, or constitution, of the material; that is, given the same overall dimensions, a piece of rubber will respond differently than a piece of metal to the same forces because of the marked differences in their internal makeup—long-chain molecules that are held together via covalent and van der Waals bonds versus collections of atoms that are held together by metallic bonds. Indeed, even different metals and metal alloys respond differently because of differences in their internal makeup and so too for collagenous tissues such as tendons and the cornea, each of which consist largely of type I collagen, albeit with very different microstructural arrangements. Likewise, the conditions of interest must be specified. Rubber, for example, behaves very differently below its glass transition temperature than it does at room temperature or above its melting point. Quantifying, via constitutive relations, the different (solid-like and fluidlike) behaviors of molecules, cells, tissues, organs, biomaterials, and other materials under conditions of importance in biomedical engineering are critically important to both analysis and design.

As noted in Section 1.7, there are five general steps in a constitutive formulation, which can be easily remembered via the acrostic DEICE. First, we must *delineate* general characteristics of the behavior. For example, we must determine if the behavior is solidlike or fluidlike. The former is said to admit a shear stress in equilibrium, no matter how small the shear; the latter is said to be incapable of supporting a shear stress in equilibrium, which is to say that it will flow as long as the shear is applied. We emphasize that although one generally thinks of solids and fluids as phases of matter, in continuum mechanics we really seek to delineate solidlike versus fluidlike *behaviors*. For example, most people would classify glass to be a solid at room temperature, and indeed it exhibits solidlike behaviors at these temperatures. Yet, over many years to centuries, one also finds that glass flows at room temperature, as evidenced by the vertical variations in the thickness of glass window panes in Gothic churches in Europe. Hence, it is really the behavior under the condition of interest, including timescales, that is most important, and we may equally well model glass as a solid or a fluid at room temperature depending on the problem at hand. Inasmuch as this is clear, we can loosely talk about solids versus fluids, as most do; we will discuss particular constitutive behaviors and relationships for biofluids in Chapter 7.

It is also very important to determine if a material's response to an applied load is *linear* or *nonlinear*. For example, if we apply increasingly greater loads (stresses), do we observe proportionate or disproportionate increases in extension (strain). Metals and bone tend to exhibit a linear stress–strain response under small strains (i.e., strains that do not cause permanent changes in the microstructure and, thus, properties). In contrast, elastomers and soft tissues tend to exhibit nonlinear stress–strain responses under large strains without a permanent change in structure, as seen in Figure 2.22. Nonlinear behavior is much harder to quantify. Indeed, note that elastomers and soft tissues exhibit behaviors very different from those of traditional engineering materials (e.g., metals) because of their long-chain polymeric structure. In particular, much of the behavior of such polymers depends on changes in the underlying conformations of the molecules (i.e., their inherent order or disorder). Their mechanical behaviors are thus said to be governed by *entropic* mechanisms in contrast to *energetic* mechanisms that govern the lattice atomic structure in metals. It is, of course, the biopolymers (proteins) elastin and collagen that dominate soft tissue behavior—entropic changes in which complicate the associated quantification.

Another important characteristic exhibited by some solids under certain conditions is a so-called *elastic* behavior. By elastic, it is meant that the material does not dissipate any energy as it deforms. In other words, the path followed by the material in a stress–strain plot is the same during

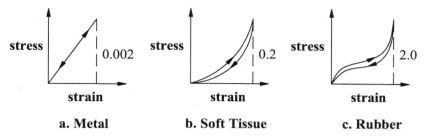

FIGURE 2.22 Qualitative comparison of the stress–strain behavior of three classes of materials: metals, soft tissues, and elastomers. Note the different order of magnitudes of the associated strains (from 0.002 to 2.0) and that the soft tissues and elastomers not only exhibit nonlinear behaviors, but they also reveal a slight hysteresis (i.e., noncoincident loading and unloading curves). The values of stress would obviously be very different as well, but we simply emphasize the general character of the curves here.

loading and unloading and the material will recover its original size and shape when all loads are removed. Moreover, an elastic behavior suggests that a material responds instantaneously to an applied load (again, the importance of timescale). Whereas metals exhibit an elastic response under small strains, tissues and rubber only exhibit a "nearly" elastic behavior under many normal conditions. That the behavior is not purely elastic is evidenced, in part, by the small differences between the loading and unloading curves (hysteresis) in Figure 2.22, the dissipation being due, in part, to moving the structural proteins within the viscous, proteoglycan-dominated ground substance matrix. Fung calls the nearly elastic behavior of soft tissues *pseudoelastic* and offers some ideas to simplify the quantification (Fung, 1990). Constitutive relations for such behavior are discussed in Chapter 6.

If the behavior of a material is independent of the position within the body/structure from which it was taken, we say that the material is *homogeneous*. Obviously, a fiber-reinforced composite like steel-reinforced concrete would not be homogeneous because the steel and surrounding matrix exhibit very different behaviors. In contrast, many metals and rubberlike materials are often homogeneous or at least nearly so, notwithstanding impurities. Although soft tissues are also composites, consisting of elastin, various collagens, proteoglycans, water, and so forth, there are cases in which it is reasonable to consider an associated homogenized behavior. Examples may include describing the behavior of skin, lung parenchyma, myocardium, bone, or even brain tissue under certain circumstances. In other cases, however, accounting for the heterogeneity due to layering (e.g., intima, media, and adventitia in blood vessels or even cortical versus cancellous bone) is essential.

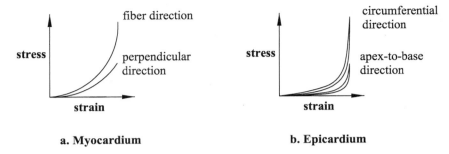

a. Myocardium   b. Epicardium

FIGURE 2.23 Schema of typical stress–strain data from a thin slab of noncontracting myocardium and associated epicardium. Both exhibit nonlinear anisotropic behaviors over finite strains, but the epicardium is more strongly nonlinear because of the initially very compliant behavior that is thought to arise due to the highly undulated collagen (cf. Fig. 1.8) in the unloaded state. Also shown is the slight hysteresis exhibited by the primarily collagenous epicardium; muscle tends to exhibit greater hysteresis (not shown).

Finally, if the behavior of a material is independent of its orientation within the body/structure, we say that its response is *isotropic*. Whereas many metals exhibit isotropy under small strains and rubber exhibits isotropy under large strains, tendons (with axially oriented type I collagen) and the stalks of plants clearly would not exhibit an isotropic response. Indeed, most tissues exhibit *anisotropic* responses (see Fig. 2.23), which again are more difficult to quantify in general. Later in this section, we will discuss two different anisotropies and compare their quantification to that of isotropy for the case of small strains.

Whereas we seek to characterize the responses of materials in terms of concepts such as linearity, elasticity, homogeneity, and isotropy, we emphasize again that these are but descriptors of behavior; no material is linear, elastic, homogeneous, or isotropic. Rather, material behaviors and the constitutive relations that describe them depend on the conditions of interest. Water, for example, behaves differently depending on the temperature; it can behave as a gas (steam), liquid (fluid), or solid (ice), each of which requires a different constitutive descriptor. Common metals also exhibit markedly different behaviors under different conditions. Under the action of a shear stress, the atoms comprising the lattice structure of a metal move relative to one another. If the shear is small (remember, even if the shear is zero relative to one coordinate system, shears will exist relative to other coordinate systems except in the very special case of a hydrostatic pressure as discussed in Example 2.4), the atoms maintain their bonds with their original neighbors, and upon the release of the loads, they return to their original positions (i.e., deform elastically). Under larger strains, however, the atoms cannot maintain bonds with their original neighbors and they slip

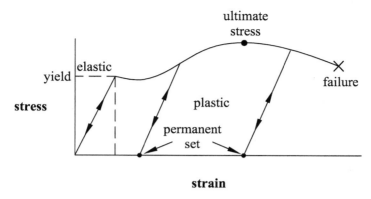

FIGURE 2.24 Schema of the stress–strain behavior of a metal that exhibits a linearly elastic response over small strains but a plastic (i.e., nonrecoverable) response thereafter. In particular, the loading and unloading curves in the plastic domain have a similar character as those in the elastic domain except that the subsequent "yield point" increases with increased plastic deformation (a so-called hardening) up to a point called the ultimate stress. Yield and failure occur due to excessive shear stresses in such ductile materials.

relative to one another and form new bonds with new neighbors (this process is called yielding). Thus, when the load is released, they remain "permanently" displaced rather than going back to their original positions (Fig. 2.24). This is called plastic set, and this inelastic behavior is called *plasticity*. For constitutive relations in plasticity, see Khan and Huang (1995). We see, therefore, that the strain level of interest can also dictate the constitutive behavior of a material. Although it is important to analyze plastic deformations in many fields of engineering (e.g., in metal forming), we seldom design implant biomaterials to exceed their yield point under the action of in vivo loads. Hence, in this book, we will focus on elastic behavior. Finally, note that soft tissues behave differently depending on whether they are hydrated, heated excessively, or exposed to certain medications. *Because constitutive relations describe material behavior, not the material itself*, the bioengineer must always be mindful of the specific conditions under which the material will perform, knowing that multiple constitutive relations may be necessary to describe the behaviors of the same material under different conditions.

2.6.1 Illustrative Characteristic Behaviors

Figure 2.25 shows illustrative data from a uniaxial test on a bovine chordae tendineae. This tissue connects the heart valve to the papillary muscle within the ventricular cavity of the heart; it consists primarily of uniaxially oriented type I collagen having only a slight undulation when unloaded. As

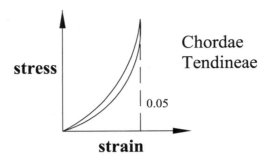

FIGURE 2.25 Schema of stress–strain data from a uniaxial test on an excised chordae tendineae, the thin stringlike tissue that connects the heart valve to the papillary muscle. Note the small, but not infinitesimal, strain. Many ligaments and tendons exhibit similar behavior.

seen in the figure, chordae (similar to tendons and ligaments of the joints) exhibit a nonlinear stress–stretch response over finite (not infinitesimal) but moderate strains. Because of the highly oriented collagen fibers, chordae are strongly anisotropic; because of the slight hysteresis upon cyclic loading/unloading, there is slight energy dissipation; because of the presence of a thin membranous covering (sheath), the tissue is not homogeneous. *Nonlinearity, inelasticity, anisotropy, and heterogeneity are common characteristics of soft tissues.* Figure 2.23 shows similar responses by excised noncontracting myocardium and epicardium. The latter is a thin collagenous membrane that covers the outer surface of the heart. Whereas the myocardium consists primarily of locally parallel muscle fibers embedded in a 3-D plexus of collagen and a ground substance matrix, the epicardium consists primarily of a 2-D plexus of collagen and elastin embedded in its ground substance matrix (proteoglycans and bound water). The collagen fibers tend to be highly undulated in both tissues in an unloaded configuration, hence the initially very compliant, perhaps isotropic response by the epicardium that is followed by a rapid stiffening (due to the straightening of the fibers). The initially greater stiffness of the myocardium is due to the presence of myofibers. Although the chordae and epicardium consist of very similar constituents, their behaviors are very different because of the different microarchitectures. Histology, the study of the fine structure of tissues, thus plays an important role in constitutive formulations. We will consider soft tissue constitutive relations in Chapter 6.

In contrast, Figure 2.26 shows results from a uniaxial test on bone. Note the much smaller range of strain and the near linear behavior. Although not shown, bone exhibits anisotropy and it is heterogeneous—cortical and cancellous bone being very different, as discussed in Chapter 4. Quantification of the stress–strain behavior of bone is discussed in Section 2.7. Although we could discuss much more about the characteristic behaviors

FIGURE 2.26 Schema of stress–strain data from bone prior to yield, which reveals an initially linear, nearly elastic response. Note that bone (type I collagen impregnated with hydroxyapatite) is much stiffer and less extensible than the chordae (Fig. 2.25; primarily type I collagen).

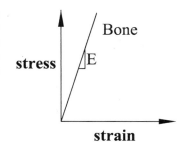

of these and other solids, we refer the student to texts on material science and biomaterials (e.g., Askeland, 1994; Ratner, 2003), which emphasize the need for biomechanics and material science to go hand-in-hand. Here, we simply note that we will focus primarily on a class of material behaviors that we refer to as LEHI:

Linear: linear stress–strain behavior
Elastic: no dissipation and the loading/unloading curve coincide
Homogeneous: same material behavior everywhere in the material/body
Isotropic: same material response in all directions at a point

2.6.2 Hookean LEHI Behavior

Due largely to A.L. Cauchy, S.D. Poisson, G. Lamé, L.M.H. Navier, and G. Green in the early to mid nineteenth century, a constitutive relation was established for LEHI behavior under small strains. It is,

$$\varepsilon_{xx} = \frac{1}{E}[\sigma_{xx} - v(\sigma_{yy} + \sigma_{zz})] + \beta \Delta T, \qquad \varepsilon_{xy} = \frac{1}{2G}\sigma_{xy},$$

$$\varepsilon_{yy} = \frac{1}{E}[\sigma_{yy} - v(\sigma_{xx} + \sigma_{zz})] + \beta \Delta T, \qquad \varepsilon_{xz} = \frac{1}{2G}\sigma_{xz}, \qquad (2.69)$$

$$\varepsilon_{zz} = \frac{1}{E}[\sigma_{zz} - v(\sigma_{xx} + \sigma_{yy})] + \beta \Delta T, \qquad \varepsilon_{yz} = \frac{1}{2G}\sigma_{yz},$$

where T is the temperature and E, v, G, and β are material parameters, the specific values of which vary from material to material. In particular, E is called Young's modulus (after T. Young, a physician interested in biomechanics, who, for example, gave lectures in 1808 to the Royal Society of London on the biomechanics of arteries); E is a measure of the extensional stiffness (i.e., change of stress with respect to strain) of a material, which can be inferred by plotting normal stress versus extensional strain in a uniaxial stress test. The parameter v is called Poisson's ratio; it describes a coupling between orthogonal directions and is often defined as $v = -\varepsilon_{lateral}/\varepsilon_{axial}$, which is to say that it describes the thinning of a material that is extended.

Thermodynamics shows that $-1 < v < \dfrac{1}{2}$, the value of $\dfrac{1}{2}$ being associated with an incompressible behavior (see below). G is called the shear modulus; it provides a measure of the resistance to shear. It can be shown that $G = E/2(1 + v)$ for LEHI behavior. Finally, β is a coefficient of thermal expansion; it tells us how much the material expands/contracts due to changes in temperature from some reference temperature T_o; that is, $\Delta T = T - T_o$ and thus there is no thermal effect when the material is isothermal at T_o. Although the body regulates temperature very closely at ~37°C, clinical interventions often involve local warming (e.g., hyperthermia treatment of cancerous cells) or cooling (e.g., cryosurgery). We will focus on isothermal behavior, however. Table A2.1 lists values of these parameters for various materials.

Example 2.7 Given values of $E = 14\,\mathrm{GPa}$ and $v = 0.32$, which are reasonable values for bone, find the values of strain for a LEHI behavior and the 2-D state of stress in Example 2.1: $\sigma_{xx} = 120\,\mathrm{kPa}$, $\sigma_{yy} = 150\,\mathrm{kPa}$, and $\sigma_{xy} = 0$. Note: $1\,\mathrm{GPa} = 10^3\,\mathrm{MPa} = 10^6\,\mathrm{kPa} = 10^9\,\mathrm{Pa}$, where $1\,\mathrm{Pa} = 1\,\mathrm{N/m^2}$.

Solution: From Eq. 2.69,

$$\varepsilon_{xx} = \frac{1}{14 \times 10^6}[120 - 0.32(150 + 0)]\frac{\mathrm{kPa}}{\mathrm{kPa}} = 5.1 \times 10^{-6},$$

$$\varepsilon_{yy} = \frac{1}{14 \times 10^6}[150 - 0.32(120 + 0)] = 8.0 \times 10^{-6}$$

$$\varepsilon_{xy} = \frac{1}{2G}(0) = 0,$$

where $G = E/2(1 + v)$. Unless the problem is treated as purely two dimensional,

$$\varepsilon_{zz} = \frac{1}{14 \times 10^6}[0 - 0.32(150 + 120)] = -6.2 \times 10^{-6},$$

which is to say that the material will thin in the z direction due to the (in-plane) stresses in the x and y directions. Note, too, that each value of strain is much less than unity, consistent with the small strain requirement for Hooke's law, and positive values denote lengthening, whereas negative values denote shortening. Moreover, strain is unitless (it represents normalized changes in length and changes in angles) and a value of strain times 10^{-6} is often called a microstrain ($\mu\varepsilon$), as, for example, $5.1 \times 10^{-6} \equiv 5.1\,\mu\varepsilon$.

Finally, it is important to note that Eq. (2.69) is called Hooke's law (although it is merely a constitutive relation, not a law) to commemorate R. Hooke's profound observation relating force and extension even though Hooke had no concept of stress or strain. Hooke's law can be derived mathematically because of the assumptions of linearity, elasticity, homogeneity, and isotropy (LEHI), but we did not do so here; we merely listed the final form. Hooke's law can also be established through a comprehensive battery of laboratory observations and experiments; again, we did not provide the associated, detailed information. Herein, therefore, we will focus on its use, not its formulation. Qualitatively, it is also useful to note that for a 1-D state of stress, say $\sigma_{xx} = f/A$, under isothermal conditions, $\varepsilon_{xx} = \sigma_{xx}/E$ and $\varepsilon_{yy} = \varepsilon_{zz} = -\nu(\sigma_{xx})/E = -\nu\varepsilon_{xx}$. Hence, it is easy to see how one could/would design a uniaxial experiment to determine the values of E and ν for a LEHI behavior; the value of G could then be calculated as $E/2(1 + \nu)$ and verified via a shear test. For example, E is simply the constant slope (stiffness) in the σ_{xx} versus ε_{xx} plot (cf. Fig. 2.26); that ν is a measure of the lateral thinning relative to the axial extension is seen by taking (for this 1-D state of stress)

$$-\frac{\varepsilon_{yy}}{\varepsilon_{xx}} = -\frac{(-\nu\sigma_{xx}/E)}{\sigma_{xx}/E} = \nu = -\frac{\varepsilon_{zz}}{\varepsilon_{xx}}, \qquad (2.70)$$

which reveals how its value can be inferred from experiment.

Finally, let us consider Poisson's ratio in more detail. Noting that $\varepsilon_{xx} = \partial u_x/\partial x \cong \partial u_x/\partial X$ where $u_x = x(X) - X$, we can think of (roughly, but not rigorously) an extensional strain ε_{xx} over a small region as a change in length divided by the original length: that is, $\varepsilon_{xx} \sim (\Delta x - \Delta X)/\Delta X = \Delta u_x/\Delta X$, where Δx is the current length and ΔX is the original length. Now, if we consider a cube having initial dimensions ΔX, ΔY, and ΔZ and deformed dimensions Δx, Δy, and Δz, then the current (deformed) volume $\Delta v = \Delta x \Delta y \Delta z$ can also be computed as

$$\begin{aligned}
\Delta v &= \Delta x \Delta y \Delta z = \Delta X(1+\varepsilon_{xx})\Delta Y(1+\varepsilon_{yy})\Delta Z(1+\varepsilon_{zz}) \\
&= (\Delta X \Delta Y \Delta Z)[(1+\varepsilon_{xx})(1+\varepsilon_{yy})(1+\varepsilon_{zz})] \\
&= (\Delta V)[1+\varepsilon_{zz}+\varepsilon_{yy}+\varepsilon_{yy}\varepsilon_{zz}+\varepsilon_{xx}+\varepsilon_{xx}\varepsilon_{zz}+\varepsilon_{xx}\varepsilon_{yy}+\varepsilon_{xx}\varepsilon_{yy}\varepsilon_{zz}] \\
&= (\Delta V)[1+\varepsilon_{xx}+\varepsilon_{yy}+\varepsilon_{zz}]+\text{H.O.T.},
\end{aligned} \qquad (2.71)$$

or

$$\Delta v = \Delta V + \Delta V(\varepsilon_{xx}+\varepsilon_{yy}+\varepsilon_{zz}) \rightarrow \frac{\Delta v - \Delta V}{\Delta V} = \varepsilon_{xx}+\varepsilon_{yy}+\varepsilon_{zz}, \qquad (2.72)$$

where H.O.T. stands for higher-order terms, terms that can be neglected in comparison to other terms (i.e., given that ε_{xx}, ε_{yy}, $\varepsilon_{zz} \ll 1$, quadratic and cubic terms are negligible with respect to linear terms). Now, if we let ε_{xx}

be the axial direction strain and ε_{yy} and ε_{zz} be the lateral direction strains, then by the above definition,

$$\varepsilon_{yy} = -v\varepsilon_{xx}, \qquad \varepsilon_{zz} = -v\varepsilon_{xx}, \qquad (2.73)$$

and, thus,

$$\frac{\Delta v - \Delta V}{\Delta V} = -v\varepsilon_{xx} + \varepsilon_{xx} - v\varepsilon_{xx} \rightarrow \frac{\Delta v - \Delta V}{\Delta V} = \varepsilon_{xx}(1-2v). \qquad (2.74)$$

Hence, if there is no volume change, then $v = \frac{1}{2}$ as alluded to earlier, which is to say that the material deforms incompressibly. Determination of a value of a material parameter from a thought experiment is thus possible, albeit uncommon. In most cases, the value of a material parameter must be calculated directly from experimental data. Note, therefore, that with respect to the aforementioned steps in formulating a constitutive relation (DEICE): *delineation* of characteristic behaviors represented our defining LEHI behaviors; *establishing* a theory involved constructing a theory of elasticity in which stress σ and strain ε are related; specifying linear relations (Hooke's law) *identifies* the specific form of the relation; and finding values for E, v, and G represents the *calculation* of the material parameters. *Evaluating* the predictive capability is thus the final step, which is typically performed by comparing computed and measured values of stress or strain for situations not used to formulate the constitutive relation. For example, if we find values of E and v for a particular material from a uniaxial test, we will want to ensure that these values also provide a good description of the behavior of the material in torsion and bending, particularly if these situations are experienced in service conditions.

2.6.3 Hooke's Law for Transverse Isotropy

We emphasize that "Hooke's law" as stated in Eq. (2.69) holds if the material behavior is isotropic (i.e., the behavior is independent of the direction the force is applied at a point within the material). This can be seen, for example, by interchanging the subscripts x, y, and z in the equations, which leaves them unchanged.

Whereas many metals exhibit an isotropic behavior under small strains, many other materials do not. Wood, fiberglass, and other man-made composites as well as tendons, ligaments, skin, bone and most other biological tissues exhibit an anisotropy. Consider, for example, a piece of wood. It is clear that the mechanical response in the direction of the grain is different from that across the grain. The same is true of heart muscle (Fig. 2.23) due to the locally parallel arrangement of the muscle fibers. When a material has a different behavior in one direction compared to all directions in an orthogonal plane, the behavior is said to be *transversely isotropic* (i.e., isotropic in a plane transverse to a preferred or different direction). If the transversely isotropic behavior is otherwise linear, elastic, and homogeneous under small strains, it is describable via a transversely isotropic Hooke's law of the form

$$\varepsilon_{xx} = \frac{1}{E}(\sigma_{xx} - v\sigma_{yy}) - \frac{v'}{E'}\sigma_{zz}, \qquad \varepsilon_{xy} = \frac{1}{2G}\sigma_{xy},$$

$$\varepsilon_{yy} = \frac{1}{E}(\sigma_{yy} - v\sigma_{xx}) - \frac{v'}{E'}\sigma_{zz}, \qquad \varepsilon_{xz} = \frac{1}{2G'}\sigma_{xz}, \qquad (2.75)$$

$$\varepsilon_{zz} = \frac{1}{E'}\sigma_{zz} - \frac{v'}{E}(\sigma_{xx} + \sigma_{yy}), \qquad \varepsilon_{yz} = \frac{1}{2G'}\sigma_{yz},$$

where, again,

$$G = \frac{E}{2(1+v)}, \qquad (2.76)$$

with the z direction (arbitrarily) taken to be the preferred direction. Note that in contrast to the relation for isotropic behavior [Eq. (2.69)], which is described by two independent parameters (E, v, and G being related to these), this relation for transversely isotropic behavior is described by five independent parameters (two Young's moduli E and E', two Poisson's ratios v and v', and a shear modulus G', where G is, again, related to E and v and thus is not independent). Again, because of the linearity, this relation can be derived theoretically or determined via a complex battery of experiments. We do not focus on either here; rather, we will simply consider in subsequent chapters how one can utilize this relation.

2.6.4 Hooke's Law for Orthotropy

Given the complexity of the microstructure of many materials in their solid phase, it should not be surprising that there are many different types of anisotropy. In addition to isotropy and transverse isotropy, however, the other most common type of material symmetry is orthotropy. As the name implies, an *orthotropic* response is one that differs in three orthogonal directions. It is thought, for example, that an artery exhibits an orthotropic response: Its behavior differs in the axial (due to axially oriented adventitial collagen), circumferential (due to the nearly circumferentially oriented smooth muscle in the media), and radial directions. Bone, too, tends to exhibit an orthotropic response, albeit nearly transversely isotropic in some cases. When the response is otherwise linear, elastic, and homogeneous under small strains, Hooke's law can be generalized to account for the orthotropy via

$$\varepsilon_{xx} = \frac{1}{E_1}\sigma_{xx} - \frac{v_{21}}{E_2}\sigma_{yy} - \frac{v_{31}}{E_3}\sigma_{zz}, \qquad \varepsilon_{xy} = \frac{1}{2G_{12}}\sigma_{xy},$$

$$\varepsilon_{yy} = \frac{1}{E_2}\sigma_{yy} - \frac{v_{12}}{E_1}\sigma_{xx} - \frac{v_{32}}{E_3}\sigma_{zz}, \qquad \varepsilon_{xz} = \frac{1}{2G_{13}}\sigma_{xz}, \qquad (2.77)$$

$$\varepsilon_{zz} = \frac{1}{E_3}\sigma_{zz} - \frac{v_{13}}{E_1}\sigma_{xx} - \frac{v_{23}}{E_2}\sigma_{yy}, \qquad \varepsilon_{yz} = \frac{1}{2G_{23}}\sigma_{yz},$$

wherein there are now nine independent material parameters: three Young's moduli E_1, E_2, and E_3, three shear moduli G_{12}, G_{13}, and G_{23}, and six Poisson's ratios v_{12}, v_{21}, v_{13}, v_{31}, v_{23}, and v_{32}, only three of which are independent; that is, it can be shown that

$$\frac{v_{12}}{E_1} = \frac{v_{21}}{E_2}, \qquad \frac{v_{13}}{E_1} = \frac{v_{31}}{E_3}, \qquad \frac{v_{23}}{E_2} = \frac{v_{32}}{E_3}. \qquad (2.78)$$

2.6.5 Other Coordinate Systems

It is essential to recognize that Hooke's law relates stress to strain at each *point* with respect to a given coordinate system. Whereas Eqs. (2.69), (2.75), and (2.77) are written in terms of Cartesians, they could also be written for cylindrical coordinates. For example, Eq. (2.69) for LEHI behavior can be written as

$$\varepsilon_{rr} = \frac{1}{E}[\sigma_{rr} - v(\sigma_{\theta\theta} + \sigma_{zz})] + \beta\Delta T, \qquad \varepsilon_{r\theta} = \frac{1}{2G}\sigma_{r\theta},$$

$$\varepsilon_{\theta\theta} = \frac{1}{E}[\sigma_{\theta\theta} - v(\sigma_{rr} + \sigma_{zz})] + \beta\Delta T, \qquad \varepsilon_{rz} = \frac{1}{2G}\sigma_{rz}, \qquad (2.79)$$

$$\varepsilon_{zz} = \frac{1}{E}[\sigma_{zz} - v(\sigma_{rr} + \sigma_{\theta\theta})] + \beta\Delta T, \qquad \varepsilon_{\theta z} = \frac{1}{2G}\sigma_{\theta z}.$$

and similarly for sphericals and so forth.

Likewise, Hooke's law can be written with respect to coordinate systems that are transformed relative to one another. For example [cf. Eq. (2.69)], for isotropy we may have, relative to (x', y', z'),

$$\varepsilon'_{xx} = \frac{1}{E}[\sigma'_{xx} - v(\sigma'_{yy} + \sigma'_{zz})], \qquad \varepsilon'_{xy} = \frac{1}{2G}\sigma'_{xy},$$

$$\varepsilon'_{yy} = \frac{1}{E}[\sigma'_{yy} - v(\sigma'_{xx} + \sigma'_{zz})], \qquad \varepsilon'_{xz} = \frac{1}{2G}\sigma'_{xz}, \qquad (2.80)$$

$$\varepsilon'_{zz} = \frac{1}{E}[\sigma'_{zz} - v(\sigma'_{xx} + \sigma'_{yy})], \qquad \varepsilon'_{yz} = \frac{1}{2G}\sigma'_{yz}.$$

Indeed, see Exercise 2.23, which asks that you prove this.

Finally, note from Eq. (2.69) that a 2-D state of stress necessarily requires a 3-D state of strain and vice versa for a 2-D state of strain. Indeed, even a 1-D state of stress (e.g., an axial force which induces a stress $\sigma_{xx} = E\varepsilon_{xx}$) will generally induce a 3-D state of strain (an extensional strain ε_{xx} plus thinning in two orthogonal directions, given by ε_{yy} and ε_{zz}). Hence, we must be careful when describing the dimension of a problem. A truly 1-D or 2-D problem is thus one wherein we simply ignore the effects in certain directions, which can be useful in some cases. For example, in a purely 1-D problem, σ_{xx} and ε_{xx} alone may exist.

Here, however, let us define a state of *plane stress* as one where

$$
\begin{bmatrix} \sigma_{xx} & \sigma_{xy} & 0 \\ \sigma_{yx} & \sigma_{yy} & 0 \\ 0 & 0 & 0 \end{bmatrix}, \quad
\begin{bmatrix} \varepsilon_{xx} & \varepsilon_{xy} & 0 \\ \varepsilon_{yx} & \varepsilon_{yy} & 0 \\ 0 & 0 & \varepsilon_{zz} \end{bmatrix}, \tag{2.81}
$$

whereas a state of *plane strain* is defined by

$$
\begin{bmatrix} \sigma_{xx} & \sigma_{xy} & 0 \\ \sigma_{yx} & \sigma_{yy} & 0 \\ 0 & 0 & \sigma_{zz} \end{bmatrix}, \quad
\begin{bmatrix} \varepsilon_{xx} & \varepsilon_{xy} & 0 \\ \varepsilon_{yx} & \varepsilon_{yy} & 0 \\ 0 & 0 & 0 \end{bmatrix}. \tag{2.82}
$$

A state of plane stress is realized easily in thin planar structures that are loaded only in-plane (Fig. 2.27) whereas a state of plane strain is realized easily in long straight members that are constrained from deforming in the

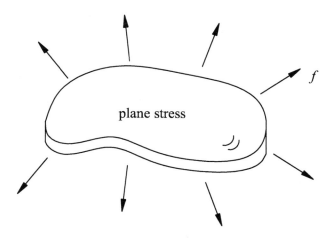

FIGURE 2.27 Schema of a state of plane stress, which is characterized by in-plane stresses only. Such states of stress also exist locally in curved membranes such as the pericardium, urinary bladder, and saccular aneurysms. Indeed, to a first approximation, many tissues (e.g., even skin in some situations) can be considered to be in a state of plane stress.

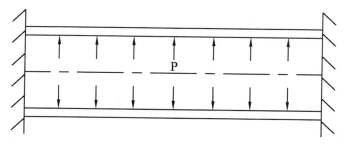

FIGURE 2.28 Schema of a state of plane strain, characterized by nonzero values in a single plane (often normal to the long axis of a prismatic structure). Although many have assumed that arteries and airways are in a state of plane strain, given that they deform primarily in the radial and circumferential directions due to internal pressurization, these tissues are actually prestretched and thus the axial strain is not zero; they are in a fully 3-D state of strain.

axial direction (Fig. 2.28). Although we will not go into these cases in detail, note that they each afford certain simplifications in formulation and solution (Timoshenko and Goodier, 1970).

2.7 Mechanical Properties of Bone

Whereas most soft tissues (e.g., skin, tendons, arteries, lung tissue, myocardium) exhibit nonlinear material behaviors over finite (large) strains, teeth and bones tend to exhibit a linearly elastic behavior over small strains. Hooke's law thus applies and the associated stress analysis is easier than that for soft tissues. Therefore, let us consider bone in some detail.

According to *Dorland's Medical Dictionary*, bone is

the hard form of connective tissue that constitutes the majority of the skeleton of most vertebrates; it consists of an organic component (the cells and matrix) and an inorganic, or mineral component; the matrix contains a framework of collagen fibers and is impregnated with the mineral component, chiefly calcium phosphate (85 percent) and calcium carbonate (10 percent).

Specifically, the type I collagen fibers tend to be organized in layers, locally parallel within a layer with the orientation varying approximately 90° from layer to layer. This layering may suggest a local transverse isotropy with the preferred direction changing from layer to layer or, more grossly, an overall orthotropic behavior at each point. Whereas the collagen endows bone with its tensile stiffness, the embedded calcium endows it with a high compressive stiffness.

Two primary cell types within mature bone are responsible for growth and remodeling: the osteoblasts, which secrete bone matrix, and the osteoclasts, which degrade it. These cells thus allow for a continuous turnover of the matrix material (Alberts et al, 2002) (i.e., a continuous maintenance or, in times of altered loading, a mechanism for adaptation). It is for this reason that bedridden patients and astronauts each suffer bone atrophy, particularly in the legs and arms, whereas athletes may have a buildup of bone. Indeed, as noted in Chapter 1, it was the work of Meyer, Wolff, and Roux in the late nineteenth century on bone that revealed a strong relationship between mechanical factors and biological growth and remodeling. For more recent work in this important and active research area, see Mow and Hayes (1991), Cowin (2001) or Carter and Beaupré (2001).

Bone typically consists of two to three layers, depending on its location within the body: an outer, dense cortical layer, a middle trabecular layer, and, in certain regions, an innermost layer of bone marrow. It is the marrow that forms blood cells. Bone thus serves several important, diverse mechanical and physiological functions: It supports and protects soft tissues and organs and it serves as a primary store of calcium and producer of blood cells. We discuss the associated microstructure further in Chapter 4.

The 206 distinct bones that constitute the adult human skeleton are often classified into 1 of 5 groups according to their shape (see Table 1.2 from Nigg and Herzog, 1994): long (e.g., femur), short (e.g., carpal), flat (e.g., sternum), irregular (pubis), and sesamoid (e.g., patella). Table 2.1 lists some of the physical properties of bone. In particular, the order of magnitude of the stiffness (Young's modulus) is ~16 GPa for cortical bone and ~1 GPa for cancellous bone. As noted earlier, however, bone does not exhibit an isotropic behavior. Rather, its linear, elastic, nonhomogeneous, and orthotropic response is better described by Eq. (2.77) with values of the parameters on the order of

$$E_1 = 6.9\,\text{GPa}, \quad \nu_{12} = 0.49, \quad \nu_{21} = 0.62, \quad G_{12} = 2.41\,\text{GPa},$$
$$E_2 = 8.5\,\text{GPa}, \quad \nu_{13} = 0.12, \quad \nu_{31} = 0.32, \quad G_{13} = 3.56\,\text{GPa}, \quad (2.83)$$
$$E_3 = 18.4\,\text{GPa}, \quad \nu_{23} = 0.14, \quad \nu_{32} = 0.31, \quad G_{23} = 4.91\,\text{GPa}$$

for the tibia and

$$E_1 = 12.0\,\text{GPa}, \quad \nu_{12} = 0.376, \quad \nu_{21} = 0.422, \quad G_{12} = 4.53\,\text{GPa},$$
$$E_2 = 13.4\,\text{GPa}, \quad \nu_{13} = 0.222, \quad \nu_{31} = 0.371, \quad G_{13} = 5.61\,\text{GPa}, \quad (2.84)$$
$$E_3 = 20.0\,\text{GPa}, \quad \nu_{23} = 0.235, \quad \nu_{32} = 0.350, \quad G_{23} = 6.23\,\text{GPa}$$

TABLE 2.1. Physical properties of bone.

Variable	Bone	Value
Density	Cortical	1700–2000 kg/m^3
	Lumbar vertebra	600–1000 kg/m^3
Mineral content	All	60–70%
Elastic modulus	Femur	5–28 GPa
Tensile strength	Femur	80–150 MPa
	Tibia	95–140 MPa
Compressive strength	Femur	131–224 MPa
	Tibia	106–200 MPa

for the femur, where 1 denotes the radial direction, 2 the circumferential, and 3 the axial (Cowin, 2001). Separate values for the cortical and cancellous portions can also be found in this reference. For more on the mechanobiology and in vivo loading of bone, see Chapter 4.

Observation 2.2 When two or more materials are bonded together, *delamination* becomes a possible mechanism of failure. Simply put, delamination is a load-induced separation between two mechanically distinct materials or layers. One way to prevent, or at least to minimize, delamination is to create a 3-D interaction (e.g., weave) at the interface. In the case of bone–metal interfaces, for example, the surface of the metal implant is often made porous to allow for in-growth of the bone. Delamination often occurs due to interfacial shear stresses and thus there is a need to design experiments that impose shear stresses. One simple experiment is a so-called "pull-out" test. Briefly, one material is bonded to the inside of a hollow sample of the second material. The outer material is then fixed in place and the inner material subjected to an axial load through its centroid. A free-body diagram of the inner material reveals that the axial load must be supported by the integrated manifestation of all the *rz* shear stresses acting on its outer surface. Although the magnitude of these shears may vary from point to point, one can determine the mean shear stress at which delamination initiates. Subsequent design would then seek to protect the bonded surface from experiencing damaging values of shear stresses.

Appendix 2: Material Properties

The properties of many materials can be found in textbooks on material science (e.g., Askeland, 1994) or biomaterials (e.g., Ratner, 2003) as well as many handbooks. Here, we simply tabulate a few of the properties that may be useful in the examination of example and exercise problems in this book.

TABLE A2.1. Physical properties of common engineering materials.

Material	Young's modulus (GPa)	Shear modulus (GPa)	Density (kg/m^3)	Yield strength (MPa)		Ultimate strength (MPa)	
				Tension	Shear	Tension	Shear
Aluminum							
2024-T4	73	27.6	2770	300	170	414	220
6061-T6	70	25.9	2770	241	138	262	165
Steel							
0.2% Carbon	200	83	7830	250	165	450	330
0.6% Carbon	200	83	7830	415	250	690	550

Exercises

2.1 Find a general relation for σ'_{xx} [Eq. (2.13)] when $\alpha = \alpha_s$.

2.2 Show that $\alpha_p = \alpha_s \pm 45°$. In addition, show that Eq. (2.25) for α_p can also be determined via $d\sigma'_{yy}/d\alpha = 0$.

2.3 Rederive the transformation equation for σ'_{yy} using the result for σ'_{xx} and the observation that σ'_{yy} exists on a face at an angle $\pi/2 + \alpha$ from the x direction.

2.4 Show that for a 2-D state of stress,

$$\sigma_{xx} + \sigma_{yy} = \sigma'_{xx} + \sigma'_{yy} \quad \forall \alpha.$$

This combination of the normal stresses is called an *invariant*; that is, its numerical value at any point is independent of the coordinate system even though its value will differ, in general, from point to point in a body and, of course, with changes in load. Invariants have been found to be useful in modeling material behavior, which, by definition, must be independent of man and his coordinate systems.

2.5 Show that Eq. (2.80) can be determined directly from Eqs. (2.69) and the transformation relations (2.13) and (2.57).

2.6 The results for the max/min normal stresses can also be found using matrix equations. Using ideas from linear algebra, show that the 2-D eigenvalue problem for the matrix equation

$$\det \begin{bmatrix} \sigma_{xx} - \Lambda_p & \sigma_{xy} \\ \sigma_{yx} & \sigma_{yy} - \Lambda_p \end{bmatrix} = 0$$

yields eigenvalues $\Lambda_1 \equiv \sigma_1$ (with $p = 1$) and $\Lambda_2 \equiv \sigma_2$ (with $p = 2$). Hint: Solve the quadratic equation for Λ_p, the two roots of which correspond to $p = 1$ and $p = 2$. Also, if familiar with linear algebra, find the eigendirections n^p, where $|n^p| = 1$, and discuss their relationship to α_p.

2.7 Given the state of stress in Example 2.1, $\sigma_{xx} = 120\,\text{kPa}$, $\sigma_{yy} = 150\,\text{kPa}$, and $\sigma_{xy} = 0\,\text{kPa}$, compute the values of σ'_{yy} for all values of α from $0°$ to $90°$ and plot as a function of α. Compare the values of α at which σ'_{yy} is max/min versus those found using the formula for α_p. Repeat for σ'_{xy} and compare the value of α at which the shear is max/min versus that using the formula for α_s.

2.8 A state of pure shear is one in which the normal stresses are zero. Consider $\sigma_{xx} = 0$, $\sigma_{yy} = 0$, and $\sigma_{xy} = \sigma_{yx} = 5\,\text{MPa}$. Find the values of the principal stresses and denote them on an infinitesimal element with orientation given by α_p.

2.9 Given a hydrostatic state of stress, $\sigma_{xx} = \sigma_{yy} = \sigma_{zz} = -p$, where p is a pressure, we computed σ'_{xx}, σ'_{yy} and σ'_{xy} for all α in Example 2.4. Likewise, compute the principal stresses σ_1 and σ_2 [i.e., $\sigma'_{xx})_{\text{max/min}}$ and

$\sigma'_{yy})_{\text{max/min}}$] as well as the maximum shear $\sigma'_{xy})_{\text{max/min}}$ using the explicit formulas in the text. Discuss your findings.

2.10 Given $\sigma_{xx} = 3\,\text{MPa}$, $\sigma_{yy} = 1\,\text{MPa}$, and $\sigma_{xy} = 2\,\text{MPa}$, find the values of the principal stresses and the maximum shear stress. What are the associated values of α_p and α_s? Draw a 2-D representation of the stress at a point p relative to each set of coordinates (x, y) and those for α_p and α_s.

2.11 Given $\sigma_{xx} = 3\,\text{MPa}$ and $\sigma_{yy} = -3\,\text{MPa}$, find the maximum shear stress and the plane on which it acts. Draw the 2-D representation of stress about a point p.

2.12 Given $u_x = (\Lambda - 1)X$ and $u_y = 0$, compute and compare the exact (E_{XX}) and the approximate/linearized (ε_{xx}) strains for $\Lambda = 1.001, 1.01, 1.1, 1.5$, and 2.0. Calculate the error introduced by the linearization in each case and determine those values of Λ for which the approximation is reasonable.

2.13 Let

$$u_x = (X + 0.001Y) - X,$$

$$u_y = Y - Y.$$

Compute the values of the components of the 2-D Green strain E_{XX}, E_{YY}, and E_{XY} and compare to those for the linearized strain ε_{xx}, ε_{yy}, and ε_{xy}. Repeat with the value premultiplying Y in the expression for u_x being 0.8.

2.14 Calculate the values of E_{YY} and E_{XY} for the rigid-body motion given by Eq. (2.54) and compare to the results for ε_{yy} and ε_{xy}.

2.15 The transformation relations for strain [Eq. (2.57)] can be found directly via coordinate transformations; recall Eqs. (2.62–2.68). Hence, if we recall from calculus that

$$x' = x\cos\alpha + y\sin\alpha, \qquad y' = -x\sin\alpha + y\cos\alpha$$

and note that similar relations hold for the displacement vector,

$$u'_x = u_x\cos\alpha + u_y\sin\alpha, \qquad u'_y = -u_x\sin\alpha + u_y\cos\alpha,$$

then show that

$$\varepsilon'_{yy} = \varepsilon_{xx}\sin^2\alpha - 2\varepsilon_{xy}\sin\alpha\cos\alpha + \varepsilon_{yy}\cos^2\alpha,$$

$$\varepsilon'_{xy} = 2\sin\alpha\cos\alpha\left(\frac{\varepsilon_{yy} - \varepsilon_{xx}}{2}\right) + (\cos^2\alpha - \sin^2\alpha)\varepsilon_{xy}.$$

Hint: Note that the angle α, which relates the two coordinate systems, is very different from the angle ϕ used in the text to represent a rigid-body rotation. Moreover, for the linearized strain,

$$\varepsilon'_{yy} = \frac{\partial u'_y}{\partial y'} = \frac{\partial u'_y}{\partial x}\frac{\partial x}{\partial y'} + \frac{\partial u'_y}{\partial y}\frac{\partial y}{\partial y'}.$$

2.16 For the delta strain gauge rosette ($\alpha_1 = 0$, $\alpha_2 = 60°$, $\alpha_3 = 120°$), show that

$$\varepsilon_{xy} = \frac{1}{\sqrt{3}}(\varepsilon_{60°} - \varepsilon_{120°}).$$

2.17 For the 0–45°–90° strain rosette of Example 2.6, find general expressions for the principal strains and maximum shear strains in terms of the measurable values ε_0, $\varepsilon_{45°}$, and $\varepsilon_{90°}$.

2.18 Whereas Eqs. (2.69), (2.75), and (2.77) are called strain–stress relations, Hooke's law can also be written as stress–strain relations. For example, for isotropy, we have

$$\sigma_{xx} = \lambda(\varepsilon_{xx} + \varepsilon_{yy} + \varepsilon_{zz}) + 2\mu\varepsilon_{xx}, \qquad \sigma_{xy} = 2\mu\varepsilon_{xy},$$

$$\sigma_{yy} = \lambda(\varepsilon_{xx} + \varepsilon_{yy} + \varepsilon_{zz}) + 2\mu\varepsilon_{yy}, \qquad \sigma_{yz} = 2\mu\varepsilon_{yz},$$

$$\sigma_{zz} = \lambda(\varepsilon_{xx} + \varepsilon_{yy} + \varepsilon_{zz}) + 2\mu\varepsilon_{zz}, \qquad \sigma_{zx} = 2\mu\varepsilon_{zx},$$

where λ and μ are called Lamé constants (material parameters), after the French scientist Lamé (1795–1870). Show that

$$\lambda = \frac{vE}{(1+v)(1-2v)}, \qquad \mu = \frac{E}{2(1+v)} \equiv G,$$

where E and v are the Young's modulus and Poisson ratio, respectively.

2.19 Note from the previous exercise and Eq. (2.69) that λ multiplies the first invariant of strain $e = \varepsilon_{xx} + \varepsilon_{yy} + \varepsilon_{zz}$, which is a measure of volume change. Show that

$$e = \frac{1-2v}{E}(\sigma_{xx} + \sigma_{yy} + \sigma_{zz})$$

for isotropy. Note that there is no change in volume (i.e., $e = 0$) if $v = \frac{1}{2}$. Moreover, if a cube of material is subjected to a hydrostatic pressure, then $\sigma_{xx} = \sigma_{yy} = \sigma_{zz} = -p$. In this case, note that

$$e = -\frac{3(1-2v)}{E}p \rightarrow -\frac{p}{e} = \frac{E}{3(1-2v)} \equiv K,$$

where K is the so called *bulk modulus*; it represents the ratio of the hydrostatic compressive stress to the decrease in volume.

2.20 For a LEHI behavior, show that a plane state of stress requires that

$$\sigma_{zz} = 0 = \frac{E}{(1+v)(1-2v)}[(1-v)\varepsilon_{zz} + v(\varepsilon_{xx} + \varepsilon_{yy})],$$

or

$$\varepsilon_{zz} = -\frac{v}{1-v}(\varepsilon_{xx} + \varepsilon_{yy}).$$

2.21 For a LEHI behavior, show that the principal stresses in a plane state of stress can be written as

$$\sigma_1 = \frac{E}{1-v^2}(\varepsilon_1 + v\varepsilon_2), \qquad \sigma_2 = \frac{E}{1-v^2}(\varepsilon_2 + v\varepsilon_1),$$

where ε_1 and ε_2 are the principal strains.

2.22 Given $\sigma_{xx} = 20\,\text{MPa}$, $\sigma_{yy} = -10\,\text{MPa}$, and $\sigma_{xy} = -20\,\text{MPa}$, find the principal stresses and principal strains with LEHI behavior and $E = 16$ GPa and $v = 0.325$.

2.23 Starting with Eq. (2.80) and using Eqs. (2.13), (2.17), (2.21), and (2.57), show that you recover Eq. (2.69). Note: We assume a rotation α about the z-axis thus $\sigma'_{zz} \equiv \sigma_{zz}$.

2.24 Given reasonable values of the material parameters for bone, estimate the axial stress in your femur due to standing, walking, and running. Toward this end, estimate the increase in the applied load (in terms of body weight and in comparison to the load due to standing) due to walking and running. Once done, note that even though we did not discuss it, bone exhibits viscoelastic, not just elastic, behavior under certain conditions. In particular, the Young's modulus increases with increases in strain rate. It has been estimated, for example, that $E \sim c\dot{\varepsilon}^d$, where $\dot{\varepsilon}$ is the extensional strain rate and c and d are material parameters. If $E = 16\,\text{GPa}$ at $\dot{\varepsilon} = 0.001\,\text{s}^{-1}$ (slow walking) and $d \sim 0.06$, find the value of c. Next, compute the value of E for vigorous activity, with $\dot{\varepsilon} \sim 0.01\,\text{s}^{-1}$, and discuss how this would effect your first estimate for stress in your femur.

2.25 Research the different constitutive relations used to describe the behavior of water in its solid, liquid, and gaseous phases (i.e., different conditions of interest). Write a two-page report on your findings, showing explicitly the different equations and discussing how the different characteristic behaviors dictate the need to establish different theoretical frameworks (DEICE).

2.26 Referring to Figure 2.22a, note that the material can return to its original configuration by releasing the energy that is stored in it due to deformation. This "strain energy" W can be computed (per initial volume) as the area under the stress–strain curve. For the 1-D test in Figure 2.22a, $\sigma_{xx} = E\varepsilon_{xx}$ and the stored energy is $\frac{1}{2}(\text{base})(\text{height})$

$= \frac{1}{2}\varepsilon_{xx}\sigma_{xx} = \frac{1}{2}\varepsilon_{xx}E\varepsilon_{xx}$. Show, therefore, that the stress can be determined as the change in energy with respect to changes in strain (i.e., $\sigma_{xx} = \partial W/\partial\varepsilon_{xx}$), whereas the stiffness can be computed as the change in stress with respect to the change in strain ($\partial\sigma_{xx}/\partial\varepsilon_{xx}$). Plot this stiffness as a function of stress and comment.

2.27 Although we chose not to derive the linearized strains directly, it is common to relate them (for illustrative purposes) to changes in length and changes in angle. The former was used to show that Poisson's ratio $v = \dfrac{1}{2}$ if the behavior is incompressible. Here, note the following for shear. Referring to Figure 2.29, let point b displace upward an amount Δu_y. With point d displacing accordingly, we call this a pure shear. Note, therefore, that the angles α and β are given by

$$\alpha = \tan\!\left(\frac{\Delta u_y}{\Delta x}\right) \approx \frac{\Delta u_y}{\Delta x}, \qquad \beta = \tan\!\left(\frac{\Delta u_x}{\Delta y}\right) \approx \frac{\Delta u_x}{\Delta y}$$

for which we used the small-angle approximation for the tangent, and thus the mean value is

$$\frac{1}{2}(\alpha + \beta) = \frac{1}{2}\left(\frac{\Delta u_y}{\Delta x} + \frac{\Delta u_x}{\Delta y}\right) \to \lim_{\Delta x\,\Delta y\to 0} \frac{1}{2}(\alpha + \beta) = \frac{1}{2}\left(\frac{\partial u_y}{\partial x} + \frac{\partial u_x}{\partial y}\right),$$

which we recognize equals ε_{xy}. Repeat this exercise for the y-z plane.

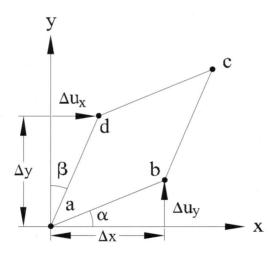

2.28 Common experimental setups include uniaxial extension or compression of a rod, biaxial extension of a sheet, tension–torsion of a cylinder, inflation–extension of a hollow cylinder, and inflation of an axisymmetric membrane (Fig. 2.30). Identify tissues that would be appropriately tested using these potential setups without excessive dissection following removal from the body.

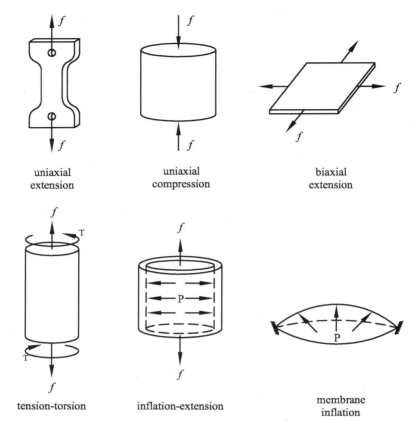

uniaxial   uniaxial   biaxial
extension   compression   extension

tension-torsion   inflation-extension   membrane
                                        inflation

FIGURE 2.30

2.29 Using Eqs. (2.13) and (2.21), it can be shown that, in two dimensions,
$\sigma_{xx} + \sigma_{yy} = \sigma'_{xx} + \sigma'_{yy}$. Show, in addition, that the principal stresses $\sigma_1 \equiv \sigma'_{xx}(\text{max})$ and $\sigma_2 \equiv \sigma'_{yy}(\text{max})$ simultaneously add to yield

$$\sigma_{xx} + \sigma_{yy} = \sigma_1 + \sigma_2.$$

2.30 A special 2-D state of stress is called an *equibiaxial stress*. It is defined
by $\sigma_{xx} = \sigma_{yy} = \sigma_o$ and $\sigma_{xy} = \sigma_{yx} = 0$. Find the principal stresses and
max/min shear stresses in this case. Note that equibiaxial stretching
tests are particularly useful in determining the anisotropy of a planar
tissue (membrane). Why?

2.31 A uniaxial test is performed on a bone specimen having a central
(gauge) region initially 6 mm long and 2 mm in diameter. Five data
points were recorded:

Axial force (N)	94	190	284	376	440
Change in length (mm)	0.009	0.018	0.027	0.050	0.094

Plot the associated stress–strain relation, calculate a Young's modulus, and show that the yield stress is ~118 MPa. Recall that the yield stress reveals the transition from elastic to plastic (cf. Fig. 2.24). Data from Ozkaya and Nordin (1999).

2.32 Data from a uniaxial tension test to failure (data point 4) for a human cortical bone are

Stress (MPa)	0	85	114	128
Strain (mm/mm)	0	0.005	0.010	0.026

Plot the data and interpret. Estimate the Young's modulus, yield stress, and ultimate stress (cf. Fig. 2.25). Clearly, much more data are useful in general. Data from Ozkaya and Nordin (1999).

2.33 Data from a uniaxial tension test in the elastic region for a bone sample are

Stress (MPa)	0	60	120	180
Strain	0	0.0034	0.0066	0.0100

Referring to Exercise 2.32, were these tests performed on the same type of bone? Compare the Young's moduli.

3
Equilibrium, Universal Solutions, and Inflation

3.1 General Equilibrium Equations[1]

Let us begin by recalling three important observations from Chapters 1 and 2. First, equilibrium requires that $\Sigma F = 0$ and $\Sigma M = 0$. Second, if a body is in equilibrium, then each of its parts are likewise in equilibrium. Third, there may exist at each point p in a body (cf. Fig. 2.4) nine components of stress, six of which are independent, which we denote as $\sigma_{(face)(direction)}$ relative to the coordinate system of choice. Because stress may vary from point to point within a body, the components at a nearby point q may have different values. (Note: It is usually convenient to refer components at different points to the same coordinate system.) Now, if we consider a small cube of material, centered about point p which is located at (x, y, z) and has stresses $\sigma_{xx}, \sigma_{xy}, \ldots, \sigma_{zz}$, then the stresses on the faces of the cube may differ from those at the center; that is, if the xx component at the center of the cube is σ_{xx}, then on the positive and negative *faces* of the cube, at distances $\pm \Delta x/2$ from the center, we may have $\sigma_{xx} + \Delta\sigma_{xx}$ and $\sigma_{xx} - \Delta\sigma_{xx}$, respectively (i.e., values slightly greater than or less than that at point p). The key question is thus: How can we evaluate this small difference $\Delta\sigma_{xx}$? As we shall do throughout this text, let us recall a Taylor series expansion from calculus $[f(x + \Delta x) = f(x) + (df/dx)\Delta x + \ldots]$ and let

$$\sigma_{xx}\left(x + \frac{\Delta x}{2}\right) = \sigma_{xx}(x) + \frac{\partial\sigma_{xx}}{\partial x}\bigg|_x \left(\frac{\Delta x}{2}\right) + \frac{1}{2}\frac{\partial^2\sigma_{xx}}{\partial x^2}\bigg|_x \left(\frac{\Delta x}{2}\right)^2 + \ldots \quad (3.1)$$

and

$$\sigma_{xx}\left(x - \frac{\Delta x}{2}\right) = \sigma_{xx}(x) - \frac{\partial\sigma_{xx}}{\partial x}\bigg|_x \left(\frac{\Delta x}{2}\right) - \frac{1}{2}\frac{\partial^2\sigma_{xx}}{\partial x^2}\bigg|_x \left(\frac{\Delta x}{2}\right)^2 - \ldots, \quad (3.2)$$

---

[1] Sections 3.1 and 3.2 may be considered optional by some instructors for a first course in biosolids.

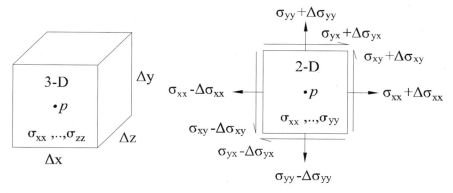

FIGURE 3.1 Representative stresses at a point p, with particular attention to those components in the x-y plane at distances $\pm\Delta x/2$ and $\pm\Delta y/2$ from point p. Note, therefore, that the components of stress may vary from point to point in general (because of the differential areas, the value of stress on an given surface area is represented by its mean over that area).

where other higher-order terms in Δx are not shown. Of course, the values of each of the components of stress on each of the six faces can likewise differ from those at the center.

For simplicity, however, let us consider a 2-D state of stress, as illustrated in Figure 3.1. We let σ_{yy} and σ_{yx}, which represent the mean value on their respective faces, vary from the bottom to the top faces (because they are y-face stresses) and thus their gradients (or changes) are with respect to y, whereas σ_{xx} and σ_{xy}, which similarly represent mean values on their respective faces, are assumed to vary from the left to the right faces (because they are x-face stresses) and their gradients are with respect to x.

Now, to ensure equilibrium, we sum forces (which requires multiplying appropriate components of stress by their respective areas) separately in the x and y directions and set them individually equal to zero. In the x direction, this yields (neglecting the $|_x$ notation because it is understood that our result will be valid at x once we shrink the cube to the point p)

$$\left[\sigma_{xx} + \frac{\partial \sigma_{xx}}{\partial x}\left(\frac{\Delta x}{2}\right) + \frac{1}{2}\frac{\partial^2 \sigma_{xx}}{\partial x^2}\left(\frac{\Delta x}{2}\right)^2 + \ldots\right]\Delta y \Delta z - \left[\sigma_{xx} - \frac{\partial \sigma_{xx}}{\partial x}\left(\frac{\Delta x}{2}\right)\right.$$

$$\left. - \frac{1}{2}\frac{\partial^2 \sigma_{xx}}{\partial x^2}\left(\frac{\Delta x}{2}\right)^2 - \ldots\right]\Delta y \Delta z + \left[\sigma_{yx} + \frac{\partial \sigma_{yx}}{\partial y}\left(\frac{\Delta y}{2}\right) + \frac{1}{2}\frac{\partial^2 \sigma_{yx}}{\partial y^2}\left(\frac{\Delta y}{2}\right)^2 + \ldots\right]\Delta x \Delta z$$

$$- \left[\sigma_{yx} - \frac{\partial \sigma_{yx}}{\partial y}\left(\frac{\Delta y}{2}\right) - \frac{1}{2}\frac{\partial^2 \sigma_{yx}}{\partial y^2}\left(\frac{\Delta y}{2}\right)^2 - \ldots\right]\Delta x \Delta z = 0. \tag{3.3}$$

Simplifying, dividing by $\Delta x \Delta y \Delta z$, and taking the limit as $\Delta x \Delta y \Delta z \to 0$, we have

$$\lim_{\Delta x \Delta y \Delta z \to 0} \frac{1}{\Delta x \Delta y \Delta z} \left(\frac{\partial \sigma_{xx}}{\partial x} + \frac{\partial^2 \sigma_{xx}}{\partial x^2} \left(\frac{\Delta x}{4} \right) + \frac{\partial \sigma_{yx}}{\partial y} + \frac{\partial^2 \sigma_{yx}}{\partial y^2} \left(\frac{\Delta y}{4} \right) \right) \Delta x \Delta y \Delta z = 0,$$

(3.4)

or

$$\frac{\partial \sigma_{xx}}{\partial x} + \frac{\partial \sigma_{yx}}{\partial y} = 0,$$

(3.5)

which is our final x-direction equation in two dimensions. Note, therefore, that all higher-order terms (H.O.T.), quadratic and above, drop out because of the limiting process. Such terms can thus be neglected in hindsight in all similar derivations. Indeed, as an exercise, show that summation of forces in the y-direction yields a similar equation:

$$\frac{\partial \sigma_{xy}}{\partial x} + \frac{\partial \sigma_{yy}}{\partial y} = 0.$$

(3.6)

Together, Eqs. (3.5) and (3.6) are the governing differential equations of equilibrium for a 2-D state of stress relative to an $(o; x, y)$ coordinate system. Note that the x equation contains only x-direction stresses and the y equation contains only y-direction stresses. This is another reason to remember the simple aid, $\sigma_{(\text{face})(\text{direction})}$.

It is easy to show, of course, that relative to a $(o; x', y')$ coordinate system, we have

$$\frac{\partial \sigma'_{xx}}{\partial x'} + \frac{\partial \sigma'_{yx}}{\partial y'} = 0 \quad \text{and} \quad \frac{\partial \sigma'_{xy}}{\partial x'} + \frac{\partial \sigma'_{yy}}{\partial y'} = 0.$$

(3.7)

Finally, if we consider a fully 3-D state of stress, relative to $(o; x, y, z)$, summation of forces separately in x, y, and z yields *our general 3-D differential equations of equilibrium*:

$$\frac{\partial \sigma_{xx}}{\partial x} + \frac{\partial \sigma_{yx}}{\partial y} + \frac{\partial \sigma_{zx}}{\partial z} + \rho g_x = 0,$$

(3.8)

$$\frac{\partial \sigma_{xy}}{\partial x} + \frac{\partial \sigma_{yy}}{\partial y} + \frac{\partial \sigma_{zy}}{\partial z} + \rho g_y = 0,$$

(3.9)

$$\frac{\partial \sigma_{xz}}{\partial x} + \frac{\partial \sigma_{yz}}{\partial y} + \frac{\partial \sigma_{zz}}{\partial z} + \rho g_z = 0.$$

(3.10)

Note that we have added a possible body force vector $\mathbf{g}$, which acts at point p and is defined per unit mass (i.e., the force per volume is $\rho \mathbf{g}$, where ρ is the mass density, and the force vector is $\rho \mathbf{g} \Delta x \Delta y \Delta z$).

At this juncture, it is instructive to review the *restrictions* associated with the derivation of Eqs. (3.5)–(3.10). First, did we specify any particular material? Actually, we did not; hence, these equations hold for *all* continua—solid or fluid, man-made or biological. Thus these equations are very general and powerful. Second, did we make any assumptions on the motion, such as the magnitude of the strains? Again, the answer is no. The only requirement was that the body be in equilibrium (i.e., not accelerating). In Chapter 8, we will relax this restriction so that we can study (accelerating) fluid flows. Third, are these equations restricted to a particular coordinate system? The answer here, of course, is yes: These equilibrium equations are valid only for Cartesian coordinate systems. Physical problems "exist" independent of coordinate systems, however, which are introduced to engender convenience when we solve a particular problem. In some cases, Cartesian coordinates will be the most convenient, such as when finding bending stresses in a cantilevered straight beam within a force transducer that is used to measure applied loads on a force plate used in gait studies. In other cases, cylindrical coordinates (e.g., when solving for stresses in an artificial artery) or spherical coordinates (e.g., when solving for stresses in an intracranial saccular aneurysm) may be preferred. Fortunately, it can be shown (although the algebra is more complex) that similar equilibrium equations exist for cylindrical and spherical coordinates (Humphrey, 2002). We merely list them here. For cylindrical coordinates, we have

$$\frac{\partial \sigma_{rr}}{\partial r} + \frac{1}{r}\frac{\partial \sigma_{\theta r}}{\partial \theta} + \frac{\partial \sigma_{zr}}{\partial z} + \frac{\sigma_{rr} - \sigma_{\theta\theta}}{r} + \rho g_r = 0, \tag{3.11}$$

$$\frac{\partial \sigma_{r\theta}}{\partial r} + \frac{1}{r}\frac{\partial \sigma_{\theta\theta}}{\partial \theta} + \frac{\partial \sigma_{z\theta}}{\partial z} + \frac{2\sigma_{r\theta}}{r} + \rho g_\theta = 0, \tag{3.12}$$

$$\frac{\partial \sigma_{rz}}{\partial r} + \frac{1}{r}\frac{\partial \sigma_{\theta z}}{\partial \theta} + \frac{\partial \sigma_{zz}}{\partial z} + \frac{\sigma_{rz}}{r} + \rho g_z = 0, \tag{3.13}$$

whereas for spherical coordinates, we have

$$\frac{\partial \sigma_{rr}}{\partial r} + \frac{1}{r}\frac{\partial \sigma_{\theta r}}{\partial \theta} + \frac{1}{r\sin\theta}\frac{\partial \sigma_{\phi r}}{\partial \phi} + \frac{1}{r}(2\sigma_{rr} - \sigma_{\theta\theta} - \sigma_{\phi\phi} + \sigma_{\theta r}\cot\phi) + \rho g_r = 0, \tag{3.14}$$

$$\frac{\partial \sigma_{r\theta}}{\partial r} + \frac{1}{r}\frac{\partial \sigma_{\theta\theta}}{\partial \theta} + \frac{1}{r\sin\theta}\frac{\partial \sigma_{\phi\theta}}{\partial \phi} + \frac{1}{r}(2\sigma_{r\theta} + \sigma_{\theta r} + (\sigma_{\theta\theta} - \sigma_{\phi\phi})\cot\theta) + \rho g_\theta = 0, \tag{3.15}$$

$$\frac{\partial \sigma_{r\phi}}{\partial r} + \frac{1}{r}\frac{\partial \sigma_{\theta\phi}}{\partial \theta} + \frac{1}{r\sin\theta}\frac{\partial \sigma_{\phi\phi}}{\partial \phi} + \frac{1}{r}(2\sigma_{r\phi} + \sigma_{\phi r} + (\sigma_{\phi\theta} + \sigma_{\theta\phi})\cot\theta) + \rho g_\phi = 0. \tag{3.16}$$

It should be obvious that solving coupled partial differential equations in three dimensions for any coordinate system is generally nontrivial and often requires sophisticated numerical methods for solution, such as the finite element method. Nonetheless, to learn the methodology of approach, we should first seek solutions that are tractable analytically; fortunately, such solutions can also be very useful, as seen in Section 3.6.

3.2 Navier–Space Equilibrium Equations

Whereas the equations of Section 3.1 are valid for all continua, it is often useful to derive specialized forms of the equilibrium equations for particular material behaviors or classes of material behaviors.[2] Here, let us do so for LEHI (Section 2.6) behaviors {i.e., materials that exhibit a linear, elastic, homogeneous, and isotropic response under small strains—these materials are described by the so-called Hooke's Law [Eq. (2.69)]}.

Let us first consider the case for a purely 2-D state of stress and strain (which is different from plane stress or plane strain, as noted in Chapter 2). From the stress–strain equation in Exercise 2.18, we note that

$$\sigma_{xx} = \lambda(\varepsilon_{xx} + \varepsilon_{yy}) + 2\mu\varepsilon_{xx}, \tag{3.17}$$

$$\sigma_{xy} = 2\mu\varepsilon_{xy} = \sigma_{yx}, \tag{3.18}$$

$$\sigma_{yy} = \lambda(\varepsilon_{xx} + \varepsilon_{yy}) + 2\mu\varepsilon_{yy}, \tag{3.19}$$

where λ and μ are material parameters (Lamé constants, with $\mu \equiv G$, the shear modulus). The linearized strains are given by Eqs. (2.44) and (2.45):

$$\varepsilon_{xx} = \frac{\partial u_x}{\partial x}, \quad \varepsilon_{yy} = \frac{\partial u_y}{\partial y}, \quad \text{and} \quad \varepsilon_{xy} = \frac{1}{2}\left(\frac{\partial u_x}{\partial y} + \frac{\partial u_y}{\partial x}\right). \tag{3.20}$$

Hence, substituting Eqs. (3.17) and (3.18) into Eq. (3.5), we have

$$\frac{\partial}{\partial x}[\lambda(\varepsilon_{xx} + \varepsilon_{yy}) + 2\mu\varepsilon_{xx}] + \frac{\partial}{\partial y}(2\mu\varepsilon_{xy}) = 0 \tag{3.21}$$

or, by using the strain–displacement relations in Eq. (3.20),

$$\frac{\partial}{\partial x}\left[(\lambda + 2\mu)\frac{\partial u_x}{\partial x} + \lambda\frac{\partial u_y}{\partial y}\right] + \frac{\partial}{\partial y}\left[2\mu\left(\frac{1}{2}\right)\left(\frac{\partial u_x}{\partial y} + \frac{\partial u_y}{\partial x}\right)\right] = 0. \tag{3.22}$$

Simplifying, note that we have

$$\lambda\frac{\partial^2 u_x}{\partial x^2} + 2\mu\frac{\partial^2 u_x}{\partial x^2} + \lambda\frac{\partial^2 u_y}{\partial x \partial y} + \mu\frac{\partial^2 u_x}{\partial y^2} + \mu\frac{\partial^2 u_y}{\partial y \partial x} = 0. \tag{3.23}$$

---

[2] Indeed, note that Eqs. (3.5) and (3.6) represent two equations in terms of three unknown stresses, whereas Eqs. (3.8)–(3.10) represent three equations in terms of six unknown stresses. In each case, we need additional equations to render the mathematical problem well posed, equations such as constitutive relations.

If we (based on hindsight, which means after first trying multiple other possibilities to no avail) note that the second term can be split into two equal parts, then by collecting like terms, we have

$$\mu\left(\frac{\partial^2 u_x}{\partial x^2} + \frac{\partial^2 u_x}{\partial y^2}\right) + (\lambda + \mu)\frac{\partial}{\partial x}\left(\frac{\partial u_x}{\partial x} + \frac{\partial u_y}{\partial y}\right) = 0. \qquad (3.24)$$

Similarly (show it), Eq. (3.6) can be written as

$$\mu\left(\frac{\partial^2 u_y}{\partial x^2} + \frac{\partial^2 u_y}{\partial y^2}\right) + (\lambda + \mu)\frac{\partial}{\partial y}\left(\frac{\partial u_x}{\partial x} + \frac{\partial u_y}{\partial y}\right) = 0. \qquad (3.25)$$

These two equations are the so-called *2-D Navier–Space equilibrium equations*; they represent two coupled partial differential equations in terms of two unknowns, the displacements u_x and u_y, as well as the Lamé constants λ and μ. Of course, relative to an (x, y) coordinate system, the displacement vector u can be written as $u = u_x \hat{i} + u_y \hat{j}$. This suggests, therefore, that these two equations merely represent the x and y components of a more general vectorial differential equation. Consequently, Eqs. (3.24) and (3.25) can be written much more compactly in vector form as

$$\mu\nabla^2 u + (\lambda + \mu)\nabla(\nabla \cdot u) = 0, \qquad (3.26)$$

where in two dimensions, relative to Cartesian coordinates, the del operator is[3]

$$\nabla = \hat{i}\frac{\partial}{\partial x} + \hat{j}\frac{\partial}{\partial y}. \qquad (3.27)$$

The advantage of writing such equations in vector form is that the equation is now very general: It can be shown that Eq. (3.26) holds for 3-D states of stress and strain [which results in three coupled partial differential equations in terms of the three components of the displacement vector (e.g., $u = u_x \hat{i} + u_y \hat{j} + u_z \hat{k}$)] as well as for any coordinate system (Cartesian, cylindrical, spherical, etc.). Because of our use of Hooke's law for linear, elastic, homogeneous, and isotropic behavior as well as the use of the linearized strains, the Navier–Space equilibrium equations are thus limited as well. It will be seen in Chapter 8 that a similar differential equation will be derived for a restricted class of (linear) fluids, an equation that is called the Navier–Stokes equation. As one might expect, solutions to coupled partial differential equations are generally challenging. Hence, rather than attempting to solve even some special cases, let us first examine a much simpler class of solutions—ones that have wide applicability despite increased restrictions due to additional assumptions. After having done so, we will return to the differential equations of equilibrium in Section 3.6.

---

[3] Vector operations are reviewed in the Appendix in Chapter 7.

3.3 Axially Loaded Rods

Perhaps the simplest problem in solid mechanics is equilibrium associated with a uniform, 1-D state of stress in an axially loaded member having one dimension much greater than the other two. For example, if the member is rectangular in cross section and x corresponds to the axial direction, then we posit that σ_{xx} is constant but nonzero, whereas all other components of stress are zero. From Eqs. (3.8)–(3.10), we see that equilibrium is thereby satisfied trivially at every point. Below, we will see how the numerical value of the σ_{xx} component of *stress is related to the applied axial force and the cross-sectional area* and, indeed, how the force must be applied to ensure an axial extension only (i.e., no bending or twisting). First, however, let us consider some biomechanical motivations for such problems.

3.3.1 Biological Motivation

Whereas many soft tissues in the body experience multiaxial states of stress (e.g., skin, the cornea and sclera of the eye, arteries and veins, the heart, the diaphragm and lungs), a few tissues experience primarily a uniaxial (or 1-D state of) stress. Consider, for example, the chordae tendineae of the heart. This thin, stringlike structure consists primarily of axially oriented type I collagen, which endows it with significant stiffness (cf. Fig. 2.25). Functionally, the chordae connects the valves in the heart to the papillary muscles, which appear as fingerlike muscular projections from the endocardial surface (Fig. 3.2). In a way then, the chordae can be thought of in

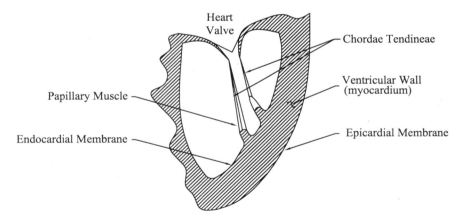

FIGURE 3.2 Schematic cross section of the left ventricle of the heart, which reveals the papillary muscle, chordae tendineae, and a heart valve. The chordae are thin, tendonlike structures that consist primarily of oriented type I collagen (recall Fig. 2.25). They function to prevent the valves from inverting as well as to augment the ejection of blood.

an analogous way as the ropes that connect a person (muscle) to a parachute (valves). Due to the action of the pressure on the valve and the resisting force in the papillary muscle, the chordae are subjected primarily to a 1-D tensile stress. The biomechanics of the chordae tendineae is obviously important in understanding valvular diseases (which affect some 96,000 Americans per year) as well as in the design of replacement valves and associated surgical procedures.

Like the chordae tendineae, many papillary muscles (some being thin and chordlike) are subjected to uniaxial loading, as are numerous tendons and ligaments in major joints. Understanding joint biomechanics is similarly important in the design of tissue repairs or replacements and, thus, surgical procedures, as well as in the design of protective devices for athletes. For example, some 170,000 athletes tear the anterior cruciate ligament (ACL) each year in the United States.

Of course, *to solve most problems in mechanics, we must know the geometry, material properties, and applied loads.* Whether it be traditional engineering materials, biomaterials, or native tissue, uniaxial stress tests are the most commonly performed test in the R&D laboratory. Figure 3.3 illustrates two common test conditions. In the situation on the right, the specimen is machined or cut into a so-called dumbbell shape so that the cross-sectional area in the central region of the specimen is much less than that at the ends. This specimen design ensures that the stresses will be highest in the central region, where failure mechanisms such as fracture or yield may be studied. In the situation illustrated on the left, the specimen is merely mounted in fixtures on each end to permit it to be coupled to the

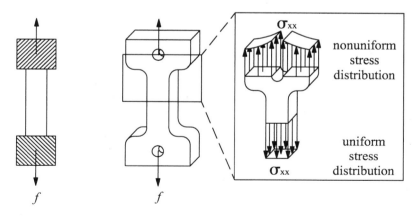

FIGURE 3.3 Schema of two typical types of specimen and mounting fixture that are used in uniaxial loading tests. Shown, too, are stresses in the dumbbell-shaped specimen, which reveal stress concentrations near the hole and a uniform distribution away from the ends. It is for this reason that stress and strain are measured away from the "ends" of the sample.

loading device. This figure is typical of many nonfailure tests on ligaments and tendons. In both cases, the state of stress can be very complex near the ends (i.e., near the loading fixtures); hence, data are collected away from the ends where the stresses are thought to be uniform (Fig. 3.3, right). This "St. Venant's Principle" can be proved mathematically for certain materials and loading conditions by solving the full differential equations of equilibrium. Below, we simply focus our attention on the uniform 1-D state of stress in the central region.

3.3.2 Mathematical Formulation

To determine the value of σ_{xx} in a generic, axially loaded member in equilibrium, a specific cutting plane is introduced to divide the member into two sections (Fig. 3.4), thereby exposing the stress of interest. For example, a free-body diagram of the left section can be drawn, where at any section, the force (magnitude) f passes through the centroid [shown below in Eqs. (3.34) and (3.36)]. The reaction force f is balanced at the cut surface by uniformly distributed normal stresses σ_{xx}, which act over differential areas at each point in the cross section; that is, the sum of each stress acting at each point, multiplied by their respective differential areas to yield a quantity having a unit of force, must balance the opposing force f to ensure equilibrium. Mathematically,

$$\sum F_x = 0 \rightarrow \int \sigma_{xx} dA - f = 0. \qquad (3.28)$$

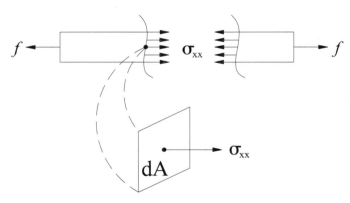

FIGURE 3.4 A free-body diagram constructed by making a fictitious cut in the central region of a test sample for isolating the x-face stresses of interest. We assume that only the normal stress σ_{xx} acts on the cut face for shear stresses are not needed to balance the applied axial load f. Moreover, although σ_{xx} is assumed to be uniform in the central region, far enough away from the ends, we must remember that stresses act over differential, not total, areas.

Because the stress σ_{xx} is distributed uniformly (i.e., it is constant in the cross section but of arbitrary value as needed to balance whatever f is applied), it can be taken outside the integral, which can then be evaluated as follows:

$$\sigma_{xx} \int dA = f \to \sigma_{xx} = \frac{f}{A}, \tag{3.29}$$

from which we see that the units of stress are clearly force per area, commonly Pa (N/m²) or psi (lb/in.²). Recall that this simple result was used extensively in Chapter 2.

That the applied force must pass through the centroid of the member to ensure that σ_{xx} is uniformly distributed over the cross section is seen easily by requiring that the sum of the moments vanish as well. Hence, let the applied force f act through an arbitrary point (y^*, z^*), which locates the line of action of f (Fig. 3.5). Considering the stress that acts over the differential area dA at (y, z), equilibrium requires that the sum of the moments about the y axis must equal zero; that is, using the right-hand rule to note the sign of each moment, we have

$$\sum M_y)_o = 0 \to \int z \sigma_{xx} dA - fz^* = 0, \tag{3.30}$$

or if the stress is uniform,

$$\sigma_{xx} \int z \, dA = fz^*, \tag{3.31}$$

where

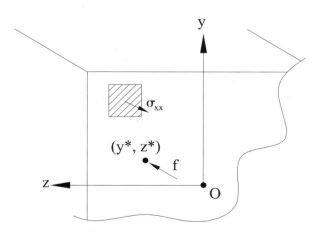

FIGURE 3.5 Cross section of the x-face area in an axially loaded member. A fictitious cut reveals σ_{xx} stresses, which act over (cross-hatched) differential areas dA [centered at point (y, z)] while the line of action of the axial load is assumed to go through the point (y^*, z^*).

$$\sigma_{xx} = \frac{f}{A} = \frac{f}{\int dA} \tag{3.32}$$

from Eq. (3.29). Hence, Eq. (3.31) can be written as

$$\frac{f}{\int dA} \int z \, dA = fz^*, \tag{3.33}$$

or

$$z^* = \frac{\int z dA}{\int dA} \to z^* = \bar{z}; \tag{3.34}$$

that is, the force must go through a point z^* that coincides with the centroid $\bar{z}$ [see the Appendix 3 for a discussion of first moments of area (centroids)]. Similarly, summing moments about the z axis,

$$\sum (M_z)_o = 0 \to -\int y\sigma_{xx} dA + fy^* = 0, \tag{3.35}$$

whereby it can be shown (do it) that

$$y^* = \frac{\int y dA}{\int dA} \to y^* = \bar{y}, \tag{3.36}$$

thus proving that the point (y^*, z^*) through which the applied force acts must coincide with the centroid $(\bar{y}, \bar{z})$ to maintain equilibrium ($\Sigma M_y = 0$ and $\Sigma M_z = 0$) in the presence of a uniform stress. We see, therefore, that *a simple analysis can help us design well a useful experiment* (i.e., to determine how the load should be applied). Finally, note that these equations hold for all cross sections—rectangular, circular, or arbitrary. Indeed, because these results were also obtained independent of the specification of particular material properties, they are called *universal solutions*. Although not emphasized in most books on the mechanics of materials, the generality of these universal solutions [Eqs. (3.29), (3.34), and (3.36)] allow them to be applied equally to problems involving the uniaxial extension of tendons, rubber bands, metallic wires, or concrete. These results are thus very important.

Example 3.1 A chordae tendineae specimen initially 10 mm long is to resist an axial tensile load f of 100 g. The specimen initially has a 1.0-mm diameter. What is the maximum axial stress that the chordae will experience?

Solution: We recall that axial stress is computed via the applied force acting over the cross-sectional area. Hence, consider

$$\frac{f}{A_o} = \frac{(100 \text{ g})(9.807 \times 10^{-3} \text{ N/g})}{\pi(0.5 \text{ mm})^2}\left(\frac{1000 \text{ mm}}{\text{m}}\right)^2 = 1.25 \text{ MPa}.$$

As seen, such calculations are very easy. A key question to ask, however, is whether a chordae tendineae can sustain a 1.25-MPa stress without tearing, which is to say, How does this value compare to the range of stresses that would be expected to exist in vivo? Toward this end, see Exercise 3.3, which should be attempted only after completion of the next two sections on inflation problems.

At this juncture, however, let us recognize another very important issue. The value of stress of 1.25 MPa in this example was computed using the applied load and the *original* cross-sectional area A_o. Such stresses are called by various names: the Piola–Kirchhoff stress (named after two nineteenth-century investigators), the nominal stress, or, sometimes, the engineering stress. The important observation though is that the derivation for $\sigma_{xx} = f/A$ in Eq. (3.29) is actually based on A, the *current* cross-sectional area over which the force actually acts. This definition is often called a Cauchy stress, after the famous mathematician/mechanicist A. Cauchy, or the true stress because one uses the actual area over which the load acts. When the deformation (and thus strain) is small, $A \sim A_o$ and the two definitions yield similar values. When the deformation is large, however, as is the case for most soft tissues, the computed values can be very different. To compute the Cauchy stress in Example 3.1, therefore, we must measure A rather than A_o. Clearly, the latter is easier experimentally; thus, the wide usage of the nominal stress $\Sigma_{(\text{face})(\text{direction})}$ by experimentalists. We will in general prefer the Cauchy stress $\sigma_{(\text{face})(\text{direction})}$ herein, however, which appears naturally in the equilibrium equations. Fortunately, we shall see in Chapter 6 that the various definitions of stress are related. Indeed, Exercise 3.4 shows how the Cauchy and nominal stress are related in the simple case of a 1-D state of stress and incompressibility.

Observation 3.1 Figures 2.23 and 2.25 reveal the characteristic nonlinear behavior exhibited by many soft tissues: initially compliant at lower strains but very stiff at higher strains. It has been suggested by many that this non-linear, perhaps exponential, behavior is due to the composite nature of such tissues and the presence of undulated collagen fibers (Fung, 1993); that is, the initially compliant, sometimes nearly linear, behavior is often ascribed to the stretching of amorphous elastin, whereas the gradual-to-rapid stiffening is ascribed to the progressive recruitment (i.e., straightening) of previously undulated collagen fibers that may exhibit a nearly linear behavior when straight. Consistent with this thinking, many early investigators in cell mechanics assumed that cytoskeletal filaments may exhibit a nearly linear behavior. However, recent technological advances now enable investigators to perform mechanical tests on single molecules, which has revealed diverse and unexpected findings.

Leckband (2000) briefly reviewed three techniques for studying the mechanics of single molecules (e.g., proteins) as well as molecular interactions. *Laser tweezers* enable force measurements over the range 1–200 pN, the *atomic force microscope* (AFM) enables measurements over 10–1000 pN, and *micropipette aspiration* enables measurements over 0.01–1000 pN (note: pico indicates 10^{-12}, whereas nano indicates 10^{-9}). The AFM is discussed further in Chapter 5. Laser tweezers, also known as optical traps, exploit the interaction of highly focused laser light (e.g., from a Nd:YAG laser, at 1065 nm) and small (1–3 µm) dielectric particles. Briefly, light can exert a "pressure" on such particles, which when directed against gravity can trap the particle in a suspended state. The net "trapping force" depends on the laser power, the speed of light, and the properties of the particle. For example, a biomolecule can be attached to and held between two functionalized (to bind to the biomolecule) polystyrene microspheres, one of which is trapped by laser light and the other is mounted on a movable micropipette. The micropipette can thus be used as an actuator and the laser tweezers as a force transducer (Fig. 3.6). Conversely, the micropipette aspiration technique allows force to be inferred from global deformations of membranous capsules having a known surface tension. The specimen is held by suction between two opposing micropipettes.

Figures 3.7a and 3.7b illustrate laser tweezer measurements of the 1-D force–extension behavior of chromatin, a DNA attached to a protein

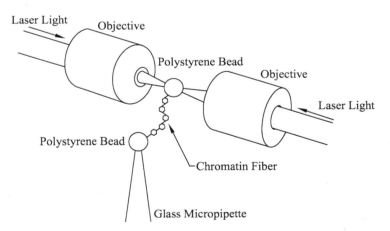

FIGURE 3.6 Schema of a laser tweezer (i.e., an optical trap) that can be used to capture and thus manipulate small particles using laser light. Functionalized particles (with appropriate ligands) can, in turn, be used to manipulate various biological molecules and thereby enable force–extension tests similar to those advocated by Hooke at the macroscale.

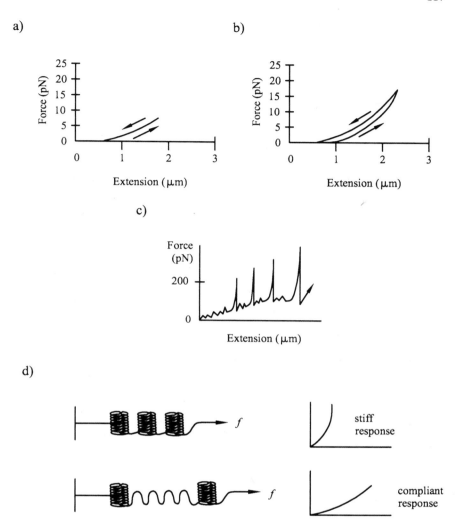

FIGURE 3.7 Schema of force–extension data for two different classes of molecules. Panels a and b show the behavior of chromatin for continuous extensions and releases, which reveal an elastic response or a pseudoelastic response depending on the degree of extension. Panel c shows the behavior of tenascin under continuous extension: a familiar nonlinear loading response is followed by an abrupt loss of force and a compliant response, only to be repeated multiple times. It is thought that this complex behavior is due to two different types of deformation of the molecule: The nonlinear response is due to a straightening of the undulated molecule, whereas the abrupt change is due to the complete unfolding of a portion of the molecule. The small oscillations represent noise in the measurement of pico-Newton (pN) level forces.

(histone) base that is the carrier of genes. Similar to data at the tissue level, one sees a nearly linearly elastic response over small extensions but a non-linearly pseudoelastic (i.e., a repeatable rate-independent hysteresis) response over larger extensions. Above ~20 pN, however, the force–extension curves were no longer reversible or repeatable (not shown).

Figure 3.7c shows a very different behavior. AFM data on tenascin, an extracellular matrix protein, reveals a "sawtooth" type of response with repeating compliant-to-stiffening curves at increasingly greater extensions. It has been suggested that such responses are due to the breaking of bonds that stabilize folded domains along the molecule; each abrupt loss of force corresponds to an unfolding event (Fig. 3.7d). It should be noted that protein unfolding exposes new chemical binding sites, thus explaining how mechanical forces or deformations can change molecular activity. Hence, albeit with sophisticated instrumentation like optical tweezers or the AFM, understanding how to interpret simple 1-D extension tests is very important, even at the molecular level. If we seek to compare behaviors from molecule to molecule, however, there is a need to go beyond Hooke's force–extension idea to concepts like stress–strain, which has not been investigated in depth in molecular biomechanics. For more on these studies, see Cui and Bustamante (2000) and Oberhauser et al. (1998), and references therein.

Example 3.2 If $f = 100\,\text{pN}$ at the maximum extension in the first cycle of loading on tenascin and if the original diameter is on the order of 10 nm, how does the stress compare with that in tissue?

Solution

Given

$$\frac{f}{A_o} = \frac{100 \times 10^{-12}\,\text{N}}{\frac{1}{4}\pi(10 \times 10^{-9}\,\text{m})^2} = 1.3 \times 10^6\,\text{N}/\text{m}^2 \cong 1\,\text{MPa},$$

we see that this value is on the order of that expected for ligaments or tendons, which are regarded as very strong tissues.

3.4 Pressurization and Extension of a Thin-Walled Tube

Let us now consider another *universal solution*—a solution in which we can find a relationship between the stresses and the applied loads and geometry without specifying a particular constitutive equation for the material [cf.

Eq. (3.29)]. Not only are such universal solutions useful because they allow the analysis of boundary value problems involving a wide variety of materials—soft tissues, rubber-like materials, metals, and so forth—but they are also extremely important in experimental design, for they provide a means of interpretation independent of the yet unknown material behavior, which is essential when one is designing an experiment to identify and quantify a yet unknown constitutive relation. As noted by Fung in 1973 and reiterated in his text (Fung, 1990), performing experiments to quantify the constitutive equations of living tissue remains one of the most important tasks in biomechanics.

3.4.1 Biological Motivation

Many soft tissues are pressurized cylindrical tubes. Examples include blood vessels, airways, and ureters. Moreover, many clinical devices include various tubes for the transport of fluids—from mechanical ventilators to heart–lung machines, from kidney dialysis units to catheters for balloon angioplasty, and even from oxygen lines to simple IV (intravenous) pumps, pressurized tubes are widespread in the hospital. Hence, quantifying the stresses in a cylindrical tube that arise from an internal pressure is fundamental to biomechanical R&D.

As a specific example, consider the saphenous vein. The great saphenous vein is one of the primary veins in the leg (along with the femoral vein); it runs along the medial aspect of the leg from the groin to the foot. Understanding well the biomechanics of the saphenous vein is not only important for understanding the normal physiology and pathophysiology of veins, but the saphenous vein has long been used as an autologous graft in coronary bypass surgery for the treatment of obstructive atherosclerotic lesions that are the cause of heart attacks. Although balloon angioplasty and intravascular stents have become widely accepted alternatives to bypass surgery (~1,069,000 angioplasty/stent procedures per year in the United States alone), saphenous vein bypass surgery continues to be performed widely (~517,000 procedures per year in the United States). As shown in Figure 3.8, a bypass procedure generally involves the suturing of the distal end of a vein segment to the ascending aorta and suturing the proximal end of the vein to a coronary artery distal to the obstruction/stenosis. Question: Why are the directions important? The treated coronary stenoses and, indeed, the continued patency of the graft are evaluated fluoroscopically via heart catheterization and dye injection into the artery. The latter requires an understanding of the biofluids, which we address in Chapters 7–10.

Whereas the first transplants of veins into the arterial system date back to the early twentieth century, coronary bypass surgery became commonplace only after basic, clinical, and biomedical engineering advances (such as the heart–lung machine) provided surgeons with sufficient time and ease

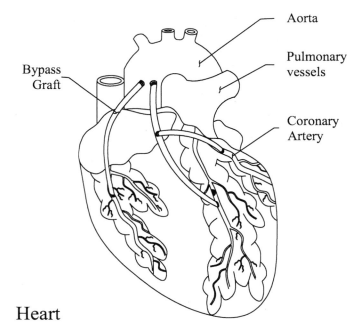

Aorta

Pulmonary
vessels

Bypass
Graft

Coronary
Artery

Heart

FIGURE 3.8 Schema of possible bypass grafts for coronary arteries originating from
the ascending aorta. Coronary artery disease (i.e., the presence of an obstructive
atherosclerotic plaque) is a leading cause of morbidity and mortality. Although
reduced flow to distal tissue can cause ischemia and thus angina, rupture of the atherosclerotic plaque and subsequent clotting at the site of rupture is responsible for
most sudden cardiac deaths. Understanding the solid mechanics (e.g., mechanisms
of rupture) and fluid mechanics (e.g., altered blood flow) are both essential to
improving clinical care.

to perform the surgery. More recently, findings in cell biology have opened
up two additional areas in which biomedical engineering can contribute significantly to bypass surgery. First, as noted in Chapter 1, we now know that
many cells in the body (the mechanocytes) alter their structure and function in response to changes in their mechanical environment. Given that
the endothelial, smooth muscle, and fibroblast cells in the venous wall (Fig.
3.9) are all very sensitive to mechanical signals and that when transplanted
into the arterial system, a vein goes from a low-pressure (5–15 mm Hg)
steady-flow environment to a high-pressure (120/80 mm Hg) pulsatile environment, understanding well the biomechanics and associated mechanobiology becomes fundamental to designing the surgical procedure. Second,
also as noted in Chapter 1, realizing the importance of mechanotransduction mechanisms in controlling cell and matrix biology suggests that we may
be able to engineer tissue replacements by applying the appropriate loads

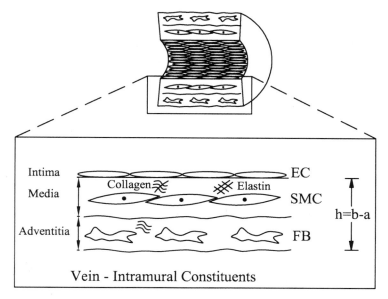

FIGURE 3.9 Portion of the wall of a vein showing the major cellular (EC = endothelial, SMC = smooth muscle, FB = fibroblast) and matrix (collagen and elastin) constituents; a is the inner radius and b the outer radius. Although veins have the same three layers as arteries (intima, media, adventitia), the thickness and composition of each layer are different between the two classes of vessels, as would be expected based on their different mechanical environments.

to vascular cells and their biodegradable scaffolds as we control their in vitro building of a replacement vessel for surgical implantation.

Given this motivation—the need to understand better both normal physiology and pathophysiology, the potential to design better surgical procedures, and the potential to engineer better replacement tissues—let us now investigate the associated stress analysis of an inflated tube.

3.4.2 Mathematical Formulation

We will see in Section 3.6 that the intramural stresses in a pressurized, thick-walled circular cylinder vary from point to point in general. For example, as illustrated in Figure 3.10, the circumferential (or hoop) stresses $\sigma_{\theta\theta}$ are often higher in the inner wall than in the outer wall due to pressurization alone.[4] It is easy to imagine, however, that if we consider only a thin

---

[4] In biological tissues and organs, a growth and remodeling process tends to introduce a so-called residual stress field that can homogenize the overall stresses (Humphrey, 2002); this is discussed briefly in Chapter 6.

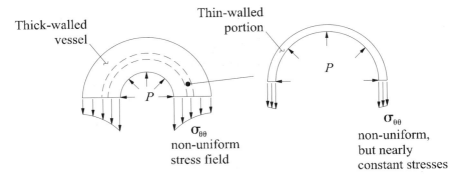

FIGURE 3.10 Schema of the radial distribution of the circumferential stress $\sigma_{\theta\theta}$ in a thick-walled cylinder under the action of an internal pressure P. Most importantly, note that the stress is nonuniform, in contrast to that in Figure 3.4 for the axially loaded member. Over a thin portion of the wall, however, the circumferential stresses can be represented well by the local mean value. This observation suggests that stresses in thin-walled pressurized cylinders may be represented well by their mean values.

segment of the thick-walled structure or, better yet, if we have a thin-walled cylinder to begin with, the normal stresses $\sigma_{(\text{theta face, theta direction})}$ (i.e., $\sigma_{\theta\theta}$), will still vary across the wall, but the difference from the inner portion to the outer portion may be so small that we can represent well this distribution by the mean value of the stress. Hence, if the cylindrical tube is *thin walled*, it is reasonable to assume that the stresses are approximately uniform across the thickness. Question: How thin is thin enough? A general rule of thumb is that the ratio of the deformed wall thickness h to the pressurized radius a should be $1/20$ or less. In this case, by making a judicious cut that separates the cylinder into halves (Fig. 3.11), we see that a force balance in the vertical direction requires the following. The sum of all internal uniform pressures acting at each point on the inner surface of the tube multiplied by their respective differential areas dA generates a net vertical force. This force is balanced by the vertical forces associated with the sum of the circumferential stresses $\sigma_{\theta\theta}$ (when we cut the tube in half) at each point multiplied by their respective differential areas; that is,

$$\sum F_v = 0 \rightarrow \int P_v dA - 2\int \sigma_{\theta\theta} dA = 0. \tag{3.37}$$

(Note: The factor 2 is necessary because we must add all of the stresses acting on each of the two cut edges.) Moreover, the "effect" of pressure in the vertical direction P_v can be written in terms of the uniform internal pressure P via $\sin\theta = P_v/P$ (Fig. 3.11). Thus, $P_v = P\sin\theta$, which acts over the differential area $r\,d\theta\,dz$, where $r \equiv a$ on the inner surface. Consequently, equilibrium requires

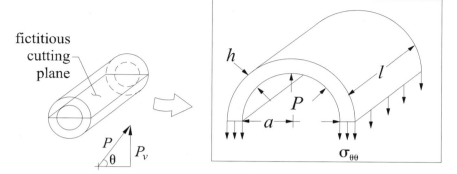

FIGURE 3.11 Free-body diagram of half of a thin cylindrical tube that is subjected to an internal pressure. Although the cut face exposes three components of stress in general ($\sigma_{\theta r}$, $\sigma_{\theta\theta}$, and $\sigma_{\theta z}$, given the face-direction nomenclature), $\sigma_{\theta\theta}$ alone is needed to balance the net vertical force due to the pressure (because we cut the tube exactly in half). Note, too, that a force balance on a finite part will yield the same result as that on an infinitesimal part because $\sigma_{\theta\theta}$ is assumed to be uniform.

$$\iint P \sin\theta a d\theta dz - 2\iint \sigma_{\theta\theta} dr dz = 0. \qquad (3.38)$$

Because the internal pressure and circumferential stress are both distributed uniformly, both can be taken outside the integrals, leaving

$$Pa\int_0^l \int_0^\pi \sin\theta d\theta dz = 2\sigma_{\theta\theta} \int_0^l \int_a^{a+h} dr dz, \qquad (3.39)$$

where l is the length of the tube, or

$$Pa\int_0^l 2dz = 2\sigma_{\theta\theta} \int_0^l h dz \to P(2a)(l) = 2\sigma_{\theta\theta}(h)(l). \qquad (3.40)$$

Thus, the basic equation for determining the *circumferential (Cauchy) stress in a thin-walled pressurized cylinder* is

$$\sigma_{\theta\theta} = \frac{Pa}{h}, \qquad (3.41)$$

where P is the uniform internal pressure, a is the inner radius of the cylinder in the pressurized configuration, and h is the thickness of the wall of the pressurized (i.e., deformed) cylinder. That a and h are values in the pressurized configuration cannot be overemphasized; numerous papers in the biomechanics literature are in error because of a failure to recognize this.

Likewise, it is very important to appreciate the implications of Eq. (3.41), which was derived independent of an explicit specification of the material properties and thus is a *universal solution*. For example, this equation says that if we have cylinders of different radii, but subjected to the same pressure and having comparable wall thickness, then the larger cylinder will have a higher stress. Although the thinness assumption may be questionable in the case of many abdominal aortic aneurysms, portions of which

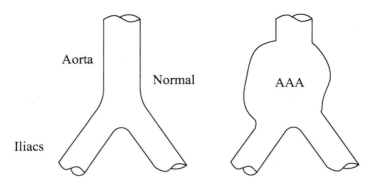

FIGURE 3.12 Schema of an abdominal aortic aneurysm (AAA). These lesions tend to be more prevalent in men, they tend to be most problematic in the elderly population, and the vast majority are fusiform in shape. AAAs may expand at different rates, often ranging from 0.1 to 1.5 cm/year. Traditional thinking has been that the rupture potential increases with overall size (~40% for lesions less than 5 cm in diameter and ~75% for lesions greater than 10 cm) although biomechanical analyses suggest that curvature may a more important predictor than size. The primary risk factors for AAAs are cigarette smoking, diastolic hypertension, and chronic obstructive pulmonary disease. Among others, Albert Einstein died from a ruptured AAA.

may be cylindrical (Fig. 3.12), many have argued based on Eq. (3.41) that the larger-diameter aneurysms are more susceptible to rupture, given comparable blood pressures, because the stress is higher.[5] Hence, this simple equation can have important clinical ramifications. One important caution, however, is that although the pressure P is assumed to be uniform and thus constant at each equilibrium state defined by each pair a and h, Eq. (3.41) does not imply that $\sigma_{\theta\theta}$ necessarily increases linearly with increases in P; that is, the radius that a tube assumes under the action of a given (equilibrium) pressure will depend on the material properties. For veins, for example, there is a highly nonlinear relation $P = P(a)$ (see Fig. 3.13), where $\sigma_{\theta\theta}$ also depends directly on the value of a for each pressure. Likewise, if the tube is subjected to an axial force f while pressurized, an increasing f will tend to decrease a at a given P. Hence, to use Eq. (3.41) correctly, we must use the correct values of a and h in the loaded configuration at each pressure P [i.e., each equilibrium state, whereby we recognize further that Eq. (3.41) can be used to compute stresses in a sequence of equilibria or increasing values of P, as long as the states are achieved quasistatically. Conversely, an illustrative example of elastodynamics is considered in Chapter 11].

---

[5] Detailed analysis of AAAs is very complex and typically requires numerical methods (Humphrey, 2002).

A

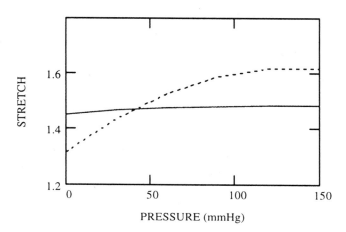

B

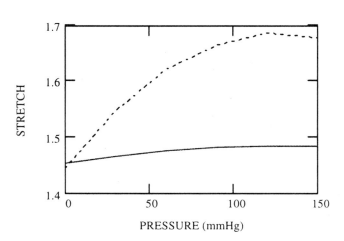

FIGURE 3.13 Pressure—stretch data for veins. Excised cylindrical segments were mounted vertically, plugged at the bottom, and inflated from the upper end. Panel A compares results in the circumferential direction between human (dashed line) and canines (solid line) vessels, whereas panel B compares data from canines in the axial (dashed line) and circumferential (solid line) directions. Note the nonlinear responses, similar to the stress–strain response of many soft tissues and of course, the species–species differences. It is also important to note that because the current radius a (where the circumferential stretch $\lambda_\theta = a/A$, with A the original radius) is dictated by the distending pressure, the pressure is thus a function of the radius [say $P = P(\text{radius})$]. This simple realization aids greatly in the interpretation of the equilibrium result for stress in veins if treated as a thin-walled tube. [Data from Wesley et al. (1975) Circ Res 37: 509–520.]

Although we will determine, below, the stresses in a tube under the action of axial loads, let us first make two additional observations. First, the integral $\iint \sin\theta\, a d\theta\, dz$ in Eq. (3.39) equaled $2al$, which is the *projected area* over which the pressure acts in the vertical direction. Recognizing this, Eq. (3.41) can be rederived easily by balancing the pressure times its projected area with the circumferential stresses times the area over which they act: $P(2al) = 2\sigma_{\theta\theta}(hl)$. When the projected area is obvious, this permits a quicker derivation. Second, if we recall that the uniform $\sigma_{\theta\theta}$ is actually the mean value that represents well the distribution of stresses across the thin wall, note that by boundary conditions, σ_{rr} equals $-P$ at the inner surface and σ_{rr} equals 0 at the outer surface (in our case although one could separately track an inner pressure P_i and outer pressure P_o). Not knowing how σ_{rr} varies from $-P$ to 0 (e.g., linearly or nonlinearly with radial location $r \in [a, a+h]$), the mean value can nonetheless be estimated simply as

$$\sigma_{rr})_{mean} = \frac{-P+0}{2} \rightarrow \sigma_{rr} \cong \frac{-P}{2}, \tag{3.42}$$

which is assumed to represent well the radial stresses within the wall of a thin-walled cylinder. Because of the thinness assumption, however, $a/h \gg 1$ and thus $\sigma_{\theta\theta}$ will be much larger numerically than σ_{rr}. It is for this reason that σ_{rr} is typically not considered in thin-walled inflated cylinders, although one must be careful to ignore effects in mechanobiology based on order of magnitude arguments alone.

Finally, let us consider two cases with regard to the axial stress σ_{zz} in an inflated thin-walled cylinder. Ask any vascular surgeon, for example, what happens to an artery or vein when it is transsected. The answer is that many vessels retract considerably when cut (e.g., the carotid artery will shorten ~50% when cut), which reveals that significant axial loads are present in vivo. These "preloads" probably arise during development, but this is speculative. Regardless, the axial stress is computed easily using the methods in Section 3.3, provided the line of action of the axial force f goes through the centroid of the overall cross section (i.e., provided f induces extension but not bending or twisting). From Figure 3.14, therefore, we have

FIGURE 3.14 Free-body diagram of a cylindrical tube that has been cut to expose z-face stresses, only one of which (σ_{zz}) is needed to balance the axial force f. Question: Under what type of loading would we also need a z-face, θ-direction stress $\sigma_{z\theta}$? Could such a stress exist in the aorta, popliteal artery, or middle cerebral artery?

FIGURE 3.15. Similar to Figure 3.14 except for the case of an internal pressure acting on a closed-ended tube. The cut exposes the stress of interest but does not depressurize the tube because it is fictitious.

$$-f + \int_0^{2\pi} \int_a^{a+h} \sigma_{zz} r\, dr\, d\theta = 0. \tag{3.43}$$

If, consistent with Eqs. (3.41) and (3.42), we consider the mean value of σ_{zz} to represent well the distribution of stresses across a thin wall, then

$$\sigma_{zz}\left[2\pi\left((a+h)^2 - a^2\right)\left(\frac{1}{2}\right)\right] = f \rightarrow \sigma_{zz} \cong \frac{f}{2\pi ah} \tag{3.44}$$

if $h \ll a$. Because the deformed cross-sectional area over which f acts is $A \cong 2\pi ah$, Eq. (3.44) is thus consistent with Eq. (3.29) for axially loaded rods of arbitrary cross section.

Conversely, if the ends of the cylinder are closed, then the internal pressure, which acts normal to and into all surfaces, will exert a net axial load as well. In this case (Fig. 3.15), we again sum the forces in the z direction. The sum of the internal pressures acting at each point in the z direction multiplied by their respective differential areas dA is balanced by the force developed by the sum of the stresses σ_{zz} acting at each point in the wall multiplied by their respective differential areas. To satisfy equilibrium, therefore,

$$\sum F_z = 0 \rightarrow \int \sigma_{zz} dA - \int P\, dA = 0, \tag{3.45}$$

or

$$\iint \sigma_{zz} r\, d\theta\, dr - \iint Pr\, d\theta\, dr = 0. \tag{3.46}$$

Again, because the internal pressure and the axial stress are assumed to be uniformly distributed, both can be taken outside the integral, thus leaving

$$\sigma_{zz} \int_0^{2\pi} \int_a^{a+h} r\, dr\, d\theta = P \int_0^{2\pi} \int_0^a r\, dr\, d\theta,$$

$$\sigma_{zz} \int_0^{2\pi} \frac{1}{2}\left[(a+h)^2 - a^2\right] d\theta = P \int_0^{2\pi} \frac{1}{2} a^2\, d\theta,$$

$$\sigma_{zz} \left\{\frac{1}{2}\left[(a+h)^2 - a^2\right]\right\}(2\pi) = P\left(\frac{1}{2} a^2\right)(2\pi), \tag{3.47}$$

$$\sigma_{zz}(2ah + h^2) = Pa^2.$$

Again assuming that the term h^2 is small compared to $2ah$ and thus that it contributes little to the overall solution, the basic equation for estimating the pressure-induced axial stress in a thin-walled closed cylinder is

$$\sigma_{zz} = \frac{Pa}{2h}, \tag{3.48}$$

where P is the internal pressure, a is the inner radius when pressurized, and h is the associated thickness of the wall. Note that this value is one-half that in the circumferential direction [cf. Eq. (3.41)].

Finally, in the case in which the ends are closed and there is an applied axial load,

$$\sigma_{zz} = \frac{Pa}{2h} + \frac{f}{2\pi ah}. \tag{3.49}$$

Superposition of solutions is allowed because the governing equations are linear. This, too, is a universal result.

Example 3.3 If a thin-walled, closed-end pressure vessel is subjected to internal pressure P (with radius a and thickness h), find the value of an additional end load f such that $\sigma_{\theta\theta} \equiv \sigma_{zz}$ (cf. Fig. 3.16).

Solution: From the above results [Eqs. (3.49) and (3.41)], the associated axial and circumferential stresses are

$$\sigma_{zz} = \frac{Pa}{2h} + \frac{f}{2\pi ah}, \quad \text{and} \quad \sigma_{\theta\theta} = \frac{Pa}{h}.$$

Now, set the two stresses equal ($\sigma_{\theta\theta} = \sigma_{zz}$) and solve for f:

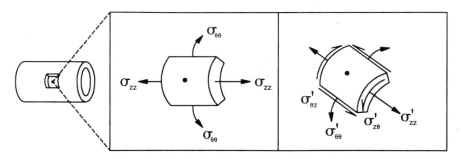

FIGURE 3.16 Schema of a small portion of a tube that could be subjected to an internal pressure and an axial force. Note that the 2-D state of stress at a point p depends on the coordinate system of choice.

$$\frac{Pa}{h} = \frac{Pa}{2h} + \frac{f}{2ah\pi} \rightarrow f = Pa^2\pi,$$

which is seen to equal the pressure times the internal projected area over which it acts.

It is important to recognize that if we neglect σ_{rr} in comparison to $\sigma_{\theta\theta}$ and σ_{zz}, then the state of stress in a thin-walled inflated tube is two-dimensional relative to (θ, z). As shown in Chapter 2, however, we know that 2-D stresses also exist relative to other coordinates, as, for example, (θ', z'). Indeed, it can be shown that,[6] perfectly analogous to Eqs. (2.13), (2.21), and (2.17),

$$\sigma'_{zz} = \sigma_{zz}\cos^2\alpha + 2\sigma_{z\theta}\sin\alpha\cos\alpha + \sigma_{\theta\theta}\sin^2\alpha, \tag{3.50}$$

$$\sigma'_{\theta\theta} = \sigma_{zz}\sin^2\alpha - 2\sigma_{z\theta}\sin\alpha\cos\alpha + \sigma_{\theta\theta}\cos^2\alpha, \tag{3.51}$$

$$\sigma'_{z\theta} = (\sigma_{\theta\theta} - \sigma_{zz})\sin\alpha\cos\alpha + \sigma_{z\theta}(\cos^2\alpha - \sin^2\alpha). \tag{3.52}$$

Hence, we can again ask questions, such as: What is the maximum normal stress? What is the maximum shear stress? At what angle α, relative to z, is the principal direction? For example, for a *closed-end* pressurized tube in the absence of an externally applied axial force,

$$\sigma'_{z\theta})_{\text{max/min}} = \sigma'_{z\theta}(\alpha = \alpha_s) = \sqrt{\left(\frac{\sigma_{\theta\theta} - \sigma_{zz}}{2}\right)^2 + (\sigma_{z\theta})^2}$$

$$= \sqrt{\frac{1}{4}\left(\frac{Pa}{h} - \frac{Pa}{2h}\right)^2 + 0} = \frac{Pa}{4h}, \tag{3.53}$$

which is to say that the maximum possible shear stress in two dimensions is one-half the axial stress. This shear occurs at

$$\alpha_s = \frac{1}{2}\tan^{-1}\left(\frac{\sigma_{\theta\theta} - \sigma_{zz}}{2\sigma_{z\theta}}\right) = \frac{1}{2}\tan^{-1}(\infty) \rightarrow \alpha_s = \pm 45°. \tag{3.54}$$

Such calculations are extremely important in design and analysis and they serve to remind us that stresses exist relative to particular coordinates.

Observation 3.2 According to Butler et al. (2000), "The goal of *tissue engineering* is to repair or replace tissues and organs by delivering implanted cells, scaffolds, DNA, proteins, and/or protein fragments at surgery." Much

---

[6] Note that z is like x, and θ is like y on a 2-D block of material.

of the early research in this field was directed toward the design of bioreactors to keep dividing cells alive ex vivo, the engineering of biodegradable
synthetic scaffolds on which these cells could adhere, migrate, and grow,
and the growing of tissue in desirable shapes such as cylindrical plugs, flat
sheets, or tubes. For obvious reasons, biochemical engineering played a significant role in the beginning of this exciting frontier. Fortunately, there
have been many successes in this regard; thus, attention is turning more
toward issues of "functionality"; that is, now that we can make tissuelike
materials in desired shapes, we must focus on making them functional.
Some tissues, like the liver, have a primarily biochemical function, but many
tissues of interest (e.g., arteries, cartilage, ligaments, heart, skin, tendons,
etc.) have mechanical as well as biological functions. Hence, biomechanics
will play a central role in functional tissue engineering. For example, early
work on arteries sought primarily to develop a nonthrombogenic tube that
would withstand arterial pressures and have sufficient suture-retention
strength. Normally functioning arteries do much more, however. For
example, they vary their smooth muscle tone to control the diameter of the
lumen and, thus, regional blood flow and they grow and remodel so as to
function well under the inevitable changes in load experienced throughout
changes in life. Tissue-engineered arteries should thus do more than pass
simple "burst" and "suture-retention" tests.

Tissue-engineered tendons for the surgical repair of damaged joints play
primarily a mechanical role, thus their material properties must likewise
mimic well those of the native tissue. Butler and Awad (1999) reported that
mesenchymal stem-cell-based tissue-engineered repairs of tendon defects
exhibited load-carrying capabilities from only 16–63% of the maximum
force experienced by the tendon during normal activity (Fig. 3.17). The
need for continued improvement in structural integrity is thus clear. The
use of biomechanical analysis will be essential, of course, in the continued
evaluation of such tissue constructs.

3.5 Pressurization of a Thin Spherical Structure

Here, we consider a situation very similar to that in Section 3.4: the uniform,
quasistatic pressurization of a thin-walled spherical structure composed of
an arbitrary material; that is, we seek another universal solution that relates
wall stresses to the applied loads and geometry. Because of the spherical
symmetry, only two stresses will be independent. The radial stress σ_{rr} at the
surfaces will, by boundary conditions, be equal and opposite the pressure
acting on either the inner or the outer surface of the sphere. Hence, as in
the case of the cylinder, the mean value of radial stress σ_{rr} will be $-P/2$ if P
is the distending (inner) pressure and the outer pressure is zero [cf. Eq.
(3.42)]. Note, too, that if we cut a sphere in half from top to bottom, we
expose intramural stresses $\sigma_{\phi\phi}$, whereas if we cut it in half from side to side,

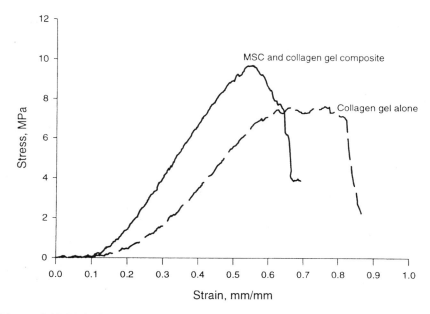

FIGURE 3.17 Uniaxial stress–strain behavior of rabbit patellar tendons wherein surgically created defects were treated with either a collagen gel filler or a collagen gel filler augmented with mesenchymal stem cells (MSCs). Data are shown 4 weeks after the repair. Clearly, augmentation of the filler with MSCs improved the stiffness and strength of the repair. Nonetheless, the repaired tendons could still support only 16% of the maximum stress borne by a native tendon. [From Butler and Awad (1999), with permission from Lippincott Williams & Wilkins.]

we expose the $\sigma_{\theta\theta}$ component. Because top to bottom and side to side are actually indistinguishable (i.e., you cannot discern a true change in orientation if you rotate a sphere), $\sigma_{\phi\phi} = \sigma_{\theta\theta}$ in a perfect sphere. Hence, our objective again is to relate this mean "hoop" stress to the applied load (distension pressure) and geometry (deformed radius and wall thickness).

3.5.1 Biological Motivation

In 1892, R. Woods presented an analysis of stresses in the wall of the heart based on the assumptions that the left ventricle is nearly spherical and thin walled. Although both of these assumptions are obviously crude, advances in cardiac mechanics came slowly: First in the late 1960s, when analyses were based on the assumption of a thick-walled sphere and material isotropy, to studies in the late 1970s to mid-1980s that focused on the nearly circular, thick-walled geometry of the equatorial region and transverse isotropy, to recent numerical studies based on more realistic geometries and material behaviors (see Humphrey, 2002). Whereas the heart cannot be modeled as a thin-walled spherical structure, many pressurized cells, tissues, and organs can be well approximated within this context. Examples may include the eye, the sphering of red blood cells, intracranial saccular aneurysms, and the urinary bladder. Indeed, in 1909, W. Osborne reported pressure–volume data on excised, intact urinary bladders that revealed the characteristic nonlinear behavior exhibited by soft tissues over large strains (cf. Fig. 3.13); his data are interpreted easily within the context of the spherical assumption. Here, however, let us consider a pathologic condition that is responsible for significant morbidity and mortality.

Intracranial saccular aneurysms are focal balloonlike dilatations of the arterial wall that often occur in or near bifurcations in the circle of Willis (the primary network of arteries that supplies blood to the brain; see Fig. 1.1). Although these lesions present in a myriad of sizes and shapes, a subclass of intracranial aneurysms can be treated reasonably well as thin-walled pressurized spherical structures (Fig. 3.18). For example, wall thickness is often on the order of 25–250 μm, whereas the pressurized inner radius is often on the order of 1.5–5 mm. Thus, $h/a \ll 1$. A dilemma faced by neurosurgeons is that the rupture potential of these aneurysms is very low, less than 0.1–1.0% per year, but when they rupture, 50% of the patients die and 50% of the survivors will have severe, lasting neurological deficits. A key question then is: How can we better predict the rupture potential of a given lesion, knowing that rupture appears to occur when wall stress exceeds strength locally? There is, therefore, a real need for biomechanical analysis in this case.

3.5.2 Mathematical Formulation

Similar to the analysis of the cylindrical tube, consider a free-body diagram in which we cut the thin-walled sphere in half to expose the internal stresses $\sigma_{\phi\phi} \equiv \sigma_{\theta\theta}$, which we will assume to be uniform (i.e., the mean values). Equilibrium thus requires force balance in the vertical direction, which is to say, a balance between all the pressures acting over their oriented differential areas and all the stresses acting over their differential areas. Mathematically (Fig. 3.19) and because the sphere is cut in half,

$$\sum F_v = 0 \rightarrow \int P_v dA - \int \sigma_{\theta\theta} dA = 0, \tag{3.55}$$

where the vertical component of the pressure is $P\cos\phi$, which acts over a differential area $dA = \rho \, d\phi \rho \sin\phi \, d\theta$, as seen in Fig. 3.19. Thus,

$$\iint P \cos\phi\rho \sin\phi d\theta\rho d\phi = \iint \sigma_{\theta\theta}\rho d\theta d\rho \tag{3.56}$$

where, in th ressure integral, $\rho = a$, the inner radius in the pressurized state. Thus,

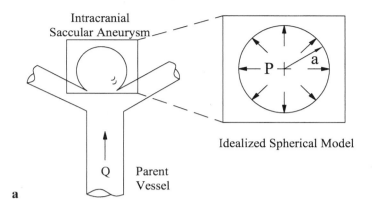

Idealized Spherical Model

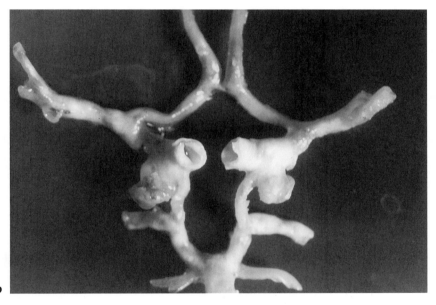

FIGURE 3.18 Schema of a subclass of intracranial saccular aneurysms (cf. Fig. 1.1) that can be modeled, to a first approximation, as a thin-walled pressurized sphere of radius a. Although pressure gradients are associated with the blood flow within the lesion (Chapter 8), these gradients tend to be small in comparison to the mean blood pressure (~93 mm Hg); hence, we can often assume a uniform internal pressure P. Also shown is a picture from the author's laboratory of a human circle of Willis with bilateral aneurysms, one ruptured and one not—the rupture being the cause of death. This reminds us that biomechanics is not just intellectually challenging and fun, it has potential to affect the lives of individuals and families.

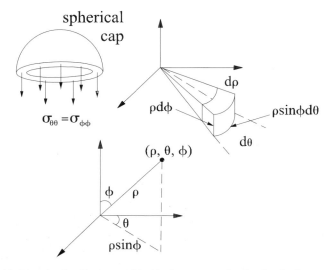

FIGURE 3.19 Free-body diagram of half of a pressurized spherical membrane, the associated coordinate system (ρ, θ, ϕ), and a convenient differential area for force balance. $\sigma_{\theta\theta} = \sigma_{\phi\phi}$ simply because of spherical symmetry—in other words, because a force balance and free-body diagram that separates the sphere into left and right halves yields the same result as one that separates it into top and bottom halves.

$$Pa^2 \int_0^{\pi/2} \int_0^{2\pi} \cos\phi \sin\phi\, d\theta d\phi = \sigma_{\theta\theta} \int_a^{a+h} \int_0^{2\pi} \rho\, d\theta d\rho,$$

$$Pa^2 \int_0^{\pi/2} 2\pi \cos\phi \sin\phi\, d\phi = \sigma_{\theta\theta} \int_a^{a+h} 2\pi\rho\, d\rho,$$

$$Pa^2 \int_0^{\pi/2} \frac{1}{2} \sin 2\phi\, d\phi = \sigma_{\theta\theta} \int_a^{a+h} \rho\, d\rho, \qquad (3.57)$$

$$Pa^2 \left(\frac{1}{2}\right)(1) = \sigma_{\theta\theta}\left[\frac{1}{2}(2ah + h^2)\right],$$

$$Pa^2 = \sigma_{\theta\theta}(2ah + h^2),$$

or because the deformed wall thickness $h \ll a$,

$$\sigma_{\theta\theta} = \frac{Pa^2}{2ah + h^2} \rightarrow \sigma_{\theta\theta} = \frac{Pa}{2h} = \sigma_{\phi\phi}. \qquad (3.58)$$

At this point, it is useful to recall that the vertical effect of the pressure is given by the pressure times the projected area over which it acts, $P\pi a^2$, which must balance the stress acting over its area, $\sim\sigma_{\theta\theta} 2\pi ah$. Together, these yield Eq. (3.58).

Example 3.4 Modeling a saccular aneurysm as a thin-walled sphere, assume that it has an inner radius of 2.5 mm and a thickness of 15 μm at a mean blood pressure of 120 mm Hg. Calculate the stress $\sigma_{\theta\theta}$ or $\sigma_{\phi\phi}$ and determine if rupture is likely if the critical stress is on the order of 5 MPa.

Solution: Given: P = 120 mm Hg $\cong$ 16000 N/m^2, where 1 mm Hg $\cong$ 133.32 N/m^2, a = 2.5 mm = 2.5 $\times$ 10^{-3} m, and h = 15 μm = 15 $\times$ 10^{-6} m, we have, by Eq. (3.58),

$$\sigma_{\theta\theta} = \sigma_{\phi\phi} = \frac{Pa}{2h} = \frac{(16000 \text{ N}/\text{m}^2)(2.5 \times 10^{-3} \text{ m})}{2(15 \times 10^{-6} \text{ m})} \cong 1,333,333 \text{ N}/\text{m}^2 \cong 1.3 \text{ MPa}$$

which, albeit less than the critical stress, is of the same order of magnitude. The factor of safety in prediction is thus only ~4. Given the 50% mortality rate associated with rupture and the sparseness of data on the mechanical behavior of saccular aneurysms, this may not be a sufficient factor of safety—one may well prefer a factor of at least 10. Note, therefore, that if the same size lesion were 60 μm in thickness rather than 15 μm, the stress decreases proportionately to 0.33 MPa, which is an order of magnitude less than the stated failure stress. In this case, because of the morbidity associated with such delicate neurosurgery (Humphrey, 2002), one may feel that such a lesion could simply be monitored over time rather than surgically treated right away. In this simple example, therefore, we see the potential utility of biomechanical analyses in surgical planning, the importance of high-resolution medical imaging (to resolve between 15- and 60-μm-thick lesions), and perhaps, most importantly, the need for better data and theories for studying saccular aneurysms (Humphrey, 2002).

3.6 Thick-Walled Cylinders

Before proceeding, let us reflect briefly on the results of the previous three sections in which we obtained universal solutions for the stresses in (1) an axially loaded member, (2) a thin-walled cylinder under axial load and pressure, and (3) a thin-walled pressurized sphere. Specifically, in each case, we were able to find *relations for the stresses in terms of the applied loads and geometry*, namely [cf. Eqs. (3.29), (3.41), (3.49), and (3.58)]:

$$\sigma_{xx} = \frac{f}{A}; \quad \sigma_{\theta\theta} = \frac{Pa}{h}, \quad \sigma_{zz} = \frac{Pa}{2h} + \frac{f}{2\pi ah}; \quad \sigma_{\theta\theta} = \frac{Pa}{2h} = \sigma_{\phi\phi}, \quad (3.59)$$

each independent of an explicit specification of the constitutive behavior. Moreover, because each is a uniform stress (i.e., they do not vary with position), the differential equations of equilibrium [Eqs. (3.8)–(3.16)] are satisfied identically. Although these results are very general, despite the associated restrictions such as loading through the centroid or thinness of the wall, in no case did we determine the associated strains or specific measures of the deformation (e.g., displacement of the end of the rod). Of course, given the value of the stress, one can use an appropriate constitutive (e.g., stress–strain) relation to find the corresponding strain. By using a constitutive relation, however, the *associated* result will not be universal, but instead will apply only to the particular class of material behaviors modeled by the constitutive relation. In this section, therefore, let us consider a case in which we determine both stress and strain for a particular material behavior by satisfying the differential equation for equilibrium.

Some veins, aneurysms, and other tissues may be analyzed by assuming that the thickness of the wall is much less than the inner radius; in other cases, the thick-walled structure must be addressed. Examples include the aorta and the equatorial region of the left ventricle of the heart. Because of the nonlinear material behavior (cf. Fig. 2.23) and large deformations, however, solving the thick-walled problem (for which universal solutions do not exist) is challenging and generally beyond the scope of an introductory text. For details on such problems, see Humphrey (2002), as well as the one example in Chapter 6.

Nonetheless, it is good to gain an appreciation of some of the complexities of the thick-walled solution and, indeed, to draw comparisons between the thick-wall and thin-wall approaches and results. Toward this end, therefore, let us consider the simplest thick-walled inflation problem: pressurization of a cylinder that exhibits a linearly elastic, homogeneous, and isotropic (LEHI) behavior under small strains. Moreover, let us assume complete axisymmetry and that there are no axial variations in stress; that is,

$$\frac{\partial}{\partial \theta} = 0 \quad \text{and} \quad \frac{\partial \left(\sigma_{(face)(direction)} \right)}{\partial z} = 0. \tag{3.60}$$

By restricting our attention to a uniform pressure, plus possibly a single axial load applied through the centroid, the only possible stresses relative to (r, θ, z) are σ_{rr}, $\sigma_{\theta\theta}$, and σ_{zz}, which may vary with radial location at most. Hence, in the absence of body forces, the cylindrical equilibrium equations [e.g., Eq. (3.11)] reduce to

$$\frac{d\sigma_{rr}}{dr} + \frac{1}{r}(\sigma_{rr} - \sigma_{\theta\theta}) = 0, \tag{3.61}$$

which is a first-order, linear ordinary differential equation. Now, for LEHI behavior [recall Eq. (2.79)],

$$\varepsilon_{rr} = \frac{1}{E}[\sigma_{rr} - \nu(\sigma_{\theta\theta} + \sigma_{zz})],$$

$$\varepsilon_{\theta\theta} = \frac{1}{E}[\sigma_{\theta\theta} - \nu(\sigma_{rr} + \sigma_{zz})], \tag{3.62}$$

$$\varepsilon_{zz} = \frac{1}{E}[\sigma_{zz} - \nu(\sigma_{rr} + \sigma_{\theta\theta})],$$

where [Eq. (2.47)]

$$\varepsilon_{rr} = \frac{\partial u_r}{\partial r}, \quad \varepsilon_{\theta\theta} = \frac{u_r}{r}, \quad \varepsilon_{zz} = \frac{\partial u_z}{\partial z}. \tag{3.63}$$

Furthermore, let the axial strain ε_{zz} be either zero or a constant. From the first two of these strain–displacement relations, note that

$$\frac{d}{dr}(r\varepsilon_{\theta\theta}) = \varepsilon_{rr}. \tag{3.64}$$

This equation is called an equation of "compatibility," which is to say that strains must be such that gaps or voids are not allowed to form (in other words, the continuum theory must be augmented to described situations such as fracture, which we do not consider, for which such incompatibilities do arise). Now, from this compatibility equation, combined with Eq. (3.62), we have

$$\frac{d}{dr}\left[r\left(\frac{1}{E}(\sigma_{\theta\theta} - \nu\sigma_{rr} - \nu\sigma_{zz})\right)\right] = \frac{1}{E}(\sigma_{rr} - \nu\sigma_{\theta\theta} - \nu\sigma_{zz}), \tag{3.65}$$

which can be written as (using the product rule)

$$r\left(\frac{d\sigma_{\theta\theta}}{dr} - \nu\frac{d\sigma_{rr}}{dr} - \nu\frac{d\sigma_{zz}}{dr}\right) + (\sigma_{\theta\theta} - \nu\sigma_{rr} - \nu\sigma_{zz}) = (\sigma_{rr} - \nu\sigma_{\theta\theta} - \nu\sigma_{zz}), \tag{3.66}$$

or

$$r\left(\frac{d\sigma_{\theta\theta}}{dr} - \nu\frac{d\sigma_{rr}}{dr} - \nu\frac{d\sigma_{zz}}{dr}\right) = (\sigma_{rr} - \sigma_{\theta\theta})(1+\nu). \tag{3.67}$$

Now, if we constrain the ends to not move in the axial direction, then $\varepsilon_{zz} = 0$ for all r and

$$\frac{d\varepsilon_{zz}}{dr} = 0 = \frac{1}{E}\left(\frac{d\sigma_{zz}}{dr} - \nu\frac{d\sigma_{rr}}{dr} - \nu\frac{d\sigma_{\theta\theta}}{dr}\right), \tag{3.68}$$

or

$$\frac{d\sigma_{zz}}{dr} = \nu\left(\frac{d\sigma_{rr}}{dr} + \frac{d\sigma_{\theta\theta}}{dr}\right). \tag{3.69}$$

Substitution of this equation as well as Eq. (3.61) (the radial equilibrium equation) into Eq. (3.67) yields

$$r\left\{\frac{d\sigma_{\theta\theta}}{dr} - v\frac{d\sigma_{rr}}{dr} - v\left[v\left(\frac{d\sigma_{rr}}{dr} + \frac{d\sigma_{\theta\theta}}{dr}\right)\right]\right\} = -r\frac{d\sigma_{rr}}{dr}(1+v), \quad (3.70)$$

or

$$(1-v^2)\frac{d}{dr}(\sigma_{rr} + \sigma_{\theta\theta}) = 0. \quad (3.71)$$

Hence, dividing through by $1 - v^2$ and integrating with respect to r, we find

$$\sigma_{rr} + \sigma_{\theta\theta} = c = \text{constant}. \quad (3.72)$$

Now, recognizing (show it) that the equilibrium equation (3.61) can be written as

$$\frac{d}{dr}(r\sigma_{rr}) = \sigma_{\theta\theta}, \quad (3.73)$$

we can obtain a single differential equation in terms of one component of stress, namely

$$\sigma_{rr} + \frac{d}{dr}(r\sigma_{rr}) = c = r\frac{d\sigma_{rr}}{dr} + 2\sigma_{rr}, \quad (3.74)$$

which reveals (show it) that we can write this equation as

$$\frac{1}{r}\frac{d}{dr}(r^2\sigma_{rr}) = c. \quad (3.75)$$

Multiplying through by r and integrating, we have

$$\int \frac{d}{dr}(r^2\sigma_{rr})dr = cr \rightarrow \sigma_{rr} = \frac{c}{2} + \frac{c_1}{r^2}, \quad (3.76)$$

which requires two boundary conditions, say $\sigma_{rr}(r = a) = -P_i$ and $\sigma_{rr}(r = b) = -P_o$. Solving the associated two algebraic equations for two unknowns, we have

$$c_1 = \frac{(P_o - P_i)a^2b^2}{b^2 - a^2}, \quad \frac{c}{2} = \frac{(P_i - P_o)b^2}{b^2 - a^2} - P_i \quad (3.77)$$

and, thus,

$$\sigma_{rr} = -P_i + \frac{(P_i - P_o)b^2}{b^2 - a^2} + \frac{(P_o - P_i)a^2b^2}{(b^2 - a^2)r^2}, \quad (3.78)$$

which can also be written as

$$\sigma_{rr} = \frac{P_ia^2 - P_ob^2}{b^2 - a^2} - \frac{(P_i - P_o)a^2b^2}{(b^2 - a^2)r^2}. \quad (3.79)$$

Note that at $r = a$ and $r = b$, we recover (show it) the boundary conditions as we should. Finally, from Eq. (3.73), the circumferential stress is

$$\sigma_{\theta\theta} = \frac{P_i a^2 - P_o b^2}{b^2 - a^2} + \frac{(P_i - P_o)a^2 b^2}{(b^2 - a^2)r^2}. \tag{3.80}$$

Together, Eqs. (3.78) and (3.80) are known as Lamé solutions. Note that, as in the previous sections, we have related the stresses to the applied loads (pressures) and geometry (radii). Remembering this common goal in each of these different problems—axially loaded rod, thin-walled cylinders and spheres, and now thick-walled cylinder—is helpful as we attempt to formulate new problems that may not be well defined. Note, too, that

$$\sigma_{rr} + \sigma_{\theta\theta} = 2\left(\frac{P_i a^2 - P_o b^2}{b^2 - a^2}\right), \tag{3.81}$$

which is constant given uniform static pressures P_i and P_o. Moreover, if $P_o = 0$, which is often the case, then

$$\sigma_{rr} = \frac{P_i a^2}{b^2 - a^2}\left(1 - \frac{b^2}{r^2}\right), \qquad \sigma_{\theta\theta} = \frac{P_i a^2}{b^2 - a^2}\left(1 + \frac{b^2}{r^2}\right), \tag{3.82}$$

which reveals that σ_{rr} increases monotonically from $-P_i$ to 0 (i.e., it is always compressive) and $\sigma_{\theta\theta}$ decreases monotonically as $1/r^2$ from its maximum value at $r = a$. See Figure 3.20 and compare to Figure 3.10 while remembering the earlier assumption of thinness to eliminate this radial dependence. It is interesting to compute (do it) the mean value of $\sigma_{\theta\theta}$, namely

$$\langle\sigma_{\theta\theta}\rangle = \frac{1}{b-a}\int_a^b \frac{P_i a^2}{b^2 - a^2}\left(1 + \frac{b^2}{r^2}\right)dr = \frac{P_i a}{b-a}, \tag{3.83}$$

which is seen to equal exactly (with wall thickness $h = b - a$) the universal result for the inflation of a thin-walled cylinder; that is, the thin-walled assumption yields the correct mean value of the stress regardless of the material properties or the thickness. Nonetheless, how well the mean value approximates the radial *distribution* of stresses must be assessed in each problem.

Finally, note that the stress σ_{zz} required to maintain the inflated cylinder at a fixed length [cf. Eq. (3.43)] is given by

$$2\pi\int_a^b \sigma_{zz} r \, dr = P_i \pi a^2 - P_o \pi b^2 + f, \tag{3.84}$$

where

$$\varepsilon_{zz} = 0 \rightarrow \frac{1}{E}[\sigma_{zz} - \nu(\sigma_{rr} + \sigma_{\theta\theta})] = 0, \tag{3.85}$$

or

$$\sigma_{zz} = \nu(\sigma_{rr} + \sigma_{\theta\theta}), \tag{3.86}$$

a constant from Eq. (3.72). Specifically, from Eqs. (3.82), we have for the case of internal pressure only:

140

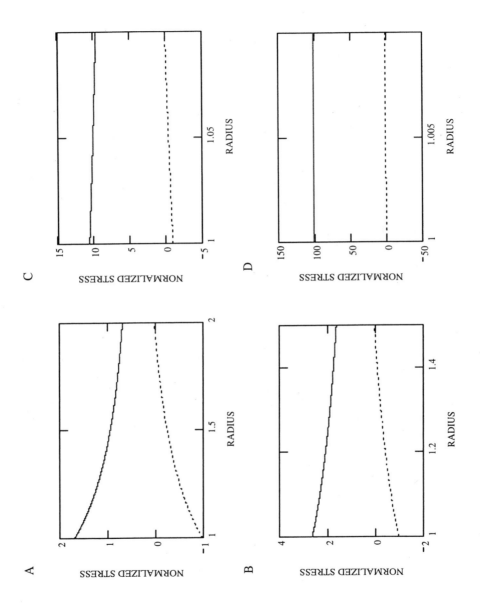

$$\sigma_{zz} = v\left(\frac{P_i a^2}{b^2 - a^2}\left(1 - \frac{b^2}{r^2}\right) + \frac{P_i a^2}{b^2 - a^2}\left(1 + \frac{b^2}{r^2}\right)\right) \tag{3.87}$$

or

$$\sigma_{zz} = \frac{2vP_i a^2}{b^2 - a^2} \equiv \frac{2vP_i a^2}{2ah + h^2}. \tag{3.88}$$

This relation, which depends on the material property v, is in contrast to Eq. (3.48) for the thin-walled case. In particular, if the wall is incompressible, $v = \frac{1}{2}$, and if $h \ll a$, then Eq. (3.88) recovers the earlier result, which did not require incompressibility. Regardless, the axial stress is uniform even though the radial and circumferential components are not.

In summary, we see that the solution of the thick-walled pressurized cylinder, even for a simple LEHI material behavior, is much more involved [Eqs. (3.61)–(3.88)] than the universal solution for the thin-walled cylinder [Eqs. (3.37)–(3.49)]. This is not surprising, of course, for we had to solve differential equations to find the pointwise distribution of stress in the thick-walled cylinder, whereas we solved simple gross force balance equations to find the uniform (mean) stresses in the thin-walled case. The decision to determine pointwise distributions versus mean values must be addressed individually in each problem, based on the desired or required detail needed to study the biomechanics or the mechanobiology. As we noted earlier, in the case of nonlinear material behavior over large strains (which is typically the case for soft tissues), solving the differential equations is nontrivial and we must often resort to numerical methods that are beyond the present scope. Moreover, because of the sensitivity of mechanocytes to changes in their mechanical environment, determination of the distribution

FIGURE 3.20 Computed transmural distributions of stress in a potentially thick-walled cylinder based on the Lamé solution with $P_o = 0$ [i.e., Eqs. (3.82)]. To evaluate the role of wall thickness, we normalize the circumferential and radial stresses by the inner pressure P_i and thus plot $\sigma_{\theta\theta}/P_i$ (solid curves) and σ_{rr}/P_i (dashed curves) as a function of r. Consider results for four different sets of inner and outer radii: panel A for $a = 1.0$ and $b = 2.0$; panel B for $a = 1.0$ and $b = 1.5$; panel C for $a = 1$ and $b = 1.1$; and panel D for $a = 1$ and $b = 1.01$. Observe the following. First, we see that the thicker the wall (e.g., panel A), the more dramatic the radial gradient in the wall stress; a corollary, therefore, is that a truly thin-walled cylinder has a nearly uniform wall stress, which is represented well by its mean value (e.g., panel D). Second, although the radial stress must always satisfy the boundary condition at the inner wall, $\sigma_{rr}(a) = -P_i$, its value becomes smaller in comparison to that of the circumferential stress as the wall gets thinner [recall the discussion near Eq. (3.42)]. Third, given the same pressure and inner radius, the thinnest-walled cylinder will have the highest stress, which is intuitive for it has less material to resist the same load. Nondimensional parametric studies such as this can often provide considerable insight and thus should be examined when possible.

of stress (i.e., the stress field) is most likely much more important than estimating the mean values.

Finally, we noted in the beginning of this section that we often desire to know the strains as well as the stresses. Having computed the values of stress, we can use the constitutive relations to determine the strains. For example,

$$\varepsilon_{\theta\theta} = \frac{1}{E}[\sigma_{\theta\theta} - \nu(\sigma_{rr} + \sigma_{zz})] \qquad (3.89)$$

is easily computed and so too the radial displacement because $u_r = r\varepsilon_{\theta\theta}$ from Eq. (3.63). The radial strain is similarly calculated easily via

$$\varepsilon_{rr} = \frac{1}{E}[\sigma_{rr} - \nu(\sigma_{\theta\theta} + \sigma_{zz})]. \qquad (3.90)$$

Of course, if the displacements were of primary interest, one could have alternatively solved the Navier–Space equilibrium equation [Eq. (3.26)] given displacement boundary conditions.

Appendix 3: First Moments of Area

Consider the cross section shown in Figure 3.21 and the associated (y', z') and (y, z) coordinate axes. The *first moments of area* with respect to the y' and z' axes are defined as

$$Q_{z'} = \iint y' dA, \qquad Q_{y'} = \iint z' dA. \qquad (A3.1)$$

These quantities may be positive, negative, or zero depending on the position of the coordinate system relative to the cross section. The *centroid* of the cross section is determined as

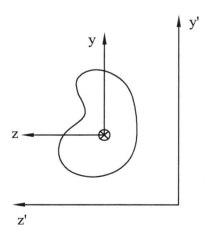

FIGURE 3.21. General cross section for the determination of the first moment of area.

FIGURE 3.22. Determine a general formula for the first moment of area for this rectangular cross section.

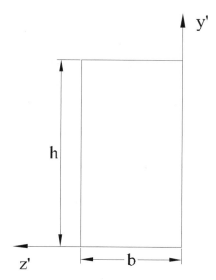

$$\overline{y}' = \frac{\iint y'dA}{\iint dA}, \qquad \overline{z}' = \frac{\iint z'dA}{\iint dA}. \tag{A3.2}$$

Once the centroid is located relative to (y', z'), it is often useful to introduce a *centroidal coordinate system* $(o; x, y, z)$ located at the centroid. Relative to this coordinate system, $(\overline{y}, \overline{z}) = (0, 0)$; that is

$$0 = \iint ydA, \qquad 0 = \iint zdA. \tag{A3.3}$$

To illustrate, consider the rectangular cross section in Figure 3.22. First, let us find $(\overline{y}', \overline{z}')$. Note, therefore, that

$$A\overline{y}' = \int_0^b \left(\int_0^h y'dy' \right) dz' = b\left(\frac{1}{2}(y')^2 \big|_0^h \right) = \frac{1}{2}bh^2, \tag{A3.4}$$

where

$$A = \int_0^b \int_0^h dy'dz' = bh. \tag{A3.5}$$

Hence, the centroid relative to (y', z') is

$$\overline{y}' = \frac{\dfrac{1}{2}bh^2}{bh} = \frac{1}{2}h, \qquad \overline{z}' = \frac{\dfrac{1}{2}b^2h}{bh} = \frac{1}{2}b. \tag{A3.6}$$

Moreover, relative to (y, z), we have

$$A\overline{y} = \int_{-b/2}^{b/2} \left(\int_{-h/2}^{h/2} ydy \right) dz = b\left(\frac{1}{2}y^2 \big|_{-h/2}^{h/2} \right) = 0, \tag{A3.7}$$

and, similarly, $A\bar{z} = 0$, as expected. With regard to first moments of area, therefore, the coordinate system of interest must be chosen carefully.

Example A3.1 Determine the centroid $(\bar{x}, \bar{y})$ for the triangle shown in Figure 3.23.

Solution: Although there are multiple ways to perform the requisite integration, let us first do so with a differential area $dA = dx\,dy$ noting that $y = (h/b)x$ (i.e., the slope, or rise over run, is h/b and the intercept is zero). Hence,

$$A = \int_0^b \left(\int_0^{hx/b} dy \right) dx = \int_0^b \frac{hx}{b} dx = \frac{h}{b} \left(\frac{x^2}{2} \Big|_0^b \right) = \frac{1}{2} bh,$$

as expected. Similarly,

$$\iint x\,dA = \int_0^b x \left(\int_0^{hx/b} dy \right) dx = \int_0^b \frac{hx^2}{b} dx = \frac{h}{b} \left(\frac{b^3}{3} \right) = \frac{1}{3} b^2 h$$

and

$$\iint y\,dA = \int_0^b \left(\int_0^{hx/b} y\,dy \right) dx = \int_0^b \frac{1}{2} \left(\frac{h^2 x^2}{b^2} \right) dx = \frac{1}{2} \frac{h^2}{b^2} \left(\frac{b^3}{3} \right) = \frac{1}{6} bh^2.$$

Consequently,

$$\bar{x} = \frac{\iint x\,dA}{\iint dA} = \frac{\frac{1}{3} b^2 h}{\frac{1}{2} bh} = \frac{2}{3} b, \qquad \bar{y} = \frac{\iint y\,dA}{\iint dA} = \frac{\frac{1}{6} bh^2}{\frac{1}{2} bh} = \frac{1}{3} h,$$

as expected. Show that the same result is obtained by considering a differential area $dA = y\,dx = (h/b)x\,dx$.

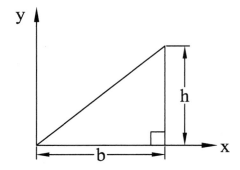

FIGURE 3.23. Determine a general formula for the first moment of area for this triangular cross section.

FIGURE 3.24. Determine the first moment of area for this circular cross section relative to two different coordinate systems.

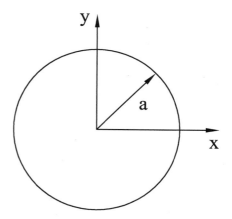

Example A3.2 Show that the centroid $(\bar{x}, \bar{y})$ for the circular region shown in Figure 3.24 is $(0, 0)$.

Solution: Knowing that $x^2 + y^2 = a^2$, or $y = \sqrt{a^2 - x^2}$, we can compute

$$A = \int_{-a}^{a} \left(\int_{-\sqrt{a^2-x^2}}^{\sqrt{a^2-x^2}} dy \right) dx = 2\left(\frac{1}{2}a^2\right)\left(\frac{\pi}{2} - \frac{-\pi}{2}\right) = \pi a^2.$$

Note from integral tables that

$$\int \sqrt{a^2 - x^2}\, dx = \frac{x}{2}\sqrt{a^2 - x^2} + \frac{a^2}{2}\sin^{-1}\left(\frac{x}{|a|}\right).$$

Alternatively, knowing that $x = r\cos\theta$ and $y = r\sin\theta$, we can compute

$$A = \int_0^a \left(\int_0^{2\pi} r\,d\theta \right) dr = 2\pi\left(\frac{r^2}{2}\bigg|_0^a\right) = \pi a^2.$$

Likewise, we can compute the centroid $(\bar{x}, \bar{y})$ in either Cartesian or cylindrical coordinates. In cylindricals,

$$\iint x\,dA = \int_0^a \left(\int_0^{2\pi} r\cos\theta\,rd\theta \right) dr = \int_0^a r^2\,dr \int_0^{2\pi} \cos\theta\,d\theta = \left(\frac{a^3}{3}\right)(\sin 2\pi - \sin 0) = 0$$

and, similarly,

$$\iint y\,dA = \int_0^a \left(\int_0^{2\pi} r\sin\theta\,rd\theta \right) dr = \int_0^a r^2\,dr \int_0^{2\pi} \sin\theta\,d\theta = \left(\frac{a^3}{3}\right)(-\cos 2\pi - \cos 0) = 0;$$

therefore,

$$\bar{x} = 0, \qquad \bar{y} = 0,$$

as expected. Repeat using Cartesians alone.

First moments of area are additive; thus, they can be used to find the centroids of composite areas. A *composite area* is simply defined as an area that can be well described via the addition or subtraction of well-defined geometric shapes (squares, rectangles, triangles, circles, ellipses, etc.).

It can be shown that the centroid for a composite area is given by

$$\bar{x} = \frac{\sum \bar{x}_i A_i}{\sum A_i}, \qquad \bar{y} = \frac{\sum \bar{y}_i A_i}{\sum A_i}, \tag{A3.8}$$

where $\bar{x}_i$ and $\bar{y}_i$ are the centroids of the individual parts $i = 1, 2, \ldots, N$, all relative to the same coordinate system. These simple formulas are best appreciated via numerical examples.

Example A3.3 Find the centroid, relative to x and y, for the cross section shown in Figure 3.25.

Solution: It is easiest to formulate these solutions in tabular form. Hence, note that

	Area	$\bar{x}_i A_i$	$\bar{y}_i A_i$
Part 1	20 (6) = 120	10 (120)	13 (120)
Part 2	2 (10) = 20	10 (20)	5 (20)

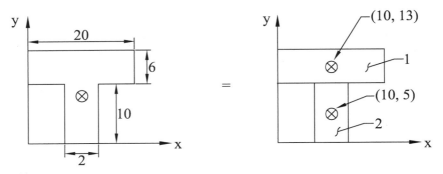

FIGURE 3.25. Compute the first moment of area of this composite section.

Therefore,

$$\bar{x} = \frac{\sum \bar{x}_i A_i}{\sum A_i} = \frac{10(120) + 10(20)}{120 + 20} = 10,$$

$$\bar{y} = \frac{\sum \bar{y}_i A_i}{\sum A_i} = \frac{13(120) + 5(20)}{120 + 20} = 11.86.$$

Note, too, that we could obtain the same result by computing values for a rectangular area 20×16 and then subtracting out small rectangular areas to yield the T-shape, namely

	Area	$\bar{x}_i A_i$	$\bar{y}_i A_i$
Part 1	20 (16) = 320	10 (320)	8 (320)
Part 2	9 (10) = 90	4.5 (90)	5 (90)
Part 3	9 (10) = 90	15.5 (90)	5 (90)

whereby

$$\bar{x} = \frac{\sum \bar{x}_i A_i}{\sum A_i} = \frac{10(320) - 4.5(90) - 15.5(90)}{320 - 90 - 90} = 10,$$

$$\bar{y} = \frac{\sum \bar{y}_i A_i}{\sum A_i} = \frac{8(320) - 5(90) - 5(90)}{320 - 90 - 90} = 11.86.$$

which is the same as found above.

Exercises

3.1 Do a literature review to find the failure strength of the anterior cruciate ligament (ACL), which is commonly torn by athletes. Given typical dimensions (e.g., cross-sectional area) of an ACL in a male college athlete, what is the maximum safe axial force that the ACL can sustain? With increased numbers of women competing in NCAA sports, however, there are increasingly more reports of ACL injuries in females than males. Why is this the case and how can this problem be addressed?

3.2 Repeat the derivation of Eqs. (3.34) and (3.36) by assuming that the line of action of the force f goes through the point $(-y^*, -z^*)$; that is, show that the final result is independent of the starting point.

3.3 Design an experiment to determine the failure strength of a chordae tendineae. Additionally, outline a method of analysis to estimate the range of stresses experienced by the chordae in vivo under physiologic values of intraventricular pressures.

3.4 Recall from Example 3.1 and the subsequent discussion that there are at least two "types" of stress: the true (Cauchy) stress, which is a measure of forces acting over a deformed oriented area A, and the nominal (Piola–Kirchhoff) stress, which is a measure of forces acting over the undeformed oriented area A_o. Note, too, that the chordae, like many soft tissues, often conserves its volume when deforming (i.e., it behaves incompressibly, which because of the large strains cannot be accounted for via $v = \frac{1}{2}$). If the original length is L and the current length is l, volume conservation requires $LA_o = lA \rightarrow A = A_o(L/l)$ assuming uniform stress and strain. Hence,

$$\sigma_{xx} = \frac{f}{A} = \frac{f}{A_o}\left(\frac{l}{L}\right) = \Lambda\Sigma_{xx},$$

where $\Lambda = l/L$ is a stretch ratio and Σ_{xx} is the nominal stress. If $L = 10$ mm, compare the values of σ_{xx} and Σ_{xx} for all l from 10.001 to 10.7 mm.

3.5   Repeat Exercise 3.4 for a rubber band, which also conserves its volume as it is extended by up to 50% of its original length. Note, therefore, the potential difference between the Cauchy and Piola–Kirchhoff stresses.

3.6   What is the maximum shear stress, in terms of the applied load f and cross-sectional area A, in a uniaxially stressed structure [Eqs. (3.29)–(3.36)] and what is the associated angle α_s relative to an axial direction x?

3.7   Based on the type of analysis in Exercise 3.6, if an investigator correlates the proliferation and migration of fibroblasts in an injured tendon with the value of axial stress during the test (e.g., no stress, nonzero but subphysiological stress, physiological stress, or supraphysiologic stress), how can one know that the mechanotransduction is induced by the experimentally convenient axial stress rather than the maximum shear stress given that both exist simultaneously (i.e., their calculation depends only on the choice of the coordinate system introduced by the investigator)?

3.8   Write a three-page summary on the history of saphenous vein bypass surgery, noting, in particular, the histological changes in the wall of the vein graft due to the biomechanics-mediated growth and remodeling.

3.9   Find the dimensions of and pressures within the human inferior vena cava. Estimate the mean wall stress $\sigma_{\theta\theta}$. Even though the aorta is not thin walled, estimate its mean wall stress $\sigma_{\theta\theta}$ in the human abdominal aorta. Discuss structural differences between the vena cava and aorta given their different functions and mechanical environments.

3.10 Compare the maximum shear stress $\sigma'_{z\theta}$ in a smooth muscle cell in the wall of an arteriole ($a \sim 75\,\mu m$ and $P \sim 45\,mm\,Hg$) of the thickness of one smooth muscle cell if the mean radial stress is ignored or included. Neglect possible $\sigma_{r\theta}$ and σ_{rz} shear stresses.

3.11 Referring to Example 3.3, show that the maximum shear stress $\sigma'_{z\theta}$ ($\alpha = \alpha_s$) is zero in a closed-end pressurized tube when $\sigma_{\theta\theta} = \sigma_{zz}$ due to the judicious choice of the applied load f and because $\sigma_{z\theta} = 0$. Indeed, note the significant ramifications of this; if the maximum shear stress, relative to $(z', \theta'$ with $\alpha = \alpha_s)$, is zero, then all shear stresses $\sigma'_{z\theta}$ are zero for any α. This is seen easily from the formula

$$\sigma'_{z\theta}(\alpha = \alpha_s) = \sqrt{\left(\frac{\sigma_{\theta\theta} - \sigma_{zz}}{2}\right)^2 + (\sigma_{z\theta})^2},$$

which reveals that if the stresses are principal *and* equal, then there is no shear stress relative to any 2-D coordinates.

3.12 If a vascular surgeon wishes to implant an arterial graft such that there will exist no shear stress $\sigma'_{z\theta}$ relative to any 2-D coordinate directions (z', θ'), find the required axial load f. Hint: See Exercise 3.11.

3.13 Let the pressure in a normal vein be denoted by P_v and likewise its normal geometry by a_v and h_v. If this vein is used as an arterial graft, its pressure will increase to value P and its radius to value a; associated thinning of the distended wall to h is expected as well. Hence, with $P > P_v$ and $a > a_v$ with $h < h_v$ we expect that $\sigma_{\theta\theta})_{\text{graft}} \gg \sigma_{\theta\theta})_{\text{vein}}$ simply due to the transplantation. Extensive laboratory evidence suggests that the vascular wall seeks to maintain the circumferential stress nearly constant. If this is so and the mechanotransduction mechanisms operative in the endothelial, smooth muscle, and fibroblast cells allow the wall to respond to the increased stress, what do you expect the vein to do?

If you said that you expect the wall to thicken, you are exactly right. In fact, consider the following data from Zwolak et al. (1987):

Tissue	EC activity	SMC activity	h (μm)	a (mm)
Artery	0.02	0.05	50	0.89
Vein	0.02	0.05	19	1.69
VG—1 week	8.10	10.30	23	1.55
VG—2 weeks	2.90	1.70	44	1.91
VG—4 weeks	1.50	0.80	77	2.36
VG—12 weeks	0.02	0.20	116	2.90
VG—24 weeks	0.10	0.20	123	2.65

where VG ≡ vein graft at the various times post-transplantation and cell activity reflects the percent turnover in cells. Plot Pa/h for the adaptation of the vein and compare it to that for the homeostatic value for the artery. Plot the cell turnover rates and discuss.

3.14 Similar to the discussion in Exercise 3.13, the arterial wall will also thicken in response to chronic hypertension: systolic/diastolic pressures such as 160/90 mm Hg versus the normal values of 120/90 mm Hg. See data on wall thickening in Fung and Liu. Discuss in terms of $\sigma_{\theta\theta}$

noting that their discussion of stress is wrong; that is, the formula $\sigma_{\theta\theta}$ = Pa/h must be based on values of a and h in the deformed (pressurized) configuration. You may note, in addition, that this formula is strictly valid only for $h/a \ll 1$. It can be shown, however, that the mean wall stress in even a thick-walled tube is estimated reasonably well by the simple formulas (cf. result in Section 3.6), which is why this relation is widely used in vascular mechanics.

3.15 The balloons used on angioplasty catheters tend to be long and cylindrical. If one performed an experiment in the laboratory in which a balloon catheter is inflated within a healthy cylindrical artery, what measurements would be needed to determine the radial stress exerted on the endothelium by the balloon?

3.16 Referring to Example 3.3, compute the maximum shear stress $\sigma'_{z\theta}$ if f = 0; that is,

$$\sigma_{\theta\theta} = \frac{Pa}{h}, \quad \sigma_{zz} = \frac{Pa}{2h}.$$

Compare it to the case when

$$f = Pa^2\pi.$$

3.17 What is the maximum shear stress in a thin-walled pressurized sphere? What does this imply with regard to the potential rupture criterion for intracranial saccular aneurysms?

3.18 Similar to the analysis of the thick-walled pressurized Hookean cylinder in Section 3.6, formulate and solve for the stresses in a thick-walled sphere. This may require library research on spherical coordinates. Note, too, that this is a non-trivial problem.

3.19 Given the solution in the previous exercise, find the average wall stress in the thick-walled Hookean sphere and compare to the results from the thin-walled analysis.

3.20 Although Eq. (3.41) was derived for a thin-walled cylinder, it can be shown that it provides a reasonable estimate of the mean circumferential stress in a thick-walled tube as well. Note, therefore, that in hypertension (i.e., a persistent increase in blood pressure), the aorta distends (i.e., a increases) and the wall thins (i.e., h decreases). Hence, $\sigma_{\theta\theta}$ increases tremendously. If the hypertensive pressure $P_H = nP$, where n is a number and the luminal radius returns to a due to smooth muscle contraction and a shear-stress-mediated vasoconstriction (see Chapter 9), how much does the aorta need to thicken to restore $\sigma_{\theta\theta}$ back to its original value? Find data in the literature on aortic morphology in hypertension to see if this is borne out by data.

3.21 In the thick-walled cylinder problem, we found that $\sigma_{zz} = v(\sigma_{rr} + \sigma_{\theta\theta})$. If we use this result in Hooke's law, then

$$\varepsilon_{rr} = \frac{1}{E}\{\sigma_{rr} - \nu[\sigma_{\theta\theta} + \nu(\sigma_{rr} + \sigma_{\theta\theta})]\},$$

$$\varepsilon_{\theta\theta} = \frac{1}{E}\{\sigma_{\theta\theta} - \nu[\sigma_{rr} + \nu(\sigma_{rr} + \sigma_{\theta\theta})]\}.$$

Show that these two equations can be inverted to yield

$$\sigma_{rr} = \frac{E}{(1+\nu)(1-2\nu)}[(1-\nu)\varepsilon_{rr} + \nu\varepsilon_{\theta\theta}],$$

$$\sigma_{\theta\theta} = \frac{E}{(1+\nu)(1-2\nu)}[\nu\varepsilon_{rr} + (1-\nu)\varepsilon_{\theta\theta}].$$

3.22 Given the result from the previous exercise that $\sigma_{rr} = f(\varepsilon_{rr}, \varepsilon_{\theta\theta})$ and $\sigma_{\theta\theta} = g(\varepsilon_{rr}, \varepsilon_{\theta\theta})$, where $\varepsilon_{rr} = du_r/dr$ and $\varepsilon_{\theta\theta} = u_r/r$, show that equilibrium requires

$$\frac{d\sigma_{rr}}{dr} + \frac{1}{r}(\sigma_{rr} - \sigma_{\theta\theta}) = 0 \rightarrow \frac{\partial^2 u_r}{\partial r^2} + \frac{1}{r}\frac{\partial u_r}{\partial r} - \frac{u_r}{r^2} = 0.$$

3.23 Verify via substitution that $u_r = C_1 r + C_2/r$ is a solution for the differential equation in the previous exercise. Moreover, show that

$$C_1 = \frac{(1+\nu)(1-2\nu)}{E}\left(\frac{P_i a^2 - P_o b^2}{b^2 - a^2}\right),$$

$$C_2 = \frac{(1+\nu)}{E}\left(\frac{P_i - P_o}{b^2 - a^2}\right)a^2 b^2.$$

Hint: Use the boundary conditions that

$$\sigma_{rr}(r = a) = -P_i, \qquad \sigma_{rr}(r = b) = -P_o,$$

with σ_{rr} and $\sigma_{\theta\theta}$ written in terms of u_r ($\equiv C_1 r + C_2/r$) via the constitutive relation (cf. Exercise 3.21) and strain–displacement relation (cf. Exercise 3.22). Verify that the final result for stress is consistent with that found in Section 3.6.

3.24 Consider the two following experimental setups, both of which are designed to impose an axial load on a thin tendon (from a laboratory rat) for purposes of studying the stress–strain behavior. Assuming a coefficient of friction μ_s between the wire and the rough cylinder, determine the axial load f that is applied to the tendon, in terms of W, in each case.

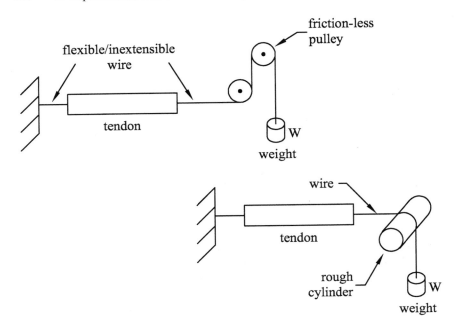

FIGURE 3.26.

3.25 If the axial first Piola–Kirchhoff stress $\Sigma = f/A_o$ in Exercise 3.24, what is the axial Cauchy stress σ if the tendon has an original cross-sectional area A_o and length L and the tendon is incompressible? Let the current area and length be A and l, respectively.

3.26 If a single protein molecule is tested in tension, what complications may arise with regard to assuming a continuum to compute the stress. See Figure 3.27.

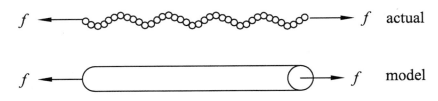

FIGURE 3.27.

3.27 Find the centroid using the method of composite sections for the following cross-sectional area.

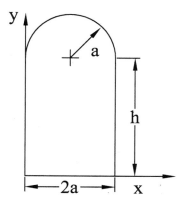

FIGURE 3.28.

4
Extension and Torsion

The deformations experienced by some biological tissues and biomaterials can be very complex. For example, we have seen that all six components of the Green strain [cf. Eq. (2.42), but relative to cylindrical coordinates] are nonzero in the wall of the heart, and each varies with position and time throughout the cardiac cycle (cf. Fig. 2.19). In such cases, we must often resort to sophisticated numerical methods to measure or compute the strain fields. Nevertheless, there are many cases in which the deformations are much simpler, as, for example, in chordae tendineae within the heart, which experience primarily an axial extension with associated lateral thinning (cf. Fig. 3.2). Indeed, as an introduction to biomechanics, it is often best to study simple motions such as extension, compression, distension, twisting, or bending, which allow us to increase our understanding of the basic approaches and which also apply to many problems of basic science or clinical and industrial importance. Whereas we considered small strains that occur during a simple inflation of a thick-walled tube in the last section of Chapter 3, here we consider in some detail small strains associated with axial extension and torsion, with an associated complete stress analysis for the latter for a linear, elastic, homogenous, and isotropic (LEHI) behavior of a circular member. Such analyses will be particularly relevant in bone mechanics.

Observation 4.1 The reader is encouraged to consult Carter and Beaupré (2001) for a description of the mechanobiology of skeletal *development*. Here, we simply recount some of their observations. For example, they write: "The flat bones of the skull and face are formed by intramembranous ossification within a condensation of cells derived from the neural crest. In the limb bones and most of the postcranial skeleton, however, mesenchymal cell condensations chondrify, creating the endoskeletal cartilage anlagen. These cartilage rudiments form the templates for the future skeleton and subsequently, in the process of growth, undergo a bony transformation." In particular, "The cartilage cells within the rudiments therefore undergo a characteristic process of cell proliferation, maturation, hypertrophy, and death, followed by matrix calcification and ossification. Variations within the cartilage growth and ossification rates in different directions within the anlage result in shape changes of developing bones ... Once a region of cartilage mineralizes and it is either resorbed or replaced by bone, further bone growth occurs by osteoblastic apposition on mineralized surfaces." As noted in Chapter 12, Developmental Biomechanics is one of the exciting frontiers of our field, one that is clearly complex.

Many factors affect the development as well as the subsequent maintenance and adaptation of bone. For example, biological factors that affect the metabolism of chondrocytes include bone-derived growth factor (BDGF), bone morphogenetic proteins (BMP), cartilage-derived morphogenetic protein (CDMP), fibroblast growth factors (FGFs), insulin-like growth factors (IGFs), interleukins (ILs), sex hormones, prostaglandins (PG), matrix metalloproteinases (MMPs) and their inhibitors (TIMPs), and even vitamins A, C, and D. In addition, of course, mechanical stimuli also play a major role in the development, maintenance, and adaptation of bone. In many cases, strains have proven convenient to correlate with the mechanotransduction. Let us now consider measures of the deformation in the simple case of axial loading.

4.1 Deformations due to Extension

4.1.1 Biological Motivation

Figure 4.1 illustrates some of the important structural and biological features of a representative mature long bone. Grossly, the three primary regions are the central long hollow shaft, the end caps, and the transitional regions between the two. These three regions are referred to respectively as the diaphysial, epiphysial, and metaphysial regions. The central core of the diaphysial region is called the medullary canal; it contains the bone marrow, which produces different types of blood cells and their precursor. Of primary concern here, however, is that there are two primary classes of bone tissue: *cortical* (or compact) and *cancellous* (or trabecular). Cortical bone constitutes most of the outer portion of a whole bone, including the majority of the wall of the diaphysis. Except in a few regions, the cortical bone is invested by a specialized covering, the periosteum, which is rich in collagen and fibroblasts and has an underlying osteogenic layer that contains active bone cells. During development and in periods of trauma and repair, cortical and cancellous bone can be of the woven type, which is often poorly structured, highly mineralized, and appears to serve as a temporary scaffolding for the development of another type of bone tissue.

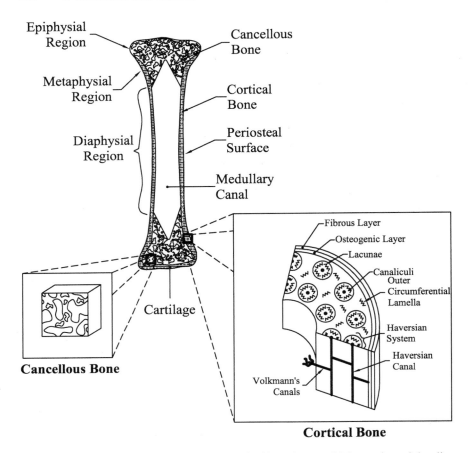

Epiphysial Region
Cancellous Bone
Metaphysial Region
Cortical Bone
Diaphysial Region
Periosteal Surface
Medullary Canal
Cartilage
Cancellous Bone
Fibrous Layer
Osteogenic Layer
Lacunae
Canaliculi
Outer Circumferential Lamella
Haversian System
Haversian Canal
Volkmann's Canals
Cortical Bone

FIGURE 4.1 Schema of the structure of a typical long bone, which consists of the diaphysial (shaft) region, the metaphysial (transition) region, and the epiphysial (end) regions. Note, too, the two primary types of bone tissue: cortical, which is found along the outer surface, and cancellous, which is found in the inner portion of the end regions. The cartilage forms as a protective covering at the end of the articulating bones; cartilage is discussed in Chapter 11.

In maturity and following healing, cortical bone consists primarily of two types of bone: *Lamellar* bone is characterized by concentrically arranged layers (or laminae), each about $20\,\mu m$ thick, with networks of blood vessels between layers; *osteonal*, or Haversian, bone is characterized by nearly cylindrical units (or osteons) $\sim200\,\mu m$ in diameter and $\sim2\,cm$ long, which contain centrally located blood vessels connected to radial channels called Volkmann's channels. Each of these channels, which allow the transport of blood and bone fluid within compact bone, contribute to an overall porosity despite the otherwise dense constitution of cortical bone (specific gravity ~2). Uniformly distributed throughout the interstitial substance of cortical bone are lenticular cavities, called lacunae, each containing a bone cell called an osteocyte. Radiating in all directions from each lacunae are anastomosing tubular passages, called canaliculi, which further contribute to the porosity and are essential to nutrient exchange.

Cancellous bone has a very different microstructure. It is much more porous, consisting of a three-dimensional lattice of branching trabeculae, which are thin-walled and of lamellar type. Cancellous bone is found, for example, near the ends of long bones. Recall from Chapter 1 that research in the late nineteenth century by von Meyer, Culmann, and Wolff suggested that the orientation of the trabeculae in the femur appeared to follow the directions of the principal stresses (Fig. 4.2). This ultimately led to "Wolff's law of bone remodeling," a topic that continues to be of interest, particularly with recent advances in mechanobiology.

In contrast to other tissues, which experience interstitial growth, bone growth occurs only via deposition on cell-laden surfaces. Such *appositional growth* thus occurs at the periosteal surface, the endosteal surface that lines the medullary canal, and on all surfaces of the tubular cavities as well as the surfaces of the trabeculae. Note, too, that trabecular growth is evidenced by an increased number of trabeculae or an increase in their thickness. Whereas skeletal development occurs over periods of years, stress- or strain-mediated adaptation occurs over months to years; fortunately, in cases of injury, such as a fracture, bone growth and thus repair can occur in weeks to months.

Although bone consists primarily of type I collagen impregnated with hydroxyapatite, $Ca_{10}(PO_4)_6(OH)_2$, an inorganic compound that endows bone with its high compressive strength, it is the bone cells that govern overall growth and remodeling. There are four primary types of cells in bone: osteoprogenator cells, osteoblasts, osteoclasts, and osteocytes. As noted earlier, like other connective tissues, most bone derives from the mesenchyme. Osteoprogenator cells are relatively undifferentiated cells found on many of the free surfaces; they are particularly active during normal development and in times of repair. Osteoblasts are responsible for forming bone, which is to say that they actively synthesize the collagen and appear to regulate the uptake and organization of the mineral component. Osteoclasts, in contrast, are responsible for the resorption of bone; they are giant cells 20–100 μm in diameter that contain many nuclei. The primary cells of fully formed bone are the osteocytes, which derive from the osteoblasts and reside in the lacunae within the interstitial space (Fig. 4.3). Once encased in calcified bone matrix, the osteocytes no longer divide; rather, they form gap junctions with neighboring osteocytes via the canaliculi, and probably participate in the control of the osteoblasts and osteoclasts. For more on the biology of bone, see Alberts et al. (2002) and Fawcett (1986).

One of the key questions in bone mechanobiology is how the embedded osteocytes or surface osteoblasts/osteoclasts sense and respond to changes

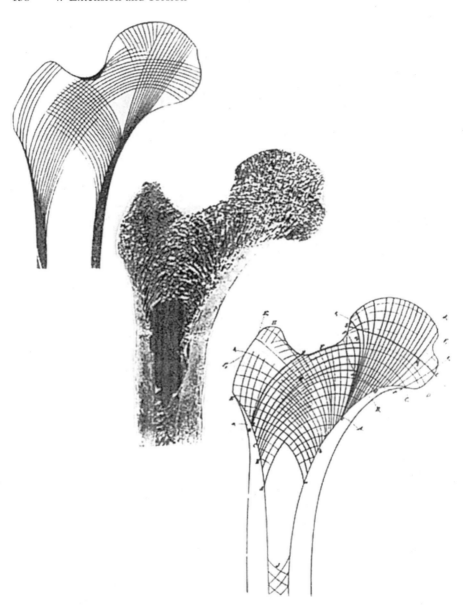

FIGURE 4.2 Correspondence between the trabecular structure in the femur and Wolff's envisioned lines of tension. [From Wolff (1986), with permission from Springer-Verlag.]

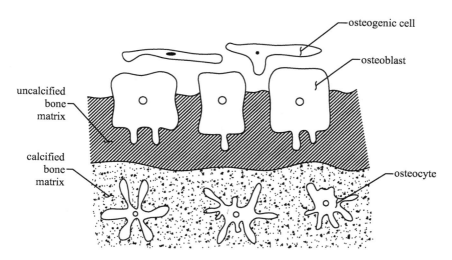

FIGURE 4.3 Schema of three of the four primary bone cells. The osteogenic cell, the osteoblast (or bone-forming cell), and the osteocytes, which are former osteoblasts that are trapped within calcified bone matrix. Not shown are the osteoclasts, which remove bone tissue.

in mechanical stimuli. We know, for example, that there is tremendous bone loss in load-bearing bones (particularly in the legs) in bedridden patients and astronauts in a microgravity environment. Conversely, there is significant increase in bone mass in athletes such as weight lifters and even tennis players (e.g., the humerus can have a 30% greater cross-sectional area in the playing versus the nonplaying arm). Such examples of decreased and increased bone mass are likewise common when applied loads are altered clinically, such as due to bone screws, plates, or implanted prostheses. For more examples, see Carter and Beaupré (2001). It is not clear, however, if the causative cellular activity correlates best with changes in stress, strain, strain rate, strain energy, or similar metric. Again, we emphasize that cells cannot directly sense these volume-averaged continuum quantities, yet they will likely be very useful for identifying such empirical correlations (Humphrey, 2001). Although strains can be measured on the outer surface of some bones, it is not possible to measure internal strains or any stresses. Hence, we must resort to the methods of mechanics to calculate the stress or strain fields experienced by the bone of interest, which, in turn, requires knowledge of the geometry, material properties, and applied loads. As noted in Chapter 2, bones can be described by Hooke's law for stress analysis in many circumstances, yet a detailed study of the mechanobiology may require structural models that account for the fine trabecular architecture or material models that account for the porosity and, indeed, the internal flow of blood or bone fluid due to applied loads. The latter necessitates

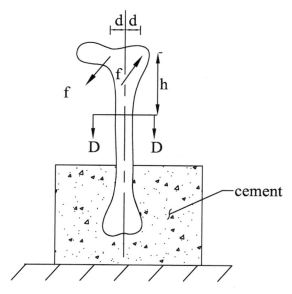

FIGURE 4.4 Schema of a portion of the femur isolated in the laboratory for mechanical testing to induce torsion via the application of a couple 2*fd*.

modeling of the solid–fluid coupling, which is addressed briefly in Chapter 11 in a different context. Solid–fluid coupling in bone is an advanced topic of current research. Here, therefore, let us consider the simplest approach, assuming on average that bone exhibits a linear, elastic, locally homogenous, and isotropic (i.e., LEHI) behavior under some circumstances. In this case, effective bone properties can be assumed to be $E \sim 15\,\text{GPa}$ and $\nu \sim 0.33$. Indeed, let us consider the stress and strain fields in the diaphysial region of a long bone, consisting of cortical bone only and subject, first, to an axial compressive load and, second, to a twisting moment as suggested by Fig. 4.4; of course, the bone could also experience bending loads, but these are considered in Chapter 5.

4.1.2 Theoretical Framework

Envision a case in which a rod of negligible weight is suspended vertically from a fixed support and loaded from the lower end by a constant force that is applied through its centroid and uniformly over the cross-sectional area. Intuitively, the axial displacement (say, u_x) will be zero at the fixed support, nonzero in the middle, and maximum at the lower end (Fig. 2.17); that is, the displacement will vary along the length of the rod (even though the stress is assumed to be constant throughout), from which we can compute the axial strain, namely

$$u_x = u_x(x) \to \varepsilon_{xx} = \frac{\partial u_x}{\partial x}. \tag{4.1}$$

Reminder: This formula for strain is restricted to small values, consistent with our desired use of Hooke's law as a descriptor of LEHI behavior. Clearly, integration of ε_{xx} with respect to x can provide the displacement at any point x, including that at the lower end $x = L$; that is,

$$\int_0^x \varepsilon_{xx} dx \equiv \int_0^x \frac{\partial u_x}{\partial x} dx = u_x(x) - u_x(0), \tag{4.2}$$

where $u_x(0) = 0$ is the displacement boundary condition (for this case) at the fixed end. Now, ε_{xx} can be related to the stress via Hooke's law [Eq. (2.69)], where the uniform 1-D state of stress in an axially loaded rod is $\sigma_{xx} = f/A$ from Eq. (3.29). Hence, we have

$$\varepsilon_{xx} = \frac{1}{E}[\sigma_{xx} - \nu(0+0)] = \frac{f}{AE} \to u_x(x) = \int_0^x \frac{f}{AE} dx, \tag{4.3}$$

where, in general, the force, cross-sectional area, and even Young's modulus could vary with x. In the special case in which all three quantities are independent of x and we seek only the value of u_x at the lower end (the so-called end deflection δ), we have the simple result

$$u_x(x = L) \equiv \delta = \frac{fL}{AE}. \tag{4.4}$$

In general, however, it is best to remember the primary result of Eq. (4.3), which determines a *deformation in terms of the applied loads, geometry, and material properties*. It can be written generally as

$$u_x(x = c) - u_x(x = a) = \int_a^c \frac{f(x)}{A(x)E(x)} dx, \tag{4.5}$$

which emphasizes that the applied axial force, cross-sectional area, and Young's modulus may each vary with x. Of course, the integral is a linear operator and, thus,

$$\int_a^c \frac{f(x)}{A(x)E(x)} dx = \int_a^b \frac{f(x)}{A(x)E(x)} dx + \int_b^c \frac{f(x)}{A(x)E(x)} dx \tag{4.6}$$

and so forth. This division of the integral over separate domains can be very helpful in cases in which $f(x)$, $A(x)$, or $E(x)$ are constant over such subdomains. Let us illustrate via a few examples how this might be useful. First, however, note some terminology: If a rod is homogeneous, then $E \neq E(x)$; if a rod has a constant cross section, then $A \neq A(x)$; and if the rod is under a constant load, then $f \neq f(x)$.

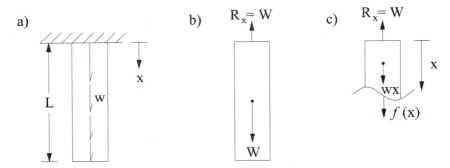

FIGURE 4.5 A vertically loaded member subject to its own weight, given as w (force per unit length) and thus a total weight $W = wL$, which acts at the center of gravity. Shown is the physical problem, a free-body diagram of the whole to isolate reaction R_x at the fixed support, and a free-body diagram of a part to isolate the internal force $f(x)$.

Example 4.1 Consider a vertically mounted, axially loaded member subject to its own distributed weight w N/m (see Fig. 4.5a). Assume that the member has a constant cross-sectional area A and a constant elastic modulus E. The total weight of the member of length L is thus $W = wL$. Find the displacement u_x at the free end [i.e., $\delta \equiv u_x(x = L)$].

Solution: First, let us construct a free-body diagram of the whole structure and ensure equilibrium to find the reactions (Fig. 4.5b):

$$\sum F_x = 0, \quad R_x - W = 0 \rightarrow R_x = W = wL,$$

$$\sum F_y = 0, \quad R_y = 0,$$

$$\sum M_z = 0, \quad M_{wall} = 0.$$

Next, construct a free-body diagram of the parts (Fig. 4.5c) recalling that if a structure is in equilibrium, then each of its parts is in equilibrium. The force $f(x)$ due to the weight of the member is $w(L - x)$ at any cross section cut at a distance x from the support; at $x = 0$, $f(0) = R_x = wL$, the entire weight, as it should. Alternatively, in terms of the total weight of the member, the force becomes $W(1 - x/L)$ and thus

$$\sum F_x = 0 \rightarrow \int \sigma_{xx} dA - f = 0 \rightarrow \sigma_{xx} = \frac{f}{A} = \frac{W}{A}(1 - x/L).$$

Note that the stress is largest at $x = 0$, where all of the weight must be borne by the material, and the stress is zero at the free end, which is free of applied loads (i.e., traction-free). Given the stress, the strain and the axial displacement can now be computed using Hooke's law and Eq. (4.5); namely

$$\varepsilon_{xx} = \frac{1}{E}[\sigma_{xx} - \nu(\sigma_{yy} + \sigma_{zz})]$$

with σ_{yy} and σ_{zz} each zero. Thus,

$$\varepsilon_{xx} = \frac{\sigma_{xx}}{E} \rightarrow \int_0^L \varepsilon_{xx} dx = u_x(x = L) - u_x(x = 0) = \int_0^L \frac{W(1 - x/L)}{AE} dx,$$

where $u_x = 0$ at $x = 0$ (a displacement boundary condition) and the end displacement is

$$\delta \equiv u_x(x = L) = \frac{W}{AE}\left(x - \frac{x^2}{2L}\right)\Big|_0^L = \frac{WL}{2AE}.$$

Of course, the displacement at any value of x is found by integrating from 0 to x rather than from 0 to L.

Example 4.2 Find the end displacement δ in each of the members illustrated in Figure 4.6.

Solution: The first structure (Fig. 4.6a) is homogenous and subject to a constant axial load P, but it does not have a constant cross-sectional area. The area changes abruptly from A_1 to A_2 at $x = L/2$. Thus, $A = A(x)$ and the end displacement is determined via

$$u_x(x) - u_x(0) = \int_0^x \frac{P}{A(x)E} dx \rightarrow \delta = u_x(0) + \frac{P}{E}\int_0^L \frac{dx}{A(x)}.$$

The integral must be separated at the point of discontinuity in the cross-sectional area to give the following results [with $u_x(0) = 0$ via a boundary condition]:

$$\delta = \frac{P}{E}\int_0^{L/2} \frac{dx}{A_1} + \frac{P}{E}\int_{L/2}^L \frac{dx}{A_2} \rightarrow \delta = \frac{P}{A_1 E}\int_0^{L/2} dx + \frac{P}{A_2 E}\int_{L/2}^L dx,$$

or

$$\delta = \frac{PL}{2A_1 E} + \frac{PL}{2A_2 E} = \frac{PL}{2E}\left(\frac{1}{A_1} + \frac{1}{A_2}\right).$$

The second structure (Fig. 4.6b) has a constant cross-sectional area and is subjected to a constant axial load P, but it is not homogenous. The material properties change at $x = L/3$ from the wall. Therefore, $E = E(x)$ and the displacement becomes

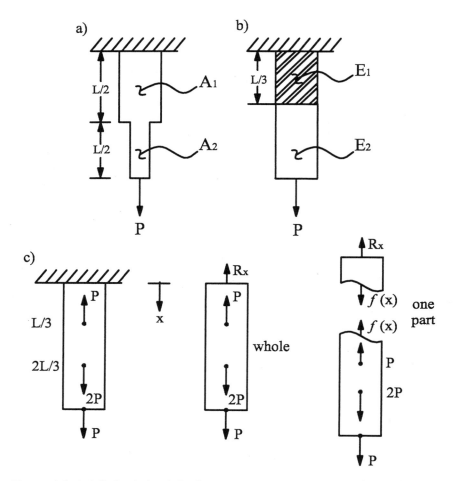

FIGURE 4.6 Axially loaded rods having a nonconstant cross section (panel a), a non-constant material composition (panel b), and multiple applied loads (panel c). Although we need to draw free-body diagrams of the whole and multiple parts for each case, we show only the free-body diagram of the whole structure and one free-body diagram for a part of the rod of panel c.

$$u_x(x) - u_x(0) = \frac{P}{A} \int_0^x \frac{dx}{E(x)} \rightarrow \delta = u_x(0) + \frac{P}{A} \int_0^L \frac{dx}{E(x)}.$$

Again dividing the integration over judicious domains, we have

$$\delta = \frac{P}{A} \int_0^{L/3} \frac{dx}{E_1} + \frac{P}{A} \int_{L/3}^L \frac{dx}{E_2},$$

where $u_x(x = 0) = 0$ again. Hence, we find

$$\delta = \frac{PL}{3AE_1} + \frac{2PL}{3AE_2} = \frac{PL}{3A}\left(\frac{1}{E_1} + \frac{2}{E_2}\right).$$

For the third problem (Fig. 4.6c), we must first solve the statics problem. Equilibrium of the whole requires that the reaction force R_x be given by

$$-R_x - P + 2P + P = 0 \rightarrow R_x = 2P,$$

whereas equilibrium of parts requires that we consider three separate cuts. For the first part,

$$-R_x + f(x) = 0 \rightarrow f(x) = 2P, \quad 0 \le x \le \frac{L}{3}.$$

Similarly, for the second part,

$$-R_x - P + f(x) = 0 \rightarrow f(x) = 3P, \quad \frac{L}{3} < x < \frac{2L}{3}.$$

Finally, for the third required part,

$$-R_x - P + 2P + f(x) = 0 \rightarrow f(x) = P, \quad \frac{2L}{3} < x \le L.$$

Indeed, the last result can be seen easily given a small part near the end. Regardless, given constants E and A and $u_x(x = 0)$, we have

$$\delta = \frac{1}{AE}\left(\int_0^{L/3} 2P\, dx + \int_{L/3}^{2L/3} 3P\, dx + \int_{2L/3}^{L} P\, dx\right) = 2\frac{PL}{AE}.$$

4.1.3 Clinical Application

Now that we have some experience with the full axial load problem for LEHI behavior, let us consider an important clinical problem. Each year in the United States, ~120,000 artificial hips are implanted surgically to relieve pain and restore ambulatory motion. Figure 4.7 shows a typical prosthesis and its insertion into the host femur. As seen at section D-D, we have nearly concentric cylindrical cross sections over part of the bone–metal interface. Although the femoral head experiences complex loads that may subject the prosthesis to compression, torsion, and bending, here let us focus on the axial load alone (other loads will be considered subsequently). This special case could be produced in the laboratory. Moreover, although the actual loads, geometry, and material properties demand a numerical (e.g., finite element) method (Fig. 4.8), let us consider a simple analysis to gain some insight into the overall problem. In particular, as a first approximation, let us assume that the bone and prosthesis each exhibit LEHI behaviors. Bone is, of course, better characterized as nonhomogeneous and

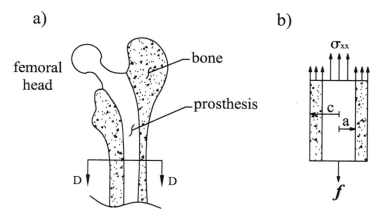

FIGURE 4.7 Schema of a metallic prosthetic hip that has been implanted to replace a damaged femoral head. One of the most common causes of femoral damage is fracture associated with osteoporosis. Defined as a reduction in bone mass, osteoporosis is a particularly debilitating disease in elderly women. If we focus on the region near section D-D in the figure and consider the action of an axial load only, then panel b shows an appropriate free-body diagram for analysis to relate the axial stress to the applied loads and geometry. Although the stress may (as a first approximation) be assumed to be uniform within each constituent, metal and bone, these mean values need not be the same.

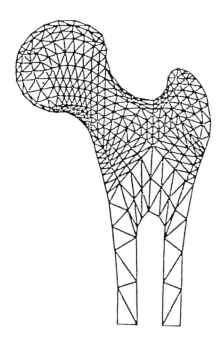

FIGURE 4.8 Illustrative finite element mesh used to analyze stresses in the femur for determining the state of stress in health, which should be mimicked as well as possible following an implant surgery. Each triangle represents a local computational domain, or element, in which equilibrium is enforced. Certain continuities, such as displacement, are also enforced from element to element. Albeit for a 2-D analysis of a normal femur, finite element studies can be conducted similarly in three dimensions, and for the case of a prosthesis, a (poly) methyl methacrylate, or PMMA, bone cement, and bone. Finite element analyses are extremely powerful, and the student is encouraged to take at least one course in this area. With permission, from Prof. B. Simon.

anisotropic, but these simplifying assumptions have been used by many and they allow us to begin to explore the problem.

Our model problem, therefore, is simply the axial loading (through the overall centroid) of a circular cylinder consisting of two LEHI materials (Figure 4.7b). Like most biological tissues, bone will grow and remodel in response to changes in mechanical stimuli. Therefore, one of the key questions with regard to prosthesis design is: How will the implant redistribute the stresses within the bone? Again, this is a complex question; we will consider the much simpler question here. On average, how does the applied load f in Figure 4.7b distribute (i.e., partition) between the metal implant and the remaining bone? Toward this end, let the radius of the prosthesis be a and the outer radius of the bone be c. If we let that part of the load carried by the prosthesis and bone be denoted as f_p and f_b, respectively, then axial equilibrium requires that $f_p + f_b = f$. The associated mean axial stresses are thus $\sigma_{xx}^p = f_p/A_p$ for the prosthesis and $\sigma_{xx}^b = f_b/A_b$ for the bone. The key question then is: What are f_p and f_b?

With $f_p + f_b = f$, we have one equation and two unknowns, thus rendering this problem statically indeterminate; that is, we cannot determine how the load or the stress partitions using statics alone—we must seek a second equation. This can be accomplished from kinematics if we simply assume that the axial displacements are the same in each component [i.e., that there is no relative movement (e.g., delamination as discussed in Observation 2.2) between the prostheses and bone as desired of a painless implant]. Hence, if $\varepsilon_{xx}^p \equiv \varepsilon_{xx}^b$, with uniform properties along the length of the prosthesis, then we have

$$u_x(x = L) - u_x(x = 0) = \int_0^L \varepsilon_{xx}^p \, dx = \int_0^L \frac{\sigma_{xx}^p}{E_p} \, dx = \int_0^L \frac{f_p}{A_p E_p} \, dx = \frac{f_p L}{A_p E_p}, \quad (4.7)$$

or, with $u_x(x = 0) = 0$,

$$\delta_p = \frac{f_p L}{A_p E_p}. \quad (4.8)$$

Similarly, for the bone,

$$u_x(x = L) - u_x(x = 0) = \int_0^L \varepsilon_{xx}^b \, dx = \int_0^L \frac{\sigma_{xx}^b}{E_b} \, dx = \int_0^L \frac{f_b}{A_b E_b} \, dx = \frac{f_b L}{A_b E_b}, \quad (4.9)$$

or

$$\delta_b = \frac{f_b L}{A_b E_b}. \quad (4.10)$$

Hence, *to ensure compatible displacements*, $\delta_p = \delta_b$ requires that

$$\frac{f_p L}{A_p E_p} = \frac{f_b L}{A_b E_b} \rightarrow f_p = \frac{A_p E_p}{A_b E_b} f_b. \quad (4.11)$$

Thus, we have a second equation in terms of the unknown "partitioned forces." From equilibrium, we have

$$f = \frac{A_p E_p}{A_b E_b} f_b + f_b = f_b \left(1 + \frac{A_p E_p}{A_b E_b}\right), \tag{4.12}$$

or

$$f_b = \frac{f A_b E_b}{A_b E_b + A_p E_p}, \tag{4.13}$$

and, similarly,

$$f_p = \frac{A_p E_p}{A_b E_b} \left(\frac{f A_b E_b}{A_b E_b + A_p E_p}\right) = \frac{f A_p E_p}{A_b E_b + A_p E_p} \tag{4.14}$$

Finally, the stresses in the prosthesis and bone are

$$\sigma_{xx}^p = \frac{1}{A_p}\left(\frac{A_p E_p f}{A_p E_p + A_b E_b}\right) = \frac{E_p f}{A_p E_p + A_b E_b},$$

$$\sigma_{xx}^b = \frac{1}{A_b}\left(\frac{A_b E_b f}{A_p E_p + A_b E_b}\right) = \frac{E_b f}{A_p E_p + A_b E_b}. \tag{4.15}$$

We see, therefore, that the load partitions according to the respective cross-sectional areas *and* the material properties. In the special case that $E_p = E_b = E$ and $A_p + A_b = A$, we recover the original homogeneous solution ($\sigma_{xx} = f/A$), as we should. Whether the bone will resorb (atrophy) or grow will depend on whether its stress (or strain) following implantation is less than or greater than the normal physiological values. Early on, artificial implants were designed primarily to be geometrically mimicking of the native femoral head and to be strong enough that they would not fail (i.e., yield, deform plastically, or fracture; cf. Fig. 2.24) under the demands of physiological loading. Yet, the associated designs failed to consider how the stress or strain in the bone redistributed and how functional adaptation might lead to a weakening of the remaining bone over time. This flaw in the analysis and design resulted in many prosthetic failures in the early days, thus necessitating much more careful biomechanical study. The interested reader is encouraged to review the current literature on prosthesis design to appreciate the development of the field. With regard to the present (simple) analysis, a take-home message is that although we were only interested in the stresses, equilibrium alone did not permit a complete solution. This is in stark contrast to the (statically determinate) universal solutions in Chapter 3. Rather, to obtain a sufficient number of equations in this statically indeterminate problem, we sought additional equations via use of strain–displacement and stress–strain relations. We will see below and in

Chapter 5 that this general approach is helpful in many different statically indeterminate problems.

4.2 Shear Stress due to Torsion

4.2.1 Introduction

Although the analysis in Section 4.1.2 was restricted to LEHI material behaviors and thus small strain, there was no restriction on the cross-sectional area; that is, the developed equations held equally well for rectangular, circular, elliptical, indeed general cross sections. As we begin our study of torsion, however, the situation is very different. It has long been known that if you subject a straight member of circular cross section to a small twist, the originally parallel cross sections remain parallel. In other words, small twisting of a circular member (shaft) does not warp the cross section. For any other cross section, such as elliptical or rectangular cross sections, torsion induces both a twist (i.e., material particles have a u_θ displacement) and a *warping motion* (e.g., u_z displacements that are nonuniform). In the next two subsections, we focus solely on small twisting motions in solid or hollow members that have a circular cross section and exhibit a LEHI behavior. As in the other problems, we will seek to relate the developed stress(es) to the applied loads and geometry [cf. Eq. (3.59)] and the deformations to the applied loads, geometry, and material properties [cf. Eqs. (3.89) or (4.5)].

4.2.2 Biological Motivation

Many biological tissues and implants are subjected to twisting loads (or torsion). Most notably, the twisting action of the heart is fundamental to the ejection of blood during each cardiac cycle; that is, consistent with Figure 2.19, the heart shortens, constricts, twists, and shears as the muscle fibers contract during the ejection phase. In particular, the twisting action comes from a unique arrangement of the cardiac muscle fibers (Fig. 4.9), which was noticed many years ago by anatomists, but not fully appreciated until the 1970s and 1980s based on biomechanical models. It is now clear that the twisting action of the heart is not only effective in aiding the ejection of blood, but it also tends to homogenize the distribution of stress across the wall of the ventricle. The latter is very important within the context of mechanobiology because a homogenized stress (or strain) field would allow the cardiac myocytes and fibroblasts to experience similar (perhaps optimal) mechanical stimuli regardless of their position within the wall of the heart. Because of the large strains and nonlinear material behavior in the heart, however, the reader is referred to Humphrey (2002) for a discussion of cardiac mechanics. Here, let us simply consider a small strain

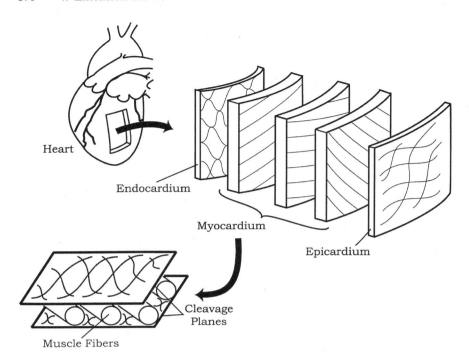

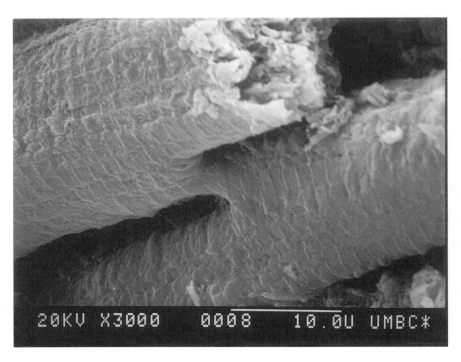

FIGURE 4.9 Schema of the heart with a cutout section from the ventricular wall showing the alternating directions of the muscle fibers (which vary smoothly throughout the wall) within the myocardium plus the delimiting connective tissue membranes on the inner (endocardial) and outer (epicardial) surfaces. The transmural splay in the muscle fibers gives rise to the twisting action of the heart upon contraction. Also shown is a scanning electron micrograph (magnification 3000×) of two connected muscle fibers that emphasize the locally parallel structure. [From Humphrey (2002), with permission.]

example. Figure 4.10 shows that a load applied can induce a small strain twisting action. Indeed, such loads occur naturally during daily activities as well as in the laboratory during material testing.

4.2.3 Mathematical Formulation

Recall from Sections 3.3–3.5 that we began each stress analysis by introducing a judicious cut to isolate (or expose) the stress $\sigma_{(\text{face})(\text{direction})}$ of interest in the free-body diagram. Once done, we enforced equilibrium and related the component of stress of interest to the applied loads and geometry. Let us take the same approach here. Consider, a solid circular cylinder that is fixed on one end and free on the other; moreover, let the free end be subjected to a positive twisting moment $M_z \equiv T$ (or torque). Equilibrium of the whole (Fig. 4.11) requires an equal and opposite reaction

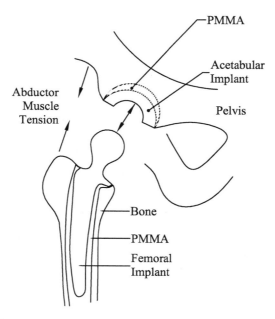

FIGURE 4.10 Schema of the femur and acetabulum, a cup-shaped cavity in which the head of the femur articulates. Because the line of action of the loads applied on the femoral head do not coincide with the long axis of the mid-shaft of the femur, these forces can cause both bending and twisting moments in addition to axial compression. Bending is addressed in Chapter 5 so we simply focus on the combined axial load and associated torque. Because of the linearity of the problem in small strain, we can use the principle of superposition and thus solve each aspect separately (compression, torsion, and bending).

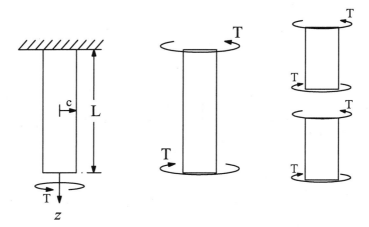

FIGURE 4.11 Schema of a solid circular cylinder (i.e., shaft) subjected to an applied torque T on the otherwise free end. (Note: The positive sign convention is consistent with the right-hand rule whereby the thumb points in the positive coordinate direction and the fingers wrap around the associated coordinate axis). Shown, too, is a free-body diagram of the whole structure to isolate the reaction at the fixed end, and a free-body diagram of two parts to isolate the internal torques. Equilibrium requires that the internal torques balance the applied and reaction torques.

torque T at the fixed wall, remembering, of course, the right-hand rule for the positive sign convention. Next, consider equilibrium of the parts. In particular, from Figure 4.12, we see that z-face, θ-direction stresses $\sigma_{z\theta}$ act on the cut face to balance the net applied torque T. Knowing that each $\sigma_{z\theta}$ acts over its respective differential area, with $dA = rd\theta dr$ in the circular cross section and that a torque is a force acting at a distance (i.e., a twisting moment), we must add up the effects of all stresses acting on their differential areas. Hence,

$$\sum M_z)_0 = 0 \rightarrow -T + \int \underbrace{r\,\sigma_{z\theta}dA}_{df} = 0 \rightarrow T = \int_0^c \int_0^{2\pi} \sigma_{z\theta}r^2\,d\theta dr. \quad (4.16)$$

Because stress can vary from point to point, in general, we must know $\sigma_{z\theta}$ as a function of position before we can evaluate the integral. Recall that we avoided this "issue" in the axially loaded rod in Section 3.3 by assuming that far enough from the ends, the stress σ_{xx} was uniform (i.e., constant) over the cross section; likewise, we avoided this issue in Sections 3.4 and 3.5 for the inflated cylinders and spheres by assuming thin walls and, consequently, that the stress was well represented by its mean (i.e., constant) value. Here, however, we will soon find that $\sigma_{z\theta}$ varies with radial location and that this spatial dependence cannot be ignored. Although we addressed this issue of nonuniform stress in the thick-walled cylinder in Section 3.6

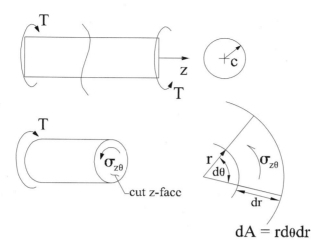

$$dA = rd\theta dr$$

FIGURE 4.12 Alternate free-body diagram for the circular cylinder shown in Figure 4.11, this time isolating a $\sigma_{z\theta}$ stress, on a cut z face, which serves to balance the applied torque. This balance is achieved, of course, via the net effect of all such stresses acting on their respective cross-sectional differential areas dA and at a distance from the axis called the moment arm.

by solving the full differential equations, here we seek an alternate, easier "strength of materials" approach. In hindsight (which means, after trying multiple approaches to no avail), it will prove convenient to employ the kinematics and constitutive relation directly.

Hence, consider the general element in Figure 4.13 in which the angle γ is introduced to measure the circumferential motion of all material particles along a line drawn along the length of the cylinder. Moreover, let $\gamma(r = c)$ be denoted by γ_c for a line drawn on the outer surface. From trigonometry,

$$\tan \gamma_c = \frac{c\Delta\theta}{\Delta z} \quad (\text{at } r = c)$$

where

$$\lim_{\Delta z \to 0} \frac{c\Delta\theta}{\Delta z} = c\frac{d\theta}{dz} \quad (\text{at } r = c). \tag{4.17}$$

Next, let us restrict our attention to small changes in angles whereby the following small-angle approximation holds:

$$\tan \gamma_c \cong \gamma_c \to \gamma_c = c\frac{d\theta}{dz} \quad (\text{at } r = c). \tag{4.18}$$

Likewise, it can be shown by the same assumptions that a similar relation holds at any radius, namely

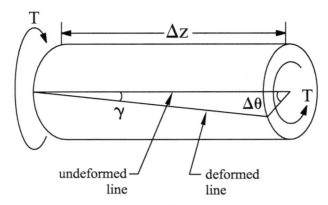

FIGURE 4.13 Schema of a circular cylinder subjected to equal and opposite end torques (assume that the torque is applied on the right end and that the torque at the left end is a reaction at a fixed boundary condition). Imagine that a straight line is drawn on the outer surface in the axial direction in the unloaded configuration. Upon the application of the torque, this line would rotate (i.e., points would displace u_θ) differently at different axial locations (cf. Fig. 2.17 for the axial load). If either the angle γ is small or the length Δz is small, then the line may be assumed to remain nearly straight and thus be describable via a single angle γ.

$$\tan \gamma_r \cong \gamma_r \rightarrow \gamma_r = r\frac{d\theta}{dz} \quad \text{(at any } r\text{)}. \tag{4.19}$$

Having these relations, obtained from simple trigonometric arguments, we should ask: What are γ_c and γ_r? As it turns out, because of the linearization of the Green strain (Section 2.5), the linearized strains are related directly to small changes in length or angle; thus, γ_c and γ_r are related to a linearized shear strain. Here, ε_{zr}, $\varepsilon_{z\theta}$, and $\varepsilon_{r\theta}$ are candidate measures of shear or angle change. Of these, the strain $\varepsilon_{z\theta}$ is the measure of interest because it alone is induced by the stress $\sigma_{z\theta}$, which is needed to resist the torque T. Recall, therefore, from Hooke's Law for LEHI behavior [Eq. (2.79)] that

$$\varepsilon_{z\theta} = \frac{1}{2G}\sigma_{z\theta}, \tag{4.20}$$

where G is the shear modulus. Moreover, it can be shown that (cf. Exercise 2.27)

$$\varepsilon_{z\theta} = \frac{1}{2}(\gamma + 0) \rightarrow \gamma = 2\varepsilon_{z\theta} \tag{4.21}$$

for any r; that is,

$$\gamma_c = 2\varepsilon_{z\theta}(r = c) \quad \text{and} \quad \gamma_r = 2\varepsilon_{z\theta}(r). \tag{4.22}$$

These results can be substituted into Eq. (4.20), and using Eqs. (4.18) and (4.19), we obtain

$$\sigma_{z\theta}(r) = 2G\varepsilon_{z\theta}(r) = G\gamma_r = Gr\frac{d\theta}{dz},$$

$$\sigma_{z\theta}(c) = 2G\varepsilon_{z\theta}(c) = G\gamma_c = Gc\frac{d\theta}{dz}.$$

(4.23)

Hindsight reveals that it is useful to take the ratio of these two stresses:

$$\frac{\sigma_{z\theta}(r)}{\sigma_{z\theta}(c)} = \frac{Gr\,d\theta/dz}{Gc\,d\theta/dz} = \frac{r}{c} \rightarrow \sigma_{z\theta}(r) = \frac{\sigma_{z\theta}(c)}{c}r.$$

(4.24)

Whereas $\sigma_{z\theta}(r)$ is still an unknown function of radius, in general, $\sigma_{z\theta}(c)$ is just the value of this function at one point, $r = c$; hence, it is just a number. Likewise, c is just a number, the value of the outer radius; hence, via kinematics and constitutive relations, we can now evaluate the equilibrium equation (4.16):

$$T = \int \frac{\sigma_{z\theta}(c)}{c}r^2 dA = \frac{\sigma_{z\theta}(c)}{c}\int r^2 dA.$$

(4.25)

By recognizing the second polar moment of area (see Appendix 4) $J = \int r^2 dA$, where $dA = rd\theta dr$, we can write,

$$T = \frac{\sigma_{z\theta}(c)}{c}J \leftrightarrow \sigma_{z\theta}(c) = \frac{Tc}{J}.$$

(4.26)

By Eq. (4.24), however, we have

$$\sigma_{z\theta}(r) = \frac{r}{c}\frac{Tc}{J} \rightarrow \sigma_{z\theta}(r) = \frac{Tr}{J}.$$

(4.27)

Note, therefore, that we have succeeded in finding the stress (relative to r, θ, z) in terms of applied load (torque T) and a measure of the geometry (second polar moment of area J). This is similar to our previous (universal) results for stress in axial loading and pressurization of a thin-walled cylinder or sphere [summary in Eq. (3.59)]:

$$\sigma_{xx} = \frac{f}{A}; \quad \sigma_{\theta\theta} = \frac{Pa}{h}, \quad \sigma_{zz} = \frac{Pa}{2h} + \frac{f}{2\pi ah}; \quad \sigma_{\theta\theta} = \frac{Pa}{2h} = \sigma_{\phi\phi}.$$

There are two significant differences between the present and prior findings, however. Whereas these prior relations for stress were universal results, good for all materials, Eq. (4.27) holds only for a small-strain LEHI behavior. Moreover, in contrast to these prior results whereby the stress was uniform (i.e., independent of position within the body), Eq. (4.27) reveals a nonuniform distribution of stress; that is, the shear stress varies linearly with radial position within a circular cylinder under torsion, the stress being zero at $r = 0$ and largest at the outer radius $r = c$. Hence, if the particular "LEHI material" of interest fails due to shear, it would be expected that failure would initiate on the outer surface.

4.3 Principal Stresses and Strains in Torsion

As in Chapter 2, the components of stress at any point relative to one coordinate system can be related to those relative to another coordinate system via transformation relations like those in Eq. (2.13):

$$\sigma'_{xx} = \sigma_{xx}\cos^2\alpha + 2\sigma_{xy}\sin\alpha\cos\alpha + \sigma_{yy}\sin^2\alpha.$$

To rewrite this equation in terms of the cylindrical-polar coordinates, let $x \to z$ and $y \to \theta$; thus,

$$\sigma'_{zz} = \sigma_{zz}\cos^2\alpha + 2\sigma_{z\theta}\sin\alpha\cos\alpha + \sigma_{\theta\theta}\sin^2\alpha, \tag{4.28}$$

where α is the now the angle between z and z' and likewise between θ and θ' (recall Eq. 3.53). For members subjected to pure torsion, σ_{zz} and $\sigma_{\theta\theta}$ equal zero, thus giving the following:

$$\sigma'_{zz} = 2\sigma_{z\theta}\cos\alpha\sin\alpha. \tag{4.29}$$

By substituting Eq. (4.27) into this transformation relation, we obtain

$$\sigma'_{zz} = 2\frac{Tr}{J}\cos\alpha\sin\alpha. \tag{4.30}$$

Similarly, from Chapter 2, Eq. (2.21),

$$\sigma'_{yy} = \sigma_{xx}\sin^2\alpha - 2\sigma_{xy}\sin\alpha\cos\alpha + \sigma_{yy}\cos^2\alpha$$

can be rewritten as

$$\sigma'_{\theta\theta} = \sigma_{zz}\sin^2\alpha - 2\sigma_{z\theta}\sin\alpha\cos\alpha + \sigma_{\theta\theta}\cos^2\alpha \tag{4.31}$$

or for our case,

$$\sigma'_{\theta\theta} = -2\sigma_{z\theta}\sin\alpha\cos\alpha \to \sigma'_{\theta\theta} = -2\frac{Tr}{J}\sin\alpha\cos\alpha. \tag{4.32}$$

Finally, Eq. (2.17) can be written as

$$\sigma'_{z\theta} = \sin\alpha\cos\alpha(\sigma_{\theta\theta} - \sigma_{zz}) + (\cos^2\alpha - \sin^2\alpha)\sigma_{z\theta}, \tag{4.33}$$

or

$$\sigma'_{z\phi} = \frac{Tr}{J}(\cos^2\alpha - \sin^2\alpha). \tag{4.34}$$

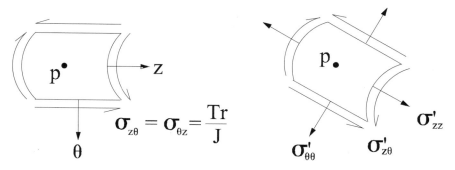

$$\sigma_{z\theta} = \sigma_{\theta z} = \frac{Tr}{J}$$

FIGURE 4.14 For pure torsion of a circular LEHI cylinder, the only nonzero component of stress at a point p is $\sigma_{z\theta}$ relative to z and θ. Relative to z' and θ', however, we may have additional components of stress, including normal and shear. We are reminded, therefore, that components of stress at a point depend on the coordinate system that is employed; they are not unique physical measurables or quantities that are "felt" directly by a cell or tissue.

See Figure 4.14. As in Chapter 2, the principal stresses can be computed as

$$\sigma_{1,2} = \begin{cases} \sigma'_{zz})_{max/min} = \dfrac{\sigma_{zz} + \sigma_{\theta\theta}}{2} \pm \sqrt{\left(\dfrac{\sigma_{zz} - \sigma_{\theta\theta}}{2}\right)^2 + \sigma_{z\theta}^2} \\[3mm] \sigma'_{\theta\theta})_{max/min} = \dfrac{\sigma_{zz} + \sigma_{\theta\theta}}{2} \pm \sqrt{\left(\dfrac{\sigma_{zz} - \sigma_{\theta\theta}}{2}\right)^2 + \sigma_{z\theta}^2} \end{cases} \tag{4.35}$$

but for members subject to pure torsion, σ_{zz} and $\sigma_{\theta\theta}$ are equal to zero; thus,

$$\sigma'_{zz})_{max/min} = \pm\sigma_{z\theta}, \quad \sigma'_{\theta\theta})_{max/min} = \pm\sigma_{z\theta}, \tag{4.36}$$

which is to say that the maximum/minimum normal stresses are numerically equal to the original value of the shear stress $\sigma_{z\theta}$:

$$\sigma_1 = +\frac{Tr}{J} \quad \text{and} \quad \sigma_2 = -\frac{Tr}{J}, \tag{4.37}$$

as seen in Figure 4.15. The plane on which the maximum normal stress acts is given by an equation similar to Eq. (2.25):

$$2\alpha_p = \tan^{-1}\left(\frac{\sigma_{z\theta}}{(\sigma_{zz} - \sigma_{\theta\theta})/2}\right), \tag{4.38}$$

where σ_{zz} and $\sigma_{\theta\theta}$ are equal to zero; hence,

$$\alpha_p = \frac{1}{2}\tan^{-1}(\infty) = \frac{\pi}{4} = 45° \tag{4.39}$$

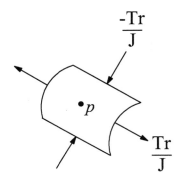

$\dfrac{-Tr}{J}$

$\dfrac{Tr}{J}$

•p

FIGURE 4.15 Principal stresses σ_1 and σ_2 at point p for the state of stress shown in Figure 4.14. Note that the principal values are equal in magnitude but opposite in direction (i.e., one is compressive and one is tensile). Moreover, note that their magnitude is equal to the magnitude of the $\sigma_{z\theta}$ shear stress. This reminds us that components of stresses can be of the same magnitude, but different because of either the different faces on which they act or the different directions in which they act.

(i.e., the maximum/minimum normal stresses will act in a direction 45° from the original z or θ axis).

Similarly, the maximum shear stress can be rewritten as (from Chapter 2)

$$\sigma'_{z\theta})_{max/min} = \sqrt{\left(\frac{\sigma_{\theta\theta} - \sigma_{zz}}{2}\right)^2 + \sigma^2_{z\theta}} \rightarrow \sigma'_{z\theta})_{max/min} = \pm\frac{Tr}{J}, \qquad (4.40)$$

which occurs at

$$\alpha_s = \frac{1}{2}\tan^{-1}\left(\frac{\sigma_{\theta\theta} - \sigma_{zz}}{2\sigma_{z\theta}}\right) \rightarrow \alpha_s = \frac{1}{2}\tan^{-1}(0) = 0; \qquad (4.41)$$

that is, the shear stress is a maximum relative to the original (z, θ) coordinate system. Finally, note that Eqs. (4.37) and (4.40) show the maximum/minimum values relative to (z, θ) and (z', θ') coordinate systems; because the stress varies with radial direction, the largest max/min values occur at $r = c$. Hence, whether the material fails first due to shear or normal stresses, we would expect failure to initiate on the outer surface $(r = c)$, in the absence of internal defects of course.

Example 4.3 A hollow LEHI cylinder has an inner radius $a = 15\,\text{mm}$, an outer radius $c = 20\,\text{mm}$, and a length $L = 0.5\,\text{m}$. The applied torque T is $600\,\text{N m}$ with an angle of twist $\Delta\theta(z = L) = 3.57°$. Calculate $\sigma'_{z\theta})_{max}$ and $\sigma'_{zz})_{max}$, find the value of the shear modulus G, and calculate $\varepsilon'_{z\theta})_{max}$ and ε_{zz}.

Solution: From Eqs. (4.37) and (4.40),

$$\sigma'_{z\theta})_{max} = \frac{Tr}{J} \quad \text{and} \quad \sigma'_{zz})_{max} = \frac{Tr}{J},$$

where

$$J = \int r^2 dA = \int\int r^2 r\,d\theta\,dr = \int_0^{2\pi}\int_a^c r^3 dr\,d\theta = \frac{\pi}{2}(c^4 - a^4).$$

Given

a = 15 mm = 0.015 m
c = 20 mm = 0.02 m
L = 0.5 m
T = 600 N m
$\Delta\theta$ = 3.57° = 0.0623 radians

first calculate $\sigma'_{z\theta})_{max}$ and $\sigma'_{zz})_{max}$:

$$\sigma'_{z\theta})_{max} \text{ and } \sigma'_{zz})_{max} = \frac{Tc}{\pi(c^4 - a^4)/2} = \frac{2(600\,\text{N m})(0.02\,\text{m})}{\pi\left[(0.02\,\text{m})^4 - (0.015\,\text{m})^4\right]}$$

$$\cong 6.98 \times 10^7 \frac{\text{N}}{\text{m}^2} = 69.8\,\text{MPa}.$$

Second, calculate G. Assuming $\gamma_c \ll 1$, we have

$$\tan\gamma_c \cong \gamma_c = c\frac{d\theta}{dz} \cong c\frac{\Delta\theta}{\Delta z} = \frac{(0.02\,\text{m})(0.0623\,\text{rad})}{(0.5\,\text{m})} = 0.00249\,\text{rad}.$$

Hence, $\sigma_{z\theta}(r = c) = 2G\varepsilon_{z\theta}(r = c) = G\gamma_c$ implies that

$$G = \frac{\sigma_{z\theta}}{\gamma_c} = \frac{6.98 \times 10^7\,\text{N/m}^2}{0.00249} = 2.80 \times 10^{10}\,\text{Pa} = 28.0\,\text{GPa}.$$

Third, calculate $\varepsilon'_{z\theta})_{max}$ and ε_{zz} using Hooke's law:

$$\varepsilon'_{z\theta})_{max} = \frac{1}{2G}\sigma'_{z\theta})_{max} = \frac{69.8\,\text{MPa}}{2(28.0\,\text{GPa})} = 0.00125$$

and, finally,

$$\varepsilon_{zz} = \frac{1}{E}[\sigma_{zz} - \nu(\sigma_{rr} + \sigma_{\theta\theta})] = 0.$$

Thus, the shaft does not extend and the maximum shear strain is indeed small, consistent with our small-strain assumption in the derivation of the governing equations and our use of a LEHI descriptor of the behavior. Also note that in reference to Table A2.1, a shear modulus $G \sim 28$ GPa suggests that the material is a 2024-T4 aluminum. The yield strength of this material is ~170 MPa in shear; hence, we would not expect that yield would have occurred.

Example 4.4 A solid circular member is to be subjected to an applied torque of 500 N m. Find the required diameter of the member so as not to exceed the maximum stress $\sigma_{z\theta}$ of 125 MPa.

Solution: Given

$$\sigma_{z\theta} = 125\,\text{MPa} = 1.25 \times 10^8 \, \frac{\text{N}}{\text{m}^2}. \quad T = 500\,\text{N m};$$

let the maximum radius $r = c$. From Eq. (4.27),

$$\sigma_{z\theta} = \frac{Tr}{J} \quad \text{or} \quad \sigma_{z\theta}(r = c) = \frac{Tc}{J},$$

where

$$J = \int r^2 dA = \int\int r^2 r\, d\theta\, dr = \int_0^{2\pi}\int_0^c r^3\, dr\, d\theta = \frac{\pi}{2} c^4.$$

Hence,

$$\sigma_{z\theta}(c) = \frac{Tc}{(\pi/2)c^4} = \frac{2T}{\pi c^3} \rightarrow c^3 = \frac{2T}{\pi \sigma_{z\theta}} \rightarrow c = \left(\frac{2T}{\pi \sigma_{z\theta}} \right)^{1/3},$$

or

$$c = \left(\frac{2(500\,\text{N m})}{\pi(1.25 \times 10^8 \, \text{N/m}^2)} \right)^{1/3} = 0.0137\,\text{m} = 13.7\,\text{mm},$$

and thus the minimum allowable diameter is $2c = 27.4\,\text{mm}$, which is just over 1 in.

Observation 4.2 Not all bones serve the same function. Some serve primarily to protect underlying soft tissue (e.g., the skull and sternum); thus, they have significant strength but carry little load most of the time. Conversely, other bones serve intermittently as load-bearing structures (e.g., the humerus, radius, and ulna of the arm), whereas still others consistently bear significant loads (e.g., the spine as well as the femur, tibia and fibula of the leg). We would expect, therefore, that the strains experienced by these different bones differ significantly throughout a normal day. Much of the attention in the mechanobiology of bone has focused on load-bearing long bones.

Regardless of their primary function, from the perspective of mechanics, bones tend to experience small strains. Hence, given that bones are also relatively stiff, standard strain gauges can be used to measure the surface strains that they experience under either in vitro or in vivo loading conditions. Given material properties, of course, stresses can then be computed from measured strains without the need to solve the equilibrium problem (for that point). Note, therefore, that the magnitude of peak *compressive* strains measured in vivo on the outer surface of load-bearing bones (e.g., cortical bone of the diaphysial region of the femur) have been reported on the order of 0.001 or less during normal walking (often ~0.0004) and between 0.002 and 0.004 during vigorous exercise. It is interesting to note, therefore, that Rubin and Lanyon (1985) reported a maintenance of cortical bone (i.e., a balanced production by osteoblasts and removal by osteoclasts) when the compressive strain is between 0.0005 and 0.0015. Above a strain of ~0.0015, there tends to be a net growth whereby production exceeds removal. Microdamage may occur, however, when the strains are greater than 0.0025 in tension or 0.004 in compression. Microdamage is also thought to stimulate a bone growth/healing response. Yield may occur at strains of ~0.006 in tension and 0.009 in compression, whereas cracks can occur when strains exceed ~0.03, which will also elicit a bone growth/healing response. Of course, sustained inactivity (e.g., bedridden patients) or gross unloading (e.g., in astronauts in a microgravity environment) leads to a net loss of bone material in bones that normally support loads. We conclude, therefore, that consistent, vigorous exercise promotes bone growth by increasing the strains (or stresses) and, through mechanotransduction mechanisms, increasing the production and organization of bone material by the osteoblasts. Let us now look at small strain deformations in torsion, one load seen daily by bones such as the femur.

4.4 Angle of Twist due to Torque

Recall from Section 4.1 that in axial load problems, it is often useful to find the maximum displacement (extension), denoted as δ, as well as the displacement vector and strain fields. So, too, with torsion, it is often useful to determine the maximum angle of twist

$$\Theta = \int \frac{d\theta}{dz} dz \quad \text{at } r = c. \tag{4.42}$$

4.4.1 Basic Derivation

From Eqs. (4.19)–(4.21), we recall that

$$\gamma_r = r \frac{d\theta}{dz}, \quad \varepsilon_{z\theta} = \frac{1}{2}\gamma_r, \quad \varepsilon_{z\theta} = \frac{1}{2G}\sigma_{z\theta}. \tag{4.43}$$

Hence, from Eq. (4.27), we have

$$\frac{d\theta}{dz} = \frac{1}{r}\gamma_r = \frac{1}{r}2\left(\frac{1}{2G}\sigma_{z\theta}\right) = \frac{1}{rG}\left(\frac{Tr}{J}\right) = \frac{T}{JG}, \tag{4.44}$$

and, consequently, the angle of twist Θ can be computed via

$$\Theta(z) - \Theta(0) = \int_0^z \frac{d\theta}{dz}dz = \int_0^z \frac{T(z)}{J(z)G(z)}dz, \tag{4.45}$$

wherein, similar to Eq. (4.5), we allow the torque, second polar moment of (cross-sectional) area, and shear modulus to vary with position z along the length in general. It is important to note, therefore, that if the shaft is homogeneous, then $G \neq G(z)$; if the shaft has a constant cross-sectional area, then $J \neq J(z)$; and if the shaft is under a constant torque, then $T \neq T(z)$. The direction of the angle of twist Θ coincides with the direction of the applied torque T.

Example 4.5 Find the total twist at a distance z in each of the members in Figures 4.16 and 4.17.

Solution

$$\Theta(z) - \Theta(0) = \int_0^z d\theta, \qquad d\theta = \frac{T}{JG}dz.$$

The first shaft is homogeneous and acted upon by a constant torque; it does not have a constant cross-sectional area however. The area changes from A_1 to A_2 at a length of $L/2$ from the wall. Therefore, $J = J(z)$ and the angle of twist becomes

$$\Theta(L) - \Theta(0) = \int_0^L \frac{T}{J(z)G}dz = \frac{T}{G}\int_0^{L/2}\frac{1}{J_1}dz + \frac{T}{G}\int_{L/2}^L \frac{1}{J_2}dz.$$

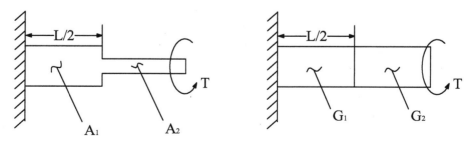

FIGURE 4.16 Two idealized circular cylinders of length L are acted upon by a single, constant end torque T. The cylinder on the left has a nonconstant cross section, whereas the one on the right is nonhomogeneous in composition.

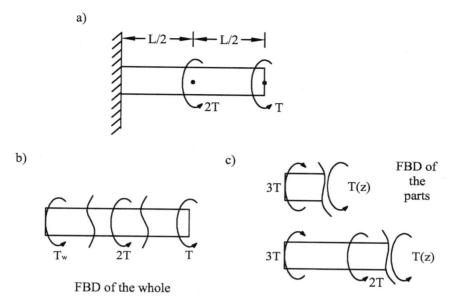

FIGURE 4.17 A LEHI circular cylinder subjected to multiple applied torques. A free-body diagram of the whole allows the reaction support T_w to be determined; free-body diagrams of judiciously selected parts allows internal torques to be determined as a function of z. Remember that judicious cuts are typically those taken between abrupt changes in applied loads.

Because the integral was broken into a sum of integrals for the discontinuity in cross-sectional area, each new integral contains terms that are constant along the range of integration and can be moved outside the integral and evaluated. Given that the twist at the fixed end is zero [i.e., $\Theta(0) = 0$], we have

$$\Theta(L) - \Theta(0) = \frac{T}{J_1 G} \int_0^{L/2} dz + \frac{T}{J_2 G} \int_{L/2}^{L} dz \to \Theta(L) = \frac{TL}{2J_1 G} + \frac{TL}{2J_2 G}.$$

The second shaft has a constant cross-sectional area and is acted on by a constant torque; it is not homogeneous however. The material properties change at a distance of $L/2$ from the wall. Therefore, $G = G(z)$ and the twist becomes

$$\Theta(L) - \Theta(0) = \int_0^L \frac{T}{JG(z)} dz = \frac{T}{J} \int_0^{L/2} \frac{1}{G_1} dz + \frac{T}{J} \int_{L/2}^{L} \frac{1}{G_2} dz,$$

or

$$\Theta(L) = \frac{TL}{2J} \left(\frac{1}{G_1} + \frac{1}{G_2} \right).$$

The third shaft is homogeneous and has a constant cross-sectional area; it is not under a constant loading however. The applied load changes at a distance of $L/2$ from the wall; thus, $T = T(z)$. Before we solve for the twist at the end of the shaft, we must determine the internal torques at each z. From equilibrium of the whole (Fig. 4.17b), we see that the reaction torque at the wall T_w must balance the combined effects of the $2T$ and the T that are applied at $z = L/2$ and $z = L$, respectively. Equilibrium of parts (note: when we have discrete changes in loads, geometry, or properties, judicious cuts are those between the abrupt changes) reveals further that the left half has an internal torque $3T$ and the right half only T. Hence, the end twist becomes

$$\Theta(L) - \Theta(0) = \frac{3T}{JG} \int_0^{L/2} dz + \frac{T}{JG} \int_{L/2}^L dz \rightarrow \Theta(L) = \frac{3TL}{2JG} + \frac{T}{JG}\left(\frac{L}{2}\right) = 2\frac{TL}{JG}.$$

4.4.2 Statically Indeterminate Problems

Just as in the case of the axially loaded rods, cases in which we do not have a sufficient number of equations from statics for the number of unknowns arise naturally and frequently in torsion problems. Such cases are called *statically indeterminate* because all quantities cannot be determined from statics alone. Here, let us return to the bone–prosthesis experiment of Section 4.1.3, but now focus on shear stresses induced by torsion. Referring to Figure 4.18, we know that if we assume separate LEHI behaviors for the prosthesis and bone that

$$\sigma_{z\theta}^p = \frac{T_p r}{J_p}, \quad 0 \le r < a, \quad \text{and} \quad \sigma_{z\theta}^b = \frac{T_b r}{J_b}, \quad a < r \le c, \quad (4.46)$$

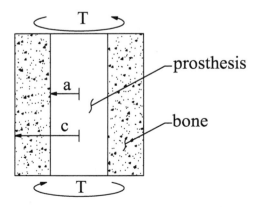

FIGURE 4.18 Similar to Figure 4.7, at section D-D, except that the bone–prosthesis system is subjected to a constant end torque T.

where T_p and T_b are those portions of the overall torque T carried by the prosthesis and bone, respectively, and

$$J_p = \frac{\pi}{2}a^4, \qquad J_b = \frac{\pi}{2}(c^4 - a^4). \tag{4.47}$$

From equilibrium, we know that $T = T_p + T_b$, but we do not yet know how the torque partitions. For a painless prosthesis, we require that there be no relative motion and, consequently, that all overall rotations, including the total end rotation, be equal; that is,

$$\Theta_p = \frac{T_p L}{J_p G_p} = \frac{T_b L}{J_b G_b} = \Theta_b \rightarrow T_p = \frac{T_b J_p G_p}{J_b G_b}, \tag{4.48}$$

which, with $\Theta(0) = 0$, yields our second equation for our second unknown. Hence,

$$T_b = \frac{T J_b G_b}{J_b G_b + J_p G_p}, \qquad T_p = \frac{T J_p G_p}{J_b G_b + J_p G_p}, \tag{4.49}$$

and, therefore,

$$\sigma_{z\theta}^p = \frac{T G_p r}{J_b G_b + J_p G_p}, \quad 0 \le r < a,$$

$$\sigma_{z\theta}^b = \frac{T G_b r}{J_b G_b + J_p G_p}, \quad a < r \le c. \tag{4.50}$$

In summary, we see again that if statics alone does not provide sufficient information, we should appeal to remaining equations (e.g., kinematics, constitutive, and boundary conditions).

Example 4.6 Consider the simple shaft shown in Figure 4.19, which has uniform LEHI properties and is fixed on both ends. Find the torque T in each section.

Solution: Because the shaft is fixed at both ends, the problem is statically indeterminate. If we let the end torques be denoted by T_A and T_C, overall equilibrium requires that $T_0 + T_C + T_A = 0$ (where T_0 is the known, applied torque). We need another equation to find the reactions however. Note, therefore, that

$$\Theta(L/2) - \Theta(0) = \int_0^{L/2} -\frac{T_A}{J_1 G}dz = -\frac{T_A L}{2 J_1 G}, \qquad \Theta(L) - \Theta(L/2) = \frac{T_C L}{2 J_2 G}$$

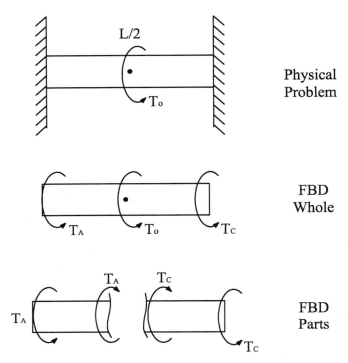

Physical
Problem

FBD
Whole

FBD
Parts

FIGURE 4.19 Statically indeterminate shaft, fixed on both ends, and subjected to a
single applied torque T_o at $z = L/2$. Free-body diagrams of the whole structure and
the parts allow the reaction and internal torques to be isolated but not determined
because we have only one nontrivial equation (the sum of the twisting moments
equals zero) for the two reaction torques T_A and T_C. There is, therefore, a need for
an additional equation.

where $\Theta = 0$ at both $z = 0$ and $z = L$. Moreover, $\Theta(L/2)$ is but a single value;
thus,

$$\Theta(L/2) = -\frac{T_A L}{2J_1 G} = -\frac{T_C L}{2J_2 G} \rightarrow T_A = T_C \frac{J_1}{J_2}$$

and therefore, having two equations and two unknowns, we can solve for
the two reactions

$$T_C = -T_0 \left(\frac{J_2}{J_1 + J_2} \right), \qquad T_A = -T_0 \left(\frac{J_1}{J_1 + J_2} \right).$$

As a special case, note that if $J_1 = J_2$ (i.e., the shaft has a constant cross
section), then $T_C = -T_0/2$ and $T_A = -T_0/2$, as expected.

Observation 4.3 One of the main complications with metallic implants (i.e., prostheses) is a gradual loosening of the device over time. Although infection and the associated degradation of bone material can cause loosening, aseptic loosening (i.e., mechanical failure) remains the most common cause of failure. PMMA, or (poly)methylmethacrylate, is commonly used as a cement to fix metallic implants within bone. Because loosening is often due to the development of microcracks within the PMMA, there is a need to understand the associated mechanics. PMMA has a stiffness (i.e., Young's modulus) of 2–3 GPa, a Poisson's ratio of 0.35, a mass density of $1220 \, kg/m^3$, a yield stress of about 28 MPa, and a tensile strength of about 83 MPa. Tensile strength is the maximum stress attained by a material on a σ versus ε curve. A particularly important characteristic, however, is the *fracture toughness* of the PMMA bone cement (i.e., its ability to withstand applied loads in the presence of flaws, including cracks). Whereas the load-carrying capability of a material containing defects or cracks is not compromised much in compression, which tends to close the defect, the behavior in tension and shear is very important. Indeed, excessive shear stresses at the bone–cement interface are thought to play a key role in the loosening of a hip implant.

A typical fracture toughness test consists of applying known axial stresses on a uniaxial sample that has a well-defined flaw in the central region (Fig. 4.20); this flaw experiences increased stress at its tip, which serves to nucleate and possibly to propagate a crack. For this simple test, a stress intensity factor K is often defined as $K = f\sigma\sqrt{\pi a}$, where f is a geometric factor for the specimen and flaw, σ is the applied axial stress, and a is a measure of the width of the flaw. The critical value of K at which the flaw begins to propagate is known as the fracture toughness K_c. Values of K_c for PMMA are around $990 \, MPa\sqrt{m}$. Whereas increased

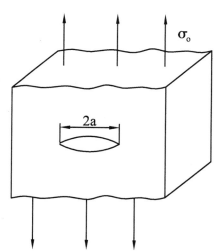

FIGURE 4.20 Schema of a specimen used for determining the fracture toughness of a material.

rates of loading or the presence of large defects reduce the fracture toughness, increasing the presence of very small inclusions tends to increase K_c. Such inclusions, including grains in metals, tend to arrest the propagation of a microcrack because more energy is needed to divert the crack around the inclusion. For this reason, small whiskers of titanium are sometimes added to the bone cement. These whiskers not only increase the fracture toughness of the bone cement, they also improve its radio-opacity and thus permit an easier examination of the integrity of the cement with X-rays. We have not considered fracture mechanics or the associated material science herein, but the student must know that many real-life problems require advanced methods and the expertise of many to understand fully the clinical problem and its most effective solution. We emphasize again, therefore, that this text is but an introduction; the interested student must pursue advanced courses in applied mechanics and biomechanics.

4.5 Experimental Design: Bone Properties

We recall from Chapter 2 that bones are typically heterogeneous (cortical and cancellous bone being very different); here, we consider a simple experiment to determine a first-order approximation for the shear modulus G in the diaphysial region of the femur based on the assumptions of homogeneity and isotropy. Pretend, however, that we do not have access to a tension–torsion device, which would allow us to perform a torsion test on a cylindrical sample and thereby to measure the end rotation $\Theta(z = L)$, applied torque T_0, length L, and second polar moment of area J that are needed to calculate $G = TL/J\Theta$. Rather, assume that we have available a much less expensive axial load device. We are thus faced with the dilemma of determining the value of the shear modulus G via an axial load experiment; let us employ our theoretical framework for help.

Actually, there are various ways to overcome this problem. First, we could recall that for isotropy, $G = E/2(1 + \nu)$ and therefore we simply need to determine E and ν. If we perform a uniaxial load test, we can infer σ_{xx} and ε_{xx} (with $\sigma_{xx} = E\varepsilon_{xx}$) from measurables: $\sigma_{xx} = f/A$, which can be determined by measuring the applied load and the cross-sectional area, and ε_{xx}, which can be obtained directly from an axially oriented strain gauge (because the bone will experience small strains in its elastic range). Indeed, if $\varepsilon_{xx} \equiv \varepsilon_{\text{axial}}$, then a second strain gauge placed orthogonal to the first would yield $\varepsilon_{\text{lateral}}$ whereby

$$G = \frac{E}{2(1 + \nu)} = \frac{\sigma_{xx}/\varepsilon_{xx}}{2(1 - \varepsilon_{\text{lateral}}/\varepsilon_{xx})} = \frac{f/A\varepsilon_{\text{axial}}}{2(\varepsilon_{\text{axial}} - \varepsilon_{\text{lateral}})/\varepsilon_{\text{axial}}}, \quad (4.51)$$

or

$$G = \frac{f}{2A(\varepsilon_{\text{axial}} - \varepsilon_{\text{lateral}})}. \tag{4.52}$$

Alternatively, we could recall our transformation equations for stress and strain (Chapter 2). For example,

$$\sigma'_{xy})_{\max} = \sqrt{\left(\frac{\sigma_{xx} - \sigma_{yy}}{2}\right)^2 + \sigma_{xy}^2}, \quad \varepsilon'_{xy})_{\max} = \sqrt{\left(\frac{\varepsilon_{xx} - \varepsilon_{yy}}{2}\right)^2 + \varepsilon_{xy}^2}, \tag{4.53}$$

where, for a uniaxial test, $\sigma_{yy} = 0$, $\sigma_{xy} = 0$, and $\varepsilon_{xy} = 0$. Thus, we simply need to invoke the constitutive relation relative to the primed coordinates, namely

$$\sigma'_{xy})_{\max} = 2G\varepsilon'_{xy})_{\max} \rightarrow G = \frac{1}{2}\frac{\sigma_{xx}/2}{(\varepsilon_{xx} - \varepsilon_{yy})/2}, \tag{4.54}$$

or with $\sigma_{xx} = f/A$, $\varepsilon_{xx} \equiv \varepsilon_{\text{axial}}$, and $\varepsilon_{yy} \equiv \varepsilon_{\text{lateral}}$,

$$G = \frac{f}{2A(\varepsilon_{\text{axial}} - \varepsilon_{\text{lateral}})}, \tag{4.55}$$

which is the same result as obtained earlier. We see again, therefore, that theory helps us to determine what to measure—that is to say, how to design a good experiment. If we were working in industry, our boss would be particularly pleased if our knowledge of theory would allow the desired result (here, the value of G) to be determined using available instrumentation (a standard axial load frame) rather than necessitating the expense and delay associated with the purchase of more specialized equipment.

4.6 Experimental Design: Papillary Muscles

4.6.1 Biological Motivation

The wall of the heart consists primarily of myocardium, which is delimited on its inner and outer surfaces by thin endocardial and epicardial membranes (Fig. 4.9). Whereas these delimiting membranes consist primarily of a 2-D plexus of collagen with admixed elastin, the myocardium consists primarily of locally parallel cardiomyocytes that are embedded in a 3-D collagenous matrix. Clearly, then, the myocardium and delimiting membranes exhibit very different mechanical behaviors (recall Fig. 2.23) consistent with their very different biomechanical functions. Fundamental to understanding overall cardiac function, therefore, is a detailed knowledge of the mechanical properties of the various tissues that constitute the heart. Quantification of the mechanical properties of the myocardium is complicated, however, by its ability to contract as a muscle and the observation that it experiences multiaxial finite extensions, shortening, and shears throughout the normal cardiac cycle (recall Fig. 2.19). There is a need, therefore, for tests that address both of these complexities.

The papillary muscles are thin, fingerlike projections within the ventricles of the heart (cf. Fig. 3.2). They consist of locally parallel myocardial fibers that are oriented along the axial direction, plus a thin delimiting endocardial membrane. Because some papillary muscles (e.g., from the right ventricle of the rabbit) are thin, nearly circular in cross section, and of modest taper along a significant portion of their length, they have proven to be ideal specimens for experiments that seek to quantify behavior in extension (i.e., axial loading) and shear (i.e., torsion) in both active and passive states; that is, the thinness of such specimens allows one to induce muscular contraction by bathing the papillary muscle in an appropriate solution, such as a normal physiologic solution augmented with barium to induce contracture or, likewise, to induce relaxation by changing the bathing solution to one containing an appropriate cardioplegic (e.g., high potassium and 2,3-butanedione 2-monoxime, or BDM). From the perspective of mechanics, therefore, one can design a tractable experiment: the combined axial extension and torsion of a cylindrical specimen having either active or passive properties. Given that we have derived formulas for axial extension and torsion, it may seem that it would be easy to design and interpret such an experiment to determine the stress–strain behavior of a papillary muscle and, thus, myocardium. Here, however, *we must be very careful*: Whereas the formula for Cauchy stress in an axially loaded member ($\sigma_{zz} = f/A$) is a universal solution and thus applicable to any material and any degree of strain, the analogous formula for Cauchy stress in the torsion of a circular member ($\sigma_{z\theta} = Tr/J$) holds only for LEHI behavior and small strains. Likewise, the formulas for end deflection (e.g., $\delta = fL/AE$) and that for end rotation ($\Theta = TL/JG$) are both restricted to small strains. The characteristic nonlinear, inelastic, heterogeneous, and anisotropic behavior exhibited by myocardium thus prohibits the use of three of our otherwise four seemingly applicable formulas.

Although we discuss some aspects of the quantification of nonlinear material behavior in Chapter 6, here let us see that how our simple results can still be used to design an appropriate experiment on a complex soft tissue.

4.6.2 Experimental Design

Consider Figure 4.21, which illustrates a possible setup for an extension–torsion test on a papillary muscle. In particular, we need actuators to induce both extension and torsion; this can be accomplished with computer-controlled stepper motors, which are commercially available at the appro-

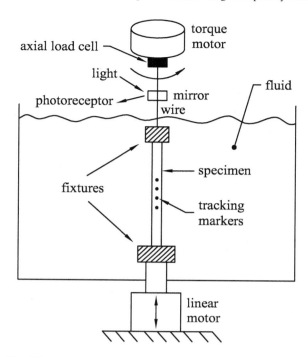

FIGURE 4.21 Possible experimental setup for performing an extension–torsion test on a thin, long, circular soft tissue. Shown are two actuators (a linear motor and torque motor to induce the extension and twisting, respectively), a standard axial load cell, a custom laser lever for measuring the torque, and a specimen in a physiologic solution. Note that the specimen has markers affixed to its surface to allow noncontacting measurements of displacements and then, via interpolation, calculation of displacement gradients and thus strains.

priate resolution in motion. We also need a method to measure the strain in the central region; although standard strain gauges cannot be used, strains can be inferred by affixing small markers to the surface of the specimen and tracking their motion with a video camera and computer image analysis system. From marker displacements, of course, we can compute the requisite displacements and their gradients (by introducing interpolation functions) to compute surface strains as discussed in Chapter 2. Although papillary muscles are small and thus subject to relatively small axial loads, commercial load cells are available with the requisite resolution. Measurement of the applied torque is not so simple however because a torque is a force acting at a distance. For a papillary muscle from the right ventricle (RV) of the rabbit or rat heart, this means a small force acting at a very small distance; hence, the applied torque will be very small. Therefore, let

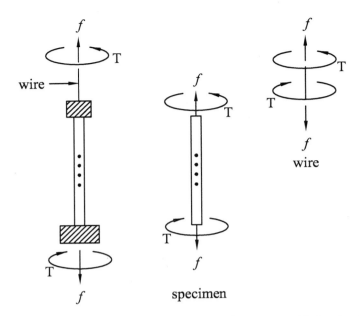

FIGURE 4.22 Free-body diagrams of the specimen–fixture assembly for the device in Figure 4.21 as well as of the isolated specimen and the wire that connects the upper fixture to the torque motor. Although the metallic wire and soft tissue have very different material properties (recall Fig. 2.22), equilibrium and Newton's third law require continuity in the applied loads from one to the other.

us see how the results of this chapter can be used to design an appropriate torque transducer.

Figure 4.22 is a free-body diagram of the bottom fixture, specimen, upper fixture, and thin connecting wire assembly. Because equilibrium of the whole implies equilibrium of the parts, each member of this assembly has a similar free-body diagram. In particular, there must be continuity of the applied loads throughout each member of this specimen–fixture assembly. In other words, if we can measure the torque acting on either the bottom or the upper fixture, we will know the torque that acts on the papillary muscle. In a Ph.D. dissertation, Sten-Knudsen (1953) recognized this and suggested that the upper fixture be connected to a thin metallic wire that exhibits a LEHI behavior. Consequently, if one measures the rotation at two points along the wire, say Θ_A and Θ_B, and if one knows the radius c and the shear modulus G of the wire, then the torque on the wire is [from Eq. (4.45)]

$$T = \frac{(\Theta_B - \Theta_A)JG}{L}, \qquad J = \frac{\pi}{2}c^4, \qquad (4.56)$$

where Θ_A and Θ_B are the rotations at points A and B and L is the distance between A and B. If the load cell in Figure 4.22 is rigidly attached to the torque motor, then Θ_A simply equals the rotation of the torque motor, which is generally available as a digital output signal. How then do we measure Θ_B? One possibility is to measure the angle of reflection of a beam of light (i.e., a laser) using a mirror that is attached rigidly to the wire at B and a photoreceptor. The resolution and range of the torque transducer is thus controlled largely by the position of the mirror at B, the radius of the wire, and the shear modulus of the wire G. Each of these quantities are easily measured.

In summary, we sought a tractable experiment to reveal the nonlinear extensional and shear behaviors of myocardium in active and passive states. Nature provided a nearly ideal sample in the thin and nearly circular papillary muscle. Whereas commercially available stepper motors, video cameras, frame-grabber boards, axial load cells, and A/D boards allow one to control and measure most of the requisite quantities, the unavailability of a commercial torque transducer having sufficient resolution (in 1953 and today) necessitated a custom design. We saw, therefore, that our simple strength of materials solution restricted to LEHI behavior could be used to design such a transducer for measuring torques in a tissue that exhibits a nonlinearly, inelastic, heterogeneous, and anisotropic material behavior. Knowing not only the restrictions but also the applications of each derivation is thus fundamental to creative analysis and design. Whereas we have considered only the design of the transducer here, Humphrey (2002) addressed the complete problem via nonlinear mechanics.

4.7 Inflation, Extension, and Twist

Because the stress boundary value problems associated with the distension of a thin-walled circular tube, the small strain axial extension of a rod, and the small strain twist of a circular shaft are each linear, their solutions can be superimposed to consider more complex loading conditions. In particular, relative to (r, θ, z) coordinates, recall the following results:

$$\sigma_{\theta\theta} = \frac{Pa}{h}, \quad \sigma_{zz} = \frac{Pa}{2h} + \frac{f}{2\pi ah}, \quad \sigma_{z\theta} = \frac{Tr}{J} \tag{4.57}$$

wherein we emphasize that each result relates the stress to the applied load and geometry. Referring to Figure 4.23, therefore, we see a potentially complex 2-D state of stress. From a design perspective, one could ask questions such as: What are the maximum principal or shear stresses and at what orientation α do they act? Knowing the value α_p for the principal values would be useful, for example, in the placement of strain gauges on the specimen. Fortunately, such questions are answered easily because the formulas for stress in Eq. (4.57) can be superimposed.

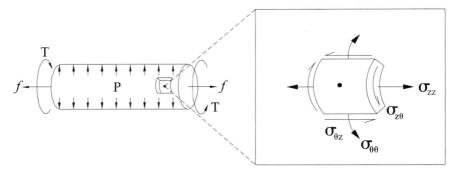

FIGURE 4.23 Complex state of stress in a cylindrical tube. Relative to z and θ, the axial stress σ_{zz} is induced by the axial load f, the shear stress $\sigma_{z\theta}$ is induced by the torque T, and the circumferential stress $\sigma_{\theta\theta}$ is induced by the pressure P. Superposition applies because the problem is linear.

In summary, as we noted in the Introduction, considering separately the mechanics of simple problems not only gives us intuition and reinforces the general method of approach, but it also yields direct applications and in some cases it allows us to consider more complex situations consisting of multiple types of applied load. In any event, we must always be mindful of the derivations (i.e., of the embodied assumptions).

Appendix 4: Second Moments of Area

In Appendix 3 in Chapter 3, we defined the first moment of area and showed how it can be used to determine a centroid. Whereas

$$\iint x^1 dA, \qquad \iint y^1 dA, \qquad \iint z^1 dA \qquad \text{(A4.1)}$$

are called *first moments of area* (given that x, y, and z are raised to the power 1),

$$\iint x^2 dA, \qquad \iint y^2 dA, \qquad \iint z^2 dA \qquad \text{(A4.2)}$$

are called *second moments of area* for obvious reasons. (Note: The word "moment" is used because of the analogy of a force acting at a distance compared to the case here of an area "acting" at a distance or a distance squared. In many books, the second moments of area are called moments of inertia, but this is incorrect, for inertia must involve a mass. Moments of inertia arise in dynamics and are equally important, but different.)

Because of the quadratic form in Eq. (A4.2), additional second moments of area are possible:

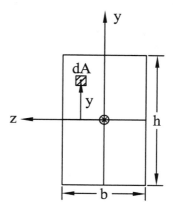

FIGURE 4.24 Schema of a rectangular cross section for purposes of determining a second moment of area.

$$\iint xy\, dA, \qquad \iint yz\, dA, \qquad \iint zx\, dA \qquad (A4.3)$$

and similarly for yx, zy, and xz terms. Clearly, these additional cross moments would have the same value as their paired result in Eq. (A4.3). Hence, like the Cauchy stress and linearized strain, there are nine components of the second moment of area, six of which are independent with respect to a particular coordinate system.

Herein, however, we shall typically focus on the x face and, thus, moments of area in the y-z plane. We typically denote these quantities by

$$I_{zz} = \iint y^2 dA, \quad I_{yy} = \iint z^2 dA, \quad I_{yz} = -\iint yz\, dA. \qquad (A4.4)$$

The minus sign in I_{yz} is introduced for convenience; we will not detail this. Rather, let us focus on I_{zz} and then I_{yy}. I_{zz} is perhaps best appreciated by calculating its value for a rectangular cross section. Referring to Figure 4.24 and locating the centroid $(\bar{y}, \bar{z})$ at $(h/2, b/2)$, we have

$$I_{zz} = \int_{-b/2}^{b/2} \int_{-h/2}^{h/2} (y^2 dy) dz = \int_{-b/2}^{b/2} \left(\frac{1}{3} y^3 \Big|_{-h/2}^{h/2} \right) dz$$

$$= \int_{-b/2}^{b/2} \frac{1}{3}\left(\frac{h^3}{8} + \frac{h^3}{8} \right) dz = \frac{1}{12} h^3 \left(z \Big|_{-b/2}^{b/2} \right) = \frac{1}{12} bh^3. \qquad (A4.5)$$

A general equation for rectangular cross sections can thus be written as

$$I_{zz} = \frac{1}{12}(\text{base})(\text{height})^3, \qquad (A4.6)$$

where base is the width of the cross section in the z direction and height is the length of the cross section in the vertical direction.

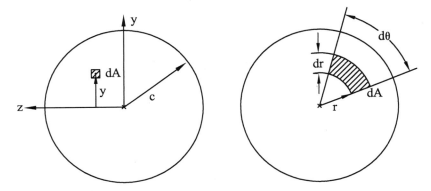

FIGURE 4.25 Schema of a circular cross section for purposes of determining a second moment of area relative to Cartesian or cylindrical coordinate systems; the latter is called the second polar moment of area and commonly denoted by J.

Next, let us consider a circular cross section (Fig. 4.25). Noting that

$$y = r\cos\theta, \qquad z = r\sin\theta \tag{A4.7}$$

and

$$dA = dy\,dz = r\,d\theta dr, \tag{A4.8}$$

then

$$I_{zz} = \iint y^2 dy\,dz = \int_0^{2\pi}\int_0^c (r^2\cos^2\theta)r\,dr\,d\theta$$

$$= \int_0^{2\pi}(\cos^2\theta)d\theta \int_0^c r^3 dr = \left(\frac{1}{2}\theta + \frac{1}{4}\sin 2\theta\right)\bigg|_0^{2\pi}\left(\frac{1}{4}r^4\right)\bigg|_0^c = \frac{\pi}{4}c^4. \tag{A4.9}$$

Here, observe two things. First, the derivation for the cylindrical cross section was easier in cylindrical coordinates, reminding us that coordinate systems should be selected to facilitate analysis. Second, it is easily shown (do it) that $I_{yy} = \pi c^4/4$ also. Indeed, let us note that

$$I_{yy} + I_{zz} = \iint z^2 dA + \iint y^2 dA = \iint (z^2 + y^2)dA, \tag{A4.10}$$

where $z^2 + y^2 = r^2$ in cylindricals. Thus,

$$I_{yy} + I_{zz} = \iint r^2 r\,dr\,d\theta \equiv J, \tag{A4.11}$$

the so-called *polar second moment of area*. For the circular cross section, therefore,

$$J = I_{yy} + I_{zz} = 2\left(\frac{\pi}{4}c^4\right) = \frac{\pi}{2}c^4, \qquad \text{(A4.12)}$$

a result that we have found to be very useful in this chapter on torsion.

Exercises

4.1   Find $\sigma_{z\theta}$, $\sigma'_{z\theta}$, and $\sigma'_{zz})_{max}$ given a positive torque T_o applied at the free end of a constant-diameter solid shaft of radius c and length L and having a shear modulus G. Assume the shaft is fixed at the left end.

4.2   Given the shaft in the following figure, (a) find the maximum shear stress $\sigma'_{z\theta})_{max}$ and note its location, (b) find the angle of twist Θ at the end of the shaft, and (c) find the maximum normal stress $\sigma'_{zz})_{max}$. Assume a LEHI behavior, with a shear modulus G, as well as a length L and radius c.

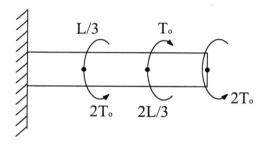

FIGURE 4.26

4.3   A laboratory test involves "potting" part of the femur in cement and then applying two loads f, each at a distance d from the centerline (cf. Fig. 4.4); this yields a couple or twisting moment $T_o = 2fd$. A strain gauge is placed at an angle α on the bone a distance $L/2$ from the fixed end. Derive a formula for the strain in the gauge that can be related to experimentally measurable quantities. Assume the bone exhibits a LEHI behavior and that it has a cross section at $z = L$ given by inner and outer radii of a and c. Discuss for what such an experiment could be utilized. Discuss why the strain gauge is not applied at length L and angle $\alpha = 0$.

4.4   For a solid shaft of diameter d for $z \in [0, 2L/3)$ and diameter nd for $z \in (2L/3, L]$, subject to torques T_o at $2L/3$ and $2T_o$ at L, (a) find the value of n such that the maximum shear stress $\sigma_{z\theta}$ is the same in each segment and (b) find the twist Θ at the free end if $n = 1$.

4.5   Some papillary muscles in the heart (which connect the valves to the endocardium through the chordae) are nearly cylindrical. We wish to perform a torsion test on such a tissue, but measuring the applied torque is difficult because of the small size. Assume that we can use the device in Figure 4.20, that the wire is made of copper, and that the distance between points A and B is 15 mm. Also assume Θ_A and Θ_B are measurable, their difference being ~90°. If the maximum torque achieved is ~0.5 mN mm, find an appropriate diameter for the wire.

4.6   Carter and Beaupré (2001) discuss an interesting finding by Lanyon and Rubin in 1984. It was suggested that the number of cycles of loading per day and the maximum achieved strain both serve as mechanobiological stimuli for bone growth. In particular, they found that bone mass was maintained (i.e., production and removal were balanced) given a strain history of 4 cycles/day at a maximum value of 0.002 or similarly at 100 cycles/day at a maximum value between 0.0005 and 0.001 (assume 0.0008). They suggested that these combined effects can be accounted for via a "daily bone stimulus" parameter ξ that is computed via the following formula, where n is the number of cycles/day, ε is the maximum strain attained per cycle, and m is an empirically determined material parameter. Given the data listed here, find a value for m.

$$\xi = \left(\sum_{day} n\varepsilon^m \right)^{1/m}$$

4.7   Based on the results of the previous exercise, determine the number of cycles that one should walk per day if the strain during normal walking is $400\,\mu\varepsilon$ (i.e., microstrain, where $1\,\mu\varepsilon = 1 \times 10^{-6}$). Carter and Beaupré (2001) suggest that 10,000 cycles of walking per day will maintain bone mass. If the normal person advances 3 ft per stride, how far should he/she walk per day to maintain bone mass?

4.8   Based on the previous exercise and an assumed Young's modulus $E = 16$ GPa and Poisson's ratio $\nu = 0.325$ for bone, compute the axial load necessary to cause a strain of $400\,\mu\varepsilon$ in the normal adult diaphysial region of the femur. Express your results in terms of percent body weight, assuming a weight of 70 kg. What would the associated axial compressive stress be? Similarly, estimate the load on the femur during running and the associated compressive stress and strain. Based on these values and the previous exercise, if a person advances 4 ft per stride when running, how far should he/she run per day?

4.9   If a 17.2-N m torque induces a maximum shearing strain of $1132\,\mu\varepsilon$ at the periosteal surface in the diaphysial region of the femur, what is the associated value of the shearing stress if the shear modulus is 3.3 GPa?

4.10 The ratio of the cortical thickness to the outer radius of most human bones is between 0.33 and 0.4. Assume that the cross section of a segment of a long bone is circular and that the periosteal and endosteal radii are 15 mm and 9 mm, respectively. Assume, too, that a 17.2-N m torque is applied for 10,000 cycles. What is the maximum extensional (principal) strain and, from the equation in Exercise 4.6, what is the value of the daily bone stimulus parameter ξ?

4.11 According to Carter and Beaupré (2001), "bone cross-sections that are formed are very dependent on the full history of loading throughout life. In the age range of 30–60 years, the normal bone has a diameter of about 32 mm and a cortical thickness of 5 mm. When the loads are reduced to 40 percent of normal at the age of 20, the bone in later adulthood has diameter of about 30 mm and a cortical thickness of about 2 mm. The bone that forms while loads are reduced to 40 percent throughout development has an adult diameter of about 22 mm and a thickness of 4 mm." What are the implications of such observations with regard to space travel, especially a voyage to Mars?

4.12 Referring to the previous exercise, note that the strains in adapting bones are generally the same regardless of the applied loads and associated cross-sectional radii. What does this suggest with regard to growth and remodeling?

4.13 Carter and Beaupré (2001, p. 81) suggest a phenomenological descriptor of bone growth (actually the rate of increase of the outer radius) of the form, $\dot{r} = \dot{r}_b + \dot{r}_m = ce^{-0.9t} + \dot{r}_m$, where $\dot{r}$ is the time rate of change of the radius, having units of microns per day; subscripts b and m denote an intrinsic biological rate and an adaptive mechanobiologic rate, respectively, and t denotes time measured in days. They suggest further that the intrinsic rate becomes relatively small shortly after birth or in early childhood, thus its representation as an exponential decay; that is, they assume that most growth and remodeling occur due to mechanobiologic factors in adolescence and maturity. In simulations, the maximum rate of biological growth was varied from 1 to 20 μm/day. Given these numbers, what would the radius be due to biological growth alone at 6 years of age? Is this value consistent with data on long bones such as the femur?

4.14 Referring to Exercise 4.13, Carter and Beaupré (2001, p. 151) note that the mass density of cancellous bone (usually ρ from 570 up to 1200 kg/m^3) is nearly constant from early adolescence to early adulthood. They suggest that this implies that in the absence of bone diseases, the intrinsic biological rate of growth is negligible with respect to the mechanobiological rate during this period. If this is true, what are the implications with regard to the modeling of bone adapatation?

4.15 Galileo thought that long bones were hollow because this afforded maximum strength with minimum weight. Discuss this in terms of the ability of a hollow versus a solid cylindrical bone of the same mass to resist a torque; assume the bone is cortical, which has a mass density of ~1700 kg/m^3. Alternatively, is the "hollowness" of a long bone consistent with a stress- or strain-based growth model wherein a maximum compressive strain of 1000 $\mu\varepsilon$ is homeostatic—assume that the bone is either subjected to a torque alone or to a combined torque and axial load wherein the stresses due to torsion exceed those due to axial loading?

4.16 A long bone is subjected to a torsion test. Assume that the inner diameter is 0.375 in. and the outer diameter is 1.25 in., both for a circular cross section. If $E = 16$ GPa and $\nu = 0.325$, find the largest torque that can be applied prior to yield, where $\sigma_{yield} = 1.25$ ksi (i.e., a maximum normal stress).

4.17 A solid circular cylinder 10 cm long and 2 cm in outer radius behaves as a LEHI material with $G = 10$ GPa. If the twisting moment (torque) applied at the free end is 3 kN m, show that $J = 25.13 \times 10^{-8}$ m^4, $\Theta = 6.84°$ at the free end, $\sigma_{z\theta}(r = c) = 238.76$ MPa, and $2\varepsilon_{z\theta}(r = c) = 0.02388$. Assume one end is fixed.

4.18 A rectangular bar $2 \times 2 \times 20$ cm in dimension is subjected to an axial force (uniform) of 4×10^6 N. Assuming $E = 100$ GPa and $\nu = 0.30$, find σ_{xx}, ε_{xx}, $\varepsilon_{yy} = \varepsilon_{zz}$ and the deformed dimensions (assuming homogenous strains).

4.19 A human femur is mounted in a torsion testing device and loaded to failure. Assuming that one end is fixed and the other rotated, failure (fracture) occurs when $T = 180$ N m and $\Theta(L) = 20°$. Assume that $L = 37$ cm and that the failure occurs at 25 cm from the fixed end, where the inner and outer radii are 7 mm and 13 mm, respectively. Find the value of the shear stress at which fracture occurs; estimate the shear modulus G. Finally, note that "torsional fractures are usually initiated at regions of the bones where the cross-sections are the smallest. Some particularly weak sections of human bones are the upper and lower thirds of the humerus, femur, and fibula; the upper third of the radius; and the lower fourth of the ulna and tibia" (Özkaya and Nordin, 1999).

4.20 A rectangular aluminum bar (~1.5 $\times$ 2.1 cm in cross section) and a circular steel rod (~1 cm in radius) are each subjected to an axial force of 20 kN. Assuming that both are 30 cm long in their unloaded configuration, find (a) the stress in each, (b) the extensional strain in each, and (c) the amount of lengthening in each. Let $E = 70$ GPa for aluminum and 200 GPa for steel.

4.21 A *brittle* behavior is characterized by an abrupt fracture soon after the elastic limit is exceeded. In contrast, a *ductile* behavior is characterized by a plastic behavior, including strain hardening, following yield. Recall Figure 2.24. We know that yield and the subsequent plastic behavior are governed by shear stresses, which cause atoms to "slip" past one another irreversibly. Hence, it is important to compute the maximum shear stress. Although a shear stress at which a material yields is easy to determine in a torsion test, tensile tests are more common. Recall from Chapter 2, therefore, that the maximum shear stress in a 1-D tension test $(\sigma_{xx} = \sigma_1, \sigma_{yy} = 0, \sigma_{xy} = 0)$ is

$$\left(\sigma'_{xy}\right)_{max} = \sqrt{\left(\frac{\sigma_1 - 0}{2}\right)^2 + 0^2} = \frac{\sigma_1}{2}.$$

This value of σ_1 at yield is called σ_y, the *yield stress*. Hence, a yield criterion in uniaxial tension is as follows: If $|\sigma_{xx}| \le \sigma_y$, then the material has not yielded. In multiaxial states of stress, more general yield criteria are needed. Two common yield theories are the *Tresca* yield condition and the *von Mises* yield conditions. Research these two yield theories and submit a two-page summary.

5
Beam Bending and Column Buckling

Although we have not emphasized it, we now note some standard terminology. Generally, a structural member having one dimension much greater than the other two is called a *rod* if it is subjected to a tensile axial load, it is called a *column* if it is subjected to a compressive axial load, it is called a *shaft* if it circular in cross section and subjected to a torque, and it is called a *beam* if it is subjected to moments or transverse loads that induce bending.[1] In this chapter, we focus on beams as well as columns that buckle (i.e., structural members having one dimension much greater than the other two that are subjected to bending loads). As in Chapter 4, we limit our examination to structural members that exhibit a linearly elastic, homogeneous, and isotropic (LEHI) behavior over small strains. Hence, again, the primary biomedical applications are (long) bones as well as select biomaterials. In addition, just as in Chapter 4, we will see that the topics herein are essential to the design of many different load cells, which, in turn, are important to many different areas of biomedical engineering, from gait analysis to studying mechanotransduction in cells. As in prior chapters, however, the most important thing is the deepening of one's understanding of the general approach of mechanics, not the specific (textbook) applications or solutions.

Whereas most engineering students learn about "bending moment and shear force diagrams" in a first course on engineering, here we briefly review these ideas because of their importance and because of the nonunique sign conventions and approaches used in different textbooks. In other words, we need to be on the same page when we begin our analysis of stress and deformation in subsequent sections.

---

[1] For completeness, note that structural members having two dimensions much greater than the third are called plates or shells if they are initially flat or curved, respectively. An example of the latter is the skull. If a plate or shell does not resist bending, it is called a membrane. The pericardium is an example of a curved membrane.

5.1 Shear Forces and Bending Moments

Given a generic straight beam (Fig. 5.1), we extract a differential element of length Δx and note the exposed shear forces V and bending moments M_z (i.e., bending moments about the z axis). We recall from statics that V and M_z can vary with position x; hence, the values on the left exposed face need not equal the values on the right face. Question: By how much do they differ? Because the right face will only Δx from the left, we expect that moments and shears on the right to differ from those on the left by only some small amount, say $M_z + \Delta M_z$ and $V + \Delta V$ on the right x face relative to M_z and V on the left face. Recall, too, that if a body is in equilibrium, then each of its parts is in equilibrium. Hence, assuming the possible existence of a uniformly distributed load $q(x)$, let us enforce equilibrium for the differential element in Figure 5.1b. Force balance requires

$$\sum F_x = 0 \rightarrow 0 = 0, \tag{5.1}$$

$$\sum F_y = 0 \rightarrow V - (V + \Delta V) - q(x)\Delta x = 0, \tag{5.2}$$

or

$$\Delta V = -q(x)\Delta x \rightarrow \frac{\Delta V}{\Delta x} = -q(x). \tag{5.3}$$

If we take the limit as Δx approaches zero, we obtain the general differential equation

$$\lim_{\Delta x \to 0} \frac{\Delta V}{\Delta x} = \frac{dV}{dx} \rightarrow \frac{dV}{dx} = -q(x). \tag{5.4}$$

Note that in this derivation, we assumed that the resultant of $q(x)$, [i.e., $\int_x^{x+\Delta x} q(x)dx$] is well approximated by $q(x)\Delta x$ over a small length Δx, which is to say that although q may vary with x, it will not vary much over the length Δx and, thus, it can be taken out of the integral. In other words, we invoke the mean value theorem for integrals from calculus.

Finally, let us enforce moment balance for the differential element in Figure 5.1b:

$$\sum M_z)_A = 0 \rightarrow -M_z + (M_z + \Delta M_z) - q(x)\Delta x\left(\frac{\Delta x}{2}\right) - (V + \Delta V)(\Delta x) = 0, \tag{5.5}$$

which reduces to

$$\Delta M_z = V\Delta x + \Delta V\Delta x + \frac{1}{2}q(x)(\Delta x)^2, \tag{5.6}$$

or

$$\frac{\Delta M_z}{\Delta x} = V + \Delta V + \frac{1}{2}q(x)\Delta x. \tag{5.7}$$

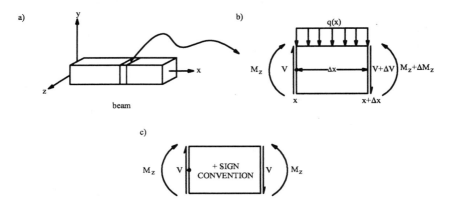

FIGURE 5.1 Panel a shows a generic initially straight beam and convenient coordinate system, with a representative section of length Δx isolated for removal via fictitious cuts. Panel b shows the removed differential element with an applied distributed load $q(x)$ and isolated internal shear forces V and bending moments M_z, each of which may vary as a function of location x. Panel c shows the positive sign convention that is adopted herein. Note: Although it is useful to consult other books for additional illustrative examples or alternate explanations and derivations, the sign convention differs considerably from book to book, which changes the governing equations and values of the computed quantities of interest. Paying careful attention to sign conventions is thus very important.

Finally, in the limit as we shrink Δx to a point (whereby ΔV and Δx go to zero), we have

$$\lim_{\Delta x \to 0} \frac{\Delta M_z}{\Delta x} = \lim_{\Delta x \to 0}\left[V + \Delta V + \frac{1}{2}q(x)\Delta x\right] = V \qquad (5.8)$$

or our final result

$$\frac{dM_z}{dx} = V(x). \qquad (5.9)$$

Equations (5.4) and (5.9) will prove very useful when we seek to draw the shear and bending moment diagrams. Next, let us consider a full analysis of a general beam problem.

When all of the forces are applied in one plane, we see that only three of the six equations of statics are available for the analysis. These are $\Sigma F_x = 0$, $\Sigma F_y = 0$, and $\Sigma M_z = 0$. In addition to providing our general differential equations, these equations also allow us to determine reaction forces at the supports of the beam. Indeed, the analysis of any beam or frame (i.e., finding internal forces and moments) should begin with a free-body diagram of the whole structure that shows both the applied and the reactive loads, which must satisfy equilibrium. The reactions can be computed using the equations of equilibrium provided that the system is statically determinate; in the case of statically indeterminate problems, which we consider below, additional equations are needed. Here, however, let us focus on the former, simpler case.

The next step in a general analysis uses the concept that if a body is in equilibrium, then each of its parts is also in equilibrium. We thus repeat the free-body diagram/equilibrium procedure for each judicious cut of interest. Consider an imaginary cut normal to the axis of the beam that separates the beam into two segments. Each of these segments is also in equilibrium. The conditions of equilibrium require the existence of internal forces and moments at each cut section of the beam. In general, at a section of such a member, at location x, a shear force $V(x)$, a horizontal force $f(x)$, and a moment $M_z(x)$ are necessary to maintain the isolated part in equilibrium. To illustrate this, consider initially straight, constant-cross-section LEHI beams in the following examples.

Example 5.1 For the beam in Figure 5.2, find $V(x)$ and $M_z(x)$ and draw the resulting shear force and bending moment diagrams. P, a, b, and L are assumed to be known.

Solution: The first step in solving this problem is to draw a free-body diagram of the whole beam and then to determine the reaction forces at the supports using the equations of (statics) equilibrium; that is,

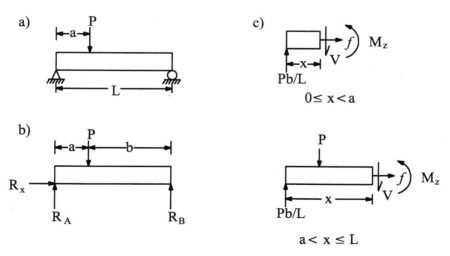

FIGURE 5.2 Shown here is a transversely loaded, simply supported beam (panel a), the free body of the whole structure (panel b), and free-body diagrams of two parts of the structure (panel c) that isolate the internal loads and moments in two regions of interest. Note: Remember that judicious cuts are often best taken between abrupt changes in loads, as done here.

$$\sum F_x = 0 \rightarrow R_x = 0,$$
$$\sum F_y = 0 \rightarrow R_A + R_B - P = 0,$$
$$\sum M_z)_A = 0 \rightarrow R_B L - Pa = 0,$$

where, from moment balance, we have

$$R_B = \frac{Pa}{L},$$

and with $L - a = b$, we now have, from vertical force balance,

$$R_A = P - \frac{Pa}{L} = P\left(1 - \frac{a}{L}\right) = P\left(\frac{L-a}{L}\right) = \frac{Pb}{L}.$$

The next step is to construct a free-body diagram for each part of interest, with judicious choices of cuts. In general, *we make cuts between any abrupt changes in load, geometry, or material properties* (see Fig. 5.2c). First, for the section cut to the left of the applied load P (i.e., for $0 < x < a$),

$$\sum F_x = 0 \rightarrow f = 0,$$
$$\sum F_y = 0 \rightarrow \frac{Pb}{L} - V = 0 \rightarrow V = \frac{Pb}{L},$$
$$\sum M_z)_A = 0 \rightarrow M_z - Vx = 0 \rightarrow M_z(x) = \left(\frac{Pb}{L}\right)x.$$

Similarly, for the section cut to the right of the applied load P (i.e., for $a < x < L$),

$$\sum F_x = 0 \rightarrow f = 0,$$
$$\sum F_y = 0 \rightarrow \frac{Pb}{L} - P - V = 0,$$

or

$$V = \frac{Pb}{L} - P = P\left(\frac{b}{L} - 1\right) = P\left(\frac{b-L}{L}\right) \rightarrow V = -\frac{Pa}{L},$$

and, finally,

$$\sum M_z)_A = 0 \rightarrow M_z - Pa - Vx = 0,$$
$$M_z = Pa - \frac{Pa}{L}x \rightarrow M_z(x) = Pa\left(1 - \frac{x}{L}\right).$$

Now that we have the functions $V(x)$ and $M_z(x)$ for each section of interest, we can construct the shear force and bending moment diagrams, recalling that

$$V(x) = \begin{cases} \dfrac{Pb}{L} & \text{for } x \in (0, a) \\ -\dfrac{Pa}{L} & \text{for } x \in (a, L) \end{cases} \quad \text{and} \quad M_z(x) = \begin{cases} \dfrac{Pb}{L}x & \text{for } x \in (0, a) \\ Pa\left(1 - \dfrac{x}{L}\right) & \text{for } x \in (a, L). \end{cases}$$

It is convenient to construct our shear and bending moment diagrams directly below the free-body diagram of the beam, using the same horizontal scale for the length of the beam (Fig. 5.3). Note, in particular, that the internal shear force is constant within each of the two sections of inter-,est. Consistent with Eq. (5.9), the associated moments are each linear in these sections of interest, the slope of which is given by the value of V. Finally, note that there is a discontinuity (jump) in $V(x)$ where the transverse load P is applied, but there is no jump in $M_z(x)$ given the absence of any applied (concentrated) moment.

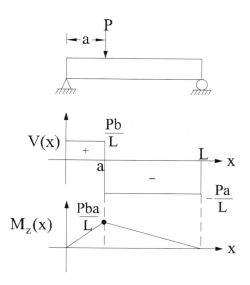

FIGURE 5.3 Shear and bending moment diagrams for the beam in FIGURE 5.2. It is best to draw such diagrams as done here: Show the physical problem with the shear and bending diagrams directly below. This will help you to develop some intuition with regard to how such diagrams should look for various physical problems.

Example 5.2 The beam in Figure 5.4 has a uniformly, or evenly, distributed load $q(x) = q_o$. Find V and M_z at all x and the resulting shear force and bending moment diagrams.

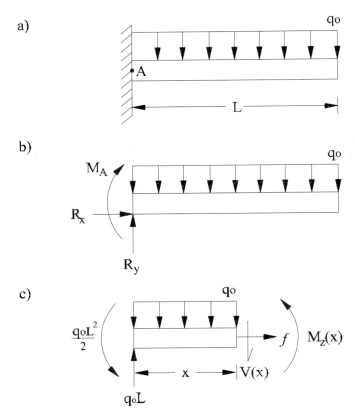

FIGURE 5.4 Shown is a uniformly loaded [i.e., $q(x) = q_o$] cantilevered beam (panel a) as well as free-body diagrams of the whole (panel b) and one part of the beam (panel c). Note that only one fictitious cut is needed to expose the internal loads and moments because there are no abrupt changes in geometry, properties, or loading between $x = 0$ and $x = L$.

Solution: Again, the first step is to draw a free-body diagram of the whole structure and to determine the reaction forces at the supports using the equations of statics; that is,

$$\sum F_x = 0 \rightarrow R_x = 0,$$

$$\sum F_y = 0 \rightarrow R_y - \int_0^L q_o dx = 0 \quad \text{or} \quad R_y = q_o L.$$

Finally,

$$\sum M_z)_A = 0 \rightarrow -M_A - \int_0^L q_o x\, dx = 0 \quad \text{or} \quad M_A = -\frac{q_o L^2}{2}.$$

Before proceeding to equilibrium of parts, let us note the following. In your first course on statics, you may have solved this problem by considering force and moment balance in terms of the *resultant* force and moment due to the uniformly applied load $q(x) = q_o$. These resultants are

$$R_F = q_o L, \quad M_F = q_o L\left(\frac{L}{2}\right),$$

which are seen easily. The reactions can thus be solved in terms of these resultants via equilibrium. Whereas this procedure is simple in simple problems, integration (which yields these results) of the applied loads directly in the equilibrium equations is preferred in general.

Next, let us consider internal forces and moments via the introduction of judicious cuts. Because of the uniform loading, however, one cut will suffice here. From Figure 5.4c, we have for all $0 < x < L$,

$$\sum F_x = 0 \rightarrow f = 0,$$

$$\sum F_y = 0 \rightarrow q_o L - \int_0^x q_o \, dx - V = 0,$$

or $V(x) = q_o(L - x)$. Similarly,

$$\sum M_z)_A = 0 \rightarrow \frac{q_o L^2}{2} - \int_0^x q_o x \, dx - Vx + M_z = 0,$$

or

$$M_z(x) = Vx + \frac{q_o}{2}(x^2 - L^2)$$

where V is known from above; thus,

$$M_z(x) = q_o(L - x)x + \frac{q_o}{2}(x^2 - L^2).$$

Given these two results, we can now plot the desired shear and bending moment diagrams. First, however, let us observe an alternate but equivalent approach. Instead of the direct approach of cutting a beam and determining the internal shear forces and bending moments at a section by statics, an efficient alternative procedure can be used if the distributed external force $q(x)$ is known and integrated easily. Recall the basic differential equations derived earlier, namely Eqs. (5.4) and (5.9):

$$\frac{dV}{dx} = -q(x) \quad \text{and} \quad \frac{dM_z}{dx} = V(x).$$

Integrating Eq. (5.4) yields the shear force V, whereas integrating Eq. (5.9) yields the bending moment M_z. These ordinary differential equations can be used to solve for the shear forces and bending moments in beam problems. Hence, consider the following example.

Example 5.3 Find $V(x)$ and $M_z(x)$ for the beam in Example 5.2 using the governing ordinary differential equations, where $q(x) = q_o$.

Solution: By Eq. (5.4),

$$\frac{dV(x)}{dx} = -q_o \rightarrow \int \frac{d}{dx}(V(x))dx = -\int q_o\, dx,$$

or

$$V(x) = -q_o x + c_1.$$

By Eq. (5.9), we have

$$\frac{dM_z(x)}{dx} = -q_o x + c_1 \rightarrow \int \frac{d}{dx}(M_z(x))dx = \int (-q_o x + c_1)dx,$$

or

$$M_z(x) = -\frac{q_o x^2}{2} + c_1 x + c_2.$$

Applying the boundary conditions for a *free end*, $V(x = L) = 0$ and $M_z(x = L) = 0$, we get

$$0 = -q_o L + c_1 \rightarrow c_1 = q_o L$$

and

$$0 = -\frac{q_o L^2}{2} + q_o L^2 + c_2 \rightarrow c_2 = -\frac{q_o L^2}{2}.$$

Therefore,

$$V(x) = -q_o x + q_o L \rightarrow V(x) = q_o(L - x)$$

and

$$M_z(x) = -\frac{q_o x^2}{2} + c_1 x + c_2 \rightarrow M_z(x) = q_o L x - \frac{q_o}{2}(x^2 + L^2).$$

These results are the same as found in the previous example, as they should be. The shear and bending moment diagrams are in Figure 5.5.

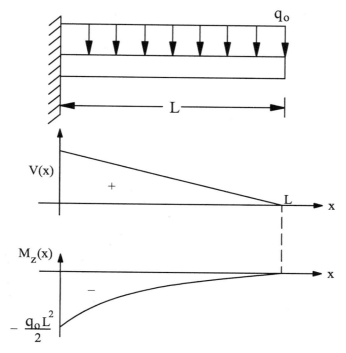

FIGURE 5.5 Shear and bending moment diagrams for the beam in Figure 5.4.

5.2 Stresses in Beams

5.2.1 Biological Motivation

Recall from Chapters 1 and 4 that it was probably first appreciated in bone that the local state of stress (or strain) influences greatly the underlying microstructure through growth and remodeling processes. Moreover, bone and teeth are among the few tissues in the body that exhibit elastic (and viscoelastic) behaviors under small strains. Finally, it is easily imagined that bones are routinely subjected to applied loads that tend to induce bending. A prime example is daily loading of the femur during walking and running because the line of action of the applied load does not go through the centroid of the diaphysis (see Figs. 4.1 and 4.7 as well as Section 3.3.2). Taken together, these observations reveal the importance of studying the bending of beams that exhibit a linear material behavior under small strains. Indeed, whether we are interested in understanding the maximum allowable transverse loads that an athlete's tibia or fibula can withstand without fracturing, in designing a prosthesis for implantation, or studying mechanotransduction in osteoblasts and osteoclasts, knowledge of simple beam bending is of paramount importance.

5.2.2 Theoretical Framework

Despite the existence of specific examples that motivate the need to study particular problems, we emphasize that continuum mechanics is an approach to solving a broad class of problems; it is not a collection of specialized solutions. Recall from Chapters 3 and 4, therefore, that in finding the *relation between stress and the applied loads and geometry* for the axially loaded rod, inflated thin-walled cylinders and spheres, and the torsion of a circular shaft, we first introduced a judicious cut to expose the stress of interest and then we enforced equilibrium (of the parts). Based on our examination of shear force and bending moment diagrams, it is clear that a fictitious cut will, in general, expose two types of stress: a normal stress σ_{xx} that serves to balance the moment M_z and a shear stress σ_{xy} that serves to balance the shear force V (Fig. 5.6). Nevertheless, let us begin our analysis by considering *pure bending,*—bending in the absence of transverse loads and a σ_{xy} stress.

Normal Stress

To determine the normal stress in a beam subjected to pure bending, consider the differential element in Figure 5.7. The forces in the x direction must balance, and so too for the moments (i.e., externally applied and the internal resisting moments). Hence,

$$\sum F_x = 0 \rightarrow \int -\sigma_{xx}\, dA = 0 \tag{5.10}$$

and

$$\sum M_z)_A = 0 \rightarrow -M_z + \int y(-\sigma_{xx})\, dA = 0. \tag{5.11}$$

Our two governing equilibrium equations are thus

$$\int -\sigma_{xx}\, dA = 0, \tag{5.12}$$

where $dA = dy\,dz$ and

$$M_z = -\int y\sigma_{xx}\, dA, \tag{5.13}$$

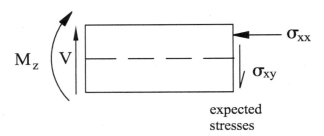

expected
stresses

FIGURE 5.6 Schema of the need for x-face normal and shear stresses to balance a bending moment M_z and shear force V.

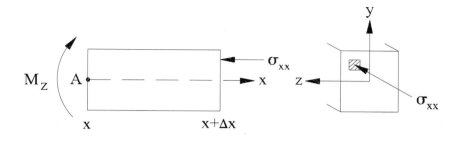

OR

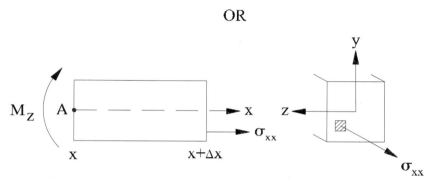

FIGURE 5.7 Side view and oblique view of a fictitiously cut beam that exposes the normal stress σ_{xx}, which acts over a differential area $dA = dydz$. Given the applied moment shown, the stress is expected to be compressive in the upper portion of the beam ($y > 0$) and tensile in the lower portion ($y < 0$). Either free-body diagram (top or bottom) is sufficient for purposes of a force balance.

the second of which relates the stress to the applied load (pure bending moment) and geometry (cross-sectional area). Similar to the torsion problem, however, the stress may vary in a yet unknown way: $\sigma_{xx} = \sigma_{xx}(y)$. Indeed, we expect a compressive stress in the upper portion of the cross section, where the (positive) bending moment tends to shorten the beam, and a tensile stress in the bottom portion, where the moment tends to lengthen the beam. Equation (5.12) requires that these compressive and tensile stresses must self-equilibrate.

As in the torsion problem, let us turn to kinematics and constitutive relations to find the function $\sigma_{xx}(y)$, which will allow us to integrate Eqs. (5.12) and (5.13) as needed. Thus, consider the differential element extracted from Figure 5.7 wherein we exaggerate the degree of bending (quantified by the radius of curvature ρ) to allow visualization (Fig. 5.8). Moreover, let us locate the coordinate system at the level wherein the width of the element is denoted $\overline{NA}$. Indeed, let us pick that level where $\overline{NA}$ equals the original width Δx; hence,

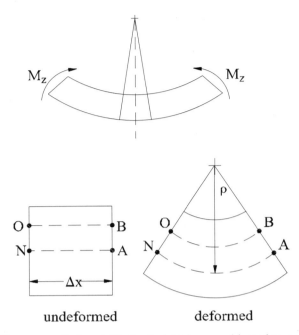

FIGURE 5.8 We assume that an initially straight beam subjected to a pure moment (i.e., no transverse loads) has a constant curvature at any depth (in the y direction). It will prove convenient, therefore, to consider line elements $\overline{OB}$ and $\overline{NA}$ in the x direction in both undeformed and deformed configurations. Furthermore, let the radius of curvature for the deformed line element $\overline{NA}$ be ρ. The arc length of $\overline{NA}$ is thus given by ρ times the subtended angle.

$$\overline{NA} = \Delta x = \rho \Delta \theta. \tag{5.14}$$

That one can find a location in the y direction where a line segment does not change length due to bending is revealed by the observation that line elements shorten in the compressive portion and lengthen in the tensile portion—somewhere between, one line segment must remain at a constant length, which thus serves as a convenient reference point.

Likewise, consider the line segment $\overline{OB}$, located a distance y above the level containing $\overline{NA}$. Whereas $\overline{OB} = \Delta x$ before deformation, after deformation we have

$$\overline{OB} = \Delta \theta (\rho - y). \tag{5.15}$$

The segment $\overline{OB}$ can also be rewritten in terms of Δx, which we prefer in light of Eqs. (5.4) and (5.9). From Eq. (5.14),

$$\overline{OB} = \Delta \theta (\rho - y) = \rho \Delta \theta - y \Delta \theta = \Delta x - y \frac{\Delta x}{\rho} = \Delta x \left(1 - \frac{y}{\rho}\right). \tag{5.16}$$

Recall from Chapter 2, therefore, that the linearized extensional strain can be thought of (over infinitesimal line segments) as

$$\varepsilon_{xx} = \frac{\text{Current length} - \text{Original length}}{\text{Original length}}. \tag{5.17}$$

Hence, the strain associated with the change in length of $\overline{OB}$ can be approximated as

$$\varepsilon_{xx} = \lim_{\Delta x \to 0} \frac{\Delta x(1 - y/\rho) - \Delta x}{\Delta x} = -\frac{y}{\rho} \to \varepsilon_{xx} = -\frac{y}{\rho}, \tag{5.18}$$

where the reciprocal of the radius of curvature ρ defines the curvature κ, with ρ and κ both constant with respect to x. It is important to note that if $y > 0$, then the strain is compressive, and if $y < 0$, then the strain is extensional, each consistent with the assumed compressive and tensile stresses σ_{xx} discussed earlier. Moreover, at $y = 0$ the strain is zero consistent with our selection of the location of $\overline{NA}$.

From Hooke's law for isotropic behavior [cf. Eq. (2.69)], $\sigma_{xx} = E\varepsilon_{xx}$ when $\sigma_{yy} = 0$ and $\sigma_{zz} = 0$, as assumed here. Hence, from Eq. (5.18),

$$\sigma_{xx} = E\left(-\frac{y}{\rho}\right). \tag{5.19}$$

Now, back to equilibrium, Eq. (5.12) becomes

$$\int -\left(\frac{Ey}{\rho}\right) dA = 0 \to \frac{E}{\rho} \underbrace{\int y\, dA}_{\bar{y}A} = 0 \to \frac{EA}{\rho}\bar{y} = 0, \tag{5.20}$$

where $\bar{y}$ is the distance from the origin of our (x, y, z) coordinate system to the centroid of the cross-sectional area A (recall Eq. A3.2). Because this integral equals zero and the Young's modulus E, radius of curvature ρ, and cross-sectional area A are each nonzero, the distance $\bar{y}$ must be set equal to zero; that is, the z axis must pass through the centroid of the cross-section. This means that the coordinate system, at $\overline{NA}$, must be located at the centroid. Next, from Eq. (5.13),

$$-\int y\left(-\frac{Ey}{\rho}\right) dA = M_z \to \frac{E}{\rho} \underbrace{\int y^2\, dA}_{I_{zz}} = M_z \tag{5.21}$$

wherein we recognize the second moment of area I_{zz} (Appendix 4 of Chapter 4), thus moment balance requires

$$M_z = \frac{EI_{zz}}{\rho}. \tag{5.22}$$

This equation is called the *moment-curvature relation*; it will prove critical in our subsequent discussion of beam deflections in Section 5.3. Here, however, note that by rearranging Eq. (5.22) and using Eq. (5.19), we obtain our desired relation for the normal stress in terms of the applied load and geometry:

$$\sigma_{xx} = -\frac{M_z(x)y}{I_{zz}}. \tag{5.23}$$

This equation is called the *flexure formula*; it is one of the most important relations in elementary solid mechanics. Before we explore its use, however, let us observe the following. In bending, the locus of all centroids is called the *neutral axis*, $\overline{NA}$, for it is where $\varepsilon_{xx} = -y/\rho$ and $\sigma_{xx} = -M_z y/I_{zz}$ both equal zero (i.e., are neutral). The neutral axis for any elastic beam of homogeneous composition can thus be determined easily by finding the centroid of the cross-sectional area of the beam.

Finally, Eq. (5.23) allows us to compute the normal stress σ_{xx} at any x [due to $M_z(x)$ dependence] or y; we assume that σ_{xx} does not vary with z. Because of our small strain, small slope of the deflection curve, and use of Hooke's law, this flexure formula is not a universal result; it is restricted to small-strain LEHI behavior. Let us now consider the case in which transverse loads exist, which give rise to shear stresses σ_{xy}.

Shear Stress

Whereas we derived the flexure formula for σ_{xx} due to pure bending (moment only), we will assume that the same formula [Eq. (5.23)] holds equally well for bending due to transverse loads and that this stress can be considered simultaneously with other stresses that arise from the transverse loads. This is tantamount to assuming a superposition of solutions in a linear problem as we did when considering combined internal pressurization and axial loading of a thin-walled cylinder.

In the presence of transverse loads, we must also account for the shear stresses that act to balance V (Fig. 5.6). Here, therefore, let us derive a general relation that relates the shear stress σ_{xy} to the applied load and geometry. Hindsight reveals that an approach different from that used to derive σ_{xx} (wherein we used the sum of the effects of all σ_{xx} stresses acting over their differential areas to balance directly the applied bending moment) will prove useful. Hence, consider the following.

Let us extract a small rectangular piece of a generic beam as shown in Figure 5.9 such that we expose the stresses σ_{yx} at the point p (i.e., in the limit as $\Delta x \to 0$). Moreover, whereas we must ensure force balance in x and y as well as moment balance, here we shall focus only on force balance in x. Note, therefore, that

$$\sum F_x = 0 \to -\int \sigma_{xx}(x)\,dA + \int \sigma_{xx}(x+\Delta x)\,dA - \int \sigma_{yx}\,dA_s = 0, \tag{5.24}$$

where dA_s simply denotes the area over which the shear stress σ_{yx} acts, and consistent with our approach in the previous section,

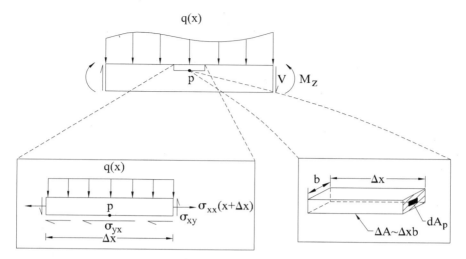

FIGURE 5.9 Consider a small rectangular portion of a beam taken above point p and on which σ_{yx} shear stresses act on the bottom y face; recalling the requirement from angular momentum balance that the stress be symmetric (i.e., $\sigma_{xy} = \sigma_{yx}$), this is consistent with the need for σ_{xy} shear stresses on the x face to balance the transverse loads. Note that the shear stress is shown according to its standard positive sign convention, not for the adopted positive sign convention for the shear force in Figure 5.6; note, too, that we neglect σ_{yy} in general and thus require σ_{xy} to balance all of the vertical load.

$$\sigma_{xx}(x) = -\frac{M_z y}{I_{zz}}, \qquad \sigma_{xx}(x+\Delta x) = -\frac{(M_z + \Delta M_z)y}{I_{zz}}. \qquad (5.25)$$

Hence, we see that force balance requires

$$+\frac{M_z}{I_{zz}}\int y\,dA_p - \frac{M_z}{I_{zz}}\int y\,dA_p - \frac{\Delta M_z}{I_{zz}}\int y\,dA_p - \int \sigma_{yx}\,dA_s = 0. \qquad (5.26)$$

Noting that the first two terms cancel and that the $\int y dA_p$ is *the first moment of area for the exposed cross section above point p*, we have

$$-\frac{\Delta M_z}{I_{zz}}Q = \int \sigma_{yx}\,dA_s, \qquad (5.27)$$

where, for notational simplicity, we let $Q \equiv \int y dA_p$. Finally, note that to complete the derivation, we must either consider the average value of σ_{yx} (as in the thin-walled tube problem) or find how σ_{yx} varies with x so that we can integrate over $dA_s = dx\,dz$. Knowing that we seek to shrink Δx to a point in the limit and assuming that σ_{yx} does not vary with z (similar to our implicit assumption for σ_{xx}), we can write

$$-\frac{\Delta M_z}{I_{zz}}Q = \sigma_{yx}\big)_{\text{ave}}\,\Delta x b \rightarrow \sigma_{yx}\big)_{\text{ave}} = -\frac{\Delta M_z}{\Delta x}\left(\frac{Q}{I_{zz}'b}\right), \qquad (5.28)$$

which in the limit becomes

$$\sigma_{yx}\big)_{\text{ave}} = -\frac{dM_z}{dx}\left(\frac{Q}{I_{zz}b}\right), \qquad (5.29)$$

whereby we recall from Eq. (5.9) that $V(x) = dM_z/dx$. Our derivation is thus complete except for one observation. Note that we denoted σ_{yx} and σ_{xy} on the isolated part of the beam according to our general sign convention for stress (cf. Fig. 2.4). We must be consistent with our sign convention for $V(x)$ and $M_z(x)$ for beam bending, whereby we defined a positive $V(x)$ on a positive x face to be in the negative direction (cf. Fig. 5.1). Hence, just as in statics, when we obtain a negative value in an analysis, this tells us that our quantity acts opposite to the direction assumed. Thus, the negative sign in Eq. (5.29) tells us that $\sigma_{yx}\big)_{\text{ave}}$ acts opposite to the direction assumed and so too for σ_{xy}, which is numerically equal to σ_{yx} at a point due to moment balance [Eq. (2.7)]. Hence, given these observations, we have the desired result

$$\sigma_{xy}\big)_{\text{ave}} = \frac{V(x)Q(y)}{I_{zz}b}, \qquad (5.30)$$

where $Q = \int y\,dA_p$ and b is the width of the beam at point p where the stress is evaluated. Note, too, that $\sigma_{xy}\big)_{\text{ave}}$ acts in the direction of the shear $V(x)$ and that this relation relates the stress to the applied shear force and geometry (measures being Q, I_{zz}, and b). Knowing that σ_{xy} is an average value over b, we will drop the notation $\sigma_{xy}\big)_{\text{ave}}$ and simply write σ_{xy}.

As a final observation, note that because Q is a first moment of area, it can be computed simply as (recall Appendix 3 of Chapter 3)

$$Q = \bar{y}_p A_p, \qquad (5.31)$$

where $\bar{y}_p$ locates the centroid of the cross-sectional area *above* point p relative to the overall centroid, and A_p is the area above point p. To better appreciate this, consider the following example.

Example 5.4 Find the value of Q for the rectangular cross section in Figure 5.10 at the following points $y = h/2,\ h/4,\ 0,\ -h/4,\ -h/2$, three of which are emphasized in the Figure.

Solution: Here, it will prove useful to compute $Q = \bar{y}_p A_p$ for each point of interest and to do so in tabular form.

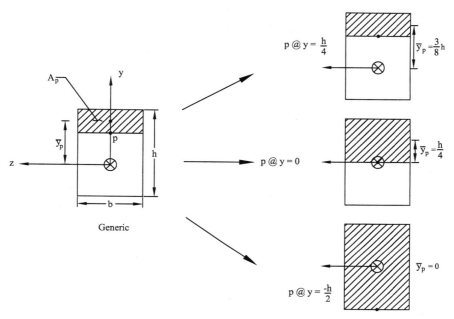

FIGURE 5.10 Illustration of the method for determining the value of $Q = \bar{y}_p A_p$ in a rectangular cross section. The subscript p reminds us that these quantities are defined for the cross-sectional areas above a point p rather than over the entire cross section. Forgetting this is a common error. Q is seen to be zero at the top and bottom surfaces for different reasons.

y-Location of p	$\bar{y}_p$	A_p	Q
$\dfrac{h}{2}$	$\dfrac{h}{2}$	0	0
$\dfrac{h}{4}$	$\dfrac{h}{4}+\dfrac{h}{8}$	$\dfrac{h}{4}b$	$\dfrac{3}{32}bh^2$
0	$\dfrac{h}{4}$	$\dfrac{h}{2}b$	$\dfrac{1}{8}bh^2$
$-\dfrac{h}{4}$	$\dfrac{h}{8}$	$\dfrac{3h}{4}b$	$\dfrac{3}{32}bh^2$
$-\dfrac{h}{2}$	0	hb	0

From this example, therefore, we see that Q varies with y, being smallest (zero) at the top and bottom and largest at the overall centroid. Indeed, for the simple case of a rectangular cross section, we can obtain a general formula for Q, namely

$$Q = \int_{-b/2}^{b/2}\int_y^{h/2} y\,dy\,dz = b\left(\frac{1}{2}y^2\Big|_y^{h/2}\right) = \frac{b}{2}\left(\frac{h^2}{4} - y^2\right), \tag{5.32}$$

which we see recovers the results in the above table. In this case, Q varies quadratically with the location of interest p in the y direction. From the relation $\sigma_{xy} = VQ/I_{zz}b$, therefore, we see that the shear stress would likewise vary quadratically with y, being zero on the top and bottom surfaces but largest at the centroid. This is in contrast to the normal stress $\sigma_{xx} = -M_z y/I_{zz}$, which varies linearly with y and is zero at the centroid and largest at the top and bottom surfaces. The stress field in a beam subject to bending will thus be complex in general.

Question: Does it make sense that σ_{xy} is zero at values of y that correspond to the top and bottom surfaces? To answer this question, it is useful to recall two things. First, $\sigma_{xy} = \sigma_{yx}$ at any point due to moment balance as revealed in Chapter 2. Second, we are only considering beams that are subjected to bending moments or transverse loads. Hence, there are no x-directed forces on the top and bottom y faces and thus no σ_{yx} ($=\sigma_{xy}$) stresses on these faces. Thus, our result for $\sigma_{xy} = VQ/I_{zz}b$ does make sense on these top and bottom surfaces.

Finally, although our approach for deriving the relation for σ_{xy} differed from the direct force balance used to derive the flexure formula for σ_{xx}, force balance must be respected nonetheless. Hence, consider the following example.

Example 5.5 Show that $\int\sigma_{xy}dA = V$ at any x.

Solution: Although the result can be obtained more generally, let us consider the rectangular cross section in Figure 5.10. Using Eq. (5.30), we have

$$\int\sigma_{xy}\,dA = \int_{-b/2}^{b/2}\int_{-h/2}^{h/2}\frac{V(x)Q(y)}{I_{zz}b}\,dy\,dz = \frac{V}{I_{zz}b}\int_{-b/2}^{b/2}dz\int_{-h/2}^{h/2}Q(y)\,dy$$

which from Eq. (5.32) can be written as

$$\int\sigma_{xy}\,dA = \frac{V}{I_{zz}}\int_{-h/2}^{h/2}\frac{b}{2}\left(\frac{h^2}{4} - y^2\right)dy,$$

where $I_{zz} = (1/12)bh^3$ for the rectangular cross section. Hence,

$$\int\sigma_{xy}\,dA = \frac{12V}{bh^3}\left(\frac{b}{2}\right)\left(\frac{h^2}{4}y - \frac{1}{3}y^3\right)\Big|_{-h/2}^{h/2} = \frac{6V}{h^3}\left[\frac{h^2}{4}\left(\frac{h}{2}+\frac{h}{2}\right) - \frac{1}{3}\left(\frac{h^3}{8}+\frac{h^3}{8}\right)\right]$$

$$= \frac{6V}{h^3}\left(\frac{h^3}{4} - \frac{h^3}{12}\right) = V,$$

which proves that vertical force balance is respected in this case, as it should be.

Principal Values and Maximum Shear

If we want to find maximum normal stress or maximum shear stress (at each point), we recall from Chapter 2 that

$$\sigma'_{xx})_{max/min} = \frac{\sigma_{xx} + \sigma_{yy}}{2} \pm \sqrt{\left(\frac{\sigma_{xx} - \sigma_{yy}}{2}\right)^2 + \sigma^2_{xy}} \tag{5.33}$$

and

$$\sigma'_{xy})_{max/min} = \pm \sqrt{\left(\frac{\sigma_{xx} - \sigma_{yy}}{2}\right)^2 + \sigma^2_{xy}}. \tag{5.34}$$

For beam bending, these equations for the maximum values become

$$\sigma'_{xx})_{max} = -\frac{M_z(x)y}{2I_{zz}} + \sqrt{\left(-\frac{M_z(x)y}{2I_{zz}}\right)^2 + \left(\frac{V(x)Q(y)}{I_{zz}b}\right)^2} \tag{5.35}$$

$$\sigma'_{xy})_{max} = \sqrt{\left(-\frac{M_z(x)y}{2I_{zz}}\right)^2 + \left(\frac{V(x)Q(y)}{I_{zz}b}\right)^2}, \tag{5.36}$$

where each varies with (x, y) in general. Let us now consider a few illustrative examples.

5.2.3 *Illustrative Examples*

Example 5.6 Find σ_{xx}, the maximum normal stress $\sigma'_{xx})_{max}$, and the maximum shear stress $\sigma'_{xy})_{max}$ for the beam in Figure 5.11 with an applied bending moment M_o. Neglect the weight of the beam.

Solution: A free-body diagram of the whole structure reveals that we have

$$\sum F_x = 0 \rightarrow R_x = 0,$$
$$\sum F_y = 0 \rightarrow R_y = 0,$$
$$\sum M_z)_B = 0 \rightarrow M_o - M_w - R_y L = 0,$$

and, thus,

$$M_w = M_o.$$

A free-body diagram of part of the beam similarly requires

a)

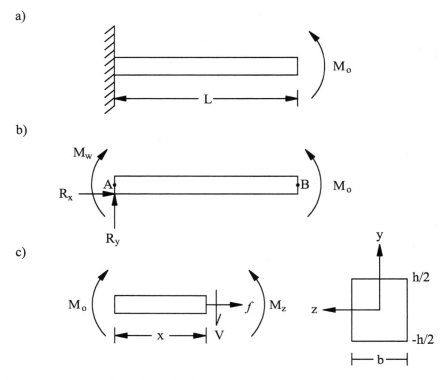

b)

c)

FIGURE 5.11 A cantilever beam having a rectangular cross section and subjected to an applied moment at the end. Free-body diagrams of the whole and the parts isolate reaction and internal forces and moments as needed. One cut is sufficient because the loads are applied only at the ends.

$$\sum F_x = 0 \rightarrow f = 0,$$
$$\sum F_y = 0 \rightarrow V(x) = 0 \quad \forall x$$
$$\sum M_z)_A = 0 \rightarrow -M_o + M_z(x) - Vx = 0,$$

and thus, as expected,

$$M_z(x) = M_o.$$

Next, we could draw the shear force and bending moment diagrams, which are trivial in this case of pure bending. From Eq. (5.23), therefore, we have

$$\sigma_{xx} = -\frac{M_0 y}{I_{zz}} \quad \forall x \in [0, L], y \in \left[-\frac{h}{2}, \frac{h}{2}\right], z \in \left[-\frac{b}{2}, \frac{b}{2}\right],$$

where

$$I_{zz} = \int y^2 \, dA = \int_{-b/2}^{b/2} \int_{-h/2}^{h/2} y^2 \, dy \, dz = \int_{-b/2}^{b/2} \frac{h^3}{12} \, dz = \frac{h^3 b}{12}.$$

Because σ_{xx} balances the applied load at each cross section, there is no need for any other component of stress relative to (x, y, z). The largest compressive and tensile loads, relative to (x, y, z), occur at $y = \pm h/2$; hence,

$$\sigma_{xx}\left(y = \pm\frac{h}{2}\right) = \mp\frac{12M_o}{bh^3}\left(\frac{h}{2}\right) = \mp\frac{6M_o}{bh^2}.$$

Considering only a 2-D state of stress here (e.g., σ_{xx}, σ_{xy}, and $\sigma_{xy} = \sigma_{yx}$), we recall from Chapter 2 that the maximum/minimum normal stresses are called principal stresses. They are computed via

$$\sigma'_{xx})_{\substack{\max \\ \min}} = \frac{\sigma_{xx} + \sigma_{yy}}{2} \pm \sqrt{\left(\frac{\sigma_{xx} - \sigma_{yy}}{2}\right)^2 + \sigma_{xy}^2},$$

where $\sigma_{yy} = 0 = \sigma_{xy}$. Using the largest value of σ_{xx} (i.e., at $y = h/2$), we find that

$$\sigma'_{xx})_{\max} = \frac{\sigma_{xx}}{2} + \sqrt{\frac{\sigma_{xx}^2}{4}} = \frac{6M_o}{bh^2},$$

$$\sigma'_{xx})_{\min} = \frac{\sigma_{xx}}{2} - \sqrt{\frac{\sigma_{xx}^2}{4}} = 0.$$

Hence, x and y are principal directions ($\alpha_p = 0$).

The maximum/minimum shear stress, however, is

$$\sigma'_{xy})_{\substack{\max \\ \min}} = \pm\sqrt{\left(\frac{\sigma_{xx} - \sigma_{yy}}{2}\right)^2 + \sigma_{xy}^2} = \pm\frac{\sigma_{xx}}{2}$$

and, therefore,

$$\sigma'_{xy})_{\substack{\max \\ \min}} = \pm\frac{3M_o}{bh^2},$$

which occurs at $\alpha_s = \pi/4$ or $45°$ and $y = \pm h/2$, not at $y = 0$ (the centroid), where σ_{yx} is largest in general. We see, therefore, that we must pay particularly close attention to the coordinate system to which quantities are referred as well as possible failure mechanisms (e.g., ductile materials yield in response to shear stresses and brittle materials fracture in response to normal stresses).

Example 5.7 Find general relations for σ_{xx} and σ_{xy} in terms of P, x, y, b, and $\bar{y}_p$ for the beam shown in Figure 5.12. Next, find the values of each of these components of stress at the following (x, y) locations: $(0, h/2)$, $(0, 0)$, $(L/2, h/2)$, $(L/2, 0)$, $(L, h/2)$, $(L, 0)$, $(L, -h/2)$.

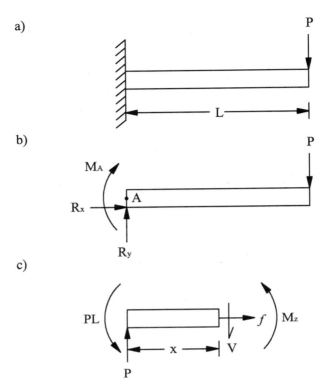

FIGURE 5.12 Similar to Figure 5.11 except for a cantilever beam having a rectangular cross section and subjected to an applied transverse load P at the end.

Solution: From the free-body diagram of the whole structure, we see that

$$\sum F_x = 0 \rightarrow R_x = 0,$$
$$\sum F_y = 0 \rightarrow R_y = P,$$
$$\sum M_z)_A = 0 \rightarrow -M_A - PL = 0,$$
$$M_A = -PL.$$

Next, let us find the internal forces and moments. From a free-body diagram of the parts, we obtain

$$\sum F_x = 0 \rightarrow f = 0,$$
$$\sum F_y = 0 \rightarrow V = P,$$
$$\sum M_z)_A = 0 \rightarrow PL - Vx + M_z = 0,$$
$$M_z = Vx - PL = P(x - L).$$

From Eq. (5.23), therefore,

$$\sigma_{xx} = -\frac{M_z y}{I_{zz}} = -\frac{P(x-L)y}{I_{zz}} \quad \forall x, y, z,$$

where

$$I_{zz} = \int_{-b/2}^{b/2} \int_{-h/2}^{h/2} y^2 \, dy \, dz = \frac{1}{12} bh^3.$$

From Eq. (5.30),

$$\left.\sigma_{xy}\right)_{ave} = \frac{VQ}{I_{zz}b} = \frac{PQ}{I_{zz}b},$$

where

$$Q = \int y \, dA_p = \int_{-b/2}^{b/2} \int_y^{h/2} y \, dy \, dz = \int_{-b/2}^{b/2} \left(\frac{h^2}{8} - \frac{y^2}{2}\right) dz = \frac{b}{2}\left(\frac{h^2}{4} - y^2\right).$$

Hence,

$$\sigma_{xx} = -\frac{12P(x-L)y}{bh^3} \quad \text{and} \quad \sigma_{xy} = \frac{12P[b(h^2/4 - y^2)/2]}{b^2 h^3}.$$

Now, we can compute σ_{xx} and σ_{xy} at various points of interest p. For example, for the points indicated, let the point $(x, y) = (0, h/2)$; $(0, 0)$; $(L/2, h/2)$; $(L/2, 0)$; $(L, h/2)$; $(L, 0)$; $(L, -h/2)$.

(x, y)	σ_{xx}	σ_{xy}
$\left(0, \dfrac{h}{2}\right)$	$-\dfrac{P(0-L)(h/2)}{(1/12)bh^3} = \dfrac{6PL}{bh^2}$	$\dfrac{P[(b/2)(h^2/4 - (h/2)^2)]}{(1/12)b^2h^3} = 0$
$(0, 0)$	$-\dfrac{P(0-L)0}{(1/12)bh^3} = 0$	$\dfrac{P[(b/2)(h^2/4 - 0^2)]}{(1/12)b^2h^3} = \dfrac{3P}{2bh}$
$\left(\dfrac{L}{2}, \dfrac{h}{2}\right)$	$-\dfrac{P(L/2-L)(h/2)}{(1/12)bh^3} = \dfrac{3PL}{bh^2}$	$\dfrac{P[(b/2)(h^2/4 - (h/2)^2)]}{(1/12)b^2h^3} = 0$
$\left(\dfrac{L}{2}, 0\right)$	$-\dfrac{P(L/2-L)0}{(1/12)bh^3} = 0$	$\dfrac{P[(b/2)(h^2/4 - 0^2)]}{(1/12)b^2h^3} = \dfrac{3P}{2bh}$
$\left(L, \dfrac{h}{2}\right)$	$-\dfrac{P(L-L)(h/2)}{(1/12)bh^3} = 0$	$\dfrac{P[(b/2)(h^2/4 - (h/2)^2)]}{(1/12)b^2h^3} = 0$
$(L, 0)$	$-\dfrac{P(L-L)0}{(1/12)bh^3} = 0$	$\dfrac{P[(b/2)(h^2/4 - 0^2)]}{(1/12)b^2h^3} = \dfrac{3P}{2bh}$
$\left(L, -\dfrac{h}{2}\right)$	$-\dfrac{P(L-L)(-h/2)}{(1/12)bh^3} = 0$	$\dfrac{P[(b/2)(h^2/4 - (h/2)^2)]}{(1/12)b^2h^3} = 0$

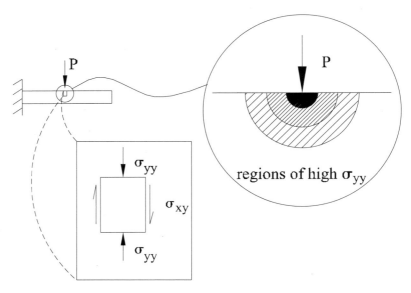

FIGURE 5.13 Schema of y-face, y-direction stresses σ_{yy} in the neighborhood of a concentrated load P. The contour lines show that the value of σ_{yy} decreases (shown by cross-hatching that becomes less dense) as one gets farther from the source of the concentrated load. No real load is truly concentrated, of course, for it must act over a finite, albeit possibly small, area.

Whereas we have determined the values of the components σ_{xx} and σ_{xy} at the indicated points, we emphasize that σ_{xx} and σ_{xy} can (should) be found at all (x, y). Indeed, at each (x, y), we should also compute the maximum normal stress $\sigma'_{xx})_{max}$ if we expect failure to occur due to normal stresses (e.g., in brittle materials) or the maximum shear stress $\sigma'_{xy})_{max}$ if we expect failure to occur in shear (e.g., in ductile materials). Because the principal and maximum shear stresses both depend on σ_{xx} and σ_{xy}, it is clearly more challenging to identify the point in the structure where an absolute maximum stress exists. For this reason and given the availability of computers and color graphics, many simply create color contour plots of the principal or maximum shear stresses to aid failure analysis.

At this juncture, we should emphasize that we have neglected the σ_{yy} component of stress. Clearly, in response to transverse loads applied to the top or bottom surfaces $y = \pm h/2$, σ_{yy} stresses will exist. Indeed, in the case of a concentrated load P (Fig. 5.13), the σ_{yy} stress can be very large close to the applied load. Inclusion of such effects generally requires numerical methods, however, and thus are beyond the present scope. Because real loads are applied over finite, not infinite, areas and because σ_{xx} and σ_{xy} tend to dominate, we will focus on these components throughout.

Observation 5.1 Although we have ignored potential "stress concentrations" due to concentrated transverse loads in our simple beam theory, the issue of stress concentrations is nevertheless very important in mechanics. Basically, a *stress concentration* can arise due to an abrupt change in geometry, material properties, or applied load; it is characterized by a significantly increased value of stress locally, with associated steep gradients with respect to stresses in the surrounding area. Stress concentrations may arise at holes, sharp edges, interfaces between materials of dissimilar stiffness, and, of course, at concentrated loads. Hence, methods to combat the potentially deleterious effects of stress concentrations include rounding edges, functionally grading the stiffness of a material (e.g., a metallic intravascular stent), and distributing a load over a broader area. Because stress concentrations are characterized by steep gradients in stress, analytic solutions are often not possible. One must often resort to numerical methods, such as finite elements, or perhaps experimental methods, such as photoelasticity. Fortunately, many general problems have been solved and categorized for reference. For example, see Roark and Young (1975) wherein stress concentration factors are given for many geometries. A stress concentration factor K is defined simply as a ratio of the maximum expected stress to the mean stress in that region (e.g., $\sigma_{max} = K\sigma_{avg}$), typically in reference to normal stresses. For example, in a LEHI material, the stress in a uniaxially loaded member is higher near a centrally placed hole by a factor of 2–3 depending on the ratio of the radius of the hole a to the width of the uniaxial sample d: $K = 3$ if $2a/d \sim 0$, but $K = 2$ if $2a/d \sim 1$. Holes are introduced in skin by dermatologists when taking a skin biopsy, in the lens capsule of the eye by ophthalmologists when implanting an intraocular device, in bone when an orthopedic surgeon puts in a bone screw, and so on. Understanding the effects of stress concentrations is thus very important in biomechanics even though our discussion here is very brief.

5.3 Deformation in Beams

5.3.1 Biological Motivation

Mechanics is the study of motions and the applied loads that cause them. Fundamental to the study of beams, therefore, is the quantification of strains and deflections. For example, if we seek to quantify the material parameters of Hooke's law for bone, we must know both stresses and strains at representative points. Likewise, recall from Chapter 3 that one type of failure can be excessive deformation. Hence, if we are to design an orthotic device to maintain two ends of a severely fractured bone in close proximity during the healing process, we must ensure that the orthotic device does not deform excessively under loads experienced during daily activity. Indeed, as we will see, measuring or computing deformations in beams is extremely important in many different situations in biomechanics.

5.3.2 Theoretical Framework

Recalling from the definition of Poisson's ratio that $v = -\varepsilon_{\text{transverse}}/\varepsilon_{\text{axial}}$, Eq. (5.18) yields

$$v = -\frac{\varepsilon_{yy}}{\varepsilon_{xx}} \rightarrow \varepsilon_{yy} = -v\left(-\frac{y}{\rho}\right) = v\left(\frac{y}{\rho}\right). \tag{5.37}$$

The "same" relation holds for ε_{zz} because

$$\varepsilon_{yy} = \frac{1}{E}[\sigma_{yy} - v(\sigma_{xx} + \sigma_{zz})] = -\frac{v}{E}\sigma_{xx} \tag{5.38}$$

and

$$\varepsilon_{zz} = \frac{1}{E}[\sigma_{zz} - v(\sigma_{xx} + \sigma_{yy})] = -\frac{v}{E}\sigma_{xx}. \tag{5.39}$$

Hence, because the Poisson's ratio v is typically non-negative, the beam widens in z where it shortens in x ($y > 0$ for $M_z > 0$) and it narrows in z where it lengthens in x ($y < 0$ for $M_z > 0$). This phenomenon is called *anticlastic* bending.

Finally, to get more information on the deformation, recall the moment-curvature relation, [Eq. (5.22)], which we now write as

$$\frac{1}{\rho} = \frac{1}{EI_{zz}}M_z. \tag{5.40}$$

Recall, too, from calculus that the curvature κ is defined as

$$\kappa = \frac{1}{\rho} = \frac{d^2v/dx^2}{\sqrt[3]{1 + (dv/dx)^2}} \tag{5.41}$$

where $v(x)$ denotes the vertical *deflection* of the neutral axis (i.e., v is a vertical displacement u_y of points along the neutral axis only). Clearly then, if we let $dv/dx \ll 1$ [i.e., if we consider beam deflections $v = v(x)$ having small slopes], then

$$\frac{1}{\rho} \cong \frac{d^2v}{dx^2}. \tag{5.42}$$

By substituting this result into the moment-curvature relation, we obtain a general differential equation for the beam deflection:

$$EI_{zz}\frac{d^2v}{dx^2} = M_z(x). \tag{5.43}$$

In summary, Eq. (5.18) allows us to compute the extensional strain ε_{xx} at any location y provided that we know the radius of curvature ρ, which is evaluated at the neutral axis $\overline{NA}$. Eq. (5.43) similarly provides information on the deflection $v(x)$ of the neutral axis given information on the applied load $M_z(x)$, geometry I_{zz}, and material property E. Compare this to the results for axial extension δ and rotation Θ in the axial load and torsion problems, respectively.

Because Eq. (5.43) is a second-order differential equation, we will need two boundary conditions for its full solution. These conditions will be on either the deflection v at a particular value of x or the slope dv/dx at a particular x. Recall, therefore, that the deflection v will be zero at a pin or a fixed end; the deflection at a roller is zero if the beam is pushed toward the roller. In contrast, a free end or an end on a slider cannot resist a deflection (Fig. 5.14). Conversely, the slope dv/dx is zero at a fixed end or a slider, but it cannot be specified at a roller, a pin, or a free end. An easy way to remember these *kinematic* boundary conditions is to remember the associated *traction* boundary conditions; that is, to restrict a vertical deflection, the support must be able to supply a resisting vertical reaction force R_y, and to restrict a rotation (i.e., a slope), the support must be able to supply a resisting bending moment M_z.

Let us note that Eqs. (5.4) and (5.9), in combination with Eq. (5.43), allow us to formulate alternative differential equations for the deflection curve $v(x)$; that is,

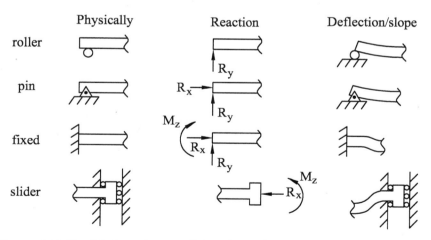

FIGURE 5.14 Possible boundary conditions for various supports showing both the reactions and the possible motions. Note that an applied force is capable of preventing (or limiting) a displacement in the same direction, whereas an applied moment is capable of preventing (or limiting) a rotation. Thus, one can prescribe at a support either an applied load or the resulting motion, but not both.

$$\frac{dM_z}{dx} = V(x) \rightarrow \frac{d}{dx}\left(EI_{zz}\frac{d^2v}{dx^2}\right) = V(x) \tag{5.44}$$

or

$$\frac{dV}{dx} = -q(x) \rightarrow \frac{d^2}{dx^2}\left(EI_{zz}\frac{d^2v}{dx^2}\right) = -q(x). \tag{5.45}$$

If the beam is of constant cross section and E does not vary with x, we see further that these third- and fourth-order differential equations can be written as

$$EI_{zz}\frac{d^3v}{dx^3} = V(x) \quad \text{and} \quad EI_{zz}\frac{d^4v}{dx^4} = -q(x). \tag{5.46}$$

In a given problem, therefore, we have the option to solve any *one* of the governing differential equations. The choice of which equation to attempt to solve should be dictated by our knowledge of the loading functions, $M_z(x)$, $V(x)$, or $q(x)$ as well as boundary conditions. Indeed, the higher-order equations require us to know moments or shears at the boundaries. At a "free end," for example, the moment or shear will be that which is applied; if it is truly a free end, with no physical support or applied load, we will have zero moment and zero shear. Various boundary conditions will be examined in the following examples.

5.3.3 *Illustrative Examples*

Example 5.8 Find the deflection curve of the beam in Example 5.6

Solution: Let $v(x)$ be the deflection curve (i.e., shape of the neutral axis in the deformed configuration). Recalling the governing differential equation $EI_{zz}d^2v/dx^2 = M_z(x)$ from Eq. (5.43) and that $M_z(x) = M_o$ for this beam, integrating once yields an expression for the slope of the deflection curve:

$$\int \frac{d}{dx}\left(\frac{dv(x)}{dx}\right)dx = \int \frac{1}{EI_{zz}}M_o\,dx = \frac{M_o}{EI_{zz}}\int dx,$$

or

$$\frac{dv(x)}{dx} = \frac{M_o}{EI_{zz}}x + c_1.$$

Integrating again yields an expression for the deflection curve $v(x)$:

$$\int \frac{d}{dx}[v(x)]dx = \int\left(\frac{M_o}{EI_{zz}}x + c_1\right)dx,$$

or

$$v(x) = \frac{M_o}{2EI_{zz}} x^2 + c_1 x + c_2.$$

Applying the boundary conditions for a fixed end, we get

$$\frac{dv}{dx}(x = 0) = 0 \rightarrow 0 = \frac{M_o}{EI_{zz}}(0) + c_1 \rightarrow c_1 = 0,$$

$$v(x = 0) = 0 \rightarrow 0 = \frac{M_o}{2EI_{zz}}(0)^2 + c_1(0) + c_2 \rightarrow c_2 = 0,$$

Therefore, with $I_{zz} = bh^3/12$ for a rectangular cross section, we have

$$v(x) = \frac{6M_o x^2}{Ebh^3},$$

with the deflection at the end (often denoted by δ) where $x = L$, being

$$\delta = v(x = L) = \frac{6M_o L^2}{Ebh^3}.$$

Although this was a simple example, with $M_z = M_o$ and $V = 0$ for all x, it illustrates the general approach, which is our primary goal.

Example 5.9 Find the deflection curve $v(x)$ and the maximum deflection $\delta = v(x = L)$ for the beam in Example 5.7.

Solution: Recalling that the moment $M_z(x) = P(x - L)$, Eq. (5.43) can be integrated once to obtain the slope:

$$EI_{zz} \int \frac{d}{dx}\left(\frac{dv}{dx}\right) dx = \int M_z \, dx = \int P(x - L) \, dx,$$

or

$$EI_{zz} \frac{dv}{dx} = P\left(\frac{x^2}{2} - xL\right) + c_1.$$

Integrating again, we have

$$EI_{zz} \int \frac{d}{dx}[v(x)] \, dx = \int \left[P\left(\frac{x^2}{2} - xL\right) + c_1\right] dx$$

$$EI_{zz} v(x) = P\left(\frac{x^3}{6} - \frac{x^2 L}{2}\right) + c_1 x + c_2.$$

Applying the boundary conditions for a fixed end, we get

$$EI_{zz}\frac{dv}{dx}(x=0)=0=P\left(\frac{(0)^2}{2}-(0)L\right)+c_1 \to c_1=0$$

and

$$EI_{zz}v(x=0)=0=P\left(\frac{(0)^3}{6}-\frac{(0)^2 L}{2}\right)+c_1(0)+c_2 \to c_2=0.$$

Thus, the deflection curve for a rectangular cross-section is,

$$v(x)=\frac{P}{EI_{zz}}\left(\frac{x^3}{6}-\frac{x^2 L}{2}\right)=\frac{12P}{Ebh^3}\left(\frac{x^3}{6}-\frac{x^2 L}{2}\right).$$

The maximum deflection is obviously at $x=L$; hence,

$$\delta=v(L)=\frac{12P}{Ebh^3}\left(\frac{L^3}{6}-\frac{L^3}{2}\right)=\frac{-4PL^3}{Ebh^3},$$

or, for any shaped cross section,

$$\delta=-\frac{PL^3}{3EI_{zz}}.$$

Note the minus sign, which indicates that the beam deflects downward given the downward transverse load.

Example 5.10 Find the deflection curve for the beam in Figure 5.15.

Solution: The free-body diagram for the whole structure reveals that

$$\sum F_x=0 \to R_x=0,$$
$$\sum F_y=0 \to R_y-P=0,$$
$$R_y=P,$$
$$\sum (M_z)_A=0 \to -M_w+M_o-PL=0,$$
$$M_w=M_o-PL.$$

Similarly, a free-body diagram of the segment from $0 \le x < L/2$ reveals that

$$\sum F_x=0 \to f=0,$$
$$\sum F_y=0 \to P-V(x)=0,$$
$$V(x)=P \quad \text{for } 0 \le x < \frac{L}{2},$$
$$\sum (M_z)_A=0 \to M_z-Vx-M_o+PL=0,$$
$$M_z=P(x-L)+M_o \quad \text{for } 0 \le x < \frac{L}{2},$$

FIGURE 5.15 A cantilevered beam with a concentrated moment applied in the middle and a concentrated transverse load at the end. Because of these concentrated loads, two cuts are necessary for analysis.

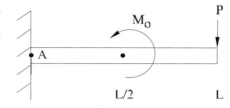

whereas a free-body diagram for the segment from $L/2 < x \leq L$ reveals that

$$\sum F_x = 0 \rightarrow f = 0,$$

$$\sum F_y = 0 \rightarrow P - V(x) = 0,$$

$$V(x) = P \quad \text{for} \quad \frac{L}{2} < x \leq L,$$

$$\sum M_z)_A = 0 \rightarrow M_z + M_o - Vx - M_o + PL = 0,$$

$$M_z(x) = P(x - L) \quad \text{for} \quad \frac{L}{2} < x \leq L,$$

Now, to find the deflections, we appeal to the moment-curvature relation

$$EI_{zz} \frac{d^2 v}{dx^2} = M_z(x),$$

but because of the discontinuity at $x = L/2$, the moment-curvature relation should be considered via two equations. Hence,

(1) $EI_{zz} \dfrac{d^2 v_1}{dx^2} = P(x - L) + M_o, \quad 0 \leq x \leq \dfrac{L}{2},$

(2) $EI_{zz} \dfrac{d^2 v_2}{dx^2} = P(x - L), \quad \dfrac{L}{2} \leq x \leq L.$

For (1), we integrate two times and denote the associated deflection curve as v_1, namely

$$EI_{zz} \int \frac{d}{dx}\left(\frac{dv_1}{dx}\right) dx = \int [P(x - L) + M_o] \, dx,$$

$$EI_{zz} \frac{dv_1(x)}{dx} = P\left(\frac{1}{2}x^2 - Lx\right) + M_o x + c_1,$$

$$EI_{zz} \int \frac{d}{dx}[v_1(x)] \, dx = \int \left[P\left(\frac{1}{2}x^2 - Lx\right) + M_o x + c_1\right] dx,$$

$$EI_{zz} v_1(x) = P\left(\frac{1}{6}x^3 - \frac{L}{2}x^2\right) + \frac{M_o}{2}x^2 + c_1 x + c_2.$$

For (2), we likewise integrate two times and denote the associated deflection curve as v_2:

$$EI_{zz} \int \frac{d}{dx}\left(\frac{dv_2}{dx}\right) dx = \int P(x-L)\, dx,$$

$$EI_{zz} \frac{dv_2(x)}{dx} = P\left(\frac{1}{2}x^2 - Lx\right) + c_3,$$

$$EI_{zz} \int \frac{d}{dx}[v_2(x)]\, dx = \int \left[P\left(\frac{1}{2}x^2 - Lx\right) + c_3\right] dx,$$

$$EI_{zz} v_2(x) = P\left(\frac{1}{6}x^3 - \frac{L}{2}x^2\right) + c_3 x + c_4.$$

Applying the boundary conditions for a fixed end,

$$\frac{dv_1}{dx}(x=0) = 0 \rightarrow c_1 = 0,$$

$$v_1(x=0) = 0 \rightarrow c_2 = 0,$$

we thus have

$$v_1(x) = \frac{1}{EI_{zz}}\left[P\left(\frac{1}{6}x^3 - \frac{L}{2}x^2\right) + \frac{M_o}{2}x^2\right] \quad \left(\text{for } 0 \le x \le \frac{L}{2}\right).$$

Two more boundary conditions are needed to find c_3 and c_4. Unfortunately, the boundary conditions for a free end do not provide anything useful. Hence, let us look for other conditions, like continuity of slope and deflection at $x = L/2$; that is, the two solutions should match at $x = L/2$. For example,

$$\frac{dv_1(L/2)}{dx} = \frac{dv_2(L/2)}{dx} \rightarrow$$

$$\frac{1}{EI_{zz}}\left\{P\left[\frac{(L/2)^2}{2} - L\left(\frac{L}{2}\right)\right] + M_o\left(\frac{L}{2}\right)\right\} = \frac{1}{EI_{zz}}\left[P\left(\frac{(L/2)^2}{2} - L\left(\frac{L}{2}\right)\right) + c_3\right].$$

Simplifying, we have

$$c_3 = M_o\left(\frac{L}{2}\right).$$

Similarly, at $x = L/2$, $v_1(L/2) = v_2(L/2)$, and therefore

$$\frac{1}{EI_{zz}}\left[P\left(\frac{(L/2)^3}{6} - \frac{L(L/2)^2}{2}\right) + \frac{M_o}{2}\left(\frac{L}{2}\right)^2\right]$$

$$= \frac{1}{EI_{zz}}\left[P\left(\frac{(L/2)^3}{6} - \frac{L(L/2)^2}{2}\right) + M_o\left(\frac{L}{2}\right)^2 + c_4\right],$$

or

$$\frac{M_o L^2}{8} = \frac{M_o L^2}{4} + c_4 \rightarrow c_4 = M_o L^2 \left(\frac{1}{8} - \frac{1}{4}\right) = -\frac{M_o L^2}{8}.$$

Therefore,

$$v_2(x) = \frac{1}{EI_{zz}}\left[P\left(\frac{x^3}{6} - \frac{Lx^2}{2}\right) + \frac{M_o L}{2}x - \frac{M_o L^2}{8}\right] \quad \left(\text{for } \frac{L}{2} \le x \le L\right).$$

Finally, we are interested in $v(x = L/2) = \delta_c$ and $v(x = L) = \delta_b$. Using the second solution, which is good for $L/2 \le x \le L$,

$$\delta_c = \frac{1}{EI_{zz}}\left[P\left(\frac{(L/2)^3}{6} - \frac{L(L/2)^2}{2}\right) + \frac{M_o L}{2}\left(\frac{L}{2}\right) - \frac{M_o L^2}{8}\right] = -\frac{5PL^3}{48EI_{zz}} + \frac{M_o L^2}{8EI_{zz}}$$

and

$$\delta_b = \frac{1}{EI_{zz}}\left[P\left(\frac{L^3}{6} - \frac{L^3}{2}\right) + \frac{M_o L^2}{2} - \frac{M_o L^2}{8}\right] = -\frac{PL^3}{3EI_{zz}} + \frac{3M_o L^2}{8EI_{zz}}.$$

Note: If $M_o = 0$, then the end deflection δ_b is the same as that calculated in Example 5.9 as expected.

Finally, as a check, note that we can also use the solution for $v_1(x)$ to find δ_c:

$$\delta_c = \frac{1}{EI_{zz}}\left\{P\left[\frac{1}{6}\left(\frac{L}{2}\right)^3 - \frac{L}{2}\left(\frac{L}{2}\right)^2\right] + \frac{M_o}{2}\left(\frac{L}{2}\right)^2\right\} = -\frac{5PL^3}{48EI_{zz}} + \frac{M_o L^2}{8EI_{zz}}$$

which matches that from $v_2(x)$, as it should.

5.4 Transducer Design: The AFM

5.4.1 Introduction

Recall from Chapter 1 that biomechanics emerged as a distinct field of study in the mid-1960s due, in large part, to parallel advances in both theory (e.g., continuum mechanics and numerical methods) and technology (e.g., computers). Indeed, scientific advances often result from the development of either a new *enabling technology* or a clever application of existing technology in a new way (e.g., X-ray crystallography aided in the discovery of the basic structure of DNA).

The history of biology reveals, for example, the important role of microscopy in our continuing understanding of the structure and behavior of living things. It was via a primitive two-lens light microscope that Robert Hooke (1635–1703) first observed remnant walls in cork, which led him to introduce the term cell, a word coming from the Latin meaning "little room." Likewise, it was through the use of a light microscope that Malpighi (1628–1694) first observed capillaries in lung tissue, which provided evidence for Harvey's (1578–1657) bold idea of "porosities in the flesh" that allowed blood to flow from arteries to veins. Using the light microscope, Schleiden (1804–1881) and Schwann (1810–1882) suggested that cells are the fundamental unit of life. Indeed, throughout the history of biology, one finds the important role of microscopy (e.g., see Harris, 1999; Lodish et al., 2000).

Although by the word "microscopy" we typically think of an optical instrument that increases, via a series of lenses, the apparent size of an object, there are now a host of technologies available: the scanning electron microscope, transmission electron microscope, confocal microscope, and two-photon microscope to name a few. The advantage of having multiple technologies is that one can exploit their particular advantages as needed. For example, the light microscope (LM) can resolve only on the order of $0.2\,\mu m$ or $(200\,nm)$, but the transmission electron microscope (TEM) has a resolution of $0.1\,nm$. The latter allows one to probe subcellular components as needed in cell mechanics. Whereas the LM and TEM provide information within cross sections, the scanning electron microscope (SEM) provides information on the 3-D surface structure to a resolution of $10\,nm$. For more information on these and related microscopic techniques, see Lodish et al. (2000). Here, however, let us consider a recent technology having a particular utility in biomechanics.

5.4.2 The Atomic Force Microscope

First reported in 1986, the atomic force microscope (AFM) has become a widely used tool in the study of protein and cell structure and properties (Binnig et al., 1986; Radmacher et al., 1992). Briefly, the AFM is similar in concept to a profilometer, a device that measures surface contours on hard materials via a moving stylus. As shown in Figure 5.16, the AFM consists of three primary components: a flexible cantilever beam with a rigid end tip that can be dragged across the surface of a soft sample or used to indent the sample; an optical lever, consisting of a precision laser and photodetector that can measure changes in the angle of the laser light that are associated with the deflection of the cantilever; and a precision piezoelectric x-y-z stage that can either move the cantilever beam or move the sample relative to a fixed beam. The cantilever beams are very small, typically 100–$400\,\mu m$ in length with a tip having a 10–50-nm radius of curvature; they are made using silicon-based nanofabrication techniques. The structural stiffness of the beam is usually quoted as a "spring constant," often on the order of 0.004–$1.85\,N/m$. It is because of this small stiffness that the AFM can resolve forces exerted by atoms, as, for example, van der Waals forces, Coulomb interactions, and hydration forces. The AFM is typically used in one of a few different modes. In the constant-force mode, the cantilever tip is dragged

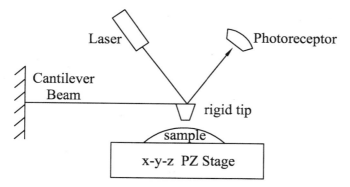

FIGURE 5.16 Schema of the basic components of an atomic force microscope (AFM). The deflection of the cantilevered beam (i.e., probe) is measured with an optical lever consisting of a laser source and photoreceptor (cf. Fig. 4.21). The sample is placed on a piezoelectric (PZ) stage that can move in x-y-z. In some systems, the PZ stage moves the probe rather than the sample, but the net effect is the same.

across the surface of a sample while the x-y-z piezoelectric stage moves the sample (or cantilever) so as to keep the tip-to-sample contact force constant. By tracking the x-y-z changes in position, one can construct a topological map of the surface of the sample. In this mode, one can resolve positions on the order of nanometers (nm). In the indentation mode, the tip can be used to indent the surface of the sample, with measured indentation force–depth data providing information on the local mechanical properties of the sample. In this case, indentation depths are usually on the order of 50–500 nm, often in cells that are less than $2\,\mu$m thick. Costa and Yin (1999) showed, therefore, that the associated sample strains are not small and thus one often should not use a linearized analysis (although most do) to infer the mechanical properties of cells using the AFM. Indeed, the associated boundary value problem is very complex and is not discussed here. Rather, we simply note that because the AFM is based on a cantilevered beam subjected to bending, we can examine the design of the device using the methods found in Sections 5.2 and 5.3.

5.4.3 Illustrative Example

Let us assume that an AFM device is constructed of a cantilevered LEHI beam of length L, with Young's modulus E, and a second moment of area $I_{zz} = I$. Moreover, let us assume that the end deflection $\delta \equiv v$ at $x = L$ is inferred from a measure of the end slope $\phi = dv/dx$ at $x = L$, as determined by the laser. If the end load P is directed upward and is transverse to the beam, we note that (cf. Example 5.9)

$$EI_{zz}\frac{d^2v}{dx^2}=M_z(x)=P(L-x),$$

from which upon two integrations and evaluation of boundary conditions ($v(x=0)=0$ and $dv(x=0)/dx=0$), we have

$$v(x)=\frac{Px^2}{6EI}(3L-x)\rightarrow v(x=L)\equiv\delta=\frac{PL^3}{3EI},$$

or

$$P=\left(\frac{3EI}{L^3}\right)\delta\rightarrow P=k\delta,$$

where k is an effective stiffness for the device having units of force per length; given its analogy with a spring wherein $f=kx$, the k in $P=k\delta$ is called the AFM spring constant. Question: If $L=400\,\mu m$ and the beam is made of silicon ($E\sim166\,GPa$), what value of I would yield a typical value of $k=1.0\,N/m$. Moreover, if the beam is rectangular in cross section with a width $b=5h$, where h is the height, and $I=bh^3/12$, what is the required thickness of the cantilever? Clearly, these and similar questions can be answered (do it) using ideas discussed herein.

In summary, the AFM has become a widely used device to study both the geometry and properties of living cells (e.g., see Figs. 5.17 and 5.18).

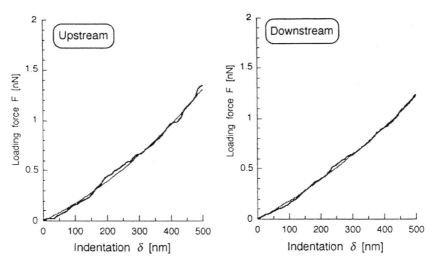

FIGURE 5.17 Measurement of the mechanical response of an isolated endothelial cell to an AFM indentation force. Shown are the force–indentation data, which reveal a nonlinear character and that the upstream portion of the cell tends to be slightly stiffer than the downstream portion when subjected to a flow-induced shear stress. [From Sato et al. (2000) J Biomech 33: 127–135, with permission from Elsevier.]

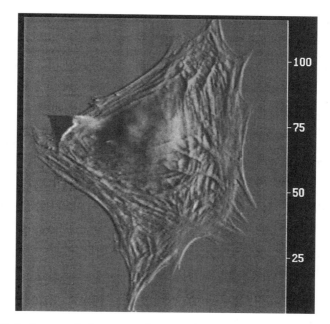

100

75

50

25

FIGURE 5.18 In the so-called constant-force mode, the AFM can measure the surface topography. Shown here is a single cell, with the region of the nucleus very clear. Such cells tend to adopt very different shapes in vivo wherein they are embedded in a 3-D plexus of extracellular matrix material and have extensive cell-to-cell junctions. There is much that we can learn from tests on isolated cells, and similarly from tests on cell cultures, yet we must remember that it is because of their extreme sensitivity to changes in applied loads (which we seek to measure) that their response in an artificial environment will be different from that in vivo. (Courtesy of Dr. G. Meininger, Texas A&M Health Science Center.)

Although some inappropriately interpret cell properties using a linearized analysis, we see that our linearized beam theory can again be used in the design of the device itself. We must always remember, therefore, under what conditions derived relations apply.

5.5 Principle of Superposition

Simply put, this principle asserts that, under certain conditions, one may add the solutions of multiple "simpler" problems to obtain the solution of a more complex problem. Superposition is particularly useful, therefore, when a complex problem can be analyzed in terms of simpler solutions which are well known, such as, the stresses due to extension/compression of an axial rod and the inflation of a thin-walled cylindrical tube (cf.

Example 3.3). Indeed, we have already used this principle in many different ways, including the use of the flexure formula for problems involving shear due to transverse loads. Here, therefore, we wish to emphasize further that it is very important to know when this principle applies and when it does not. Whereas universal solutions can be superimposed, for they are valid for all materials and levels of strain, it is also important to recognize that solutions to linear problems can also be superimposed. The latter should be familiar to those who have had a course in ordinary differential equations wherein we often exploit superposition; the so-called *homogenous* and *particular* solutions can be added when the differential equation is linear.

In this section, therefore, let us explore the utility of the principle of superposition in beam problems wherein the governing differential equations for the deflection are linear [cf. Eqs. (5.43) and (5.46)]; by linear, of course, we mean linear in the deflection $v(x)$, not that the right-hand side of the equation is linear in x. In particular, we shall see that this approach is very useful in two different classes of problem. Let us now illustrate this utility via two examples.

Example 5.11 Find the deflection curve $v(x)$ for a cantilevered beam subjected to a linearly increasing distributed load $q(x)$ and an applied moment M_o at the end. Assume that the beam exhibits a LEHI behavior, is of length L, is initially straight, and has a constant rectangular cross section.

Solution: Let us divide this "complex" problem into two simpler problems: a cantilever subjected to a uniformly increasing load $q(x)$ and an identical cantilever subjected to an end moment M_o. For "beam 1," our governing differential equation is

$$EI_{zz}\frac{d^4v_1}{dx^4} = -q(x) = -q_o\left(\frac{x}{L}\right),$$

where q_o is the value of $q(x)$ at the end. Integrating this equation four times yields

$$EI_{zz}\frac{d^3v_1}{dx^3} = -\frac{q_o}{L}\left(\frac{x^2}{2}\right) + c_1,$$

$$EI_{zz}\frac{d^2v_1}{dx^2} = -\frac{q_o}{L}\left(\frac{x^3}{6}\right) + c_1 x + c_2,$$

$$EI_{zz}\frac{dv_1}{dx} = -\frac{q_o}{L}\left(\frac{x^4}{24}\right) + c_1\left(\frac{x^2}{2}\right) + c_2 x + c_3,$$

$$EI_{zz}v_1(x) = -\frac{q_o}{L}\left(\frac{x^5}{120}\right) + c_1\left(\frac{x^3}{6}\right) + c_2\left(\frac{x^2}{2}\right) + c_3 x + c_4$$

for which we need four boundary conditions:

(a) $v_1(x = 0) = 0,$

(b) $\dfrac{dv_1}{dx}(x = 0) = 0,$

(c) $M_z(x = L) = 0 = EI_{zz}\dfrac{d^2v_1(x = L)}{dx^2},$

(d) $V(x = L) = 0 = EI_{zz}\dfrac{d^3v_1(x = L)}{dx^3}.$

Hence, from (a), we have $c_4 = 0$, and from (b), we have $c_3 = 0$. Similarly, from (d), we have $c_1 = q_oL/2$, and thus from (c), we have $c_2 = -q_oL^2/3$. Our first solution is

$$EI_{zz}v_1(x) = -\frac{q_o}{L}\left(\frac{x^5}{120}\right) + \frac{q_oL}{2}\left(\frac{x^3}{6}\right) - \frac{q_oL^2}{3}\left(\frac{x^2}{2}\right),$$

or

$$v_1(x) = \frac{1}{EI_{zz}}\left(\frac{-q_o}{120L}\right)(x^5 - 10L^2x^3 + 20L^3x^2).$$

Next, for "beam 2," we have

$$EI_{zz}\frac{d^2v_2}{dx^2} = M_o,$$

which can be integrated twice to yield

$$EI_{zz}\frac{dv_2}{dx} = M_ox + c_5,$$

$$EI_{zz}v_2(x) = M_o\left(\frac{x^2}{2}\right) + c_5x + c_6,$$

for which we need but two boundary conditions:

(e) $v_2(x = 0) = 0,$

(f) $\dfrac{dv_2(x = 0)}{dx} = 0.$

Hence, $c_5 = 0$ and $c_6 = 0$ and

$$v_2(x) = \frac{1}{2EI_{zz}}(M_ox^2).$$

The solution for our original problem is thus

$$v(x) = v_1(x) + v_2(x)$$

by superposition.

Clearly, we could obtain solutions to even more complicated problems by simply adding together the solutions of multiple (appropriate) simpler problems. Here, however, let us consider the second primary utility of the principle of superposition in problems of beam bending. Recall that a statically indeterminate problem is one that cannot be solved via statics alone. In traditional problems of beam bending, we recall further that we have but three general equilibrium equations ($\Sigma F_x = 0$, $\Sigma F_y = 0$, $\Sigma M_z = 0$) to find the reactions. Hence, in cases in which there are four or more reactions (i.e., a statically indeterminate problem), we must seek additional equations to solve the problem. Let us illustrate how the principle of superposition can be useful in this regard.

Example 5.12 Find the reactions for the beam in Figure 5.19 assuming that L, E, I_{zz}, and q_o are all known.

Solution: First, note that equilibrium of the whole requires that

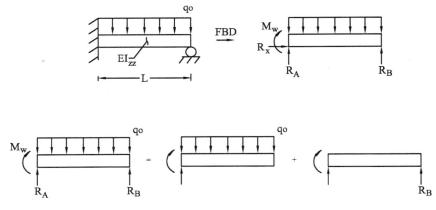

FIGURE 5.19 Statically indeterminate beam, cantilevered on one end and supported by a roller on the other. The three equations of static equilibrium are thus insufficient to determine the four reactions: M_w, R_A, R_B, and R_x at A; that is, whereas R_x can be shown to be zero via axial force balance, the remaining equations (vertical force balance and moment balance) are not sufficient to find the remaining three unknowns. Also shown are free-body diagrams of two convenient subproblems: a cantilevered beam subjected to a uniformly distributed load and a cantilevered beam subjected to a transverse end load. In the latter case, we can treat the reaction R_B as we would an applied load and thus solve the problem as usual.

$$\sum F_x = 0 \rightarrow R_x = 0,$$

$$\sum F_y = 0 \rightarrow R_A + R_B - \int_0^L q_o \, dx = 0,$$

$$\sum M_z)_A = 0 \rightarrow R_B L - \int_0^L (q_o x) \, dx - M_w = 0,$$

which yields three equations in terms of four unknowns (R_x, R_A, R_B, M_w). To generate a fourth equation, let us divide the problem into two problems (Fig. 5.19): a cantilever subjected to a uniformly distributed load q_o and a cantilever subjected to an end load R_B (whose value is as yet unknown). Clearly, we know the solutions for the deflection curves for each of these "simpler" problems. From an analysis similar to that in the previous example, show that

$$EI_{zz} v_1(x) = -q_o \left(\frac{x^4}{24} \right) + q_o L \left(\frac{x^3}{6} \right) - \frac{q_o L^2}{2} \left(\frac{x^2}{2} \right),$$

whereas from Example 5.9 (with $P = -R_B$), we have

$$EI_{zz} v_2(x) = -R_B \left(\frac{x^3}{6} \right) + R_B L \left(\frac{x^2}{2} \right).$$

Note: The direction of R_B is opposite the previously considered end load P, but otherwise the present problem is no different than that considered earlier. Hence, the solution to our original problem is

$$v(x) = v_1(x) + v_2(x),$$

subject to the constraint that

$$v_1(x = L) + v_2(x = L) = 0$$

because the roller at $x = L$ does not allow a deflection [i.e., $v(x = L) = 0$ is a boundary condition for the full problem]. This constraint provides an additional equation in terms of one of the original four unknowns; thus, we have succeeded in identifying four equations (three from equilibrium and one from a kinematic constraint condition) for our four unknowns and the problem can be solved. In particular, from the constraint condition, we find that

$$0 = -q_o \left(\frac{L^4}{24} \right) + q_o L \left(\frac{L^3}{6} \right) - \frac{q_o L^2}{2} \left(\frac{L^2}{2} \right) + (-R_B) \left(\frac{L^3}{6} \right) + R_B L \left(\frac{L^2}{2} \right),$$

$$0 = q_o L^4 \left(-\frac{1}{24} + \frac{1}{6} - \frac{1}{4} \right) + R_B L^3 \left(-\frac{1}{6} + \frac{1}{2} \right),$$

or

$$R_B = \frac{3}{8} q_o L,$$

hence finding the reaction at B in terms of the known values of q_o and L. Returning to the three equilibrium equations, we can now find R_A and M_w, which, in turn, will allow a full stress analysis. This is left for the reader to complete.

In closing, we emphasize yet again that mechanics is not a subject consisting of solutions to individual problems; rather, it is a subject in which a common method is used to solve diverse problems. Note, therefore, that we have used kinematic constraint conditions earlier to render a problem well posed: in Section 4.1.3, we used the condition that the end deflection δ was the same for the bone and the metal prosthesis in an axial load problem, and in Section 4.4.2, we used the condition that an angle of twist Θ was likewise the same in a bone–prosthesis torsion problem. *Kinematic constraints*, in the present case matching the deflections from two solutions at a single point in a beam, are thus very useful to impose in many problems and should be considered in problems wherein statics alone does not yield a sufficient number of equations.

Observation 5.2 We have noted that materials and structures can fail via a variety of mechanisms. They can deform excessively and thus cease to fulfill the intended function; they may yield and thus experience a permanent set which prevents them from returning to an original shape or location when unloaded; or they can fracture (i.e., rupture) and thus fail catastrophically. In each of these cases, failure may occur the first time that the applied loads exceed safe values (e.g., the yield stress). Another type of failure that is potentially problematic in many biomechanical problems is *fatigue failure*. In material science, the term "fatigue" denotes a loss in strength of a material due to repeated loading. Fatigue often occurs in three stages: the initiation of small cracks, the propagation of these cracks, and, finally, fracture due to the development of large cracks. A common method to test a material's resistance to fatigue is the "rotating cantilever test." In this test, a cylindrical specimen is loaded, via a bearing, by a transverse load at its end while the specimen is rotated many (sometimes millions) times. Because the specimen experiences tension on the top and compression on the bottom, material away from the neutral axis experiences a sinusoidal cycle from maximum to minimum tensile and compressive stresses. Tests are performed to failure, with the number of cycles to failure noted. Similar tests at multiple levels of applied load (i.e., maximum stress) reveal differences in the number of cycles to failure at different stresses. When the number of cycles to failure (abscissa) is plotted against the stress during the test (ordinate), one obtains a so-called S–N (or stress–number) curve. As one might expect, the number of cycles to failure is greater for lower values of applied stress and, conversely, it is lower for higher values of stress. Given that prosthetic hips and knees must survive millions of cycles due to daily walking or running and, likewise, artificial heart valves must survive over 30 million cycles per year, fatigue failure is an important concern in the biomechanical design of prosthetic devices. Question: Why is fatigue failure less of an issue for biological tissues? Answer: Tissues are continually replaced via a balanced synthesis and degradation of material; hence, the "same" material does not experience the thousands to millions of cycles needed to cause fatigue failure. Of course, repeated surgical replacement of prosthetic devices to renew the material is not a viable option for the biomedical engineer; thus, there is a need to decrease the likelihood of fatigue failure.

Let us note a few additional terms: The *fatigue life* tells us how long a particular component is expected to survive at a particular stress under normal conditions and the *fatigue strength* is the maximum stress for which failure will not occur for a prescribed number of cycles (e.g., 300 million). Fatigue testing is obviously a very important and yet potentially time-consuming activity. For this reason, one often seeks to perform accelerated tests whereby the requisite number of cycles can be achieved in much less time than would be required at the physiological rate. For a heart valve, for example, a 10-year equivalent fatigue test can be performed in 1 year if the tests are performed at 10 Hz rather than the physiological ~1 Hz. Yet, 1 year is still a long time to wait for experimental results and one might be tempted to perform the test at 100 Hz and thus obtain results in ~5 weeks instead of 1 year. One must ask, however, whether the behavior of the material of interest is sensitive to the rate of deformation, because this could adversely affect the results. We shall see in Chapter 11, for example, that strain-rate sensitivity is one of the characteristics of a viscoelastic behavior.

For many polymers, one can alternatively use a concept of time–temperature equivalency (Ferry, 1980), which states that similar behaviors occur much faster at higher temperature. Thus, by performing tests at temperatures above service conditions (e.g., at 70°C rather than 37°C), one can collect data over much shorter periods. Temperature can have very different effects on other materials, however, including tissue; thus, one must be very careful when employing this equivalency for experimental expediency. Fatigue testing is nevertheless often time-consuming. Because of its importance, including Food and Drug Administration (FDA) specifications in many cases, the biomechanical engineer must investigate this deeply. We refer the interested reader to books on material science.

5.6 Column Buckling

Recall from Section 5.1 that a column is any structural member having one dimensions greater than the third and subjected to a compressive axial load. In some cases, the column may fail due to an excessive load simply by frac-

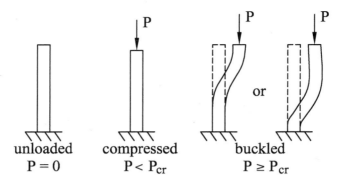

FIGURE 5.20 A cantilevered beam is subjected to a compressive end load P. Initially the beam-column will simply compress and the stress $\sigma_{xx} = P/A$, as in Chapter 3. After a critical value P_{cr} is reached, the beam column will buckle (i.e., bend abruptly) and the analysis of stress and strain becomes much more complex. Hence, rather than computing these complex states of stress or strain, let us focus simply on that value of P_{cr} that induces buckling.

ture, plastic deformation, or excessive compression. In other cases, however, the primary concern may be the possibility that the column may become unstable and buckle. A simple example of such buckling can be appreciated by taking a plastic ruler and compressing it along its long axis— the sudden bending out, or *buckling*, occurs when a critical value of the compressive load P_{cr} is achieved (Fig. 5.20). Let us now consider the general concept of stability as well as the specific example of column buckling.

5.6.1 Concept of Stability

Consider the two structural members in Figure 5.21. In each case, statics tells us that the reaction force at the pin is $R_y = W$, the weight of the member. Indeed, in each case, the pin is exerting an upward directed force and we might say that the two problems are statically equivalent. From the perspective of stability, however, these two problems are very different. If we subject member B to a small lateral *disturbing* load, or perturbation, we expect the member to move initially in the direction of the load, but then, like a pendulum, to swing back and forth until it regains its original position (assuming a frictionless pin but resistance to motion due to the air). Conversely, we expect member A to respond very differently to the same lateral disturbing force—we expect it to swing down and eventually gain the position of member B. Note: This experiment is accomplished easily by holding your pen loosely between two fingers in each of the original configurations and subjecting it to a small lateral disturbing force. Although both members A and B are initially in equilibrium, we say that A is unsta-

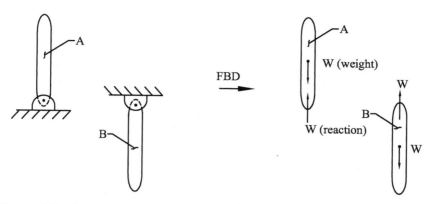

FIGURE 5.21 Illustration of the concept of stability. Although structural members A and B are both in equilibrium, which is to say that the reaction force $R_y = W$ for both, structure A is unstable—a small transverse (i.e., disturbing) load will cause it to swing down and assume a position similar to that of B. Structure B, on the other hand, will simply swing back and forth if disturbed by a small transverse load, until it regains its original equilibrium position (assuming air friction or friction in the pin, otherwise with no energy dissipation the member could swing back and forth about the original equilibrium indefinitely).

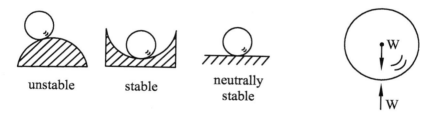

FIGURE 5.22 Another simple illustration of the concept of stability. Although each ball is initially in equilibrium (if centered) and thus has the same initial free-body diagram, a small lateral force will cause the ball on the hill to roll off, whereas a small lateral force will cause the ball in the trough to simply roll back and forth until it comes to rest in its original position (again, assuming some friction in the system). These are called unstable and (asymptotically) stable, respectively. The ball on the flat plate may move only slightly when disturbed; thus, this is called neutrally stable—it need not experience an abrupt change in equilibrium position, but it also need not regain its original position.

ble and B is stable. *Mechanical stability*, then, is the ability to resist a small disturbing force, which is a very important structural characteristic.

Another good illustration of the concept of stability is seen in Figure 5.22. In this case, imagine three otherwise identical balls on low friction surfaces. Moreover, imagine the response of each initially centered ball if it is subjected to a small lateral disturbing force. In case A, we easily imagine the ball "rolling off the hill," which is to say, moving in such a way that it cannot regain its original position. We would say that this ball is unstable because the disturbing force caused the ball to find another equilibrium position. Conversely, in case B, we can easily imagine that, provided the disturbing force is not too large, the ball will first move in the direction of perturbation, but then roll back and forth until it regains its original position. This ball would be said to be (asymptotically) stable. Thus, in cases A and B in Figures 5.21 and 5.22, we see that mechanical stability is an ability to resist a small disturbing force, which is to say, an ability to regain the original position or configuration following the disturbance. Case C in Figure 5.22 illustrates one final possibility. In this case, the ball will not regain its original position, but it may not move far from that position. Such cases are called neutrally stable; they are, in fact, a cause for concern, for they may easily degenerate into an instability given slight imperfections (e.g., if the flat surface is at a slight incline). We will consider the static stability of an elastomeric balloon in Chapter 6 and the dynamic stability of an aneurysm in Chapter 11. We should be very mindful, therefore, that stability is an important consideration in biomedical design, analysis, and experimentation with regard to both biomaterials and native tissues. Let us now consider the generic case of column buckling, the classical introduction to stability in engineering mechanics and another subject in mechanics that was touched by the genius of Euler.

5.6.2 Buckling of a Cantilevered Column

Consider the initially straight but buckled column in Figure 5.23, which is assumed to exhibit a linear, elastic, homogenous, and isotropic (LEHI)

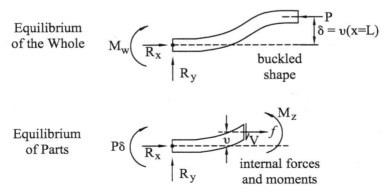

FIGURE 5.23 Detailed free-body diagrams of the whole and part of a cantilevered beam column as shown in Figure 5.20.

response. If the axial load P is applied through the centroid, we expect one of two possibilities. First, consider the case wherein the column just compresses as an axial rod. In this case, $\sigma_{xx} = -P/A$ (compressive) and the displacement is given by

$$u_x(x) - u_x(0) = \int_0^x -\frac{P}{AE}dx = -\frac{P}{AE}x, \qquad (5.47)$$

assuming a constant cross section and homogenous constitution. These results are valid as long as $|P| < |P_{cr}|$, the so-called critical buckling load.

If we continued to load the column until $|P| = |P_{cr}|$, however, the situation is very different. As can be seen, the buckled beam appears "bent" and, consequently, as in the prior sections of this chapter, we must consider the bending. The main difference, however, is that it is a compressive load, not a transverse load or applied bending moment, that gives rise to the buckling of the beam. Equilibrium of the whole reveals that the reactions at the wall are (Fig. 5.23)

$$\sum F_x = 0 = R_x - P \rightarrow R_x = P,$$
$$\sum F_y = 0 \rightarrow R_y = 0,$$
$$\sum M_z)_A = 0 = P\delta - M_w \rightarrow M_w = P\delta, \qquad (5.48)$$

where $v(x = L) = \delta$. Equilibrium of the parts thus reveals (Fig. 5.23) that

$$\sum F_x = 0 = P + f \rightarrow f = -P,$$
$$\sum F_y = 0 \rightarrow V = 0,$$
$$\sum M_z)_A = 0 = \rightarrow M_z - fv - P\delta = 0, \qquad (5.49)$$

or

$$M_z = fv + P\delta = P\delta - Pv. \qquad (5.50)$$

Assuming that the moment-curvature relation [Eq. (5.43)] holds in this case of bending, we have

$$EI_{zz}\frac{d^2v}{dx^2} = P\delta - Pv, \qquad (5.51)$$

or

$$\frac{d^2v}{dx^2} + \frac{P}{EI_{zz}}v = \frac{P}{EI_{zz}}\delta. \qquad (5.52)$$

We recognize from our study of differential equations (reviewed in Appendix 8 of Chapter 8) that this is a second-order, linear, nonhomogenous differential equation with a constant coefficient (for each value of P). It will prove useful, therefore, to let this coefficient be denoted by $k^2 \equiv P/EI_{zz}$, thus yielding our final governing differential equation

$$\frac{d^2v}{dx^2} + k^2 v = k^2 \delta. \tag{5.53}$$

Because this is a linear equation, let us first seek its homogenous and then its particular (i.e., nonhomogenous) solutions whereby $v(x) = v_h(x) + v_p(x)$. First, for the homogenous equation, note that it can be written in operator form as

$$\frac{d^2 v_h}{dx^2} + k^2 v_h = 0 \leftrightarrow (D^2 + k^2) v_h = 0, \tag{5.54}$$

whereby we have a solution if $D = \pm ki$, where $i = \sqrt{-1}$. We know that the solution of such equations can be assumed to be of the form

$$v_h(x) = e^{(a+bi)x} = e^{ax}(c_1 \cos bx + c_2 \sin bx). \tag{5.55}$$

Hence, for our problem, we have $a = 0$ and $b = k$; thus,

$$v_h(x) = c_1 \cos kx + c_2 \sin kx. \tag{5.56}$$

As an exercise, verify that this solution does in fact satisfy the homogenous differential equation; this is accomplished easily by taking the second derivative and substituting back into Eq. (5.54).

Next, for the particular solution, note that the right-hand side of Eq. (5.53) is constant and thus let

$$v_p(x) = A, \tag{5.57}$$

from which we see that this is a solution of the nonhomogenous equation provided that $A = \delta$. Thus, our full solution is

$$v(x) = c_1 \cos kx + c_2 \sin kx + \delta. \tag{5.58}$$

The boundary conditions are

$$v(x = 0) = 0 \rightarrow 0 = c_1 + \delta \rightarrow c_1 = -\delta.$$
$$\frac{dv}{dx}(x = 0) = 0 \rightarrow 0 = c_2 k \rightarrow c_2 = 0. \tag{5.59}$$

Hence,

$$v(x) = \delta(1 - \cos kx), \tag{5.60}$$

where $\delta \equiv v(x = L)$ provides the constraint condition that

$$\delta = \delta(1 - \cos kL), \tag{5.61}$$

which, in turn, requires that $\cos kL = 0$ for all k. The cosine function equals zero, of course, at $\pi/2, 3\pi/2, 5\pi/2, \ldots$; hence, we must have

$$kL = n\frac{\pi}{2}, \quad n = 1, 3, 5, \ldots. \tag{5.62}$$

Now, recalling that $k^2 = P/EI_{zz}$, this says that

$$\sqrt{\frac{P}{EI_{zz}}}L = \frac{n\pi}{2} \tag{5.63}$$

or that a value (magnitude) of the compressive axial load P for which we have buckling is

$$P = \frac{n^2\pi^2}{4L^2}EI_{zz}. \tag{5.64}$$

We are interested, of course, in the smallest buckling load, called P_{cr} or the *critical buckling load*, which is given by $n = 1$, and therefore

$$P_{cr} = \frac{\pi^2}{4L^2}EI_{zz} \tag{5.65}$$

for this case of a cantilevered column subjected to an axial end load P. Note that the critical buckling load is increased by an increased stiffness of the material E and increased second moment of area I_{zz}. Conversely, P_{cr} is reduced as the length of the column is increased. All of these effects are intuitive; for example, if we try to buckle a plastic ruler, we can make it more difficult to do so by simply supporting it in such a way that its effective length is reduced. Try it. Likewise, if we increase the stiffness (e.g., use a wooden ruler rather than a plastic one), it is harder to induce buckling, and so too if we increase the cross-sectional area. Indeed, note that the result for P_{cr} depends on I_{zz}. Actually, it is somewhat arbitrary how we define the y and z directions in the cross section, so note that a ruler tends always to buckle in one direction (i.e., in the direction of least thickness), the one associated with the smallest second moment of area.

Finally, a few words about our solution $v(x)$. It may be tempting to draw the buckled shape of the column using our solution for $v(x)$ and k and, indeed, some seek to explain the different buckling modes (shapes) via different values of n (i.e., different curves defined by sines and cosines). One knows, for example, that a buckled plastic ruler could assume various sinusoidal shapes depending on how strongly one pushes on the ends. To try to explain such buckled shapes based on our analysis is ill advised, however, because our solution was based on the moment-curvature relation, which, in turn, was based on the assumption of a small slope ($dv/dx \ll 1$). This assumption is not respected by the buckled shape in general. Hence, we can only use this formulation to find P_{cr}, which is *the load at which buckling is imminent but not realized*. This example serves to remind us again that it is essential to remember and respect all assumptions. To determine the buckled shape, we must first derive and then solve a nonlinear differential equation. This is beyond the present scope.

Finally, note that our governing differential equation (5.52) is not a general equation; it is valid only for a column with a free end. Other boundary conditions will thus modify both the general equation and the associated unknown coefficients in the solution. Each case is solved similarly, but they are different. Consider the following example.

Example 5.13 Find the critical buckling load for the fixed–pinned column in Figure 5.24.

Solution: First, consider a free-body diagram for the whole structure:

$$\sum F_x = 0 \rightarrow R_x = P,$$
$$\sum F_y = 0 \rightarrow R_y = -N,$$
$$\sum M_z)_o = 0 \rightarrow -M_w + NL = 0,$$

and, thus,

$$M_w = NL \quad \text{and} \quad R_y = -\frac{M_w}{L}.$$

Second, consider a free-body diagram for the part:

$$\sum F_x = 0 \rightarrow f = -P,$$
$$\sum F_y = 0 \rightarrow -\frac{M_w}{L} - V = 0,$$
$$V = -\frac{M_w}{L},$$
$$\sum M_z)_o = 0 \rightarrow -M_w + M_z - Vx - v(x)f = 0,$$
$$M_z = M_w - \frac{M_w}{L}x - Pv(x).$$

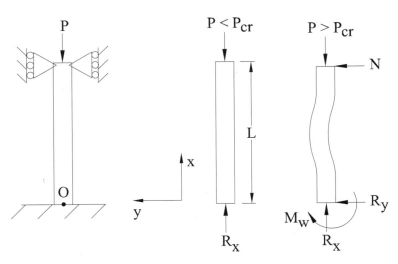

FIGURE 5.24 Solutions of beam-column problems are very sensitive to the boundary conditions. Shown here is a fixed-pinned column.

Thus, from the linearized moment-curvature relation [Eq. (5.43)], we have

$$EI_{zz}\frac{d^2v}{dx^2} = -Pv + M_w\left(1 - \frac{x}{L}\right).$$

Rearranging this relation into standard form, we have

$$\frac{d^2v}{dx^2} + \frac{P}{EI_{zz}}v = \frac{M_w}{EI_{zz}}\left(1 - \frac{x}{L}\right).$$

Consistent with the previous example, the homogeneous solution is

$$v_h(x) = c_1\cos kx + c_2\sin kx.$$

For a particular solution, given that the right-hand side is linear in x, assume

$$v_p(x) = c_3 + c_4 x,$$

whereby

$$\frac{dv_p}{dx} = c_4 \quad \text{and} \quad \frac{d^2v_p}{dx^2} = 0.$$

Hence, substituting into the governing differential equation for $v_p(x)$,

$$0 + \frac{P}{EI_{zz}}(c_3 + c_4 x) = \frac{M_w}{EI_{zz}}\left(1 - \frac{x}{L}\right) \rightarrow c_3 + c_4 x = \frac{M_w}{P} - \frac{M_w}{PL}x$$

and, consequently,

$$c_3 = \frac{M_w}{P} \quad \text{and} \quad c_4 = -\frac{M_w}{PL}.$$

The full solution then becomes $v(x) = v_h(x) + v_p(x)$, or

$$v(x) = c_1\cos kx + c_2\sin kx + \frac{M_w}{P} - \frac{M_w}{PL}x,$$

from which

$$\frac{dv(x)}{dv} = c_1(-k\sin kx) + c_2(k\cos kx) - \frac{M_w}{PL}.$$

Enforcing the boundary conditions at the fixed end, $v(x = 0) = 0$ and $dv(x = 0)/dx = 0$,

$$0 = c_1 + \frac{M_w}{P} \rightarrow c_1 = -\frac{M_w}{P},$$

$$0 = c_2 k - \frac{M_w}{PL} \rightarrow c_2 = \frac{M_w}{kPL}.$$

Enforcing the boundary conditions at the pinned end, $v(x = L) = 0$,

$$0 = -\frac{M_w}{P}\cos kL + \frac{M_w}{kPL}\sin kL + \frac{M_w}{P} - \frac{M_w}{PL}L,$$

or

$$\frac{1}{kL}\sin kL - \cos kL = 0.$$

Hence,

$$\frac{1}{kL} = \frac{\cos kL}{\sin kL} \rightarrow kL = \tan kL.$$

This is a transcendental (nonlinear) equation, which does not admit a direct solution. However, one can use an iterative numerical method to show that the smallest root is

$$kL \approx 4.4935 \text{ (radians)}$$

from which

$$P_{cr} \approx \frac{20.19EI_{zz}}{L^2} = \frac{2.05\pi^2 EI_{zz}}{L^2},$$

the latter of which permits an easier comparison to the previous result.

Appendix 5: Parallel Axis Theorem and Composite Sections

Recall from Appendix 4 of Chapter 4 that the second moment of area I_{zz} is given by

$$I_{zz} = \iint y^2 \, dy \, dz, \tag{A5.1}$$

where y and z are taken here to be the in-plane coordinates (i.e., cross sectional) and x is directed along the long axis of a beam. Moreover, because of our need to locate the origin of our $(o; x, y, z)$ coordinate system at the centroid [recall Eq. (5.20)], this I_{zz} must likewise be computed relative to the centroid. For simple geometries, such as rectangular or circular, such computations are straightforward, as seen in Appendix 4. In many beam-bending problems, however, the cross section of the beam is often complex, whether it is the cross section of a long bone or the cross section of a beam used as a transducer. For this reason, the so-called *parallel axis theorem* is very useful.

Consider the centroidal coordinate system (y,z) and general cross section shown in Figure 5.25. Clearly,

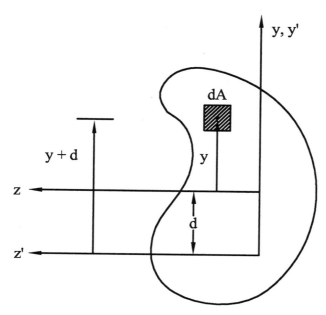

FIGURE 5.25 Coordinate systems and generic cross section for deriving the parallel axis theorem; (y, z) are centroidal coordinates, which are very useful in beam theory.

$$I_{zz} = \iint y^2 \, dy \, dz$$

is computed with respect to the centroidal coordinates (y, z), whereas

$$I'_{zz} = \iint (y+d)^2 \, dA = \iint y^2 \, dA + 2d \iint y \, dA + d^2 \iint dA \qquad (A5.2)$$

is computed with respect to another coordinate system (y', z') oriented parallel to our centroidal system. In addition to recognizing the second moment of area with respect to the centroidal system, $\iint y^2 dA$, we also recognize the first moment of area $\bar{y}A = \iint y dA$, also with respect to the centroidal system (y, z). The value of $\bar{y}$ relative to the centroidal system (i.e., the distance the centroid is from the centroid) is zero, however; thus, we have

$$I'_{zz} = I_{zz} + d^2 A, \qquad (A5.3)$$

which is known as the parallel axis theorem. It allows us to compute the second moment of area of a cross section of area A given its "centroidal second moment of area" and the distance d between the centroidal axis z and any parallel axis z' of interest. Clearly, a similar, more general result can be obtained if y and y' do not coincide. Regardless, we emphasize that the parallel axis theorem is very useful for determining the value of the second moment of area of a "composite" cross section relative to the overall centroid. To illustrate this, consider the I-beam cross section shown in Figure 5.26.

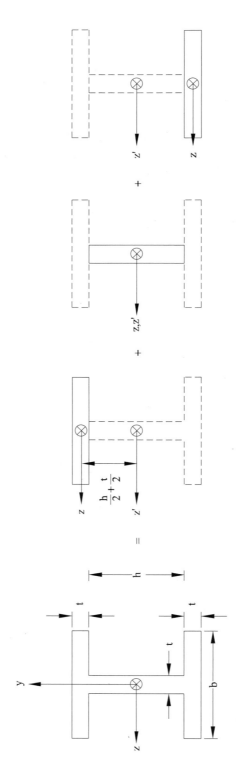

FIGURE 5.26 Illustration that, like first moments of area (cf. Appendix 3 of Chapter 3), second moments of area can be determined using a method of "composite" sections shown by solid lines. The parallel axis theorem is fundamental to such determinations.

To compute the *overall* centroidal second moment of area, we can use the parallel axis theorem three times to transform the easily computed *individual* centroidal values of rectangles ($bh^3/12$) to the overall centroid; that is, for the top, middle, and bottom parts respectively, we have

$$I_{zz} = \left[\frac{1}{12}bt^3 + \left(\frac{h}{2}+\frac{t}{2}\right)^2 bt\right]_{\text{top}} + \left(\frac{1}{12}th^3 + 0^2 bt\right)_{\text{middle}}$$

$$+ \left[\frac{1}{12}bt^3 + \left(-\frac{h}{2}-\frac{t}{2}\right)^2 bt\right]_{\text{bottom}}, \tag{A5.4}$$

which can be simplified algebraically if desired. The two key observations are that the distance squared term $d^2 A$ provides a positive contribution regardless of the location of the small part relative to the centroid and that $d = 0$ recovers our standard relation.

Hence, for a composite section

$$I'_{zz}(\text{whole}) = \sum (I_{zz} + d^2 A)_{\text{parts}}, \tag{A5.5}$$

where $(\ldots)_{\text{parts}}$ is computed relative to the centroidal coordinate system for each part. Note, too, that one can use this idea to compute the second moment of area of a hollow cross section. For example, the simple case in Figure 5.27 has the solution

$$I_{zz} = \left[\frac{1}{12}BH^3 + 0^2(BH)\right] - \left[\frac{1}{12}bh^3 + 0^2(bh)\right] = \frac{1}{12}(BH^3 - bh^3). \tag{A5.6}$$

Hence, as in the case of composite sections and centroids (Appendix 3 of Chapter 3), we can easily add or remove the scalar second moment of areas.

Finally, it should be noted that second moments of areas, like stress and strain, obey coordinate transformation relations. Hence, if we know I_{yy}, I_{zz}, and I_{yz} in two-dimensions, then values with respect to (y', z') can be computed as (Fig. 5.28)

$$I'_{yy} = I_{yy}\cos^2\alpha + 2I_{yz}\cos\alpha\sin\alpha + I_{zz}\sin^2\alpha,$$
$$I'_{zz} = I_{yy}\sin^2\alpha - 2I_{yz}\cos\alpha\sin\alpha + I_{zz}\cos^2\alpha,$$
$$I'_{yz} = (I_{zz} - I_{yy})\cos\alpha\sin\alpha + I_{yz}(\cos^2\alpha - \sin^2\alpha), \tag{A5.7}$$

although we will not prove these results here. Clearly, though, given the values of the second moments of area with respect to convenient centroidal axes, one can determine components for any other related coordinate system via the parallel axis theorem and/or transformation relations.

Example A5.1 Determine the second moment of area I_{zz} about the horizontal axis for the I-beam in Figure 5.29.

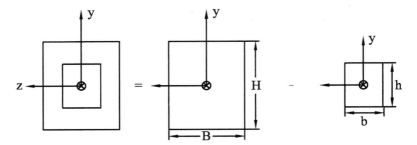

FIGURE 5.27 Another example of the method of composite sections to determine a second moment of area—this time for a cross section with a hole.

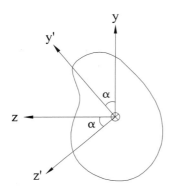

FIGURE 5.28 Similar to components of stress and strain, second moments of area relative to one coordinate system can be related easily to those of an associated coordinated system via a simple transformation.

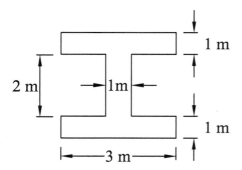

FIGURE 5.29 A dimensioned I-beam.

Solution: For this problem, the centroid of the entire area has been located for you. To obtain I_{zz} for the whole cross section, the parallel axis theorem is used; that is,

$$I'_{zz}(\text{whole}) = \sum (I_{zz} + d^2 A)_{\text{parts}},$$

where the cross section can be broken into three parts as in Figure 5.26 and the coordinate system located at the overall centroid. Because each part is rectangular, Eq. (A4.5) can be used to calculate $(I_{zz})_c$ for each piece of the cross section.

For part 1:

$$I'_{zz} = (I_{zz})_c + Ad^2 = \frac{1}{12}(\text{base})(\text{height})^3 + (\text{base})(\text{height})d^2,$$

$$I'_{zz} = \frac{1}{12}(3\,\text{m})(1\,\text{m})^3 + (3\,\text{m})(1\,\text{m})(1.5\,\text{m})^2 = 7\,\text{m}^4.$$

For part 2:

$$I'_{zz} = (I_{zz})_c + Ad^2 = \frac{1}{12}(\text{base})(\text{height})^3 + (\text{base})(\text{height})d^2,$$

$$I'_{zz} = \frac{1}{12}(1\,\text{m})(2\,\text{m})^3 + (1\,\text{m})(2\,\text{m})(0)^2 = \frac{2}{3}\,\text{m}^4.$$

For part 3:

$$I'_{zz} = (I_{zz})_c + Ad^2 = \frac{1}{12}(\text{base})(\text{height})^3 + (\text{base})(\text{height})d^2,$$

$$I'_{zz} = \frac{1}{12}(3\,\text{m})(1\,\text{m})^3 + (3\,\text{m})(1\,\text{m})(-1.5\,\text{m})^2 = 7\,\text{m}^4.$$

For the composite section, therefore:

$$I'_{zz} = 7\,\text{m}^4 + \frac{2}{3}\,\text{m}^4 + 7\,\text{m}^4 = 14.667\,\text{m}^4.$$

Example A5.2 Determine the second moment of area I_{yy} about the horizontal x axis for an area that is shaped like a "C". Let the overall width and height be 3.5 in. and 6.0 in., respectively. Let the "cut-out" be centered vertically but toward the right and 3 in. in width and 5 in. in height.

Solution: First, sketch the cross section. Second, find the centroid of the area. To do this, visualize breaking the cross section into parts. One way to visualize the cross section is by a sum of multiple parts that are joined together. Another way to visualize it is by subtracting the hollow interior from a solid cross section. Using the second method, we must find the centroid of the entire cross section. The best way to do this is to organize a chart of the parts in order to locate the centroid relative to (x, y).

Part	Area (A)	$\bar{x}$	$\bar{y}$	$A\bar{x}$	$A\bar{y}$
1	6 in. $\times$ 3.5 in.	1.75 in	3 in	36.75 in^3	63 in^3
2	$-$(3 in. $\times$ 5 in.)	2 in	3 in	$-$30 in^3	$-$45 in^3
Σ	6 in.^{2}			6.75 in^3	18 in^3

Thus,

$$\bar{y} = \frac{\sum A\bar{y}}{\sum A} = \frac{18\,\text{in.}^3}{6\,\text{in.}^2} = 3\,\text{in.}, \qquad \bar{x} = \frac{\sum A\bar{x}}{\sum A} = \frac{6.75\,\text{in.}^3}{6\,\text{in.}^2} = 1.12\,\text{in.}$$

To calculate $I_{xx} = \iint y^2 dA$ let us use the parallel axis theorem. This is most easily done by considering a solid rectangular cross section and subtracting the hollow interior from it. Once the overall centroid has been located, originate the coordinate system there. Because each of the cross sections is rectangular, the general formula, $(I_{xx})_c = (1/12)(\text{base})(\text{height})^3$, can be used. For the solid area:

$$I'_{xx} = (I_{xx})_c + Ad^2 = \frac{1}{12}(\text{base})(\text{height})^3 + (\text{base})(\text{height})d^2,$$

$$I'_{xx} = \frac{1}{12}(3.5\,\text{in.})(6\,\text{in.})^3 + (3.5\,\text{in.})(6\,\text{in.})(0)^2 = 63\,\text{in.}^4.$$

For the hollow interior:

$$I'_{xx} = (I_{xx})_c + Ad^2 = \frac{1}{12}(\text{base})(\text{height})^3 + (\text{base})(\text{height})d^2,$$

$$I'_{xx} = \frac{1}{12}(3\,\text{in.})(5\,\text{in.})^3 + (3\,\text{in.})(5\,\text{in.})(0)^2 = 31.25\,\text{in.}^4.$$

For the composite section, therefore,

$$I'_{xx} = 63\,\text{in.}^4 - 31.25\,\text{in.}^4 = 31.75\,\text{in.}^4$$

Now, various quantities dependent on I_{xx}, such as stress or the critical buckling load, can be calculated for this particular cross section.

Exercises

5.1   Find σ_{xx} and σ_{xy} for the following beam.

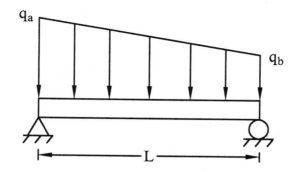

FIGURE 5.30

5.2   Find σ_{xx} and σ_{xy} for the following beam.

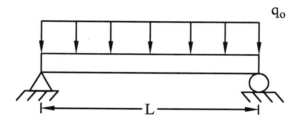

FIGURE 5.31

5.3   Find σ_{xx} and σ_{xy} for the following beam.

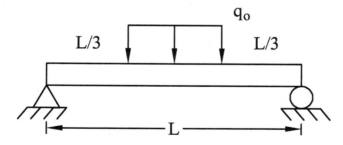

FIGURE 5.32

5.4   Find σ_{xx} and σ_{xy} for the following beam.

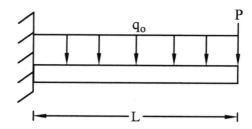

FIGURE 5.33

5.5 Find σ_{xx} and σ_{xy} for the following beam.

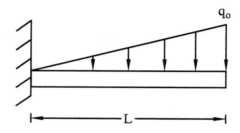

FIGURE 5.34

5.6 Find the maximum value of σ'_{xx} and σ'_{xy} for the beam in Exercise 5.1.
5.7 Find the maximum value of σ'_{xx} and σ'_{xy} for the beam in Exercise 5.2.
5.8 Find the maximum value of σ'_{xx} and σ'_{xy} for the beam in Exercise 5.3.
5.9 Show that $(\sigma_{xy})_{\text{ave}}$ in a beam having a rectangular cross section (area = bh) has as its largest value $1.5\,V/A$, which is at the centroid.
5.10 Find the deflection curve for the beam in Example 5.1.
5.11 Find the deflection curve for the beam in Example 5.2.
5.12 Find the deflection curve for a simply supported beam (a pin and roller at the two ends) with a constant distributed load $q(x) = -q_o$. Note whether the beam deflects up or down.
5.13 Find the deflection curve for the beam in Exercise 5.5.
5.14 Find the deflection curve for the beam in the previous example except with a distributed load of $q(x) = (q_o/L^2)x^2$.
5.15 You are to design a force transducer based on a cantilever beam subject to an end load. Assume the beam is rectangular in cross section and that redundant strain gauges are placed at $(x = L/2, y = \pm h/2)$. Find a formula for selecting the value of Young's modulus E if the maximum allowable measured strain ε_{xx} is ε_o (i.e., find E in terms of, possibly, ε_o, L, h, b, P, etc.). Note that Popov (1999) is a nice introduction to mechanical engineering applications of strength of materials such as this problem.

5.16 Radmacher et al. (1992) pointed out that if one "drags" the AFM probe across a surface, the tip of the probe experiences both a normal force and a tangential force. The latter will contribute to the bending. Given the probe shown below, find the end deflection $\delta = v(x = L)$.

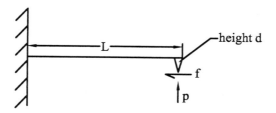

FIGURE 5.35

5.17 Using the principle of superposition, find the displacement vector $u(x)$ of the neutral axis for the beam shown below. Hint: Let $u(x) = v(x)\hat{j} + w(x)\hat{k}$.

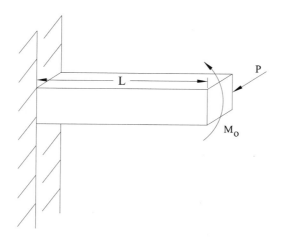

FIGURE 5.36

5.18 Use the principle of superposition to find the deflection curve $v(x)$ for the neutral axis for the beam in Exercise 5.4.

5.19 Use the principle of superposition to find the reactions for a beam that is fixed on both ends and subjected to a uniformly distributed load. Hint: Assume that there is no axial load and divide the problem into three cantilever beams: one with the distributed load, one with an end load R_B, and one with an end moment M_B. Use the kinematic constraint conditions that

$$v(x = L) = 0 = v_1(x = L) + v_2(x = L) + v_3(x = L)$$

and

$$\frac{dv}{dx}(x = L) = 0 = \frac{dv_1}{dx}(x = L) + \frac{dv_2}{dx}(x = L) + \frac{dv_3}{dx}(x = L),$$

from which we see that we have the requisite five equations (three equilibrium and two constraints) for the five unknowns (R_x, R_A, M_A, R_B, M_B).

5.20 Two potential experiments for determining the (effective) Young's modulus E for a bone sample are the so-called three-point and four-point bending tests, shown schematically here Figure 5.37. Assuming that three strain gauges (A, B, C) are applied equidistantly to the bottom surface of each beam sample and that their lengths are $L/50$ each, note that the desired value of the Young's modulus can be determined via

$$\sigma_{xx}(x, y) = E\varepsilon_{xx}(x, y) \rightarrow E = -\frac{M_z(x)y}{I_{zz}\varepsilon_{xx(\text{gauge})}},$$

where x and y must correspond to the placement of one or more of the strain gauges. Via a mechanical analysis, show why the preferred measurement (of the three sites shown) would be via gauge B in the four-point bending test. Hint: Recall that a gauge is of a finite, albeit small, length, whereas strain is defined at a point.

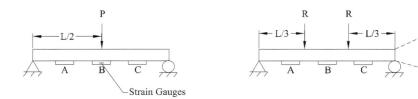

FIGURE 5.37

5.21 For the three-point and four-point bending tests shown in Exercise 5.20, find the value of R in terms of P (assume given) such that the maximum value of σ_{xx} is the same in each beam. Note the value of (x, y) at which σ_{xx} is maximum.

5.22 Assume that a LEHI gate is designed as a "dam." Find the deflection curve assuming that the bottom support can be modeled as a pin and the top support as a roller (pushing opposite the force of the fluid). Hint: First determine the uniform loading on the beam gate given the differential equation for fluid statics

$$\frac{dp}{dx} = \rho g,$$

where ρ is the density of the fluid and g is the gravitational constant. Note the boundary condition that $P = P_{atm}$ at $x = 0$.

5.23 Noting that the flexure formula $\sigma_{xx} = -M_z y/I_{zz}$ was determined via an approximate, linear theory, superposition of stresses holds. Hence, for a combined axial load and bending,

$$\sigma_{xx} = \frac{f}{A} - \frac{M_z y}{I_{zz}}.$$

In like fashion, note that "symmetrical" bending due to moments applied with respect to both the z and y axes will induce a superimposed stress (Boresi et al., 1993)

$$\sigma_{xx} = -\frac{M_y z}{I_{yy}} - \frac{M_z y}{I_{zz}},$$

the first contribution of which can be obtained directly by deriving the flexure formula due to a moment M_y alone. Show that this is the case for the beam of length L.

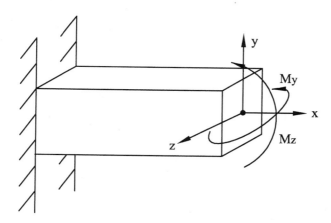

FIGURE 5.38

5.24 Find σ_{xy} for the beam shown in the figure.

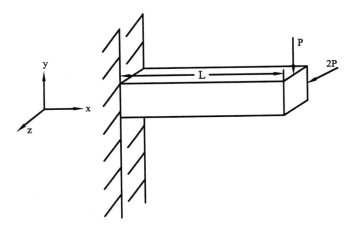

FIGURE 5.39

5.25 Whereas a moment M_z induces a bending in the x-y plane and thus a y-direction displacement, a moment M_y induces bending in the x-z plane and thus a z-direction displacement. For a combined symmetrical bending, therefore, the displacement vector $\boldsymbol{u}$ for points along the neutral axis are given by

$$\boldsymbol{u}(x) = v(x)\hat{\boldsymbol{j}} + w(x)\hat{\boldsymbol{k}},$$

where

$$EI_{zz}\frac{d^2v}{dx^2} = M_z, \qquad EI_{yy}\frac{d^2w}{dx^2} = M_y.$$

Hence, find $\boldsymbol{u}$ for a cantilevered beam of length L, rectangular cross section h deep and b wide and subjected to a shear force at $x = L$ given by $-P\hat{\boldsymbol{j}} + 2P\hat{\boldsymbol{k}}$ using the same figure as Exercise 5.24

5.26 Throughout this chapter we have assumed that the cross sections are homogenous. This need not be the case. Consider the layered beam shown below (Fig. 5.40), which consists of two materials characterized by LEHI properties $E^{(1)}$, $v^{(1)}$ and $E^{(2)}$, $v^{(2)}$, respectively. If we assume a continuous strain, then $\varepsilon_{xx} = -y/\rho$ as earlier, where ρ is the radius of curvature of the neutral axis, which need not be at the centroid as in the case of a homogenous beam. Indeed, the neutral axis can be found from the axial force balance equation

$$\sum F_x = 0 = -\iint \sigma_{xx}\,dA = -\left[\iint \sigma_{xx}^{(1)}\,dA^{(1)} + \sigma_{xx}^{(2)}\,dA^{(2)}\right],$$

where

$$\sigma_{xx}^{(1)} = E^{(1)}\varepsilon_{xx}, \qquad \sigma_{xx}^{(2)} = E^{(2)}\varepsilon_{xx}.$$

Hence, the neutral axis is located by

$$E^{(1)} \iint y \, dA^{(1)} + E^{(2)} \iint y \, dA^{(2)} = 0,$$

wherein the $-1/\rho$ was factored out. Show that this recovers the result for a homogeneous beam if $E^{(1)} = E^{(2)}$ and if $A^{(1)} + A^{(2)} = A$, the total cross-sectional area. Note, too, that from moment balance, we get

$$-M_z + \iint -\sigma_{xx} y \, dA = 0 \rightarrow M_z = -\iint -\frac{E^{(1)}}{\rho} y^2 \, dA^{(1)} - \iint -\frac{E^{(2)}}{\rho} y^2 \, dA^{(2)}.$$

Show that this leads to the following moment-curvature relation:

$$\frac{1}{\rho} = \frac{M_z y}{E^{(1)} I_{zz}^{(1)} + E^{(2)} I_{zz}^{(2)}}$$

and thus

$$\sigma_{xx}^{(1)} = -\frac{M_z y E^{(1)}}{E^{(1)} I_{zz}^{(1)} + E^{(2)} I_{zz}^{(2)}}, \qquad \sigma_{xx}^{(2)} = -\frac{M_z y E^{(2)}}{E^{(1)} I_{zz}^{(1)} + E^{(2)} I_{zz}^{(2)}},$$

where $I_{zz}^{(1)} + I_{zz}^{(2)} = I_{zz}$.

5.27 Locate the neutral axis for the composite cross section (Fig. 5.41). Hint: Assume that the neutral axis (i.e., origin of y-z axes) is at the centroid, but then find its true value relative to the interface between materials 1 and 2. Toward this end, note that

$$\iint y \, dA^{(1)} = \int_{-b/2}^{b/2} \int_{-H}^{h^{(1)}-H} y \, dy \, dz = \frac{b}{2}\left[(h^{(1)})^2 - 2Hh^{(1)}\right],$$

$$\iint y \, dA^{(2)} = \int_{-b/2}^{b/2} \int_{-(h^{(2)}-H)}^{-H} y \, dy \, dz = -\frac{b}{2}\left[(h^{(2)})^2 + 2Hh^{(2)}\right],$$

where H is the distance (assumed down) from the centroid to the interface between material 1 and material 2. Thus, axial force balance yields

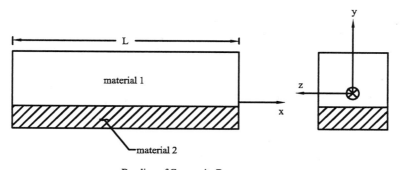

Bending of Composite Beams

FIGURE 5.40

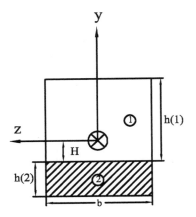

FIGURE 5.41

$$H = \frac{1}{2}\left(\frac{E^{(1)}\left(h^{(1)}\right)^2 - E^{(2)}\left(h^{(2)}\right)^2}{E^{(1)}h^{(1)} + E^{(2)}h^{(2)}} \right).$$

Note that if $E^{(1)} = E$, $E^{(2)} = 0$, $h^{(1)} = h$, and $h^{(2)} = 0$, then $H = h/2$, thus locating the neutral axis at the centroid, as it should for a homogenous beam.

5.28 In cardiopulmonary resuscitation (CPR), one seeks to augment cardiac output by pressing down on the sternum. This increases blood flow by direct compression of the heart between the sternum and spine as well as via changes in intrathoracic pressure. Typically, the sternum is compressed 1.5–2 in. with each compression. One concern in CPR, however, is that excessive force may fracture the ribs. Referring to Figure 5.42, we see that the transversely applied load P induces bending stresses in the rib. If you are biomedical engineer charged with designing an automatic device to load the sternum, find the induced stresses in the ribs as a function of the applied load and geometry. Hint: The rib can be assumed to exhibit a LEHI behavior and it is a structure having one dimension much larger than the other two and subjected to bending. The ribs are clearly not initially straight beams, thus our flexure formula does not apply. It can be shown, however, that the bending stress in a curved beam can be computed via (Boresi et al., 1993)

$$\sigma_{\theta\theta} = \frac{f}{A} + \frac{M(A - rA_m)}{Ar(RA_m - A)},$$

where f is a force applied normal to the θ-face cross section of area A, M is the bending moment, $A_m = \int (1/r)dA$, where r is the radial location of the point of interest in the cross section, and R is the radial dis-

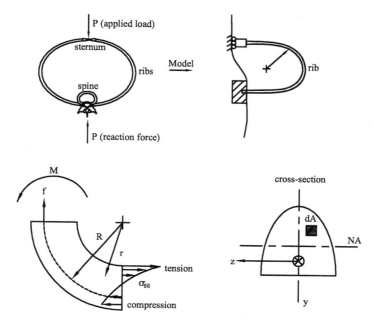

FIGURE 5.42

tance from the center of curvature of the beam to the centroid of the cross section (see figure). Because we have merely listed, not derived, this formula, we must note the assumptions/restrictions. First, plane sections are assumed to remain plane; the radial stress σ_{rr} and shear stress $\sigma_{\theta r}$ are assumed to be small in comparison to $\sigma_{\theta\theta}$, the cross section is assumed to be symmetric about the vertical y axis shown in the figure; the applied loads all lie in the plane of symmetry; and R/h < 5. In other words, if R/h > 5, the flexure formula [Eq. (5.23)] is often used even for a curved beam. Finally, see Table 5.1 for formulas for A and A_m for common cross sections.

5.29 Show that the critical load for a fixed–fixed column subjected to an end load P is

$$P_{cr} = \frac{4\pi^2 EI_{zz}}{L^2}.$$

5.30 Show that the critical load for a pinned–pinned column subjected to an end load P is

$$P_{cr} = \frac{\pi^2 EI_{zz}}{L^2}.$$

TABLE 5.1. Formulae for A and A_m for curved beams having different cross-sections. Note the coordinate directions. See Boresi et al. (1993).

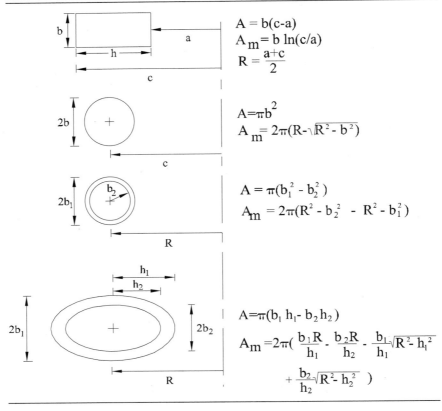

$$A = b(c-a)$$
$$A_m = b \ln(c/a)$$
$$R = \frac{a+c}{2}$$

$$A = \pi b^2$$
$$A_m = 2\pi(R - \sqrt{R^2 - b^2})$$

$$A = \pi(b_1^2 - b_2^2)$$
$$A_m = 2\pi(\sqrt{R^2 - b_2^2} - \sqrt{R^2 - b_1^2})$$

$$A = \pi(b_1 h_1 - b_2 h_2)$$
$$A_m = 2\pi(\frac{b_1 R}{h_1} - \frac{b_2 R}{h_2} - \frac{b_1}{h_1}\sqrt{R^2 - h_1^2}$$
$$+ \frac{b_2}{h_2}\sqrt{R^2 - h_2^2})$$

6
Some Nonlinear Problems

Although we have derived universal results that relate stresses to the applied loads and geometry for axially loaded rods and inflated thin-walled cylinders and spheres (see Sections 3.3–3.5 of Chapter 3), all other results for stresses and deformations have been restricted to linear, elastic, homogenous, and isotropic (LEHI) behaviors under small strains. Such results for LEHI behaviors are very important in bone mechanics, many applications involving biomaterials, and the design of experimental systems. Nonetheless, as noted in Chapter 2, the behavior of cells, soft biological tissues, and elastomers used in biomedical applications typically exhibit a nonlinear stress–strain behavior over large deformations and, consequently, there is a need for additional theoretical frameworks. In particular, LEHI behavior is characterized by two linearities: a linear relation between stress and strain [e.g., Eq. (2.69)] and a linear relation between strain and displacement gradients [e.g., Eq. (2.45)]. We need to avoid both linearizations in soft tissue mechanics. The interested reader is referred to Humphrey (2002) for a more complete treatment of the nonlinear theories, but here we provide a simple, brief introduction.

6.1 Kinematics

Recall from Section 2.5 of Chapter 2 that displacements alone cannot be related effectively to stresses to describe material behavior. Rather, combinations of displacement gradients, called strains, are more useful in formulating constitutive equations for stress. Indeed, in Eq. (2.42), we listed six independent components of the Green strain, which is an exact measure for large or small deformations that is insensitive to rigid-body motions. There is, however, a more fundamental measure of motion that is useful in large deformation problems; it is called the *deformation gradient*. Recall, therefore, that we said the displacement vector u is a measure of where we (a material particle) are minus where we were. In Cartesians, we have components

$$u_x = x(X,Y,Z) - X, \quad u_y = y(X,Y,Z) - Y, \quad u_z = z(X,Y,Z) - Z, \quad (6.1)$$

where (x, y, z) locates a point of interest in the current configuration and (X, Y, Z) locates the same point in its original (reference) configuration. In particular, it is because the location of a point in a current configuration depends on where it started that we need the functional dependence $x(X, Y, Z)$, and so too for y and z. Moreover, the displacement gradients can be written as

$$\frac{\partial u_x}{\partial X} = \frac{\partial x}{\partial X} - 1, \quad \frac{\partial u_y}{\partial Y} = \frac{\partial y}{\partial Y} - 1, \quad \frac{\partial u_z}{\partial Z} = \frac{\partial z}{\partial Z} - 1, \quad (6.2)$$

and so forth. Given that there are nine such terms, it is convenient to write these results in a matrix form.[1] Indeed, if we denote the displacement gradient matrix by $[H]$ and the deformation gradient matrix by $[F]$, then we see that

$$[H] = [F] - [I] \leftrightarrow [F] = [I] + [H], \quad (6.3)$$

where the components of the deformation gradient can be calculated (with respect to Cartesian coordinates) via

$$[F] = \begin{bmatrix} \dfrac{\partial x}{\partial X} & \dfrac{\partial x}{\partial Y} & \dfrac{\partial x}{\partial Z} \\[2mm] \dfrac{\partial y}{\partial X} & \dfrac{\partial y}{\partial Y} & \dfrac{\partial y}{\partial Z} \\[2mm] \dfrac{\partial z}{\partial X} & \dfrac{\partial z}{\partial Y} & \dfrac{\partial z}{\partial Z} \end{bmatrix}. \quad (6.4)$$

$[I]$ is the so-called identity matrix, with components

$$[I] = \begin{bmatrix} 1 & 0 & 0 \\ 0 & 1 & 0 \\ 0 & 0 & 1 \end{bmatrix}. \quad (6.5)$$

This matrix is called an identity matrix for $[A][I] = [I][A] = [A]$ for any 3×3 matrix $[A]$; that is, when $[I]$ operates on a matrix $[A]$, it returns $[A]$ unaltered.

It is the deformation gradient that plays the key role in nonlinear analyses—it is the fundamental measure of finite deformation for it includes both the deformation and the rigid-body motion. For example, if we denote the Green strain via the matrix $[E]$, where

---

[1] Matrix operations are reviewed briefly in Appendix 6.

$$[E] = \begin{bmatrix} E_{XX} & E_{XY} & E_{XZ} \\ E_{YX} & E_{YY} & E_{YZ} \\ E_{ZX} & E_{ZY} & E_{ZZ} \end{bmatrix} \tag{6.6}$$

are the Cartesian components that are listed in Eq. (2.42), it is easy to show that

$$[E] = \frac{1}{2}\left([F]^T[F] - [I]\right), \tag{6.7}$$

where the superscript T denotes a transpose of the matrix (i.e., the interchanging of rows and columns as discussed in Appendix 6). Without going into details (see Humphrey, 2002), the operation of $[F]^T[F]$ removes the rigid-body information and the subtraction of $[I]$ renders $[E] = [0]$ in the absence of deformation/strain; both of these features are desirable of a strain measure. It can be shown further that the linearized strain is given by

$$[\varepsilon] = \frac{1}{2}\left([F] + [F]^T - 2[I]\right). \tag{6.8}$$

Example 6.1 Compute the components of $[F]$ for the motions associated with Eqs. (2.51), (2.53), and (2.56), and then calculate the associated values of $[E]$ and $[\varepsilon]$.

Solution: First, consider the simple 1-D stretching motion given by $x = \Lambda X$, $y = Y$, and $z = Z$, where Λ is a stretch ratio (i.e., just a number for each equilibrium stretch). Clearly,

$$[F] = \begin{bmatrix} \dfrac{\partial x}{\partial X} & \dfrac{\partial x}{\partial Y} & \dfrac{\partial x}{\partial Z} \\ \dfrac{\partial y}{\partial X} & \dfrac{\partial y}{\partial Y} & \dfrac{\partial y}{\partial Z} \\ \dfrac{\partial z}{\partial X} & \dfrac{\partial z}{\partial Y} & \dfrac{\partial z}{\partial Z} \end{bmatrix} = \begin{bmatrix} \Lambda & 0 & 0 \\ 0 & 1 & 0 \\ 0 & 0 & 1 \end{bmatrix}$$

and therefore

$$[E] = \frac{1}{2}\left([F]^T[F] - [I]\right) = \begin{bmatrix} \frac{1}{2}(\Lambda^2 - 1) & 0 & 0 \\ 0 & 0 & 0 \\ 0 & 0 & 0 \end{bmatrix},$$

whereas

$$[\varepsilon] = \frac{1}{2}\left([F]+[F]^T - 2[I]\right) = \begin{bmatrix} \Lambda - 1 & 0 & 0 \\ 0 & 0 & 0 \\ 0 & 0 & 0 \end{bmatrix},$$

as we found earlier. Again, for $\Lambda \sim 1$ (small strain), the numerical values of $[E]$ and $[\varepsilon]$ differ little, but for larger values typically experienced by soft tissues (stretches often on the order of 10–100%), the difference becomes pronounced. For example, if $\Lambda = 1.5$, a 50% extension, then $E_{11} = 0.625$ (exact), whereas $\varepsilon_{11} = 0.5$ (approximate), thus revealing a 20% error in the computation of the strain.

Second, consider a simple shear motion given by $x = X + \kappa Y$, $y = Y$, and $z = Z$, where κ is just a number for each equilibrium stretch. Hence,

$$[F] = \begin{bmatrix} 1 & \kappa & 0 \\ 0 & 1 & 0 \\ 0 & 0 & 1 \end{bmatrix}$$

and, therefore,

$$[E] = \frac{1}{2}\left(\begin{bmatrix} 1 & 0 & 0 \\ \kappa & 1 & 0 \\ 0 & 0 & 1 \end{bmatrix}\begin{bmatrix} 1 & \kappa & 0 \\ 0 & 1 & 0 \\ 0 & 0 & 1 \end{bmatrix} - \begin{bmatrix} 1 & 0 & 0 \\ 0 & 1 & 0 \\ 0 & 0 & 1 \end{bmatrix}\right) = \frac{1}{2}\begin{bmatrix} 0 & \kappa & 0 \\ \kappa & \kappa^2 & 0 \\ 0 & 0 & 0 \end{bmatrix},$$

whereas

$$[\varepsilon] = \frac{1}{2}\left(\begin{bmatrix} 1 & \kappa & 0 \\ 0 & 1 & 0 \\ 0 & 0 & 1 \end{bmatrix} + \begin{bmatrix} 1 & 0 & 0 \\ \kappa & 1 & 0 \\ 0 & 0 & 1 \end{bmatrix} - \begin{bmatrix} 2 & 0 & 0 \\ 0 & 2 & 0 \\ 0 & 0 & 2 \end{bmatrix}\right) = \frac{1}{2}\begin{bmatrix} 0 & \kappa & 0 \\ \kappa & 0 & 0 \\ 0 & 0 & 0 \end{bmatrix}.$$

This comparison reveals a significant conceptual difference between $[E]$ and $[\varepsilon]$. Note that the extensional strain in the Y direction $E_{YY} = \kappa^2/2$ whereas $\varepsilon_{yy} = 0$; that is, shear and extension are coupled in the (exact) nonlinear theory, whereas the linearization of $[\varepsilon]$ loses this coupling. Although $\kappa^2/2$ will be negligible in comparison to $\kappa/2$ if $\kappa \ll 1$, this will not be the case for large shears, as experienced by the heart during each cardiac cycle. Again, therefore, the exact (nonlinear) theory must be used when the deformations or rigid rotations are large. The latter is revealed by considering the third case, the rigid-body motion associated with Eq. (2.56):

$$x = X \cos\phi + Y \sin\phi, \qquad y = -X \sin\phi + Y \cos\phi, \qquad z = Z.$$

In this case,

$$[E] = \frac{1}{2}\left(\begin{bmatrix} \cos\phi & -\sin\phi & 0 \\ \sin\phi & \cos\phi & 0 \\ 0 & 0 & 1 \end{bmatrix}\begin{bmatrix} \cos\phi & \sin\phi & 0 \\ -\sin\phi & \cos\phi & 0 \\ 0 & 0 & 1 \end{bmatrix} - \begin{bmatrix} 1 & 0 & 0 \\ 0 & 1 & 0 \\ 0 & 0 & 1 \end{bmatrix}\right) = \begin{bmatrix} 0 & 0 & 0 \\ 0 & 0 & 0 \\ 0 & 0 & 0 \end{bmatrix},$$

as it should, for $[E]$ is insensitive to rigid-body motion, but

$$[\varepsilon] = \frac{1}{2}\left(\begin{bmatrix} \cos\phi & \sin\phi & 0 \\ -\sin\phi & \cos\phi & 0 \\ 0 & 0 & 1 \end{bmatrix} + \begin{bmatrix} \cos\phi & -\sin\phi & 0 \\ \sin\phi & \cos\phi & 0 \\ 0 & 0 & 1 \end{bmatrix} - \begin{bmatrix} 2 & 0 & 0 \\ 0 & 2 & 0 \\ 0 & 0 & 2 \end{bmatrix}\right)$$

$$= \begin{bmatrix} \cos\phi - 1 & 0 & 0 \\ 0 & \cos\phi - 1 & 0 \\ 0 & 0 & 0 \end{bmatrix},$$

as found in Eq. (2.56), which reveals that $[\varepsilon]$ is inappropriately sensitive to a rigid-body rotation unless the rotation is small (i.e., as $\phi \to 0$, $\cos\phi \to 1$). Although these three motions are very simple, they serve to illustrate the use of $[F]$ as a fundamental measure of the motion.

Observation 6.1 The cell is the fundamental structural and functional unit of living things and, as noted in Chapter 1, understanding mechanotransduction therein is vital to many areas of biomechanical analysis and design. As one might expect, many different types of tests have been performed on cells in an attempt to correlate changes in cell structure and function with mechanical stimuli. These tests include micropipette aspiration, indentation tests, atomic force microscopy (AFM, both indentation and pulling), and magnetic bead cytometry (Fig. 6.1). In micropipette aspiration, one infers the bending stiffness of the cell membrane by monitoring the amount of cell membrane that is drawn into a pipette of known radius by a known pressure gradient. In indentation tests, one measures the force that is required to indent the cell a known amount and interprets this relation in terms of homogenized properties of the cell membrane and cytoplasm. The AFM was discussed in Chapter 5. One can use the AFM to indent the cell or to pull on focal adhesion complexes by functionalizing the tip of the AFM with an appropriate ligand. In magnetic bead cytometry, one similarly functionalizes a magnetic microsphere that can be moved within a magnetic field and then measures torque–twist responses, often over small twists. This test is thus useful for interrogating time-dependent shearing (viscoelastic) behaviors, which are discussed in Chapter 11. Clearly, therefore, advances in technology permit such empirical studies to be performed, but the interpretation of the data requires a biomechanical analysis of the associated initial boundary value problem. Given the complex geometries and loading, many in cell mechanics have assumed linear material behaviors and small strains to facilitate analysis. Such assumptions should be

Empirical Studies on Cells

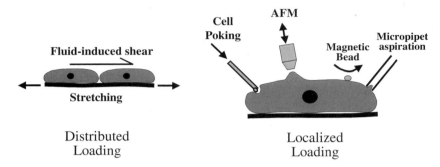

FIGURE 6.1 Possible tests for interrogating the mechanical properties or responses of cells. Micropipette aspiration, magnetic bead cytometry, atomic force microscopy (AFM), and cell poking induce localized loads, whereas stretching sheets on which cells are adhered or subjecting a monolayer to a fluid-flow-induced shear stress induce distributed loads. Flow-induced shears are discussed in Chapter 9 on biofluid mechanics. (Courtesy of R. Gleason.)

based on the physics, however, not the ease of solution, and the experiments should be designed based on theoretical frameworks, not just the availability of a new technology.

Associated with these many experiments has been a variety of attempts to model the mechanics of cells. Among various models, one finds the following: tensegrity models, which emphasize the importance of prestress within a cell and the possibility of mechanical stresses acting at a distance; percolation theories that emphasize dynamic changes in cytoskeletal interconnectiveness; soft glassy rheological models that suggest that the cytoskeleton is metastable, able to transform instantaneously from more solidlike to more fluidlike behaviors; and continuum models, based on cells as inclusions within a matrix that allow study of cell–matrix interactions (see Mow et al., 1994; Stamenovic and Ingber, 2002; Humphrey, 2002 and references therein). No single model enjoys wide acceptance, however, even for a particular class of mechanocytes; thus, there remains a pressing need for much more research on cell mechanics. Cell mechanics is essential, for example, for explaining basic processes such as cell adhesion, contraction, division, migration, spreading, and even phagocytosis (the engulfing and digestion of extracellular material). Likewise, it appears that cellular apoptosis (i.e., programmed cell death), the synthesis and degradation of matrix, and the production of growth regulatory molecules, cytokines, and cell surface receptors are also influenced greatly by the mechanics. Each of these activities manifests itself at the tissue and organ level, of course, and thereby are linked to development, tissue maintenance, wound healing, growth and remodeling, and pathogenesis. Hence, whether one seeks to understand normal physiology, disease, injury, interactions between medical devices and tissues, or even the engineering of tissue or organ replacements, there is a need to understand the mechanics of cells. Given the diversity of cell types and the various environments in which they function, we should probably expect that multiple approaches will be equally useful in modeling the many different aspects of cell mechanics. Although not emphasized in the past, *large strain analyses must be used to describe well the finite displacement gradients and finite rotations experienced by the cell membrane and cytoskeletal constituents*, which together endow the cell with much of its structural integrity. The interested reader is referred to the collection of papers in Mow et al. (1994) and a special issue of the *Journal of Biomechanics* (Vol. 28, pp. 1411–1572, 1995) for a discussion of some of these issues.

6.2 Pseudoelastic Constitutive Relations

Figure 6.2 shows a typical 1-D stress–stretch behavior of a soft tissue. As in Chapter 2, note the nonlinear response, which is initially compliant but then becomes very stiff, and the hysteresis, which reveals an inelastic character. Although the exact source of the inelasticity is not clear, it is thought to be due in part to the movement of structural proteins (primarily elastin and types I and III collagen) within the so-called ground substance matrix that consists largely of proteoglycans and water; that is, one source of energy

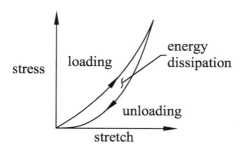

FIGURE 6.2 Typical stress–stretch response of a soft tissue. The nonlinear behavior is usually over finite strains, which disallows the use of Hooke's law and linearized strains, and thus necessitates the formulation of more general constitutive relations. Note, too, that the hysteresis is a characteristic of a viscoelastic behavior, or perhaps if small, a pseudoelastic behavior as defined by Fung.

dissipation revealed by the noncoincident loading and unloading curves may be a viscous dissipation and, in particular, a solid–fluid interaction at the molecular level. We will briefly discuss approaches to model *viscoelastic* responses in Chapter 11, but let us make a further observation here.

Most viscoelastic responses depend not just on the amount of the deformation but also the rate of deformation. A very simple example of this is revealed by a "kindergarten experiment." If one pulls his or her fingers very slowly through a solution of cornstarch (a solid–fluid mixture), the resistance is very small; in contrast, if the fingers are pulled through rapidly, the resistance increases considerably; that is, the response by the cornstarch solution to the applied load depends strongly on the rate of deformation, a characteristic common for viscoelastic behavior (discussed more in Chapter 7). Although most soft tissues exhibit a viscoelastic character under many conditions, Fung reported in the late 1960s that the behavior of soft tissues tends not to depend strongly on strain rate (Fig. 6.3) (see Fung, 1990). Indeed, Fung suggested that if one cyclically loads and unloads various soft tissues, there tends to be repeatable (but separate) loading and unloading responses. Because the theory of viscoelasticity is more complex to implement than is the theory of elasticity, Fung suggested that in some cases it may be reasonable to treat separately the loading and unloading behaviors as elastic; that is, although one would use the same function to relate stress to strain in loading and unloading, one would use separate values of the associated material parameters. To remind us that the behavior is not truly elastic, Fung called such an approach *pseudoelasticity*, which is now used frequently in many areas of biomechanical design and analysis.

At this juncture, let us note that Fung's concept of pseudoelasticity appears to be particularly applicable to tissues that are subjected in vivo to consistent loading and unloading, such as the arteries, diastolic heart, and lungs. Indeed, Fung also showed that pseudoelastic responses (i.e., separate but repeatable loading and unloading behaviors) were obtained in the laboratory only after a sufficient number of loading cycles (Fig. 6.4), usually

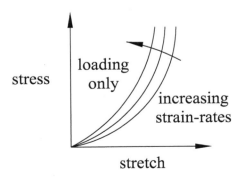

FIGURE 6.3 Possible sensitivity of the stress–stretch response to changes in strain rate (loading curves only). In many cases, however, there is little change in the response with three orders of magnitude changes in strain rate.

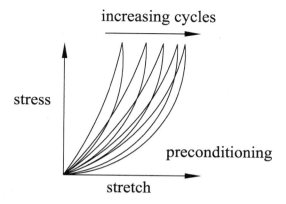

FIGURE 6.4 Preconditioning response exhibited by many soft tissues. Note that the response becomes more extensible and eventually nearly repeatable after a sufficient number of cycles of loading.

3–10; that is, it appears that following excision, whereby the tissue is removed from its normal dynamic loading environment, the tissue must be "conditioned" to obtain a repeatable pseudoelastic response. Given such conditioning, the tissue tends to dissipate less energy upon cyclic loading and to become less stiff, both of which appear to be teleologically favorable. Fung called this experimental process *preconditioning*. Whereas few have sought to understand the underlying mechanisms of preconditioning, most simply exploit it to obtain pseudoelastic responses which are easier to describe mathematically and which appear to be more physiologic in many cases because of the periodic loading experienced by many soft tissues.

Inasmuch as linear elasticity is easily described mathematically, we typically do not think deeply about its implications. Yet, one of the remarkable consequences of a linear stress–strain relation is that it is unique: There is but one way to draw a straight line (i.e., $y = mx + b$). Nonlinear behavior, on the other hand, not only requires the use of a more complex function to fit the stress–strain data, but it is not necessarily unique. Referring to the loading response in Figure 6.2, for example, one investigator may "see" a parabola and thus suggest a quadratic relationship between stress and stretch, whereas another investigator may see a trigonometric relationship and postulate a tangent function to describe the data. Indeed, because of the inherent scatter in experimental data, multiple functions may be found to describe the data similarly, thus raising the question: What is the best constitutive descriptor for this behavior?

Again, Fung offered a very helpful approach. Fung suggested that instead of plotting stress versus strain (or stretch), we could plot the stiffness as a function of stress. Strictly speaking, *stiffness* is defined as a change in stress with respect to a change in a conjugate strain (or stretch); thus, it is the slope of a stress–strain or stress–stretch curve. To appreciate this, let us consider a simple, 1-D, linearly elastic response. If $\sigma = E\varepsilon$, where E is the Young's modulus, then the stiffness $K = d\sigma/d\varepsilon = E$ for all σ; that is, if we plotted K versus σ, we would obtain a constant value. Integrating then, we would obtain

$$\int \frac{d\sigma}{d\varepsilon}\, d\varepsilon = \int E\, d\varepsilon \rightarrow \sigma = E\varepsilon + c_0, \tag{6.9}$$

where $c_0 = 0$ if the stress is zero at zero strain, thus yielding the (previously unknown) stress–strain relation. Let us now see what Fung observed.

Fung performed one-dimensional extension tests on excised strips of mesentery, a thin collagenous membrane found in the abdomen. Recalling from Chapter 3 that there are actually multiple definitions of stress (σ is the so-called Cauchy stress, which is a measure of forces acting over current oriented areas, whereas Σ is the so-called nominal or Piola–Kirchhoff stress, which is a measure of the force acting over an original oriented area), Fung chose to use Σ as his measure of stress and Λ as his measure of extension (Λ being a stretch ratio, which is a component of the deformation gradient as seen in Example 6.1). Doing so, Fung obtained a result similar to that shown in Figure 6.5: a near-linear relation between the stiffness $d\Sigma/d\Lambda$ and the first Piola–Kirchhoff stress Σ; that is, the data appeared to be well described by

$$\frac{d\Sigma}{d\Lambda} = \alpha + \beta\Sigma, \tag{6.10}$$

which is a first-order, nonhomogenous, linear differential equation with a constant coefficient. This equation is solved easily using either standard methods for differential equations (Appendix 8 of Chapter 8) or a direct integration. For example, for this class of differential equations, we expect

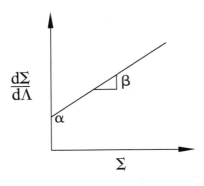

FIGURE 6.5 Fung's idea to plot stiffness (i.e., a change in stress with respect to stretch, $d\Sigma/d\Lambda$ in one dimension) versus the first Piola–Kirchhoff stress (Σ), which, for many tissues, results in a nearly linear relationship.

a homogenous solution to be of an exponential form and a particular solution to be a constant; that is, let our trial solution be of the form

$$\Sigma = c_1 e^{c_2 \Lambda} + c_3. \tag{6.11}$$

To find the values of the unknown constants, note that

$$\frac{d\Sigma}{d\Lambda} = c_1 e^{c_2 \Lambda} c_2 = (\Sigma - c_3)c_2 = c_2 \Sigma - c_3 c_2 \tag{6.12}$$

whereby from Eq. (6.10), we find that

$$c_2 \Sigma - c_3 c_2 = \beta \Sigma + \alpha \rightarrow \begin{cases} c_2 = \beta \\ c_3 = \dfrac{-\alpha}{\beta}. \end{cases} \tag{6.13}$$

To find c_1, we need another condition. For example, if we require the stress Σ to be zero when the strain is zero (i.e., the stretch $\Lambda = 1$), then

$$0 = c_1 e^{\beta} - \frac{\alpha}{\beta} \rightarrow c_1 = \frac{\alpha}{\beta} e^{-\beta} \tag{6.14}$$

and, therefore, the solution to the *experimentally* obtained stress–stretch relation is

$$\Sigma = \frac{\alpha}{\beta} e^{-\beta} e^{\beta \Lambda} - \frac{\alpha}{\beta} \rightarrow \Sigma = \frac{\alpha}{\beta} (e^{\beta(\Lambda-1)} - 1), \tag{6.15}$$

where α and β are the experimentally measurable intercept and slop, respectively, in Figure 6.5. Hence, rather that *guessing* functional forms for a stress–stretch relation (e.g., quadratic or trigonometric), Fung used a clever way of replotting the data as stiffness versus stress that revealed a linear relation, which, in turn, unambiguously suggested an exponential form of the constitutive relation. Although plots of stiffness versus stress are not perfectly linear for all tissues over all ranges of stress of interest, years of experience have revealed that exponential constitutive relations often provide good descriptions of the data for certain conditions. See, for example, the fit of a multiaxial exponential relation to data on epicardium in Figure 6.6.

Given the existence of so many different soft tissues in the body, each having unique structure and function and subjected to different multiaxial stresses and deformations, it should not be surprising that there is a wide variety of proposed constitutive relations in the literature; that is, despite the success of Fung's exponential as well as other functional relations, there is still no general agreement on the "best" relations for any given soft tissue. This situation is in stark contrast to the general acceptance since the mid-1800s of Hooke's law as a descriptor of LEHI behavior. There is, therefore, a pressing need for continued research into constitutive relations for soft

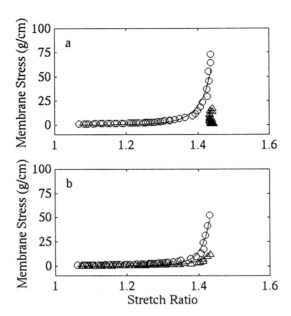

FIGURE 6.6 Fit to biaxial data for excised epicardium based on a Fung-type exponential relation. [From the author's laboratory (courtesy of J. Harris)].

tissues. For more discussion of this need and comparisons of other suc-
cessful nonlinear constitutive relations to experimental data, see Humphrey
(2002). Here, let us conclude our discussion with two observations.

First, although there is a pressing need for improved, well-accepted nonlinear constitutive functions for many soft tissues, we must remember what
a constitutive relation is: It is a mathematical descriptor of particular behaviors exhibited by a material under well-defined conditions, which is to say,
it is *not* a descriptor of the material. All materials, including soft tissues,
exhibit different behaviors under different conditions; hence, we should
expect that multiple constitutive relations will be needed for each tissue
depending on the condition of interest. For example, a pseudoelastic relation may be sufficient to describe the cyclic behavior of an artery under
physiological conditions, but a viscoelastic descriptor may be needed to
describe the artery's response during balloon angioplasty, a thermomechanical relation may be needed to analyze the thermal ablation of an atherosclerotic lesion–artery complex, and a growth and remodeling relation
may be needed to describe the long-term response of an artery to hypertensive conditions. Thus, when we say that there is a pressing need for
improved, well-accepted relations, this does not suggest that we should seek
a single relation that describes well all behaviors under all conditions.
Rather, we will still need multiple improved relations for specific conditions
of interest. Constitutive formulations, which combine theory and experi-

mentation, thereby remain one of the most important and challenging aspects of biomechanics.

Second, it is important to note that although there are many different measures of stress, they are not independent. It can be shown, for example, that the Cauchy and the first Piola–Kirchhoff stresses are related through the deformation gradient (which, as a fundamental measure of motion, relates undeformed and current areas over which a force must act), namely (Humphrey, 2002)

$$[\sigma] = \frac{1}{\det[F]}[F][\Sigma] \leftrightarrow [\Sigma] = \det[F][F]^{-1}[\sigma], \tag{6.16}$$

where $\det[F]$ denotes the determinant and $[F]^{-1}$ denotes the inverse of $[F]$. The latter is defined such that $[F]^{-1}[F] = [I] = [F][F]^{-1}$. Recall that matrix operations are discussed in Appendix 6.

For example, let us consider a simple 1-D stress test wherein a sample is extended in the "1" direction as in the test by Fung on mesentery. It can be shown that the motion is well described by

$$x = \Lambda X, \qquad y = \lambda Y, \qquad z = \lambda Z \tag{6.17}$$

and, thus, from Eq. (6.4)

$$[F] = \begin{bmatrix} \Lambda & 0 & 0 \\ 0 & \lambda & 0 \\ 0 & 0 & \lambda \end{bmatrix}. \tag{6.18}$$

Now, if the behavior is incompressible or nearly so, as is the case for many tissues, including soft tissues, then volume is conserved and $\det[F] = 1$. Hence $\lambda = 1/\sqrt{\Lambda}$. Show that this is true by computing the undeformed and deformed volumes and setting them equal. If the associated first Piola–Kirchhoff stress is measured as

$$[\Sigma] = \begin{bmatrix} \Sigma_{11} & 0 & 0 \\ 0 & 0 & 0 \\ 0 & 0 & 0 \end{bmatrix} = \begin{bmatrix} \dfrac{f}{A_o} & 0 & 0 \\ 0 & 0 & 0 \\ 0 & 0 & 0 \end{bmatrix}, \tag{6.19}$$

where f is the applied load and A_o is the undeformed area over which the load "acts," then the Cauchy stress is

$$[\sigma] = \frac{1}{1} \begin{bmatrix} \Lambda & 0 & 0 \\ 0 & \dfrac{1}{\sqrt{\Lambda}} & 0 \\ 0 & 0 & \dfrac{1}{\sqrt{\Lambda}} \end{bmatrix} \begin{bmatrix} \dfrac{f}{A_o} & 0 & 0 \\ 0 & 0 & 0 \\ 0 & 0 & 0 \end{bmatrix} = \begin{bmatrix} \Lambda\dfrac{f}{A_o} & 0 & 0 \\ 0 & 0 & 0 \\ 0 & 0 & 0 \end{bmatrix}, \tag{6.20}$$

where $\det[F] \equiv 1$. From Figure 6.7, we note further that $\Lambda = l/L$ and incompressibility requires that $lA = LA_o$ or $A = A_o/\Lambda$. Hence, we see that the only non–zero component of the Cauchy stress is

$$\sigma_{11} = \frac{\Lambda f}{A_o} = \frac{f}{A_o/\Lambda} = \frac{f}{A};$$

(6.21)

that is, the value of the Cauchy stress is computed as the force acting over the current area A as expected. Perhaps more importantly, however, Eq. (6.16) allows us to compute the Cauchy stress for Fung's mesentery sample in terms of his exponential constitutive relation derived for the first Piola–Kirchhoff stress [Eq. (6.15)]. For the uniaxial test on mesentery, therefore, we have

$$\sigma_{11} = \Lambda\Sigma_{11} = \Lambda\frac{\alpha}{\beta}\left(e^{\beta(\Lambda-1)} - 1\right).$$

(6.22)

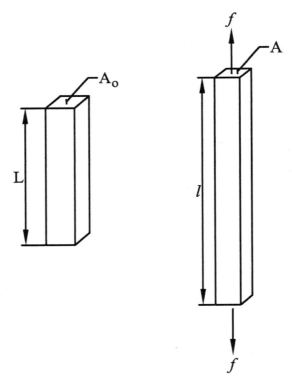

FIGURE 6.7 Uniaxial sample subjected to a uniform axial stress; note the dimensions in the undeformed and deformed configurations.

Whereas the first Piola–Kirchhoff stress is often preferred by experimentalists, for they only have to measure the cross-sectional area A_o (in uniaxial tests) once, in contrast to each A at each f, the Cauchy stress is often preferred in the solution of boundary value problems [cf. Eqs. (3.8)–(3.10)]. Fortunately, relations such as Eq. (6.16) allow us to compute one type of stress from another, thus allowing us to work with that one which is more convenient. Indeed, this is a comparable situation to our use of stress transformation equations in Chapter 2, which allow us to determine components relative to the most convenient coordinate system and then to calculate from them the components relative to any coordinate system that is desired. We conclude, therefore, by noting yet another definition of stress, the second Piola–Kirchhoff stress $[S]$ that is often useful in constitutive formulations in nonlinear elasticity. It is related to the Cauchy and (first) Piola–Kirchhoff stress via

$$[\sigma] = \frac{1}{\det[F]}[F][S][F]^T, \qquad [\Sigma] = [S][F]^T. \qquad (6.23)$$

Albeit related to $[\sigma]$ and $[\Sigma]$, $[S]$ is very different—it does not have a physical interpretation. $[S]$ can be shown to be a measure of a well-defined but fictitious force acting on the actual undeformed area. The utility of $[S]$ lies in its mathematical relationship to the Green strain $[E]$. It can be shown, for example, that $[S]$ can be determined by differentiating a scalar "strain-energy function" W with respect to $[E]$, which is to say that each component of $[S]$ is found by taking derivatives of W with respect to the associated components of $[E]$, as, for example,

$$S_{11} = \frac{\partial W}{\partial E_{11}}, \qquad S_{12} = \frac{\partial W}{\partial E_{12}}, \qquad \text{etc.} \qquad (6.24)$$

Here, E_{11} represents E_{XX} in Cartesians or even E_{RR} in cylindricals. This situation is comparable to that discussed in Exercise 2.26, in which it was noted that a similar strain-energy W can be thought of as the area under the linear stress–strain curve for a 1-D LEHI behavior, with $\sigma_{xx} = E\varepsilon_{xx}$ and, thus, $W = \frac{1}{2}(\varepsilon_{xx})\sigma_{xx} = \frac{1}{2}E\varepsilon_{xx}^2$, and, finally, $\sigma_{xx} = \partial W/\partial\varepsilon_{xx}$. Equation (6.24) and its related constitutive relations for W are much more general, however, being valid for nonlinear elastic behavior over large strains. For example, motivated by (but not directly derivable from) the 1-D exponential result of Eq. (6.15), Fung (1990) postulated that a potentially useful strain-energy function may be

$$W = \frac{1}{2}c(e^Q - 1), \qquad (6.25)$$

where, for orthotropy,

$$Q_{\text{orth}} = c_1 E_{11}^2 + c_2 E_{22}^2 + c_3 E_{33}^2 + 2c_4 E_{11} E_{22} + 2c_5 E_{22} E_{33} + 2c_6 E_{33} E_{11}$$
$$+ c_7 (E_{12}^2 + E_{21}^2) + c_8 (E_{23}^2 + E_{32}^2) + c_9 (E_{31}^2 + E_{13}^2) \tag{6.26}$$

and c and c_1–c_9 are material parameters and E_{11}–E_{33} are the nine components of $[E]$ relative to a particular coordinate system. Hence, for example,

$$S_{11} = \frac{\partial W}{\partial E_{11}} = \frac{1}{2} ce^Q (2c_1 E_{11} + 2c_4 E_{22} + 2c_6 E_{33})$$
$$= ce^Q (c_1 E_{11} + c_4 E_{22} + c_6 E_{33}) \tag{6.27}$$

with other components of $[S]$ computed similarly. This relatively simple multiaxial relation has been shown to provide reasonable fits to data for various soft tissues, including myocardium, arteries, and skin (Fung, 1990), but, again, there remains a need to search for improved relations in many cases. One advantage of Eqs. (6.25)–(6.26) is that, similar to a generalized Hooke's law (which can be found from a W that is quadratic in the components of $[\varepsilon]$), one can specify simplifications in the constants in Q for different material symmetries [cf. Eqs. (2.69), (2.75), and (2.77)]. Whereas one needs all nine values of c_1–c_9 for orthotropy, this number reduces to five for transverse isotropy and only two for isotropy. Specifically, for transverse isotropy, with the preferred direction being the (3) direction, $c_1 = c_2$, $c_5 = c_6$, $c_8 = c_9$; thus,

$$Q_{\text{trans}} = c_1 (E_{11}^2 + E_{22}^2) + c_3 E_{33}^2 + 2c_4 E_{11} E_{22} + 2c_5 (E_{11} + E_{22}) E_{33}$$
$$+ c_7 (E_{12}^2 + E_{21}^2) + c_8 (E_{23}^2 + E_{32}^2 + E_{13}^2 + E_{31}^2), \tag{6.28}$$

where $c_7 = 2(c_1 - c_4)$. Finally, for isotropy, $c_1 = c_3$, $c_4 = c_5$, $c_7 = c_9$; thus,

$$Q_{\text{iso}} = c_1 (E_{11}^2 + E_{22}^2 + E_{33}^2) + 2c_4 (E_{11} E_{22} + E_{22} E_{33} + E_{33} E_{11})$$
$$+ c_7 (E_{12}^2 + E_{21}^2 + E_{23}^2 + E_{32}^2 + E_{13}^2 + E_{31}^2). \tag{6.29}$$

Note that for isotropy, we can alternatively write Q as

$$Q_{\text{iso}} = \alpha (E_{11} + E_{22} + E_{33})^2 + \beta (-E_{12}^2 - E_{21}^2 - E_{23}^2 - E_{32}^2 - E_{13}^2$$
$$- E_{31}^2 + E_{11} E_{22} + E_{22} E_{33} + E_{33} E_{11}), \tag{6.30}$$

where $\alpha = c_1$ and $\beta = -c_7$ or, recognizing that these combinations of strains are invariant under coordinate transformations,

$$Q_{\text{iso}} = \alpha I_E^2 + \beta II_E, \tag{6.31}$$

where

$$I_E = E_{11} + E_{22} + E_{33},$$
$$II_E = E_{11} E_{22} + E_{22} E_{33} + E_{33} E_{11} - 2E_{12} E_{21} - 2E_{23} E_{32} - 2E_{31} E_{13}, \tag{6.32}$$

and $E_{12} = E_{21}$, $E_{13} = E_{31}$, and $E_{23} = E_{32}$ by definition. This form for Q was found to describe the behavior of lung parenchyma reasonably well (Fung, 1993). Let us now explore a useful experiment and the utility of Fung's exponential (pseudo)strain-energy function.

6.3 Design of Biaxial Tests on Planar Membranes

6.3.1 Biological Motivation

A membrane is defined differently in biology and mechanics. In biology, a membrane is a thin layer of tissue that covers a surface or separates a space; examples include the cell membrane, the basement membrane in the arterial wall, the pleural membrane which covers the lung, the epicardial membrane which covers the heart, and the mesentery within the abdomen. Consideration of the important structural roles played by these membranes, as well as membranes such as the urinary bladder and saccular aneurysms, reveals the need to understand the associated mechanics. In mechanics, a membrane is defined as a structure having two dimensions much greater than the third and, in particular, a structure that offers negligible resistance to bending; that is, the in-plane load-carrying capability is most important in membranes, a simple example being that a soap bubble resists a distension pressure solely through its (in-plane) surface tension. Experience has revealed that most biological membranes behave mechanically as membranes within the context of mechanics; hence, in many cases, we are interested primarily in their in-plane properties.

Many biological membranes consist primarily of a 2-D plexus of elastin and collagen embedded in a viscous ground substance matrix consisting of proteoglycans and water. Moreover, in most cases, the elastin and collagen fibers have complex orientations (Fig. 6.8) that give rise to anisotropic responses. For this reason, it is not sufficient to study the material behavior using a single uniaxial test. Rather, it is useful to employ in-plane biaxial stretching tests to interrogate the mechanical behavior. Question: Why would bending tests such as those in Chapter 5 not be useful?

6.3.2 Theoretical Framework

In most cases, biaxial tests are designed such that multiple, individual loading fixtures are applied to each of the four sides of the sample (Fig. 6.9). The primary reason for this is that stretching in one direction will induce an associated shortening (i.e., Poisson-like effect) in the orthogonal direction. Whereas single clamps on each side would impede such deformations, individual loading fixtures do not, hence their use. Note, however, that individual (point) loads introduce *stress concentrations* at the points of application, where the tissue is likely to fail first if overloaded. Thus, if the

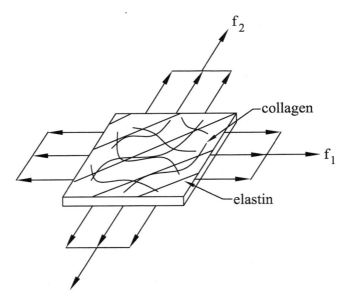

FIGURE 6.8 Schema of the complex distributions of elastin and collagen in a planar tissue such as the epicardium, pericardium, pleura, or mesentery. Such tissues are easily tested in a biaxial setting (i.e., subjected to orthogonal in-plane loads f_1 and f_2).

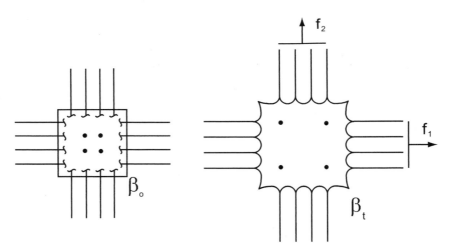

FIGURE 6.9 Specimen prepared for biaxial testing, loaded via multiple individual loading strings per side. Individual loading fixtures allow the tissue to thin in one direction when pulled in the orthogonal direction, which is important, yet they also introduce stress concentrations at the loading sites. From Humphrey (2002), with permission.

experiment is designed to investigate failure properties, one often introduces a circular or elliptical hole in the center of the specimen to control
and initiate the failure process. We will not consider such tests here, but
they have applicability to the design of incisions and wound healing, including cataract surgery and the removal of the lens, as well as to the study of
failure mechanisms. Rather, let us consider subfailure tests wherein one
focuses on the central region in which the deformation is measured (similar
to the use of a gauge length in the axially loaded rod experiment). Because
of the large deformations, however, we cannot use strain gauges as in tests
on bone. Question: How then should we measure the deformation? Recalling our relation for the deformation gradient [Eq. (6.4)], note that we
simply need to know the current positions (x, y) of points that were originally at (X, Y) in an undeformed reference configuration. One way to do
this is to *track* markers that are placed in the central region of the sample.
In general, we place multiple markers (e.g., 3, 4, 9, . . .) and use so-called
interpolation functions to obtain continuous expressions for

$$x = f(X, Y) \qquad \text{and} \qquad y = g(X, Y) \tag{6.33}$$

or, similarly, for the displacements,

$$u_x = x - X = f(X, Y) - X \qquad \text{and} \qquad u_y = y - Y = g(X, Y) - Y, \tag{6.34}$$

so that we can compute the requisite deformation or displacement gradients. For such approaches, see Humphrey (2002). Here, we will simply
assume that the deformation is homogeneous (i.e., uniform) in the central
region, thus allowing us to compute pointwise quantities like strain over
finite lengths. In particular, referring to Figure 6.10, let

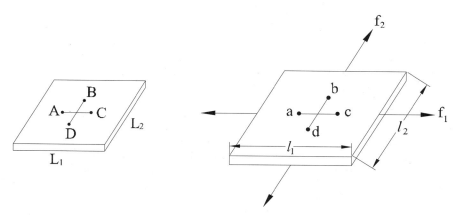

FIGURE 6.10 Dimensions of a biaxial specimen in the central region, before and after
loading and assuming a homogeneous deformation; a, b, c, and d denote tracking
markers for computing strain in the central region of the soft tissue sample. Strain
gauges cannot be used, of course, because of their extreme stiffness.

$$\overline{ac} = l_x, \qquad \overline{bd} = l_y \tag{6.35}$$

and, similarly, in the undeformed configuration

$$\overline{AC} = L_x, \qquad \overline{BD} = L_y \tag{6.36}$$

such that [cf. Eq. (2.50)]

$$x = \Lambda_1 X = \frac{l_x}{L_x} X, \qquad y = \Lambda_2 Y = \frac{l_y}{L_y} Y. \tag{6.37}$$

Although a membrane is treated two dimensionally, there will be thinning in the third direction. If we account for this via the stretch $\Lambda_3 = h/H$, where h and H are the deformed and undeformed thickness, respectively, we have

$$z = \Lambda_3 Z = \frac{h}{H} Z. \tag{6.38}$$

Hence, for this simple in-plane biaxial test, we have, from Eq. (6.4),

$$[F] = \begin{bmatrix} \dfrac{\partial x}{\partial X} & \dfrac{\partial x}{\partial Y} & \dfrac{\partial x}{\partial Z} \\[2mm] \dfrac{\partial y}{\partial X} & \dfrac{\partial y}{\partial Y} & \dfrac{\partial y}{\partial Z} \\[2mm] \dfrac{\partial z}{\partial X} & \dfrac{\partial z}{\partial Y} & \dfrac{\partial z}{\partial Z} \end{bmatrix} = \begin{bmatrix} \Lambda_1 & 0 & 0 \\ 0 & \Lambda_2 & 0 \\ 0 & 0 & \Lambda_3 \end{bmatrix}, \tag{6.39}$$

where $\Lambda_1 = l_x/L_x$, $\Lambda_2 = l_y/L_y$, and $\Lambda_3 = h/H$ are all measurable in principle, although on-line measurement of changes in thickness are problematic in many cases. It is for this reason that experimentalists often invoke the incompressibility constraint (det $[F] = 1$), for this allows the thinning to be inferred from the more easily measured in-plane quantities. Indeed, because many tissues are nearly incompressible under many cases of cyclic loading, volume conservation yields

$$l_x l_y h = L_x L_y H \rightarrow \Lambda_1 \Lambda_2 \Lambda_3 = 1, \tag{6.40}$$

where we recognize that $\Lambda_1 \Lambda_2 \Lambda_3 = \det[F]$.

Now, for the constitutive behavior. Recall from Chapters 1 and 2 that a general formulation requires five general steps: DEICE, which is to say, *delineating* general characteristic behaviors, *establishing* an appropriate theoretical framework, *identifying* a specific functional relationship between the independent and dependent constitutive parameters, *calculating* best-fit values of the associated material parameters, and *evaluating* the predictive capability of the final relation. In the present context, we assume that general characteristics are a preconditionable nonlinear, pseudoelastic, homogeneous, anisotropic, incompressible behavior of a membrane having negligible bending stiffness and subject to large in-plane deformations. One possible theoretical framework is thus the theory of large-deformation membrane elasticity, which we will employ.

In practice, identifying the specific functional form of the constitutive relation, for conditions of interest, is the most challenging of the five steps. In general, there are three ways by which this can be accomplished: via an educated guess that is based on extant observations, theoretical restrictions, and prior experience; via a formal theoretical derivation, often based on statistical mechanical arguments; or via inference directly from clever interpretations of experimental data. Fung's identification of the 1-D exponential $\Sigma = \Sigma(\Lambda)$ relation in Eq. (6.15) is a good example of an experimentally based identification. Such identifications are much more difficult in two or three dimensions however, and the interested reader is referred to Humphrey (2002) for details. Here, let us proceed by assuming (guessing) a 2-D Fung-type exponential relation of the form

$$W = c(e^{Q} - 1), \qquad Q = c_1 E_{11}^2 + c_2 E_{22}^2 + 2c_3 E_{11} E_{22}, \qquad (6.41)$$

similar to the 3-D form given in Eqs. (6.25) and (6.26). Note that we ignore shear here because only E_{11} and E_{22} are assumed to be nonzero in our biaxial test. Note, too, that the units of W are energy per volume; we could alternatively use a strain energy w defined per surface area (of the membrane), where $w \sim HW$ and H is the undeformed thickness.

From Eqs. (6.23) and (6.24), therefore, we have for Eqs. (6.39) and (6.41) the following:

$$[\sigma] = \frac{1}{\Lambda_1 \Lambda_2 (h/H)} \begin{bmatrix} \Lambda_1 & 0 & 0 \\ 0 & \Lambda_2 & 0 \\ 0 & 0 & \Lambda_3 \end{bmatrix} \begin{bmatrix} S_{11} & 0 & 0 \\ 0 & S_{22} & 0 \\ 0 & 0 & 0 \end{bmatrix} \begin{bmatrix} \Lambda_1 & 0 & 0 \\ 0 & \Lambda_2 & 0 \\ 0 & 0 & \Lambda_3 \end{bmatrix}, \qquad (6.42)$$

where

$$S_{11} = (2ce^{Q})(c_1 E_{11} + c_3 E_{22}), \qquad S_{22} = (2ce^{Q})(c_2 E_{22} + c_3 E_{11}) \qquad (6.43)$$

and, therefore,

$$\sigma_{11} = \frac{\Lambda_1}{\Lambda_2} \left(\frac{H}{h} \right) (2ce^{Q})(c_1 E_{11} + c_3 E_{22}) \qquad (6.44)$$

and

$$\sigma_{22} = \frac{\Lambda_2}{\Lambda_1} \left(\frac{H}{h} \right) (2ce^{Q})(c_2 E_{22} + c_3 E_{11}), \qquad (6.45)$$

with $\sigma_{33} = 0$. Note that this is an example of a state of plane stress, which was defined in Chapter 2. The nonzero in-plane tensions that represent the load-carrying capability of the membrane are thus,

$$T_1 = \sigma_{11}h = \left(\frac{\Lambda_1}{\Lambda_2}\right)Hce^Q[c_1(\Lambda_1^2 - 1) + c_3(\Lambda_2^2 - 1)], \tag{6.46}$$

$$T_2 = \sigma_{22}h = \left(\frac{\Lambda_2}{\Lambda_1}\right)Hce^Q[c_2(\Lambda_2^2 - 1) + c_3(\Lambda_1^2 - 1)], \tag{6.47}$$

where T_1 and T_2 are the principal tensions (force per length). Moreover, from Eq. (6.16), we know that

$$[\sigma] = \frac{1}{\Lambda_1\Lambda_2(h/H)}\begin{bmatrix} \Lambda_1 & 0 & 0 \\ 0 & \Lambda_2 & 0 \\ 0 & 0 & \Lambda_3 \end{bmatrix}\begin{bmatrix} \dfrac{f_1}{A_{1o}} & 0 & 0 \\ 0 & \dfrac{f_2}{A_{2o}} & 0 \\ 0 & 0 & 0 \end{bmatrix} \tag{6.48}$$

where A_{1o} is the easily measured original area over which the resultant force f_1 acts in the undeformed configuration and similarly for A_{2o}. Hence, we have

$$T_1 = \sigma_{11}h = \frac{H}{\Lambda_2}\left(\frac{f_1}{A_{1o}}\right), \qquad T_2 = \sigma_{22}h = \frac{H}{\Lambda_1}\left(\frac{f_2}{A_{2o}}\right), \tag{6.49}$$

which allows us to relate our theoretically predicted and experimentally determined principal *tensions* (sometimes called *stress resultants*); that is, Eqs. (6.46) and (6.47) combined with Eq. (6.49) allow us to *calculate* the values of the material parameters as demanded in the fourth step of the DEICE procedure. Although we only have four "unknown" material parameters (c, c_1, c_2, and c_3), which may imply the need for only two data points (i.e., two sets of σ_{11} versus Λ_1 and σ_{22} versus Λ_2 data, which provide four equations for our four unknowns), it is common practice to determine the values of the parameters in a least-squares sense to avoid the consequences of the inevitable experimental errors. Because $c_1, c_2,$ and c_3 appear in the exponential, one must use a *nonlinear least-squares* method to determine the best-fit values. In principle, then, we seek to minimize the sum of the squares of the differences between theoretically predicted and experimentally determined tensions at $j = 1, 2, \ldots n$ equilibrium configurations by minimizing the objective function e:

$$e = \sum_{j=1}^{n}\left\{\left[\left(\frac{\Lambda_1}{\Lambda_2}\right)cHe^Q[c_1(\Lambda_1^2 - 1) + c_3(\Lambda_2^2 - 1)] - \frac{H}{\Lambda_2}\left(\frac{f_1}{A_{1o}}\right)\right]_j^2\right.$$

$$\left. + \left[\left(\frac{\Lambda_2}{\Lambda_1}\right)cHe^Q[c_2(\Lambda_2^2 - 1) + c_3(\Lambda_1^2 - 1)] - \frac{H}{\Lambda_1}\left(\frac{f_2}{A_{2o}}\right)\right]_j^2\right\} \tag{6.50}$$

where we note that the effect of thickness H can be removed entirely as a common factor as expected of this 2-D analysis. Minimization of e is accomplished, of course, by solving the simultaneous nonlinear equations given by

$$\frac{\partial e}{\partial c} = 0, \qquad \frac{\partial e}{\partial c_1} = 0, \qquad \frac{\partial e}{\partial c_2} = 0, \qquad \frac{\partial e}{\partial c_3} = 0. \qquad (6.51)$$

Commercially available codes accomplish this easily, provided the functional form of the constitutive equation is well chosen; thus, we need not be concerned with numerical details here. Of course, once the best-fit values of the parameters are calculated, one should complete the DEICE procedure by *evaluating* the general predictive capability of the final relation. This is generally accomplished by comparing its predictions to data that were not used in the constitutive formulation. Such validations are essential for engendering confidence in the use of relations determined from relatively simple experiments. Simple experiments are sought by experimentalists, of course, for they ease performance and interpretation, but we must ensure that the associated results hold for the generally much more complex in vivo situation of interest. For example, if we determine a Young's modulus E for cortical bone via a 1-D tension test, we must verify that this value can be used in an analysis of bending wherein the moment-curvature equation (5.22) governs the response. Such issues are even more important in nonlinear relations.

6.4 Stability of Elastomeric Balloons

6.4.1 Biological Motivation

In 1963, C. Dotter forced a catheter retrograde through an occluded iliac artery in a patient to obtain a routine aortogram. In doing so, flow was improved through the previously occluded vessel; this marked the inadvertent beginning of the use of catheters as interventional rather than just diagnostic devices. Indeed, since that time, millions of balloon angioplasties have been performed to open atherosclerotic vessels, and based on the associated successes, other balloon-based procedures have arisen.

For example, Zubkov et al. (1984) were the first to report the use of a balloon dilatation to treat vasospasm in the cerebral vasculature. Simply put, a *vasospasm* is a persistent, nonphysiologic constriction of an artery that reduces distal flow and may thereby lead to ischemia or necrosis. Intracranial vasospasms occur in 30–70% of all patients who experience a subarachnoid hemorrhage, the most common cause of which is the rupture of an intracranial aneurysm. Although responsible for significant mortality and morbidity, with symptoms typically presenting 3–15 days after the bleed, vasospasm remains poorly understood. Nevertheless, because the reduced lumen associated with vasospasm compromises blood flow, Zubkov et al. suggested that a balloon dilatation could restore the lumen to near its normal value; that is, they reasoned that a controlled injury to the arterial wall could weaken it so that the normal blood pressure could distend the vessel more. Because the affected cerebral vessels are not stiffened by atherosclerosis or supported by significant perivascular tissue, early neuroangioplasty balloons were constructed of latex or silicone in contrast to the much stiffer polyethylene balloons used in traditional coronary or peripheral vessel angioplasty (Fig. 6.11). Experience revealed, however, that the dilatations by the "softer" balloons were difficult to control. In hindsight, there are at least two reasons for this, which the design engineer should have anticipated had he or she been familiar with nonlinear elasticity. First, latex rubber exhibits a preconditioning-like (cf. Fig. 6.4) softening effect referred to as the *Mullin's effect*. Hence, the pressure–volume behavior of a given balloon may change from cycle to cycle, thus complicating its control (Fig. 6.12). Many neurointerventionalists would inflate and deflate a balloon before a procedure to ensure that the device did not leak, but this was not a well-specified, repeatable part of the procedure. Second, as we know from common experience, rubber party balloons can exhibit an instability that leads to a rapid expansion at a constant pressure, which again would complicate one's attempt to control the dilatation. Such an instability is similar to that experienced by columns whereby the structure changes shape dramatically due to a small change in load. Hence, let us briefly look at this type of a material/structural instability here.

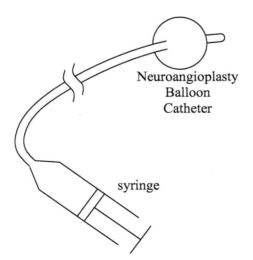

Neuroangioplasty
Balloon
Catheter

syringe

FIGURE 6.11 Schema of a neuroangioplasty balloon, which is constructed out of silicone rather than a polyethylene as in coronary angioplasty balloons.

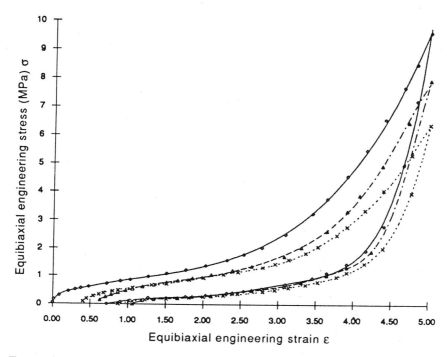

FIGURE 6.12 Pressure–volume behavior of a rubber balloon over a number of infla-
tion/deflation cycles. Note that, like preconditioning of a soft tissue (cf. Fig. 6.4), the
balloon stress-softens with repeated loading, an effect called the Mullin's effect.
(From Johnson and Beatty (1995), with permission from Elsevier).

6.4.2 Theoretical Framework

Although we will illustrate the phenomenon of an inflation instability in an
elastomeric balloon by considering the simple case of a spherical geometry, let us first consider the more general case of an axisymmetric inflation
of a membrane. Axisymmetry implies that the undeformed and deformed
shapes of the membrane can each be described by generator curves that
define the entire surface when rotated through 2π radians (Fig. 6.13). It can
be shown (Humphrey, 2002) that the two governing equilibrium equations
for an axisymmetric membrane are

$$\frac{d}{dr}(rT_1) = T_2, \qquad \kappa_1 T_1 + \kappa_2 T_2 = P, \qquad (6.52)$$

where T_1 and T_2 are the principal tensions (Cauchy stress resultants) in the
meridional and circumferential directions, κ_1 and κ_2 are the principal
curvatures in the deformed configuration, and $P > 0$ is the transmural

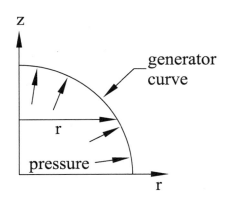

z

generator curve

r

pressure

r

FIGURE 6.13 Pressure–volume geometry for an axisymmetrically inflated membrane. By axisymmetric, we mean that a generator curve can be revolved about an axis to yield the surface of the membrane. Axisymmetry imposes certain restrictions on the material properties and applied loads.

(distension) pressure. The second of these relations is known as Laplace's equation, which is widely used (sometimes misused) in medicine and surgery. Moreover, it can be shown from differential geometry that

$$\frac{d}{dr}(r\kappa_2) = \kappa_1. \tag{6.53}$$

Our goal then is to solve for the tensions T_1 and T_2 in terms of the applied load and measures of the (deformed) geometry [cf. Eq. (3.59)]. Given two equilibrium equations in terms of our two unknown tensions, note that it is often easiest to reduce one of the equations to contain only one of the unknowns. Toward this end, let us substitute for T_2 and κ_1, from Eqs. (6.52) and (6.53), in Laplace's equation such that

$$\kappa_1 T_1 + \kappa_2 T_2 = P \rightarrow \frac{d}{dr}(r\kappa_2)T_1 + \kappa_2 \frac{d}{dr}(rT_1) = P. \tag{6.54}$$

Additionally, note that (check it) this equation can be rewritten as

$$\frac{1}{r}\frac{d}{dr}(r\kappa_2 rT_1) = P, \tag{6.55}$$

which, in turn, can be integrated as (see Appendix 8 of Chapter 8)

$$\int \frac{d}{dr}(r\kappa_2 rT_1)\, dr = \int Pr dr. \tag{6.56}$$

If P is assumed to be constant at each equilibrium state, we thus obtain

$$r^2\kappa_2 T_1 = P\frac{r^2}{2} + c_1 \rightarrow T_1 = \frac{P}{2\kappa_2} + \frac{c_1}{r^2\kappa_2}, \tag{6.57}$$

which must be valid for all r, including $r = 0$; this requirement implies that $c_1 = 0$ which gives us our result for T_1 in terms of the applied load and measure of the geometry, namely P and κ_2. Substituting back into Laplace's equation, we then find the desired relation for T_2 as well. The final results are:

$$T_1 = \frac{P}{2\kappa_2}, \qquad T_2 = \frac{P}{\kappa_2}\left(1 - \frac{\kappa_1}{2\kappa_2}\right).$$ (6.58)

In Humphrey (2002), it is shown that these equations are fundamental to designing and interpreting inflation tests on axisymmetric membranes as well as to analyzing the stresses in such membranes, such as intracranial saccular aneurysms. Here, however, let us consider but one special case—the inflation of a spherical membrane.

6.4.3 Inflation of a Neuroangioplasty Balloon

The two principal curvatures in an inflated spherical membrane (i.e., spherical in its undeformed and deformed configurations) are equal and simply given by 1 over the radius of curvature, which in this special case equals the radius of the deformed sphere a. Hence,

$$\kappa_1 = \kappa_2 = \frac{1}{a}$$ (6.59)

and, therefore,

$$T_1 = \frac{Pa}{2} \quad \text{and} \quad T_2 = Pa\left(1 - \frac{a}{2a}\right) = \frac{Pa}{2}$$ (6.60)

whereby we see that the tension is uniform (i.e., $T_1 = T_2$ independent of location r, θ, ϕ) in the sphere (which must be isotropic in-plane for a sphere to inflate into a sphere) and given by

$$T = \frac{Pa}{2}.$$ (6.61)

To compute the Cauchy stress from this tension (i.e., Cauchy stress resultant T has units of force per length), we simply divide by the deformed thickness h to obtain

$$\sigma = \frac{T}{h} = \frac{Pa}{2h},$$ (6.62)

which we recognize to be the same result as that obtained in Chapter 3 for the inflation of a thin-walled sphere [Eq. (3.58)], as it should.

One important observation from this derivation, however, is that a is, strictly speaking, not the deformed radius; rather it is the deformed radius of curvature, *curvature* being the controlling geometric feature of axisymmetric membranes [cf. Eq. (6.58)]. Indeed, this simple realization may explain, in part, a long-standing controversy in neurosurgery. If $\sigma \cong Pa/2h$ in a saccular aneurysm and if the mean blood pressure ($\sim$110 mm Hg) and wall thickness ($\sim$100 μm) are similar from patient to patient, one would expect that the larger-diameter lesions would be much more susceptible to rupture. Although larger lesions are often more lethal, many smaller lesions rupture, whereas larger lesions do not. It is suggested in Humphrey (2002) that this enigma may be due, in part, to the focusing on size rather than shape; that is, a "large" lesion can have a small radius of curvature, whereas a "small" lesion may have a large radius of curvature, curvature being the controlling factor (Fig. 6.14). Again, therefore, we are well advised to remember how our governing equations are derived so that we can interpret their implications.

That said, let us return to the problem at hand—the possible instability of a neuroangioplasty balloon, which we idealize as a sphere. It can be shown (Humphrey, 2002) that the "simplest" descriptor of the behavior of a rubberlike material over moderate stretches (up to $\sim$1.3–1.4) is the so-called neo-Hookean relation. For a membrane, it can be written as (Humphrey, 2002)

$$W = C\left(2E_{11} + 2E_{22} + \frac{1}{1 + 2E_{11} + 2E_{22} + 4E_{11}E_{22}} - 1\right), \qquad (6.63)$$

where C is a material parameter having units of stress and E_{11} and E_{22} are the principal components of the Green strain [cf. Eq. (6.41) for the Fung material]. Now, for the inflation of a thin-walled sphere, the deformation gradient can be written as

$$[F] = \begin{bmatrix} \Lambda_1 & 0 & 0 \\ 0 & \Lambda_2 & 0 \\ 0 & 0 & \Lambda_3 \end{bmatrix}, \qquad (6.64)$$

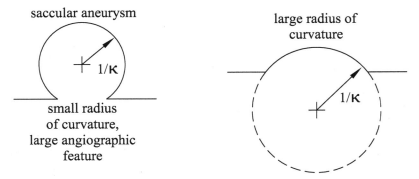

FIGURE 6.14 Radius of curvature for two aneurysms revealing that a larger lesion may have a smaller radius of curvature. Contrary to popular clinical belief, it is curvature, not size, that likely controls the associated biomechanics.

where $\Lambda_1 = 2\pi a/2\pi A = \Lambda_2$ are the principal stretches in the meridional and circumferential directions, with a and A the deformed and undeformed radii, respectively, and $\Lambda_3 = h/H$ is the principal stretch in the thickness direction, with h and H the deformed and undeformed thicknesses, respectively. Hence, let $\Lambda_1 = \Lambda_2 = a/A = \Lambda$ for convenience. From Eqs. (6.23) and (6.24), therefore, we have

$$[\sigma] = \frac{1}{\det[F]}[F]\left[\frac{\partial W}{\partial E}\right][F]^T, \tag{6.65}$$

where, for the neo-Hookean material behavior,

$$\frac{\partial W}{\partial E_{11}} = C\left(2 + 0 + \frac{(1 + 2E_{11} + 2E_{22} + 4E_{11}E_{22})(0) - 1(2 + 4E_{22})}{(1 + 2E_{11} + 2E_{22} + 4E_{11}E_{22})^2}\right). \tag{6.66}$$

For the $[F]$ given here, Eq. (6.7) reveals that

$$E_{11} = \frac{1}{2}(\Lambda_1^2 - 1) = \frac{1}{2}(\Lambda^2 - 1) \tag{6.67}$$

and similarly for E_{22}. Hence, for the spherical deformation,

$$\begin{aligned}
\left.\frac{\partial W}{\partial E_{11}}\right|_{E_{11}=E_{22}=\frac{1}{2}(\Lambda^2-1)} &= C\left(2 - \frac{2 + 4\left(\frac{1}{2}\right)(\Lambda^2 - 1)}{\left[1 + 4\left(\frac{1}{2}\right)(\Lambda^2 - 1) + 4\left(\frac{1}{4}\right)(\Lambda^2 - 1)^2\right]^2}\right) \\
&= 2C\left(1 - \frac{1 + \Lambda^2 - 1}{(1 + 2\Lambda^2 - 2 + \Lambda^4 - 2\Lambda^2 + 1)^2}\right)
\end{aligned} \tag{6.68}$$

or

$$\left.\frac{\partial W}{\partial E_{11}}\right|_{E_{11}=E_{22}=\frac{1}{2}(\Lambda^2-1)} = 2C\left(1 - \frac{\Lambda^2}{\Lambda^8}\right) = 2C\left(1 - \frac{1}{\Lambda^6}\right) \tag{6.69}$$

and similarly for $\partial W/\partial E_{22}$. Hence,

$$[\sigma] = \frac{1}{\Lambda^2(h/H)}\begin{bmatrix} \Lambda & 0 & 0 \\ 0 & \Lambda & 0 \\ 0 & 0 & \dfrac{h}{H} \end{bmatrix}\begin{bmatrix} 2C\left(1 - \dfrac{1}{\Lambda^6}\right) & 0 & 0 \\ 0 & 2C\left(1 - \dfrac{1}{\Lambda^6}\right) & 0 \\ 0 & 0 & 0 \end{bmatrix}\begin{bmatrix} \Lambda & 0 & 0 \\ 0 & \Lambda & 0 \\ 0 & 0 & \dfrac{h}{H} \end{bmatrix}$$

$$\tag{6.70}$$

from which we see that

$$\sigma_{11} = \frac{2CH}{h}\left(1 - \frac{1}{\Lambda^6}\right) = \sigma_{22} \qquad (6.71)$$

and $\sigma_{33} = 0$, or in terms of the principal tensions, the nonzero values are

$$T_1 = h\sigma_{11} = 2CH\left(1 - \frac{1}{\Lambda^6}\right) = T_2 \equiv T. \qquad (6.72)$$

Finally, appealing to equilibrium, Eq. (6.60), we have

$$T = \frac{Pa}{2} \rightarrow P = \frac{2T}{a} = \frac{2}{\Lambda A}\left[2CH\left(1 - \frac{1}{\Lambda^6}\right)\right], \qquad (6.73)$$

with $a = \Lambda A$, or

$$P(\Lambda) = \frac{4CH}{A}\left(\frac{1}{\Lambda} - \frac{1}{\Lambda^7}\right). \qquad (6.74)$$

Plotting the distension pressure as a function of stretch Λ (i.e., increase in normalized radius a/A) reveals a local maximum [i.e., a transition from a stable loading path to an unstable path, the later being characterized by a rapid increase in size even in the presence of a diminishing pressure (Fig. 6.15)]. Question: At what values of Λ does this instability occur? To answer this, recall from calculus that we find local extrema by taking the first derivative, namely

$$\frac{dP}{d\Lambda} = 0 = \frac{4CH}{A}\left(-\frac{1}{\Lambda^2} + \frac{7}{\Lambda^8}\right) \rightarrow \Lambda^6 = 7. \qquad (6.75)$$

Hence, if a near-spherical neuroangioplasty balloon exhibits a neo-Hookean behavior, it is expected to become unstable when $\Lambda = 7^{1/6} \approx 1.38309\ldots$, where $\Lambda = a/A$. This phenomenon explains, in part, why neuroradiologists had trouble controlling the expansion of the neuroangioplasty balloon based on pressure[2] and reveals yet again *the importance of a careful analysis in the design and use of a medical device.* Whereas the present analysis holds approximately for the inflation of an isolated balloon, its inflation within a vessel will change its geometry and thus complicate the analysis. With an appropriate theoretical framework and a reasonable descriptor of the material properties, however, the more complex analysis for nonspherical geometries can be conducted with the aid of a computer. The take-home message here is simply that we must often appreciate and employ the general methods of nonlinear continuum mechanics in the design of many medical devices and, furthermore, in troubleshooting clinical complications.

---

[2] Whereas pressure is not useful for feedback control, volume is. We revisit this problem in Section 10.3 of Chapter 10 in terms of a saline infused balloon.

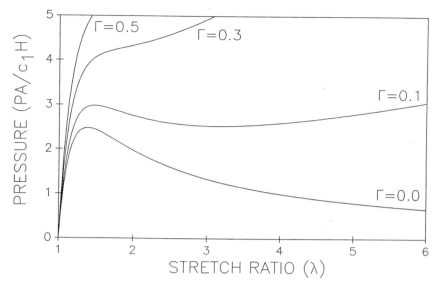

FIGURE 6.15 Pressure–stretch response of spherical rubber (party) balloons with a local maximum revealing a limit point instability (i.e., a transition from a stable to an unstable loading path). Such instabilities depend on the value of one of the material parameters, Γ, that describes the properties of a Mooney-Rivlin rubber; $\Gamma = 0$ for neo-Hookean case. From Humphrey (2002), with permission.

Observation 6.2 The eye is a remarkable organ; it collects and focuses incoming light on the retina, which, in turn, serves to convert the light into electrical signals that the brain can interpret in bold and vibrant hues of the rainbow. The eye consists of an outer shell, five-sixths of which is the collagenous, opaque sclera and the remainder is the collagenous, transparent cornea. Contained within the eye is the lens, which helps to focus the light on the retina and divides the interior into two chambers: The anterior chamber contains the aqueous humor and the posterior chamber contains the vitreous humor. The iris serves as an aperture for the lens to control the amount of incoming light. The lens is held in place by a thin membrane, the lens capsule, which consists primarily of type IV collagen and allows the curvature of the lens to be changed via contraction and relaxation of the ciliary process. Finally, the posteriorly located optic nerve conducts the signals from the eye to the brain. See Figure 6.16.

Mechanics plays many roles in ophthalmology. Glaucoma is a disease characterized by an increase in intraocular pressure; it causes pathological changes in the optical disk and concomitant defects in the field of vision. Diagnosis and treatment of this pressure-induced disease requires an understanding of the mechanics. A cataract is a disease of the lens characterized

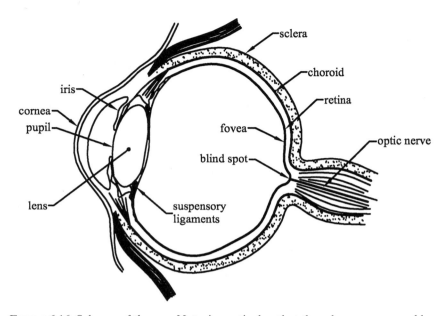

FIGURE 6.16 Schema of the eye. Note, in particular, that the sclera, cornea, and lens capsule have each been assumed to have negligible bending stiffness (i.e., to behave mechanically as membranes) in various biomechanical analyses. One must be cautious, however, in adopting assumptions from the literature; the cornea, for example, has significant bending stiffness, which is likely fundamental to any analysis of clinical procedures such as radial keratotomy or the more popular LASIK surgery. The lens capsule, on the other hand, has much less bending stiffness and may be well described as a membrane in some applications. We are reminded, therefore, that constitutive relations describe the behavior of a material under specified conditions, not the material itself. We must always be mindful of the specific application, particularly in biomechanics.

by an opacification that blurs vision and, in extreme cases, causes blindness. Surgical correction involves the removal and replacement of the lens with an intraocular device. Performed over 1.2 million times a year, cataract surgery is currently the most commonly performed surgery in the United States. The cornea is nearly 18 mm in diameter, less than 1 mm thick in the middle, and has radii of curvature of ~7.8 mm and 6.8 mm at its outer and inner surfaces, respectively. The lens power of the eye is achieved primarily by the curvature of the cornea and secondarily by the lens. In recent years, surgical alteration of the curvature of the cornea has become a popular alternative to the use of eyeglasses or contact lens. This is accomplished by introducing a series of incisions to relieve some of the corneal stress (e.g., radial keratotomy) or by locally shrinking portions of the cornea via thermal denaturation (laser thermokeratoplasty) of the type I collagen in the cornea; in either case, the net effect is that the tension in the pressurized cornea is altered and it thereby changes its curvature and thus refractive power. A rigorous biomechanical analysis of this procedure, for purposes of surgical planning, is beyond the scope of an introductory textbook because of the nonlinear material behavior of the cornea, the need for fracture-mechanics-type analyses to model the incisions, and the complex geometry. Of course, not only do we need to predict the change in the configuration of the cornea (e.g., thickness and curvature) immediately after the surgery, we must account for changes due to healing and adaptation, the latter of which will be stimulated by changes in the stress field experienced by the cells within the cornea. The interested reader is thus encouraged to research the different applications of mechanics in opthalmology.

Here, however, let us briefly consider the lens capsule. This tissue consists primarily of type IV collagen; it is very thin (~10–20 μm for the anterior lens capsule and 3–4 μm for the posterior lens capsule in the human) and contains a monolayer of epithelial cells on its anterior portion. Clearly, the stress field within the lens capsule is altered due to the surgical removal of the lens and replacement with an intraocular device, which tends to be much smaller than the native lens. One complication of cataract surgery is a secondary opacification of the posterior lens capsule, which often requires a revision, corrective procedure (e.g., thermal ablation). It is thought that this "secondary cataract" is due, in part, to the migration of epithelial cells to the central region of the posterior lens capsule and their production of excessive matrix material. Moreover, it is thought that this altered migration and synthetic activity results from the surgical perturbation of the stress or strain field in the native lens capsule. In other words, mechanotransduction mechanisms likely alter the gene expression by the epithelial cells. It is important, therefore, to quantify the native stress and strain fields in the lens capsule and how they are altered by cataract surgery. Indeed, one would hope that biomechanical analyses could identity designs for the intraocular devices that do not adversely perturb the mechanical environment of the epithelial cells. As a first approximation, one could think of the lens capsule as two hemiellipsoids that are loaded by a transmural pressure (difference in radial stress boundary conditions due to the lens on the inner surface of the lens capsule and the pressure in the aqueous or vitreous humor). Notwithstanding the tractions due to the ciliary process, Eqs. (6.52) and (6.53) could thus be used, to a first approximation, to estimate the membrane stresses in the native configuration. Hence, we see another application of the equations of membrane mechanics.

6.5 Residual Stress and Arteries

6.5.1 Biological Motivation

Hypertension, atherosclerosis, aneurysms, and stroke—these and other vascular diseases continue to be leading causes of mortality and morbidity. Although manifested differently in each case, the fundamental mechanisms by which these diseases begin and then progress relate to basic cellular functions: cell migration, replication, and apoptosis; the production of vasoactive, growth regulatory, inflammatory and degratory molecules; and the synthesis and organization of constituents of the extracellular matrix. As noted in Chapter 1, many of these cellular functions are influenced, via mechanotransduction mechanisms, by changes in the local mechanical environment. For example, an increased blood pressure nonuniformly increases intramural wall stress, which, in turn, increases cell proliferation and synthesis of the matrix in hypertension, first in the inner portion of the wall but eventually throughout the wall. Conversely, decreased blood flows may lead to decreases in wall shear stress, which, in turn, promote the production of adhesion molecules by the endothelium and the attendant adhesion of monocytes to the endothelium; these monocytes subsequently migrate into the subintimal space, transform into macrophages, and contribute to atherogenesis. There is a need, therefore, to understand arterial mechanics both in health and disease. Whereas a more detailed discussion can be found in Humphrey (2002), here we consider the first step, a simple analysis of wall stress in blood vessels.

First, however, let us briefly review some aspects of the structure of arteries. Despite the wide variety of arteries in the body—from the aorta to the renal arteries, coronary arteries, cerebral arteries, and so on—each of these vessels similarly consist of one to three concentric layers: the tunica intima, tunica media, and tunica adventitia (Fig. 6.17). Indeed, all blood vessels have an intima, which consists of a monolayer of endothelial cells and an underlying basal lamina composed primarily of mesh-like type IV collagen and the adhesion molecules fibronectin and laminin. In addition to being a smooth, nonthrombogenic interface between the blood and the contents of the vascular wall, the endothelium is biologically active. In response to chemical and mechanical stimuli, endothelial cells produce various vasoactive molecules (which dilate or constrict the vessel), growth factors (which promote cell replication or the synthesis of proteins), and factors that regulate the clotting process (e.g., heparan sulfate and the vascular cell adhesion molecule, VCAM-1). Moreover, the endothelium can modify blood-borne substances (e.g., lipids) for transport into the wall, which thereby play an important role in atherosclerosis. In contrast, the medial layer consists primarily of smooth muscle cells embedded in a plexus of elastin, various types of collagen (including types I, III, and V), and proteoglycans. In general, the closer these vessels are to the heart, the more elastin and the farther away the more smooth muscle. Regardless, wall thickness tends to increase so as to maintain the mean circumferential wall stress on the order of ~150 kPa. Whereas smooth muscle is primarily responsible for synthesizing the extracellular matrix proteins during development, it endows the mature vessel with an ability to constrict or dilate— functions that regulate blood flow locally. Most smooth muscle cells are oriented in the circumferential direction, although in some vessels, they are

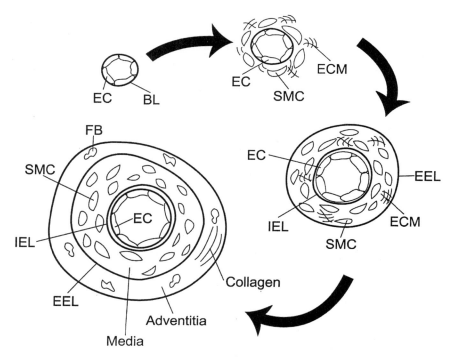

FIGURE 6.17 The three primary layers of the arterial wall: The inner layer, or intima, consists primarily of a monolayer of endothelial cells on a basement membrane; the middle layer, or media, consists largely of smooth muscle with surrounding elastin, collagen, and proteoglycans; the outer layer, or adventitia, consists primarily of collagen with abundant fibroblasts, nerves, and, in some vessels, a small vascular network called the vasa vasorum. Vascular development is seen to begin with endothelial cells (EC) on a basement layer (BL), with smooth muscle cells (SMC) attracted to the abluminal side of the BL, where they replicate and produce significant extracellular matrix (ECM). The adventitia, with abundant fibroblasts (FB), is added last. IEL and EEL denote the internal elastic lamina and the external elastic lamina, respectively, which demarcate the media in most large vessels.

oriented helically or in the axial direction. Note, too, that smooth muscle hypertrophy (increase in size), hyperplasia (increase in number), apoptosis (cell suicide), and migration play especially important roles in diseases such as aneurysms, atherosclerosis, and hypertension. Loss of matrix proteins, particularly elastin, similarly plays an essential role in the formation of aneurysms or vascular dissections. Finally, large vessels have an adventitial layer that connects with the perivascular tissue. The adventitia consists primarily of fibroblasts and axially oriented type I collagen, but also includes admixed elastic fibers, nerves and its own small vasculature, the vasa vasorum. The fibroblasts are responsible for regulating the matrix, particularly the collagen. It is thought that the adventitia serves, in part, as a protective sheath that prevents overdistension of the media (like all muscle, smooth muscle contracts maximally at a certain length, above and below which the contractions are less forceful). In summary, most arteries appear to have the same structural motif: a central parenchymal layer delimited by biologically or structurally important "membranes," the intima and adventitia. Although a detailed understanding of wall mechanics will require detailed modeling of the different properties of these layers and, indeed, the individual constituents within a layer, let us now consider a simple introduction to quantification.

6.5.2 Theoretical Framework

Like many other soft tissues, normal arteries often exhibit a nonlinear, pseudoelastic, heterogeneous, and anisotropic behavior over large physiologic strains. Moreover, they tend to behave incompressibly in many cases. It can be shown that in the case of incompressibility, the general constitutive equation embodied in Eq. (6.23) must be modified. For example, for incompressible behavior, we have

$$[\sigma] = -p[I] + [F]\left[\frac{\partial W}{\partial E}\right][F]^T, \tag{6.76}$$

where p is a scalar, pressure-like quantity (actually a Lagrange multiplier) that enforces the incompressibility constraint. That the $-p[I]$ contribution is needed is seen easily by noting that the second term on the right-hand side of Eq. (6.76) represents the stress due to deformation, which is zero in the absence of a deformation.

Imagine then a cube of incompressible material subjected to a hydrostatic pressure P (Fig. 6.18). Clearly, $\sigma_{11} = -P$, $\sigma_{22} = -P$, and $\sigma_{33} = -P$, with all shear stresses zero with respect to $(x, y, z) \equiv (1, 2, 3)$, even though there is no deformation because of incompressibility. Hence, in this case, and this case alone, the Lagrange multiplier p equals the hydrostatic pressure P and Eq. (6.76) correctly describes the state of stress in the absence of a deformation. For a formal derivation of Eq. (6.76), see Humphrey (2002).

Now, let us consider the deformation of an artery. Although arteries can have complex geometries (e.g., tapering and bifurcating), we consider here the simplest case: a short, straight, circular, excised segment of uniform thickness. Such samples are commonly tested in the laboratory. Moreover, to simulate in vivo deformations, let us consider a uniform axial extension and inflation via a constant axial load L and pressure P. If we label a generic point in the cross section as (ρ, ϑ, ζ) in an unloaded, excised configuration and (r, θ, z) in an extended and inflated configuration, we can imagine that such a point is extended uniformly by an amount $z = \lambda\zeta$, similar to that in

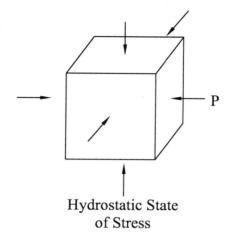

FIGURE 6.18 A small cube of incompressible material subjected to a hydrostatic pressure experiences stress but not strain. Indeed, the components of stress (equal and opposite the pressure) are the same regardless of the coordinate system. (cf. Exercise 2.9).

P

Hydrostatic State
of Stress

each direction in a biaxial test. Moreover, we might imagine that the deformation is axisymmetric, which is to say, each point may move out radially or axially, but it will maintain its angular position; that is, let $\theta = \vartheta$. Finally, we could relate r to ρ similar to the axial motion (through a constant stretch ratio), but careful consideration suggests that a point on the inner (intimal) surface may displace more radially than a point on the outer (adventitial) surface. Thus, let $r = \tilde{f}(\rho)$, in general. It can be shown that the deformation gradient $[F]$ associated with this assumed motion,

$$r = \tilde{f}(\rho), \qquad \theta = \vartheta, \qquad z = \lambda\zeta, \qquad (6.77)$$

is calculated easily and so too the associated stress $[\sigma]$ given a specific form of W. Indeed, there are many reports in the literature from the mid-1960s to the mid-1980s in which this was done. Briefly, these analyses suggested that the circumferential stress $\sigma_{\theta\theta}$ is comparatively much higher in the inner wall than in the outer wall (cf. thick-walled inflation solution in Chapter 3). Indeed, some investigators suggested that this was one cause of atherosclerosis, a disease of the inner wall. In 1983, however, it was noted that the unloaded, excised configuration is not stress-free. In particular, if one introduces a radial cut in such an arterial segment, it "springs open" into a sector. Fung suggested that this revealed the presence of a *residual stress* in arteries that was likely due to differential growth during development. Indeed, Skalak had suggested, in 1981, the possibility of residual stresses due to growth and remodeling. It was Fung and colleagues, however, who demonstrated that vascular adaptations via growth and remodeling processes, such as in hypertension, actually alter this residual stress field. One of the important consequences of residual stress is that it appears to *homogenize* the transmural distribution of stresses within the arterial wall. If this is true, this would suggest that cells at different locations within the

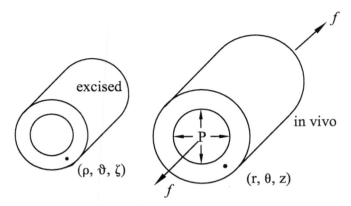

FIGURE 6.19 A segment of an artery that is removed from the body (left) and the same segment isolated fictitiously in the body via a free-body diagram.

wall tend to experience similar stresses under normal conditions. Conversely, altered conditions may render the transmural stress field less homogeneous and thus induce growth and remodeling processes that are different at different spatial locations but that together seek to restore overall normalcy. Quantification of soft tissue growth and remodeling is thus a very important current area of research. Here, however, let us return to the question of wall stress in the normal wall.

Figure 6.19 shows two configurations that occur naturally in an experiment on an artery: One is the unloaded, excised cylindrical segment and the other is a fictitious segment associated with the free-body diagram from the in vivo state. Figure 6.20 shows an additional radially cut configuration for which we label our previous generic material point via (R, Θ, Z). It can be shown that the motion from (R, Θ, Z) to (ρ, ϑ, ζ) can be approximated by

$$\rho = g(R), \qquad \vartheta = \frac{\pi}{\Theta_0}\Theta, \qquad \zeta = \delta Z, \tag{6.78}$$

where Θ_0 is a measure of how much the arterial segment springs open when cut radially and δ is a possible axial extension associated with this cutting process. From Eqs. (6.78) and (6.77), therefore, we see that

$$r = \tilde{f}(g(R)), \qquad \theta = \frac{\pi}{\Theta_0}\Theta, \qquad z = \lambda\delta Z, \tag{6.79}$$

or

$$r = f(R), \qquad \theta = \frac{\pi}{\Theta_0}\Theta, \qquad z = \Lambda Z, \tag{6.80}$$

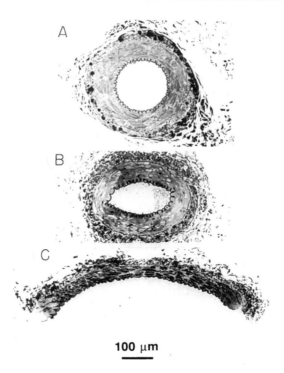

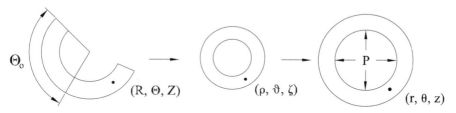

FIGURE 6.20 Three configurations of importance in arterial mechanics: a radially cut unloaded ring, which opens (panel C), an intact but unloaded ring (panel B), and an intact and loaded segment (panel A). The internal elastic lamina is seen to be waviest in the intact, unloaded configuration, consistent with the presence of compressive residual stresses in the inner wall. (From Fung and Liu (1992), reprinted with permission from the American Physiological Society.) The schema below shows coordinate systems for each configuration.

where $\Lambda \equiv \lambda\delta$. It can be shown (Humphrey, 2002) that components of the deformation gradient $[F]$, relative to cylindrical coordinates, can be computed in general via

$$[F] = \begin{bmatrix} \dfrac{\partial r}{\partial R} & \dfrac{1}{R}\dfrac{\partial r}{\partial \Theta} & \dfrac{\partial r}{\partial Z} \\[2mm] r\dfrac{\partial \theta}{\partial R} & \dfrac{r}{R}\dfrac{\partial \theta}{\partial \Theta} & r\dfrac{\partial \theta}{\partial Z} \\[2mm] \dfrac{\partial z}{\partial R} & \dfrac{1}{R}\dfrac{\partial z}{\partial \Theta} & \dfrac{\partial z}{\partial Z} \end{bmatrix}. \tag{6.81}$$

Herein, the derivation of the form for $[F]$ is not critical; rather, we will focus on its use. Nonetheless, note that each term is a nondimensional ratio of a current length to an original length. This is the reason for the presence of the r and R in the circumferential terms—a radius times an angle gives an arc length having units of length rather than radians.

From Eqs. (6.80) and (6.81), therefore, we have for our residually stressed artery,

$$[F] = \begin{bmatrix} \dfrac{\partial f}{\partial R} & 0 & 0 \\[2mm] 0 & \dfrac{r\pi}{R\Theta_0} & 0 \\[2mm] 0 & 0 & \Lambda \end{bmatrix}. \tag{6.82}$$

Recalling that incompressibility requires that $\det[F] = 1$, we have

$$\frac{\partial f}{\partial R}\left(\frac{r\pi}{R\Theta_0}\right)\Lambda = 1 \rightarrow r\frac{\partial r}{\partial R} = \frac{\Theta_0}{\pi\Lambda}R, \tag{6.83}$$

where we let $\partial f/\partial R \equiv \partial r/\partial R$ because $r \equiv f$. This equation can be integrated,

$$\int_{r_i}^{r} r\frac{\partial r}{\partial R}\,dR = \int_{R_i}^{R}\frac{\Theta_0}{\pi\Lambda}R\,dR, \tag{6.84}$$

to yield [because $dr = (\partial r/\partial R)dR$ by the chain rule]

$$r^2 - r_i^2 = \frac{\Theta_0}{\pi\Lambda}(R^2 - R_i^2) \quad \forall\, R \in [R_i, R_a], \tag{6.85}$$

where the subscripts i and a denote intimal and adventitial, respectively. Note, therefore, that incompressibility determines the previously unknown form of $r = f(R)$, which is seen to be quadratic in R. Now, the Green strains are determined easily from Eq. (6.7):

$$[E] = \frac{1}{2} \begin{bmatrix} \left(\dfrac{\Theta_0 R}{\pi \Lambda r}\right)^2 - 1 & 0 & 0 \\[2ex] 0 & \left(\dfrac{r\pi}{R\Theta_0}\right)^2 - 1 & 0 \\[2ex] 0 & 0 & \Lambda^2 - 1 \end{bmatrix}, \tag{6.86}$$

from which we see that $E_{RR} \equiv E_{11}$ and $E_{\Theta\Theta} \equiv E_{22}$ are functions of radius, whereas $E_{ZZ} \equiv E_{33}$, the axial strain, is not.

Fung's orthotropic exponential pseudostrain-energy function W [Eqs. (6.25) and (6.26)] has been shown to describe some arterial behaviors; hence, we will use it here for illustrative purposes. Note, therefore, that in cylindricals (and for principal strains), it is

$$W = \frac{1}{2}c(e^Q - 1), \tag{6.87}$$

with

$$Q = c_1 E_{RR}^2 + c_2 E_{\Theta\Theta}^2 + c_3 E_{ZZ}^2 + 2c_4 E_{RR}E_{\Theta\Theta} + 2c_5 E_{\Theta\Theta}E_{ZZ} + 2c_6 E_{ZZ}E_{RR}. \tag{6.88}$$

Hence, our nonzero components of Cauchy stress are, from Eq. (6.76),

$$\sigma_{rr} = -p + \left(\frac{\Theta_0 R}{\pi \Lambda r}\right)^2 \frac{\partial W}{\partial E_{RR}}, \tag{6.89}$$

$$\sigma_{\theta\theta} = -p + \left(\frac{r\pi}{R\Theta_0}\right)^2 \frac{\partial W}{\partial E_{\Theta\Theta}}, \tag{6.90}$$

$$\sigma_{zz} = -p + (\Lambda)^2 \frac{\partial W}{\partial E_{ZZ}}, \tag{6.91}$$

where the requisite partial derivatives are computed easily from Eqs. (6.87) and (6.88). Now, let us turn to the equilibrium equations. First, however, note that the deformation depends only on the radial direction. In the absence of body forces, Eqs. (3.11)–(3.13) (equilibrium in cylindrical coordinates) reduce to

$$\frac{d\sigma_{rr}}{dr} + \frac{\sigma_{rr} - \sigma_{\theta\theta}}{r} = 0, \qquad -\frac{\partial p}{\partial \theta} = 0, \qquad -\frac{\partial p}{\partial z} = 0, \tag{6.92}$$

the last two of which reveal that the Lagrange multiplier p, like the deformation, depends on r at most. Our only nontrivial equilibrium equation thus becomes

$$\frac{d\sigma_{rr}}{dr} = \frac{\sigma_{\theta\theta} - \sigma_{rr}}{r}, \tag{6.93}$$

which can be integrated as

$$\int \frac{d\sigma_{rr}}{dr}\, dr = \int \left(\frac{\sigma_{\theta\theta} - \sigma_{rr}}{r} \right) dr. \tag{6.94}$$

Depending on our prescription of the integration limits, this equation allows us to determine either the transmural pressure P or the Lagrange multiplier. For the first, consider

$$\int_{r_i}^{r_a} \frac{d\sigma_{rr}}{dr}\, dr = \sigma_{rr}(r_a) - \sigma_{rr}(r_i) = \int_{r_i}^{r_a} \left(\frac{\sigma_{\theta\theta} - \sigma_{rr}}{r} \right) dr \tag{6.95}$$

or

$$P_i - P_a \equiv P = \int \left(\frac{\sigma_{\theta\theta} - \sigma_{rr}}{r} \right) dr, \tag{6.96}$$

given stress boundary conditions that $\sigma_{rr}(r_i) = -P_i$ and $\sigma_{rr}(r_a) = -P_a$, the intimal and adventitial pressure, respectively. Conversely, consider

$$\int_{r_i}^{r} \frac{d\sigma_{rr}}{dr}\, dr = \sigma_{rr}(r) - \sigma_{rr}(r_i) = \int_{r_i}^{r} \left(\frac{\sigma_{\theta\theta} - \sigma_{rr}}{r} \right) dr \tag{6.97}$$

whereby σ_{rr} is given by Eq. (6.89) and thus

$$-p(r) + \left(\frac{\Theta_0 R}{\pi \Lambda r} \right)^2 \frac{\partial W}{\partial E_{RR}} + P_i = \int_{r_i}^{r} \left(\frac{\sigma_{\theta\theta} - \sigma_{rr}}{r} \right) dr \tag{6.98}$$

allows one to determine p as a function of r. Finally, note that in either case, for the Fung exponential, Eqs. (6.89) and (6.90) allow us to compute

$$\int \left(\frac{\sigma_{\theta\theta} - \sigma_{rr}}{r} \right) dr = \int \left[\left(\frac{r\pi}{R\Theta_0} \right)^2 \frac{\partial W}{\partial E_{\Theta\Theta}} - \left(\frac{\Theta_0 R}{\pi \Lambda r} \right)^2 \frac{\partial W}{\partial E_{RR}} \right] \frac{dr}{r}, \tag{6.99}$$

where

$$\frac{\partial W}{\partial E_{\Theta\Theta}} = c e^{Q} (c_2 E_{\Theta\Theta} + c_4 E_{RR} + c_5 E_{ZZ}), \tag{6.100}$$

$$\frac{\partial W}{\partial E_{RR}} = c e^{Q} (c_1 E_{RR} + c_4 E_{\Theta\Theta} + c_6 E_{ZZ}) \tag{6.101}$$

from Eqs. (6.87) and (6.88), with

$$E_{RR} = \frac{1}{2}\left(\left(\frac{\Theta_0 R}{\pi \Lambda r}\right)^2 - 1\right), \qquad E_{\Theta\Theta} = \frac{1}{2}\left(\left(\frac{r\pi}{R\Theta_0}\right)^2 - 1\right), \qquad E_{ZZ} = \frac{1}{2}(\Lambda^2 - 1).$$

$$(6.102)$$

Hence, if we know the material properties $(c, c_1, \ldots, c_6)$, the residual stress related opening angle Θ_0 and extension δ, the radially cut dimensions R_i and R_a, the stretch λ and either r_i or P, then we can solve for the stresses. The associated integrals are obviously complex but can be evaluated easily via numerical methods such as Simpson's rule or more sophisticated methods such as the Romberg method or a Gauss quadrature.

6.5.3 Illustrative Results

Panel A in Figure 6.21 shows predicted transmural stresses for the case in which we ignore the residual stress (i.e., assume $\Theta_0 = \pi$ and $\delta = 1$) but let $\lambda = 1.8$ and $P = 120$ mm Hg. Note the steep gradients in stress, which, as noted earlier, suggested to some that atherosclerosis occurs because of large intimal stresses. Panel A in Figure 6.22 shows, however, that these predicted stresses are much closer to a homogenous distribution if one includes the residual stress, here with $\Theta_0 = 71.4°$ and $\delta = 1.017695$ (Humphrey, 2002). One of the remarkable things revealed by this (simple) nonlinear analysis is that inclusion of residual stress reduces the computed stresses many fold despite the residual stresses [based on deformations from (R, Θ, Z) to (ρ, ϑ, ζ) only] actually being very small in magnitude, ~3 kPa (panel B in Fig. 6.21). This dramatic effect is due solely to the material and kinematic nonlinearities, which would simply be missed with a linear analysis [cf. Eq. (3.80)]. Moreover, it is clear that the principle of superposition of Section 5.5 of Chapter 5 does not hold in this case; subtracting a circumferential residual stress of ~3 kPa from those in Figure 6.21A clearly does not yield the computed values in Figure 6.22A. The full nonlinear analysis is thus essential here as in most problems in soft tissue biomechanics and likely cell mechanics.

In conclusion, we emphasize that this analysis was presented simply to illustrate some of the unique aspects of a nonlinear stress analysis and to provide a glimpse into methods used in cardiovascular mechanics and other areas of soft tissue mechanics. This presentation—even for a straight uniform segment—was simplified, however, for we did not consider the heterogeneity of the composition of the wall (i.e., different behavior of the media and adventitia) or the dynamical loading due to pulsatile flow. These and other effects are addressed in Humphrey (2002). Nevertheless, even in that text, the analysis is simplified. There is much more to learn about the mechanics of blood vessels, their basic constitutive relations, especially for smooth muscle, their variations in properties along the length of a vessel or

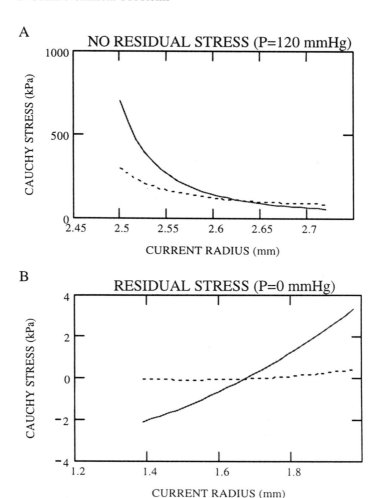

FIGURE 6.21 Panel A: Predicted transmural distribution of stresses (solid line is circumferential stress and dashed line is axial stress) when one neglects the effects of residual stress and smooth muscle activation. Panel B: Computed residual stresses alone. Note the different orders of magnitude in the stresses in the two panels and the compression in the inner wall in Panel B consistent with the histology in Figure 6.20. The following values of the parameters were used in the computations: $c = 22.4$ kPa, $c_1 = 0.0499$, $c_2 = 1.0672$, $c_3 = 0.4775$, $c_4 = 0.0042$, $c_5 = 0.0903$, and $c_6 = 0.0585$, whereas $R_i = 3.92$ mm, $R_a = 4.52$, $\delta = 1.0177$, and $\lambda = 1.767$.

through a bifurcation, their changes due to growth and remodeling, and so on. To advance our understanding in these areas, we must not only apply mechanics, but we must also develop and extend it. The challenge is great, but so is the need.

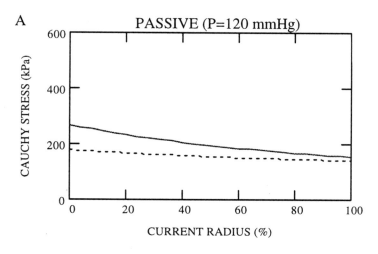

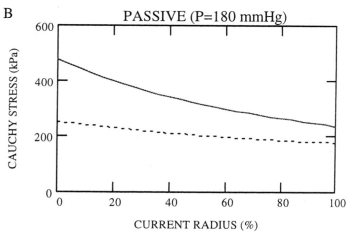

FIGURE 6.22 Panel A: Predicted transmural distribution of stress when residual stress is accounted for; in comparison to Panel A in Figure 6.21, inclusion of the residual stress tends to reduce and homogenize the stresses. Panel B: Despite the presence of residual stress, high blood pressure (acute hypertension) would tend to increase the stresses and their transmural gradients in the absence of any functional adaptation. Such deviations from normal values could set into motion various growth and remodeling processes.

6.6 A Role of Vascular Smooth Muscle

6.6.1 Muscle Basics

Although we only considered the passive (i.e., noncontracting) mechanical behavior of arteries in Section 6.5, the role of smooth muscle activation is fundamental to vascular function. Vascular smooth muscle constitutes

40–60% (by dry weight) of the medial portion of large and resistive blood vessels. It is responsible for maintaining a "basal tone," which, in turn, allows the vessel to vasoconstrict or vasodilate as needed to control local blood flow. Such regulation is fundamental, for example, for diverting blood to muscles during exercise, to the digestive system following eating, or away from the skin to minimize heat transfer when the external temperature is low.

Like all muscle (e.g., skeletal and cardiac), contraction of vascular smooth muscle depends on the concentration of intracellular free calcium and a sliding filament, cross-bridge-mediated mechanics (Chapter 1). The steady-state cytoplasmic calcium is maintained primarily via a calmodulin-regulated Ca-ATPase activity. Sources of calcium include the intracellular sarcoplasmic reticulum and transmembrane influx from the extracellular milieu; a rise in cytoplasmic free calcium triggers a contraction. Despite the similarities, vascular smooth muscle differs from skeletal and cardiac muscle in numerous ways. Smooth muscle has a much higher ratio of actin to myosin, and the actin–myosin complexes are not arranged in sarcomeres, as they are in striated muscle. Smooth muscle can also shorten more than striated muscle, albeit at a much lower rate, and it can maintain its maximum contraction at a steady level for much longer periods and at a lower energy expenditure than striated muscle. Finally, whereas striated muscle generates its greatest force at a length where the passive stress is nearly zero, smooth muscle generates its greatest force at a length where the passive stress is significant (Fig. 6.23). The ability to generate force (or stress) also depends on the contractile state, which can be governed by the concentration of a particular agonist like the neurotransmitter norepinephrine (NE). An associated sigmoidal dose–response curve is illustrated

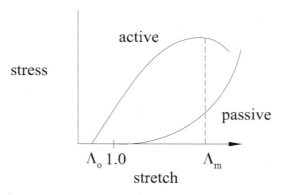

FIGURE 6.23 Schema of the length—tension response of vascular smooth muscle. This response is similar to that in skeletal muscle, with the exception that the passive tension is significant in smooth muscle at that value of stretch where the active force generation is largest.

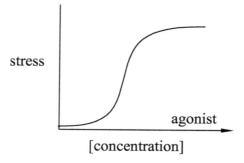

FIGURE 6.24 Dose–response curve for vascular smooth muscle. The concentration of the agonist could range from 10^{-7} to 10^{-4} for a typical test using norepinephrine (NE), a vasoconstrictor.

in Figure 6.24; that is, active force generation essentially increases with the concentration of the agonist once a threshold is exceeded, but force generation does not increase beyond a saturation value of the agonist.

6.6.2 Quantification

Despite its fundamental importance and despite the cross-bridge mechanism being proposed nearly 50 years ago (1954), we still do not have a widely accepted mathematical descriptor for active stress generation. Indeed, perhaps one of the greatest unknowns in biomechanics is the *multiaxial character* of muscle activation. Historically, muscle has been thought to generate stress only in the direction of the long axis of the actin–myosin complex (i.e., is one dimension). We now know that muscle activation is multiaxial, but the data and theory needed for quantification remain lacking. Here, therefore, let us consider the current state of the art.

It is generally assumed that the passive and active contributions to the stress are additive; thus, $\sigma = \sigma^p + \sigma^a$, where the superscripts p and a denote passive and active. Moreover, the smooth muscle is oriented in the circumferential direction in most blood vessels; hence, it is generally assumed that

$$\sigma_{rr} \equiv \sigma_{rr}^p, \qquad \sigma_{\theta\theta} = \sigma_{\theta\theta}^p + \sigma_{\theta\theta}^a, \qquad \sigma_{zz} \equiv \sigma_{zz}^p. \qquad (6.103)$$

Perhaps the best currently available relation for the active component of stress is that proposed in 1999 by Rachev and Hayashi (see Chapter 7 in Humphrey, 2002). It can be written as

$$\sigma_{\theta\theta}^a = T_0(Ca^{2+})\Lambda_\theta \left[1 - \left(\frac{\Lambda_m - \Lambda_\theta}{\Lambda_m - \Lambda_0} \right)^2 \right], \qquad (6.104)$$

where T_0 has units of stress and represents the dose–response dependency on cytoplasmic free calcium. The second part of this relation represents the stretch-dependent stress generation, where Λ_θ is the stretch in the circumferential direction, Λ_0 is the minimum value of stretch where no stress generation is possible, and Λ_m is that value of stretch where the active stress generation is maximum. Typical values of Λ_0 and Λ_m are 0.6–0.8 and 1.5–1.75, respectively. Note, too, that $T_0 \sim 50$ kPa for a basal tone, but it can range from 0 to ~100 kPa. From Section 6.5, we see that $\Lambda_\theta = r\pi/R\Theta_0$ where Θ_0 is the residual stress related opening angle and r and R are radii in the current and original (stress-free) configurations, respectively.

Recall from the previous section that including the effects of residual stress dramatically reduced the computed values of the circumferential stress, particularly in the inner wall, and its transmural gradient. Rachev and Hayashi showed that including a basal smooth muscle tone further reduces the computed circumferential stress and its gradient (compare panels A and B in Fig. 6.25). Hence, it appears that a basal smooth muscle tone not only allows vasoconstriction or vasodilation as needed, it also modifies the normal stiffness of the wall and thereby helps to homogenize the transmural distribution of stress. A homogenous stress field in normalcy seems reasonable teleologically because each cell would experience the same baseline mechanical environment.

Of course, disease, injury, and clinical intervention (e.g., balloon angioplasty) can perturb the mechanical environment, which, via mechanotransduction mechanisms, tends to set into motion various growth and remodeling responses that seek to restore normalcy or at least to arrest the damage or insult. As noted earlier, modeling biological growth and remodeling is one of the most important and exciting areas of research in biomechanics today. It is, however, mathematically complex and beyond an introductory text. We merely discuss a few of the basics of vascular growth and remodeling in Section 11.1.

In closing, however, let us see how an abrupt change in blood pressure can increase the value and gradients of the transmural stresses and how an immediate vasoactive response can tend to restore the stresses *toward* normal values; full restoration requires growth and remodeling, however. Whereas panel B in Figure 6.25 shows stresses at $P = 120$ mm Hg, panel A in Figure 6.26 shows values for the same artery at 180 mm Hg with the same basal tone. The rise in stress is due largely to the distention of the vessel and the associated stretching of the nonlinear passive components of the wall. Indeed, a vasodilatation (panel B in Fig. 6.22) allows a further distension and thus exacerbates the pressure-induced increase in wall stress. Conversely, a vasoconstriction (panel B in Fig. 6.26, with $T_0 = 100$ kPa rather than 50 kPa) tends to reduce the wall stresses toward their original values. Hence, vasoconstriction may be one early mechanism that the blood vessel uses to combat hypertension. Again, however, the detailed mechanics are

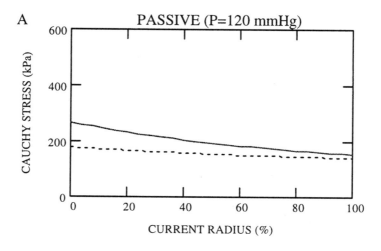

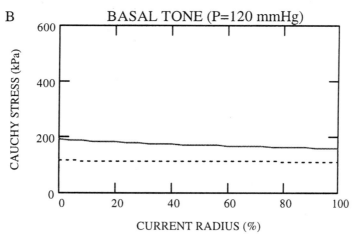

FIGURE 6.25 Panel A: We repeat panel A from Figure 6.22 to compare it directly to the case when a basal tone (i.e., smooth muscle activation modeled via $T_0 = 50\,\text{kPa}$) is included in the analysis (Panel B). It is seen that smooth muscle activation tends to reduce the stresses and their gradients.

very complex and the reader is referred to Humphrey (2002) and the archival literature for more details. Indeed, this chapter was but a brief introduction to the nonlinear mechanics of tissues and cells, hopefully an appetizer that has stimulated the reader's desire to learn more and to explore more deeply this important and fascinating area of research.

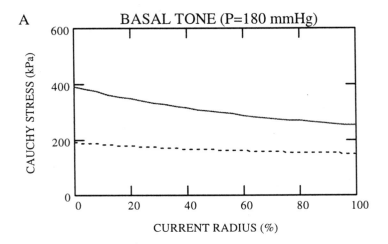

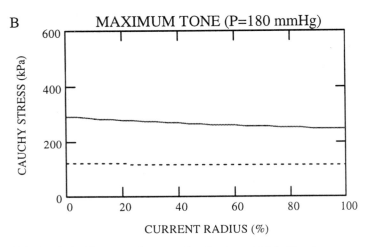

Figure 6.26 Panel A: Transmural stress distribution at high blood pressure when the smooth muscle is at its basal tone. Compared to Figure 6.22B, we see that a basal smooth muscle activation tends to decrease the stresses from those in the passive state alone. Indeed, panel B shows that further activation ($T_0 = 100$ kPa rather than the 50 kPa in the basal case) tends to decrease the stresses in the case of high blood pressure. It seems reasonable to hypothesize, therefore, that an augmented smooth muscle activation may be an early response of large vessels to hypertension. Because the stresses are not restored to normal values and because of the increased energetic demand of fully contracted muscle, however, subsequent growth and remodeling would be expected. This is consistent with the observation that hypertensive vessels tend to thicken over time via an increase in both smooth muscle and extracellular matrix, especially collagen.

Appendix 6: Matrices

A matrix is a mathematical device that is useful for manipulating arrays of numbers (or variables). In general, a $m \times n$ matrix is written with m rows of entries and n columns; it is convention to enclose these entries within brackets [ ]. For example, a 2×2 matrix has four entries, given by two rows and two columns, represented as

$$[A] = \begin{bmatrix} A_{11} & A_{12} \\ A_{21} & A_{22} \end{bmatrix}, \tag{A6.1}$$

from which we see that the indices denote the (row, column). Hence, the ith row, jth column entry can be denoted as A_{ij} (i.e., for stress, rows represent faces and columns the directions). Accepted *rules* govern the addition, subtraction, and multiplication of matrices. For example, two matrices $[A]$ and $[B]$ can be added or subtracted *if and only if* they have the same number of rows and columns. A 2×2 matrix can thus be added to another 2×2 matrix, but it cannot be added to a 3×3 matrix. In the former case, we have

$$[A] \pm [B] = \begin{bmatrix} A_{11} & A_{12} \\ A_{21} & A_{22} \end{bmatrix} \pm \begin{bmatrix} B_{11} & B_{12} \\ B_{21} & B_{22} \end{bmatrix} = \begin{bmatrix} A_{11} \pm B_{11} & A_{12} \pm B_{12} \\ A_{21} \pm B_{21} & A_{22} \pm B_{22} \end{bmatrix}, \tag{A6.2}$$

which is to say, addition and subtraction are accomplished by simply adding or subtracting like entries.

Matrix multiplication is much different. Two matrices can be multiplied *if and only if* the number of columns of the first matrix equal the number of rows of the second matrix. Thus, a $m \times n$ matrix cannot be multiplied by a $m \times n$ matrix if $m \neq n$. Rather, a $m \times p$ matrix must multiply a $p \times n$ matrix ($m = n$ allowed but not required) whereby we find that the resulting number of rows and columns of the product matrix is $(m \times p) \times (p \times n) = m \times n$. For example, if

$$[C] = \begin{bmatrix} C_{11} & C_{12} \\ C_{21} & C_{22} \\ C_{31} & C_{32} \end{bmatrix}, \qquad [B] = \begin{bmatrix} B_{11} & B_{12} \\ B_{21} & B_{22} \end{bmatrix}, \tag{A6.3}$$

then we expect $(3 \times 2) \times (2 \times 2) = 3 \times 2$, namely

$$[C][B] = \begin{bmatrix} C_{11}B_{11} + C_{12}B_{21} & C_{11}B_{12} + C_{12}B_{22} \\ C_{21}B_{11} + C_{22}B_{21} & C_{21}B_{12} + C_{22}B_{22} \\ C_{31}B_{11} + C_{32}B_{21} & C_{31}B_{12} + C_{32}B_{22} \end{bmatrix}, \tag{A6.4}$$

from which we see that each row of $[C]$ multiplies each column of $[B]$.

In addition to operations such as addition and multiplication, matrix methods introduce new operations such as the transpose and inverse. The transpose of a matrix, denoted by the superscript T, such as, transpose $([C])$ $\equiv [C]^T$, is computed by exchanging rows and columns. For example,

$$[C] = \begin{bmatrix} C_{11} & C_{12} \\ C_{21} & C_{22} \\ C_{31} & C_{32} \end{bmatrix} \rightarrow [C]^T = \begin{bmatrix} C_{11} & C_{21} & C_{31} \\ C_{12} & C_{22} & C_{32} \end{bmatrix}, \qquad (A6.5)$$

which reveals that $(m \times n)^T = n \times m$. At this point, note that matrices can have a single row or a single column. These are often called "vectors," because the associated entries can represent components of a vector in an n-space. For example, $n \times 1$ and $1 \times n$ matrices, with $n = 3$ as in Euclidean 3-space, can be written as

$$[X] = \begin{bmatrix} X_1 \\ X_2 \\ X_3 \end{bmatrix}, \qquad [Y] = [X]^T = [X_1 \quad X_2 \quad X_3]. \qquad (A6.6)$$

Hence,

$$[X]^T[X] = [X_1 \quad X_2 \quad X_3] \begin{bmatrix} X_1 \\ X_2 \\ X_3 \end{bmatrix} = [X_1^2 + X_2^2 + X_3^2], \qquad (A6.7)$$

revealing that a 1×1 matrix also exists, which, of course, can represent the value of a scalar. Note, too, that $(1 \times 3) \times (3 \times 1) = (1 \times 1)$, whereas $(3 \times 1) \times (1 \times 3) = (3 \times 3)$, namely

$$\begin{bmatrix} X_1 \\ X_2 \\ X_3 \end{bmatrix} [X_1 \quad X_2 \quad X_3] = \begin{bmatrix} X_1^2 & X_1 X_2 & X_1 X_3 \\ X_2 X_1 & X_2^2 & X_2 X_3 \\ X_3 X_1 & X_3 X_2 & X_3^2 \end{bmatrix}, \qquad (A6.8)$$

thus revealing that matrix multiplication does *not* commute in general.

A matrix can be multiplied or divided by a scalar, such as $a[A] = [B]$, where

$$a \begin{bmatrix} A_{11} & A_{12} \\ A_{21} & A_{22} \end{bmatrix} = \begin{bmatrix} aA_{11} & aA_{12} \\ aA_{21} & aA_{22} \end{bmatrix} = \begin{bmatrix} B_{11} & B_{12} \\ B_{21} & B_{22} \end{bmatrix}, \qquad (A6.9)$$

and likewise for scalar division. Hence, a scalar acts on each entry individually. This property allows one to factor out common values, such as the $\frac{1}{2}$ in the definition of the Green strain $[E]$ in terms of the deformation gradient $[F]$, as in Eq. (6.7).

Whereas a matrix can be multiplied or divided by a scalar, and a matrix can be multiplied by another matrix, *a matrix cannot be divided by another matrix*. This is similar, of course, to vector algebra: A vector can be multiplied by a scalar or a vector (e.g., dot or cross products), but a vector cannot be divided by another vector. In the case of matrices, this issue is addressed in part via the inverse operation. The inverse of a matrix, denoted by the superscript −1, is defined such that

$$[A]^{-1}[A] = [I] = [A][A]^{-1}, \qquad (A6.10)$$

where $[I]$ is the identity matrix, which has values of 0 in off-diagonal entries and 1 in diagonal entries [cf. Eq. (6.5)]. To illustrate, consider the 2×2 matrix $[A]$. Letting

$$[A] = \begin{bmatrix} A_{11} & A_{12} \\ A_{21} & A_{22} \end{bmatrix}, \quad [I] = \begin{bmatrix} 1 & 0 \\ 0 & 1 \end{bmatrix}, \qquad (A6.11)$$

it can be shown that

$$[A]^{-1} = \frac{1}{J} \begin{bmatrix} A_{22} & -A_{12} \\ -A_{21} & A_{11} \end{bmatrix}, \qquad (A6.12)$$

where the scalar J (Jacobian) is the determinant of $[A]$ given by

$$J = \det[A] = A_{11}A_{22} - A_{12}A_{21}. \qquad (A6.13)$$

Note, therefore, that

$$
\begin{aligned}
[A]^{-1}[A] &= \frac{1}{A_{11}A_{22} - A_{12}A_{21}} \begin{bmatrix} A_{22} & -A_{12} \\ -A_{21} & A_{11} \end{bmatrix} \begin{bmatrix} A_{11} & A_{12} \\ A_{21} & A_{22} \end{bmatrix} \\
&= \frac{1}{A_{11}A_{22} - A_{12}A_{21}} \begin{bmatrix} A_{22}A_{11} - A_{12}A_{21} & A_{22}A_{12} - A_{12}A_{22} \\ -A_{21}A_{11} + A_{11}A_{21} & -A_{12}A_{21} + A_{22}A_{11} \end{bmatrix} \\
&= \begin{bmatrix} 1 & 0 \\ 0 & 1 \end{bmatrix} = [I], \qquad (A6.14)
\end{aligned}
$$

as desired. In general, if we let A_{ij} represent the ith row, jth column entry, then

$$A_{ij}^{-1} = \frac{\text{cof}[A_{ij}]}{\det[A]}, \qquad (A6.15)$$

where the cofactor of the entry A_{ij} is defined by

$$\text{cof}[A_{ij}] = (-1)^{i+j} M_{ji}, \qquad (A6.16)$$

where M_{ji} is a so-called minor; it is given by the determinant of the entries left after striking out all entries in row i and column j. For example, given

Eq. (A6.1), the M_{11} minor is A_{22}, the M_{12} minor is A_{21}, the M_{21} minor is A_{12}, and the M_{22} minor is A_{11}.

Finally, the determinant of $[A]$ is defined by

$$\det[A] = \sum_{j=1}^{n} A_{ij}(-1)^{i+j} M_{ij} \tag{A6.17}$$

for any i.

Although the inverse and determinant can be difficult to compute for large matrices, they are straightforward for 2×2 and 3×3 matrices, which are particularly useful in mechanics to represent 2-D and 3-D states of stress or strain. In cases of larger matrices, computers are essential.

Exercises

6.1   Given a motion defined by

$$x = \Lambda_1 X + \kappa_1 Y, \qquad y = X + \Lambda_2 Y, \qquad z = Z,$$

find $[F]$, $[E]$, and $[\varepsilon]$ and discuss.

6.2   It can be shown (Humphrey, 2002) that $\det[F] = 1$ if the deformation is volume conserving (i.e., isochoric). If

$$x = \Lambda X, \qquad y = \Lambda Y, \qquad z = \beta Z$$

describes an "equibiaxial stretching" of amount Λ in the in-plane (x, y) directions, determine the value of β such that volume is conserved.

6.3   Assuming an isochoric motion (i.e., $\det[F] = 1$), find the value of Λ in terms of β, when

$$x = \beta X + \kappa Y, \qquad y = \kappa X + \beta Y, \qquad z = \Lambda Z.$$

6.4   It can be shown that in cylindrical coordinates (see Humphrey, 2002),

$$[F] = \begin{bmatrix} \dfrac{\partial r}{\partial R} & \dfrac{1}{R}\dfrac{\partial r}{\partial \Theta} & \dfrac{\partial r}{\partial Z} \\[2mm] r\dfrac{\partial \theta}{\partial R} & \dfrac{r}{R}\dfrac{\partial \theta}{\partial \Theta} & r\dfrac{\partial \theta}{\partial Z} \\[2mm] \dfrac{\partial z}{\partial R} & \dfrac{1}{R}\dfrac{\partial z}{\partial \Theta} & \dfrac{\partial z}{\partial Z} \end{bmatrix}$$

for a particle originally at (R, Θ, Z) that is currently at (r, θ, z), with $r = r(R, \Theta, Z)$, and so forth. Find $[F]$ and $[E]$ for the following specific motion:

$$r = \beta R, \qquad \theta = \Theta, \qquad z = \Lambda Z,$$

where β and Λ are stretch ratios. Interpret this motion.

6.5   Find the solution to Eq. (6.10) using a direct integration method. Hint: Note that

$$\frac{d\Sigma}{d\Lambda} = \alpha + \beta\Sigma \rightarrow \frac{1}{\alpha + \beta\Sigma}\frac{d\Sigma}{d\Lambda} = 1.$$

6.6   Using the stress–stretch function of the form

$$\Sigma = \frac{\alpha}{\beta}\left(e^{\beta(\Lambda-1)} - 1\right)$$

and letting values of the parameters be $\alpha = 10\,\text{MPa}$ and $\beta = 2.5$, plot the associated Σ versus Λ and $d\Sigma/d\Lambda$ versus Σ curves. Additionally, plot $\ln\Sigma$ versus Λ (for Λ up to 1.2) and discuss how such information (if Σ and Λ came from an experiment) could be used to find a constitutive function.

6.7   Given the stress–stretch relation $\Sigma = \Sigma(\Lambda)$ in Exercise 6.6, note that the relation is nonlinear in terms of the material parameter β. Because of the exponential relation, however, one may be able to determine the values of α and β from data using a linear instead of a nonlinear least-squares regression. Find the requisite equations for the linear regression. Hint: Use the natural logarithm.

6.8   Use the general solutions for axisymmetric membranes [Eq. (6.58)] to determine the stress resultants (tensions) for the inflation of an elliptical membrane, an approximate example of which is the lens capsule of the eye.

6.9   The neo-Hookean strain energy function W was used in Section 6.4 to examine the stability of an elastomeric spherical membrane. This W was written in terms of the in-plane principal components of the Green strain [Eq. (6.63)]. Show that an equivalent form is

$$W = C(\Lambda_1^2 + \Lambda_2^2 + \Lambda_3^2 - 3),$$

where $E_{11} = (\Lambda_1^2 - 1)/2$ and so forth. Moreover, note that for a membrane, W is written in terms of in-plane components only. If we enforce incompressibility kinematically, rather than constitutively, then $\det[F] = 1$ requires that $\Lambda_1\Lambda_2\Lambda_3 = 1$ if $[F] = \text{diag}[\Lambda_1, \Lambda_2, \Lambda_3]$. Rewrite W in terms of Λ_1 and Λ_2 alone.

6.10   In addition to the Fung exponential for a 2-D membrane,

$$W = c(e^Q - 1), \qquad Q = c_1 E_{11}^2 + c_2 E_{22}^2 + 2c_3 E_{11}E_{22},$$

relative to principal directions, another often used relation for biomembranes is the Skalak, Tozeren, Zarda, Chien (STZC) relation:

$$W = \frac{c}{8}\{4(E_{11}^2 + E_{22}^2) + \Gamma[4(E_{11}^2 + E_{22}^2) + 8E_{11}E_{22}$$
$$+16E_{11}^2 E_{22} + 16E_{11}E_{22}^2 + 16E_{11}^2 E_{22}^2]\},$$

where c and Γ are material parameters. If

$$E_{11} = \frac{1}{2}(\Lambda^2 - 1), \qquad E_{22} = \frac{1}{2}(\Lambda^2 - 1),$$

as in the analysis of the neuroangioplasty balloon, show that

$$T_1 = T_2 = T = \frac{c}{2}[\Lambda^2 - 1 + \Gamma\Lambda^2(\Lambda^4 - 1)].$$

6.11 Based on the previous exercise, determine if a STZC spherical membrane exhibits a limit point instability in inflation if $c > 0$ and $\Gamma > 0$. Hint: First, show that

$$P(\Lambda) = \frac{c}{A}\left(\Lambda - \frac{1}{\Lambda} + \Gamma(\Lambda^5 - \Lambda)\right).$$

6.12 Repeat the previous exercise for a Fung spherical membrane with $c > 0$, $c_1 = c_2$, and $c_1 + c_3 \equiv \Gamma > 0$.

6.13 Given the following dataset, use Eq. (6.50) to find best-fit values of the four material parameters; the units of the stress resultants are g/cm (convert them to N/m). [Data from Harris JL (2002) Thermal modification of collagen under biaxial isotonic loads. Ph.D. dissertation, Texas A&M University, College Station.]

DATA POINTS

	1	2	3	4	5	6	7	8	9	10	11	12
T_1	1.0	1.2	1.3	1.6	2.4	4.9	6.6	8.5	10.5	13.0	15.0	16.8
Λ_1	1.44	1.44	1.44	1.44	1.44	1.43	1.43	1.43	1.43	1.43	1.44	1.44
T_2	1.0	1.3	1.7	2.6	5.4	14.8	23.8	32.0	42.6	52.1	64.5	73.2
Λ_2	1.06	1.12	1.19	1.26	1.33	1.40	1.41	1.43	1.43	1.43	1.44	1.44

6.14 Plot and compare the active stress–stretch response [Eq. (6.104)] for values of $T_0 = 0$, 20, 40, 60, and 80 kPa. Use values of $\Lambda_0 = 0.7$ and $\Lambda_m = 1.5$.

6.15 Repeat Exercise 6.14 with $T_0 = 50$ kPa and (Λ_0, Λ_m) pairs of (0.6, 1.3), (0.7, 1.4), and (0.8, 1.5) and discuss.

6.16 Given the proposed 1-D descriptor for smooth muscle behavior in arteries,

$$\sigma_{\theta\theta}^a = T_0(Ca^{2+})\Lambda_\theta\left[1-\left(\frac{\Lambda_m-\Lambda_\theta}{\Lambda_m-\Lambda_0}\right)^2\right],$$

where $T_0 \sim 50\,\text{kPa}$ in the basal state, Λ_m is the circumferential stretch at which activation is maximum ($\Lambda_m \sim 1.5$), and Λ_0 is that value of stretch at which active force generation ceases ($\Lambda_0 \sim 0.6$), add this contribution to the passive stress of Eq. (6.90) and recompute and plot the stress fields in Figure 6.22. Note, too, that Λ_θ is simply the circumferential stretch and, thus, $\Lambda_\theta = \pi r/\Theta_0 R$.

6.17 P. Hunter, at Auckland New Zealand, proposed a different form of $T(Ca^{2+}, \Lambda)$ for cardiac muscle. It is $T(Ca^{2+}, \Lambda) = T_0(Ca^{2+})f(\Lambda)$, where Λ is a stretch ratio for the sarcomere. Specifically, he let

$$T(Ca^{2+}, \Lambda) = \left(\frac{[Ca^{2+}]^n}{[Ca^{2+}]+C_{50}^n}\right)T_{max}[1+\beta(\Lambda-1)],$$

where

$$C_{50} = \frac{4.35}{\sqrt{e^{4.75(L-1.58)}-1}}$$

and where L is the sarcomere length in μm (range from 1.58 to 2.2 μm), $n = 2$, $T_{max} = 100\,\text{kPa}$, $\beta = 1.45$, and $\Lambda = L/1.58\,\mu$m. Plot and compare values of T for sarcomere lengths from 1.58 to 2.2 μm for two different calcium levels: $1.8\,\mu M$ and $1.04\,\mu M$.

6.18 Using results from Exercise 6.17, compare the active stress generation (kPa) at a sarcomere length of $1.8\,\mu$m for all calcium concentrations from $1.04\,\mu M$ to $1.8\,\mu M$; that is, plot T versus $[Ca^{2+}]$.

6.19 The total axial force f on the artery is computed via

$$f = \int_0^{2\pi}\int_{r_i}^{r_a}\sigma_{zz}r\,dr\,d\theta.$$

Because of the Lagrange multiplier p, however, it proves useful to (do it) convert the integral to

$$f = \pi\int_{r_i}^{r_a}(2\sigma_{zz}-\sigma_{rr}-\sigma_{\theta\theta})r\,dr - \pi r_i^2\sigma_{rr}(r_i)+\pi r_a^2\sigma_{rr}(r_a).$$

Hint: Let $\sigma_{zz} = \sigma_{zz} - \sigma_{rr} + \sigma_{rr}$, integrate by parts, and use the radial equilibrium equation

$$\frac{d\sigma_{rr}}{dr}+\frac{\sigma_{rr}-\sigma_{\theta\theta}}{r} = 0.$$

6.20 Find the components of $[C] = [A][B]$ if

$$[A] = \begin{bmatrix} A_{11} & A_{12} & A_{13} \\ A_{21} & A_{22} & A_{23} \\ A_{31} & A_{32} & A_{33} \end{bmatrix}, \qquad [B] = \begin{bmatrix} B_{11} \\ B_{21} \\ B_{31} \end{bmatrix}.$$

6.21 If $[I]$ is the identity matrix, show that $[I][A] = [A] = [A][I]$.

6.22 If $[X]$ is a 3×1 matrix, we sometimes call it a column matrix or simply a vector. Noting from Eq. (A6.7) that $[X]^T[X]$ yields a scalar equal to the sum of the squares of the entries, compare this result to the vector dot product $X \cdot X$ if

$$X = X_1 \hat{e}_1 + X_2 \hat{e}_2 + X_3 \hat{e}_3.$$

We see, therefore, that operations in matrix methods can yield the same results as those in vector methods. Matrix methods are particularly well suited if m and/or n are >4 in a $m \times n$ matrix.

6.23 Show that $\det[A] = \det[A]^T$.

6.24 If

$$[A] = \begin{bmatrix} 2 & 4 & 2 \\ 2 & -1 & -2 \\ 4 & 1 & -2 \end{bmatrix},$$

show that

$$[A]^{-1} = \frac{1}{4} \begin{bmatrix} 4 & 10 & -6 \\ -4 & -12 & 8 \\ 6 & 14 & -10 \end{bmatrix},$$

and that $[A][A]^{-1} = [I]$.

6.25 In contrast to the neo-Hookean descriptor for rubber (Exercise 6.9), some prefer the Mooney-Rivlin relation. In terms of principal stretches, it is

$$W = C\left[(\Lambda_1^2 + \Lambda_2^2 + \Lambda_3^2 - 3) + \Gamma\left(\frac{1}{\Lambda_1^2} + \frac{1}{\Lambda_2^2} + \frac{1}{\Lambda_3^2} - 3 \right) \right]$$

where C and Γ are material parameters. Recompute results in Fig. 6.15 for $\Gamma = 0.1$. See Humphrey (2002), Chapter 4.

Part III
Biofluid Mechanics

7
Stress, Motion, and Constitutive Relations

7.1 Introduction

A fluid can be defined as any substance that flows under the action of a shear stress, no matter how small the shear. Alternatively, some define a fluid as any substance that quickly assumes the shape of the container in which it is placed. Regardless of the specific definition, common experience shows that materials in either their liquid or their gaseous phases can fall within the purview of fluid mechanics.

As noted in Chapter 1, biofluid mechanics is a very important field within biomedical engineering. For example, biofluid mechanics helps us to understand blood flow within the cardiovascular system, airflow within the airways and lungs, and the removal of waste products via the kidneys and urinary system. Biofluid mechanics is thus fundamental to understanding basic physiologic processes as well as clinical observations. We also know that the human body, and similarly each of its cells, consists largely of water (~70%). Research over the last two decades has revealed that interstitial water in tissues and organs and, likewise, the cytostolic water in cells each play key roles in many biomechanical as well as biochemical processes. For example, the extensive water within the cartilage in articulating joints, such as the knee, carries a significant portion of the early compressive loading during standing, walking, and running. These tissues are porous, however, and the fluid can flow into or out of the tissues. Such fluid flow plays a key role in governing both the mechanical properties and many of the mechanotransduction mechanisms. Hence, again, biofluid mechanics is very important. Finally, biofluid mechanics is fundamental to the design of medical devices such as artificial hearts, ventricular-assist devices, heart valves, artificial arteries, mechanical ventilators, heart–lung machines, artificial kidneys, IV pumps, and so on. Exercise 7.1 asks you to brainstorm other medical devices that can be designed only with knowledge of biofluid mechanics.

In addition to its importance in the clinical setting, biofluid mechanics plays a key role in the research laboratory. One frequently finds fluids moving through tubes, pumps, valves, and so forth in the biomedical laboratory, with research varying from studies of the effects of pharmacologic agents on the behavior of the heart, kidney, or arteries to Earth-based studies of the effects of microgravity on gene expression in cells. Again, specific examples are manifold. Let us accept, therefore, that biofluid mechanics is essential to basic and applied research as well as to clinical care, and thus begin our study.

Just as in solid mechanics, we generally divide fluid mechanics into two general categories: statics and dynamics (cf. Fig. 1.4). In the former, we are interested only in the pressure distribution within the fluid; in the latter, we are generally interested in calculating both the pressure and the velocity *fields*, for from these we can compute any quantity of interest (e.g., normal and shear stresses, acceleration, and vorticity). Whether static or dynamic, to solve any problem in mechanics, we must, in general, address five classes of relations:

Kinematics: to compute velocities, accelerations, shear rates, and so forth
Stresses and tractions: to quantify the intensity of forces acting over oriented areas
Balance relations: to ensure the balance of mass, momentum, and energy
Constitutive relations: to describe mathematically the behavior of the material
Boundary/initial conditions: for mathematical and physical completeness

Indeed, a review of our problem formulations in Part II reveals that we used these five classes of relations in each problem. Within this general five-step approach, there are multiple ways of formulating problems in biofluid mechanics: in terms of differential equations, which provide the most detail; in terms of integral equations, which provide gross or average information; and via semi-empirical methods that exploit the availability of particular data. We illustrate each approach in Chapters 8–10.

7.2 Stress and Pressure

Recall from Chapter 2 that Euler and Cauchy first showed that the concept of stress is extremely useful in studying the mechanics of continua. In particular, depending on the application, one can define stress in various ways—the Cauchy stress $[\sigma]$, the first Piola–Kirchhoff stress $[\Sigma]$, or the second Piola–Kirchhoff stress $[S]$ as discussed in Chapter 6. Whereas each of these "measures" of stress is useful in nonlinear solid mechanics, experience has shown that Cauchy's definition of stress is the most useful when studying the mechanics of fluids whether they exhibit a linear or a nonlinear behavior. Recall from Chapter 2, therefore, that the *Cauchy stress is a measure of a force acting on an oriented area in the current configuration.* Moreover, because we can resolve any force that acts on a generic cube of material in terms of three components, the three orientations of the faces of the cube yield nine components of the Cauchy stress, which can be represented using a 3×3 matrix $[\sigma]$ at each point in general. Fortunately, only six of these components are independent due to the need to respect the balance of angular momentum.

Because our discussion of Cauchy stress in Chapter 2 was independent of the material or the deformation, everything holds equally well for a fluid. Hence, we will again be interested in components denoted as

$$\sigma_{(\text{face})(\text{direction})} \tag{7.1}$$

with respect to the coordinate system of interest [Eqs. (2.8)–(2.12)]. If needed, the stress transformation relations [Eqs. (2.13)–(2.21)] similarly hold.

In contrast to solid mechanics, however, the concept of pressure is particularly important in describing the mechanical behavior of fluids as well as in the solution of initial boundary value problems of interest. The *hydrostatic pressure* is defined herein as $-1/3$ of the sum of the diagonal components of the Cauchy stress when it is written in matrix form. For example, in Cartesians,

$$[\sigma] = \begin{bmatrix} \sigma_{xx} & \sigma_{xy} & \sigma_{xz} \\ \sigma_{yx} & \sigma_{yy} & \sigma_{yz} \\ \sigma_{zx} & \sigma_{zy} & \sigma_{zz} \end{bmatrix} \tag{7.2}$$

and thus the pressure p is given by

$$p = -\frac{1}{3}(\sigma_{xx} + \sigma_{yy} + \sigma_{zz}). \tag{7.3}$$

Because this combination of components is a coordinate invariant quantity (cf. Exercise 2.4), we equivalently have

$$p = -\frac{1}{3}(\sigma'_{xx} + \sigma'_{yy} + \sigma'_{zz}) \tag{7.4}$$

and similarly for cylindricals, sphericals, and other convenient coordinate systems. Pressure is thus a scalar field quantity, its value being independent of coordinate system. We will see in Section 7.4 that pressure and stress each play important roles in the constitutive formulation of a common class of fluid behaviors.

7.3 Kinematics: The Study of Motion

Let us next consider the motion of a fluid element in a flow field. For convenience, we follow an infinitesimal fluid element having a fixed identifiable mass Δm, which occupies a region $\Delta x \Delta y \Delta z$ in its current configuration (Fig. 7.1). As Δm moves in a flow field, several things may happen: It may

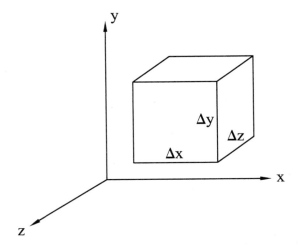

FIGURE 7.1 Schema of a differential cube that defines a fluid element of mass $\rho\Delta x\Delta y\Delta z$, where ρ is the mass density. Whereas actual cubes of solid (like an ice cube) are convenient for study, this is merely a fictitious cube of fluid.

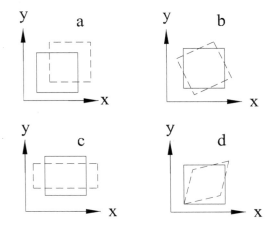

FIGURE 7.2 Four different types of motion (shown here in two dimensions for simplicity) that can be experienced by a fluid element: (a) a rigid-body translation, (b) a rigid-body rotation, (c) an extension, and (d) a pure shear.

translate, rotate, stretch, or shear. Although each of these motions may occur in three dimensions, it is useful to visualize them first in two dimensions. If the element translates (Fig. 7.2a), each particle undergoes the same displacement; for example, a particle at location (x_0, y_0, z_0) at time t_0 moves to a different location (x, y, z), at any time t, and the associated displacement or velocity vector holds for all particles of Δm. If the element rotates, it again moves as a rigid body, but the particles do not experience the same displacement. Figure 7.2b shows a rotation of the element about a z coordinate axis; clearly, the central particle does not change its location (x_0, y_0, z_0), whereas all other particles do. In general, of course, the element may rotate about all three of the coordinate axes at the same time. In addition, the fluid element may deform or "strain." Just as in solids, the deformation may be thought to consist of two types: changes in lengths (extensions) and changes in internal angles (shear). See Figures 7.2c and 7.2d. In fluids, we are often interested in the time rate of change of the deformation, not just the displacements or their spatial gradients (i.e., strains). This will be discussed in detail below. In general, therefore, a fluid element may undergo a combination of translation, rotation, extension, and shear, all in three dimensions, during the course of its motion.

7.3.1 Velocity and Acceleration

Consider a particle located at (x, y, z) at time t. Its position vector, relative to an origin at $(0, 0, 0)$, is $\boldsymbol{x} = x(t)\hat{\boldsymbol{i}} + y(t)\hat{\boldsymbol{j}} + z(t)\hat{\boldsymbol{k}}$ with respect to a Cartesian coordinate system. Velocity is defined as the time rate of change of position. Hence, we have

$$\boldsymbol{v} = \frac{d\boldsymbol{x}}{dt} = \frac{dx}{dt}\hat{\boldsymbol{i}} + \frac{dy}{dt}\hat{\boldsymbol{j}} + \frac{dz}{dt}\hat{\boldsymbol{k}} \tag{7.5}$$

Because velocity is a vector,[1] it can be written as

$$\boldsymbol{v} = v_x\hat{\boldsymbol{i}} + v_y\hat{\boldsymbol{j}} + v_z\hat{\boldsymbol{k}} \tag{7.6}$$

with respect to a Cartesian coordinate system. If so, its Cartesian components are simply

$$v_x = \frac{dx}{dt}, \qquad v_y = \frac{dy}{dt}, \qquad v_z = \frac{dz}{dt}. \tag{7.7}$$

Similarly, relative to a cylindrical coordinate system, one can have[2]

$$\boldsymbol{v} = v_r\hat{\boldsymbol{e}}_r + v_\theta\hat{\boldsymbol{e}}_\theta + v_z\hat{\boldsymbol{e}}_z. \tag{7.8}$$

The primary difference between the Cartesian and cylindrical representations is that the Cartesian bases $(\hat{\boldsymbol{i}}, \hat{\boldsymbol{j}}, \hat{\boldsymbol{k}})$ do not change with position, whereas two of the cylindrical bases $(\hat{\boldsymbol{e}}_r, \hat{\boldsymbol{e}}_\theta, \hat{\boldsymbol{e}}_z)$ do change with position; that is, the directions of $\hat{\boldsymbol{e}}_r$ and $\hat{\boldsymbol{e}}_\theta$ change with θ (Fig. 7.3). This will prove important below, when velocity gradients are calculated.

---

[1] We shall make extensive use of vectors in our discussion of fluids. A brief review is in Appendix 7.
[2] For consistency, we could let $\hat{\boldsymbol{i}} \equiv \hat{\boldsymbol{e}}_x, \hat{\boldsymbol{j}} \equiv \hat{\boldsymbol{e}}_y$, and $\hat{\boldsymbol{k}} \equiv \hat{\boldsymbol{e}}_z$ in Eq. (7.6).

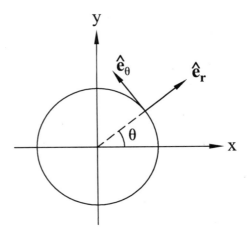

FIGURE 7.3 Illustration of the base vectors for a cylindrical-polar coordinate system. The direction of $\hat{e}_r$ changes with changes in θ and so too for the direction of $\hat{e}_\theta$. It is for this reason that spatial gradients of these bases vectors need not be zero, in contrast to the case of Cartesian bases. This proves to be important in the derivation of many equations in biofluids because of the cylindrical shape of many conducting tubes (e.g., arteries, airways, and ureters).

Acceleration is simply the time rate of change of velocity, but there are two ways of computing the acceleration. The most natural way is to take the time derivative of Eq. (7.5) (and similarly for cylindricals and other coordinate systems); that is,

$$a = \frac{dv}{dt} = \frac{d^2x}{dt^2}\hat{i} + \frac{d^2y}{dt^2}\hat{j} + \frac{d^2z}{dt^2}\hat{k}. \tag{7.9}$$

Because acceleration is also a vector, we have

$$a = a_x\hat{i} + a_y\hat{j} + a_z\hat{k}; \tag{7.10}$$

thus, the Cartesian components are

$$a_x = \frac{d^2x}{dt^2}, \qquad a_y = \frac{d^2y}{dt^2}, \qquad a_z = \frac{d^2z}{dt^2}. \tag{7.11}$$

This approach for computing a is common in solid mechanics and rigid-body dynamics; it is usually called a Lagrangian formulation. A prime example in rigid-body dynamics would be the calculation of the acceleration of a baseball after it leaves the bat; one only needs to know the (x, y, z) position of the mass center of the ball at each time as you "follow it" over the fence. Recall, too, that in solid mechanics, it is often useful to define the displacement vector $u(t)$ as the difference between where we (a material point) are and where we were, namely

$$u(t) = [x(t) - x(0)]\hat{i} + [y(t) - y(0)]\hat{j} + [z(t) - z(0)]\hat{k}, \tag{7.12}$$

from which the acceleration can alternatively be computed as

$$a = \frac{d^2 u_x}{dt^2}\hat{i} + \frac{d^2 u_y}{dt^2}\hat{j} + \frac{d^2 u_z}{dt^2}\hat{k}. \tag{7.13}$$

Equations (7.11) and (7.13) are equivalent because the original positions $x(0)$, $y(0)$, $z(0)$ at time $t = 0$ do not change with time.

A second approach for computing acceleration arises if we recognize that the velocity v is a field quantity (i.e., rather than following a single particle at each time t, we compute / measure v at each point x and each time t). For example, in Cartesians, we recognize that, in general,

$$v = v(x, y, z, t), \tag{7.14}$$

for which we must use the chain rule to compute the time derivative:

$$a = \frac{dv}{dt} = \frac{\partial v}{\partial t}\frac{dt}{dt} + \frac{\partial v}{\partial x}\frac{dx}{dt} + \frac{\partial v}{\partial y}\frac{dy}{dt} + \frac{\partial v}{\partial z}\frac{dz}{dt}. \tag{7.15}$$

From Eq. (7.7), therefore, the acceleration becomes

$$a = \frac{\partial v}{\partial t} + v_x\frac{\partial v}{\partial x} + v_y\frac{\partial v}{\partial y} + v_z\frac{\partial v}{\partial z}, \tag{7.16}$$

or, in more compact notation,

$$a = \frac{\partial v}{\partial t} + (v \cdot \nabla)v, \tag{7.17}$$

where $\partial v/\partial t$ is called the *local* acceleration and $(v \cdot \nabla)v$ is called the *convective* acceleration. A very important definition in fluids is that of a *steady flow*: A flow is said to be steady if $\partial v/\partial t = 0$ (i.e., if there is no local acceleration). Of course, in Cartesians, the del operator ∇ can be written as[3]

$$\nabla = \hat{i}\frac{\partial}{\partial x} + \hat{j}\frac{\partial}{\partial y} + \hat{k}\frac{\partial}{\partial z}. \tag{7.18}$$

Before continuing, it is instructive to show that Eqs. (7.16) and (7.17) are indeed equivalent. Given Eq. (7.16), we can factor out the common term, v, leaving us with

$$a = \frac{\partial v}{\partial t} + \left(v_x\frac{\partial}{\partial x} + v_y\frac{\partial}{\partial y} + v_z\frac{\partial}{\partial z}\right)v. \tag{7.19}$$

---

[3] Because ∇ is a differential operator, we remember that it operates on quantities to its right to form a gradient $\nabla\phi$ of a scalar ϕ, divergence $\nabla \cdot v$ of a vector v, or curl $\nabla \times v$ of a vector.

Now, resolving each term in the parentheses into its vector components, we are left with

$$a = \frac{\partial v}{\partial t} + \left[\left(v_x \hat{i} + v_y \hat{j} + v_z \hat{k} \right) \cdot \left(\hat{i} \frac{\partial}{\partial x} + \hat{j} \frac{\partial}{\partial y} + \hat{k} \frac{\partial}{\partial z} \right) \right] v, \qquad (7.20)$$

whereby we arrive at Eq. (7.17),

$$a = \frac{\partial v}{\partial t} + (v \cdot \nabla) v. \qquad (7.21)$$

In so doing, note that $v \cdot \nabla \neq \nabla \cdot v$ because the former yields a differential operator and the latter a scalar value. Finally, recognizing that a is a vector, Cartesian components are thus

$$a_x = \frac{\partial v_x}{\partial t} + v_x \frac{\partial v_x}{\partial x} + v_y \frac{\partial v_x}{\partial y} + v_z \frac{\partial v_x}{\partial z},$$

$$a_y = \frac{\partial v_y}{\partial t} + v_x \frac{\partial v_y}{\partial x} + v_y \frac{\partial v_y}{\partial y} + v_z \frac{\partial v_y}{\partial z}, \qquad (7.22)$$

$$a_z = \frac{\partial v_z}{\partial t} + v_x \frac{\partial v_z}{\partial x} + v_y \frac{\partial v_z}{\partial y} + v_z \frac{\partial v_z}{\partial z}.$$

This approach for computing acceleration is often adopted in fluid mechanics; it is called an Eulerian approach. In contrast to the aforementioned *Lagrangian approach* in which we follow material particles at different times, in the *Eulerian approach* we "watch" what happens to particles that go through multiple fixed points in space. A simple example that illustrates these approaches is to watch snowflakes fall while driving a car at night. One can either watch a particular snowflake approach the car and flow over the hood and past the windshield (a Lagrangian approach) or one could focus on a point in space and watch what happens to all the snowflakes that go through the point of interest (an Eulerian approach). The latter approach would typically be adopted by a fluid mechanicist and repeated at multiple points of interest to quantify the entire flow field $v = v(x, y, z, t)$. From Eq. (7.17), we recognize that a fluid particle moving in a flow field may accelerate for either of two reasons. For example, a fluid particle accelerates or decelerates if it is forced by an unsteady source (e.g., a pulsatile pump like the heart). This is a local acceleration. Alternatively, a fluid particle accelerates if it flows through a constriction even if acted upon by an unchanging force (e.g., gravity flow through a funnel or venous flow through a stenotic valve). See Figure 7.4. In this case, the particle is accelerated because it is convected to a region of higher velocity, hence the terminology convective part of a.

Similarly, in cylindricals, the acceleration vector can be written as

$$a = a_r \hat{e}_r + a_\theta \hat{e}_\theta + a_z \hat{e}_z, \qquad (7.23)$$

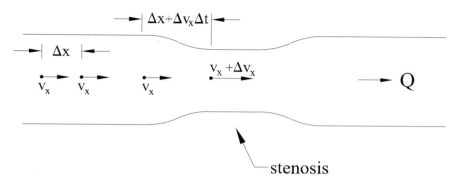

FIGURE 7.4 Schematic drawing of a stenosis (i.e., local narrowing) within a blood vessel. Stenoses often arise due to atherosclerosis, but other causes exist as well (e.g., extravascular blood due to a subarachnoid hemorrhage will cause a local vasospasm, whereas a local trauma due to a crushing injury or the overaggressive use of a vascular clamp during surgery can damage the vessel). Clearly, a fluid particle within the narrowed region will travel at a greater velocity than one in the inlet region; thus, the particles may accelerate as they enter the constriction despite no "time-dependent" change in the velocity.

where it can be shown (Humphrey, 2002) that the components are

$$a_r = \frac{\partial v_r}{\partial t} + v_r \frac{\partial v_r}{\partial r} + \frac{v_\theta}{r} \frac{\partial v_r}{\partial \theta} - \frac{v_\theta^2}{r} + v_z \frac{\partial v_r}{\partial z},$$

$$a_\theta = \frac{\partial v_\theta}{\partial t} + v_r \frac{\partial v_\theta}{\partial r} + \frac{v_\theta}{r} \frac{\partial v_\theta}{\partial \theta} + \frac{v_r v_\theta}{r} + v_z \frac{\partial v_\theta}{\partial z}, \qquad (7.24)$$

$$a_z = \frac{\partial v_z}{\partial t} + v_r \frac{\partial v_z}{\partial r} + \frac{v_\theta}{r} \frac{\partial v_z}{\partial \theta} + v_z \frac{\partial v_z}{\partial z}.$$

Note that "extra" terms arise in cylindrical coordinates [cf. Eq. (7.22)] because we are taking derivatives of the velocity (magnitude and direction) with respect to both time and direction. In contrast to Cartesians wherein $\partial(\hat{\boldsymbol{i}})/\partial x = \boldsymbol{0}$ and so forth, in cylindricals

$$\frac{\partial}{\partial \theta}(\hat{\boldsymbol{e}}_r) = \hat{\boldsymbol{e}}_\theta \quad \text{and} \quad \frac{\partial}{\partial \theta}(\hat{\boldsymbol{e}}_\theta) = -\hat{\boldsymbol{e}}_r, \qquad (7.25)$$

as suggested by Figure 7.3 and demonstrated in Appendix 7. These derivatives of the base vectors are very important to remember. Similarly, in spherical coordinates, acceleration becomes

$$\boldsymbol{a} = a_r \hat{\boldsymbol{e}}_r + a_\theta \hat{\boldsymbol{e}}_\theta + a_\phi \hat{\boldsymbol{e}}_\phi, \qquad (7.26)$$

for which it can be shown (Humphrey, 2002) that

$$a_r = \frac{\partial v_r}{\partial t} + v_r \frac{\partial v_r}{\partial r} + \frac{v_\theta}{r} \frac{\partial v_r}{\partial \theta} - \frac{v_\theta^2 + v_\phi^2}{r} + \frac{v_\phi}{r \sin \theta} \frac{\partial v_r}{\partial \phi},$$

$$a_\theta = \frac{\partial v_\theta}{\partial t} + v_r \frac{\partial v_\theta}{\partial r} + \frac{v_\theta}{r} \frac{\partial v_\theta}{\partial \theta} + \frac{v_r v_\theta}{r} - \frac{v_\phi^2}{r} \cot \theta + \frac{v_\phi}{r \sin \phi} \frac{\partial v_\theta}{\partial \phi}, \quad (7.27)$$

$$a_\phi = \frac{\partial v_\phi}{\partial t} + v_r \frac{\partial v_\phi}{\partial r} + \frac{v_\theta}{r} \frac{\partial v_\phi}{\partial \theta} + \frac{v_r v_\phi}{r} + \frac{v_\theta v_\phi \cot \theta}{r} + \frac{v_\phi}{r \sin \phi} \frac{\partial v_\phi}{\partial \phi}.$$

Again, the "extra" terms arise because of the derivatives of the base vectors, in particular

$$\frac{\partial}{\partial \theta}(\hat{e}_r) = \hat{e}_\theta, \qquad \frac{\partial}{\partial \theta}(\hat{e}_\theta) = -\hat{e}_r,$$

$$\frac{\partial}{\partial \phi}(\hat{e}_r) = \sin \theta \hat{e}_\phi, \qquad \frac{\partial}{\partial \phi}(\hat{e}_\theta) = \cos \theta \hat{e}_\phi, \qquad (7.28)$$

and

$$\frac{\partial}{\partial \phi}(\hat{e}_\phi) = -\sin \theta \hat{e}_r - \cos \theta \hat{e}_\theta. \qquad (7.29)$$

We will use spherical coordinates in Chapter 11 to investigate the interaction of cerebrospinal fluid and an intracranial saccular aneurysm.

Example 7.1 Given the following velocity field, determine if the Eulerian acceleration is zero:

$$v = Ax\hat{i} - Ay\hat{j},$$

where A is a number.

Solution: Although v is independent of time and thus does not have a local component of acceleration (i.e., it is a steady flow), we must check the convective part. Note, therefore, that the Cartesian components of v are

$$v_x = Ax, \qquad v_y = -Ay$$

and the velocity gradients are

$$\frac{\partial v_x}{\partial x} = A, \qquad \frac{\partial v_x}{\partial y} = 0, \qquad \frac{\partial v_y}{\partial x} = 0, \qquad \frac{\partial v_y}{\partial y} = -A.$$

Hence, from Eq. (7.22),

$$a = [0 + Ax(A) + (-Ay)(0) + 0]\hat{i} + [0 + Ax(0) + (-Ay)(-A) + 0]\hat{j} + 0\hat{k},$$

or

$$a = A^2\left(x\hat{\boldsymbol{i}} + y\hat{\boldsymbol{j}}\right)$$

and, consequently, the acceleration is not zero.

Before concluding this section, we emphasize that velocity and acceleration are both field quantities; that is, they depend on position as well as time in general. Consequently, it is useful to note some common terminology. A flow is said to be one, two, or three dimensional if the velocity field depends on respectively one, two, or three space variables. For example, with respect to Cartesians, a velocity field would be one dimensional if $\boldsymbol{v} = \boldsymbol{v}(x)$ but two dimensional if $\boldsymbol{v} = \boldsymbol{v}(x, y)$. Note, therefore, that one, two, or three dimensional does not have anything to do with the number of components, v_x, v_y, or v_z, that are nonzero; the dimensionality merely specifies the spatial dependence of all components. A flow in which only one component is nonzero with respect to a given coordinate system is said to be *unidirectional* regardless of its dependence on position.

Observation 7.1 Mathematically, a *field* is simply a contiguous collection of points in a region of space. When a quantity is defined at any such point, we say that it is a field quantity. Scalar quantities such as mass density, temperature, and pressure are examples of field quantities. Herein, we see too that velocity is a field quantity because it can depend on position as well as time. Figure 7.5 is a schematic drawing of part of the cardiovascular system of the dog. Clearly, the vasculature consists of a complex network of curved, tapering, and branching tubes called blood vessels. It is easy to imagine, therefore, that the velocity of the blood varies tremendously from point to point as well as throughout the cardiac cycle. Indeed, even locally within a curved region (aortic arch), bifurcation (aorto-iliac) or stenosis (Fig. 7.4), the velocity can vary tremendously in both magnitude and direction. The same is true, of course, in the airways within the lung and in the urinary tract. Quantification of the pressure and velocity fields is a prime objective in biofluid mechanics; this will require four governing equations, in general-one for pressure and one for each of the three components of velocity. We discuss these governing equations below.

7.3.2 Fluid Rotation

As noted above, we can imagine a fictitious cube of fluid (of mass Δm) rotating about all three axes x, y, and z. Such rotations can be described by an angular velocity vector, namely

$$\boldsymbol{\omega} = \omega_x\hat{\boldsymbol{i}} + \omega_y\hat{\boldsymbol{j}} + \omega_z\hat{\boldsymbol{k}}, \tag{7.30}$$

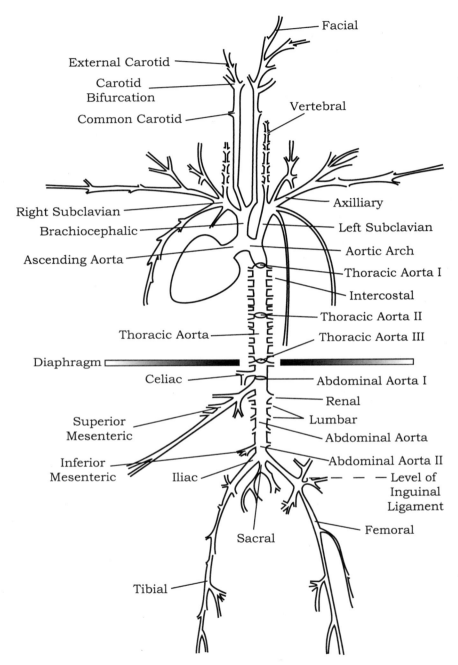

FIGURE 7.5 Schema of some of the large vessels that comprise the vasculature. There are, for example, ~1.2×10^5 arteries, 2.8×10^6 arterioles, 2.7×10^9 capillaries, 1.0×10^7 venules, and 7.0×10^5 veins in a 20-kg dog. From Humphrey (2002), with permission.

where ω_x is the rotation about the x axis, ω_y is the rotation about the y axis, and ω_z is the rotation about the z axis. To evaluate the components of this rotation vector $\boldsymbol{\omega}$, we may define the angular velocity about an axis as the average angular velocity of two initially perpendicular differential line segments in a plane perpendicular to the axis. For example, the component of rotation about the z axis is equal to the average angular velocity of two originally perpendicular infinitesimal line segments in the x-y plane (Fig. 7.6). The time rate of rotation of line segment $\overline{oa}$ of length Δx, is given by

$$\omega_{oa} = \lim_{\Delta t \to 0} \frac{\Delta \alpha}{\Delta t}, \tag{7.31}$$

where $\tan \Delta \alpha \sim \Delta \alpha$ for small-angle changes during the period Δt and $\tan \Delta \alpha$ is simply the opposite over the adjacent. (Note: In contrast to the small-angle assumption used by some to derive the small strain ε of Chapter 2, specification of a small-angle change here is not an assumption, it is merely consistent with the consideration of changes over a short interval of time Δt). If point o travels vertically at velocity v_y, then point a must have a greater velocity in order for $\Delta \alpha$ to change from 0 (at time $t = 0$) to some nonzero but small value at time $t + \Delta t$. Hence, if point a moves vertically at velocity $v_y + \Delta v_y$, then by the Taylor series expansion,

$$\Delta v_y = \frac{\partial v_y}{\partial x} \Delta x + \text{H.O.T.} \tag{7.32}$$

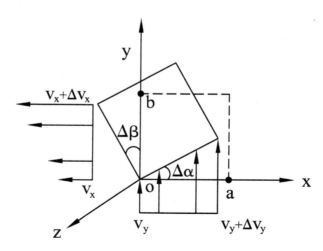

FIGURE 7.6 Rigid-body rotation of a fluid element in two dimensions. Note that such a rotation implies that the velocity must vary from point to point, such as the y-direction velocity at point a (e.g., $v_y + \Delta v_y$) must be greater than that at point o (e.g., v_y) for the line $\overline{oa}$ to rotate counterclockwise. A similar difference must exist in v_x at points b and o.

where H.O.T. stands for higher-order terms, which are negligible with respect to terms that are linear in delta such as Δx or Δy (recall our derivation in Section 3.1). Thus, in the increment of time Δt, point a moves vertically a distance (which is given by a velocity multiplied by time)

$$\left(v_y + \frac{\partial v_y}{\partial x} \Delta x - v_y \right) \Delta t \tag{7.33}$$

relative to o, and ω_{oa} becomes

$$\omega_{oa} = \lim_{\Delta t \to 0} \frac{1}{\Delta t} \left(\frac{(\partial v_y / \partial x)\Delta x \Delta t}{\Delta x} \right) = \frac{\partial v_y}{\partial x}. \tag{7.34}$$

Similarly, the angular velocity of line segment $\overline{ob}$ about the z axis is

$$\omega_{ob} = \lim_{\Delta t \to 0} \frac{\Delta \beta}{\Delta t}, \tag{7.35}$$

where b is moving in the negative x direction. If o moves leftward at velocity $-v_x$, then b must move at $-(v_x + \Delta v_x)$, where, again, we appeal to the Taylor's series. The distance along the negative x axis moved by b relative to o is thus

$$-\left(v_x + \frac{\partial v_x}{\partial y} \Delta y - v_x \right) \Delta t \tag{7.36}$$

and, therefore, ω_{ob} becomes

$$\omega_{ob} = \lim_{\Delta t \to 0} \frac{1}{\Delta t} \left(\frac{-(\partial v_x / \partial y)\Delta y \Delta t}{\Delta y} \right) = -\frac{\partial v_x}{\partial y}. \tag{7.37}$$

The mean angular velocity about the z axis, ω_z, is thus defined as the mean of ω_{oa} and ω_{ob}, or

$$\omega_z = \frac{1}{2} \left(\frac{\partial v_y}{\partial x} - \frac{\partial v_x}{\partial y} \right). \tag{7.38}$$

Similarly, for ω_x and ω_y, it is easy to show that (do it)

$$\omega_x = \frac{1}{2} \left(\frac{\partial v_z}{\partial y} - \frac{\partial v_y}{\partial z} \right), \quad \omega_y = \frac{1}{2} \left(\frac{\partial v_x}{\partial z} - \frac{\partial v_z}{\partial x} \right), \tag{7.39}$$

where $\boldsymbol{\omega} = \omega_x \,\hat{\boldsymbol{i}} + \omega_y \,\hat{\boldsymbol{j}} + \omega_z \,\hat{\boldsymbol{k}}$, and therefore

$$\boldsymbol{\omega} = \frac{1}{2} \left[\left(\frac{\partial v_z}{\partial y} - \frac{\partial v_y}{\partial z} \right)\hat{\boldsymbol{i}} + \left(\frac{\partial v_x}{\partial z} - \frac{\partial v_z}{\partial x} \right)\hat{\boldsymbol{j}} + \left(\frac{\partial v_y}{\partial x} - \frac{\partial v_x}{\partial y} \right)\hat{\boldsymbol{k}} \right]. \tag{7.40}$$

If we recognize the term in the square brackets as the curl $\boldsymbol{v} \equiv \nabla \times \boldsymbol{v}$ in Cartesian coordinates, then we can rewrite this result in a more compact (and general) vector notation as

$$\omega = \frac{1}{2} \nabla \times v. \tag{7.41}$$

Moreover, if we define a new quantity, $\zeta = 2\omega$, then

$$\zeta = \nabla \times v, \tag{7.42}$$

where ζ is called the *vorticity*—it is an alternate measure of the rotation of fluid elements. A flow in which the vorticity is zero is said to be *irrotational*. We will see later that irrotational flows allow significant simplification in the governing differential equations of motion; thus, it is useful to compute the vorticity for a given flow field.

Finally, note that in cylindrical coordinates, the vorticity is given by

$$\nabla \times v = \left(\frac{1}{r} \frac{\partial v_z}{\partial \theta} - \frac{\partial v_\theta}{\partial z} \right) \hat{e}_r + \left(\frac{\partial v_r}{\partial z} - \frac{\partial v_z}{\partial r} \right) \hat{e}_\theta + \left(\frac{1}{r} \frac{\partial (r v_\theta)}{\partial r} - \frac{1}{r} \frac{\partial v_r}{\partial \theta} \right) \hat{e}_z. \tag{7.43}$$

Likewise, in spherical coordinates, the vorticity is

$$\nabla \times v = \left(\frac{1}{r \sin \theta} \frac{\partial (v_\phi \sin \theta)}{\partial \theta} - \frac{1}{r \sin \theta} \frac{\partial v_\theta}{\partial \phi} \right) \hat{e}_r$$

$$+ \left(\frac{1}{r \sin \theta} \frac{\partial v_r}{\partial \phi} - \frac{1}{r} \frac{\partial (r v_\phi)}{\partial r} \right) \hat{e}_\theta + \left(\frac{1}{r} \frac{\partial (r v_\theta)}{\partial r} - \frac{1}{r} \frac{\partial v_r}{\partial \theta} \right) \hat{e}_\phi. \tag{7.44}$$

Here, we recognize one of the advantages of writing general results [Eq. (7.42)] in vector form. To derive Eqs. (7.43) and (7.44), we did not have to draw differential elements for cylindrical or spherical domains and determine how associated line elements rotate about each axis of interest; we merely used the del operator in the coordinate system of interest to compute general relations for the curl of the velocity. Such an approach is common in continuum mechanics: Based on the underlying physics, derive a relationship of interest with respect to Cartesian coordinates, which is generally the simplest to derive, and then extend the generalized result to other coordinate systems as needed using mathematical manipulations alone.

Example 7.2 Determine if the velocity field in Example 7.1 is irrotational.

Solution: Given that $v_x = Ax$ and $v_y = -Ay$,

$$\zeta = \frac{1}{2}(0 + 0)\hat{i} + \frac{1}{2}(0 + 0)\hat{j} + \frac{1}{2}(0 + 0)\hat{k},$$

hence the velocity is irrotational.

7.3.3 *Rate of Deformation*

During extensional deformations, a fluid element will simply change in length. Rather than using strain to quantify the extension, as in solid mechanics, it proves to be convenient in fluid mechanics to focus on the rate of extension (i. e., the stretching or rate of strain). Like strain, such rates of change can also be described by nine scalar components relative to a coordinate system of interest, six of which are independent. To measure the time rate of change of extension in one direction, consider Figure 7.4, in which a fluid accelerates, and thereby "stretches," within a constriction. A normalized rate of extension (or stretching) can be defined as the rate at which a line element lengthens divided by its original length; that is, it can be computed by knowing the difference in lengths at two times. Let the original length be Δx at time t. For this length to have increased at time $t + \Delta t$, the various particles cannot have moved at the same velocity. Hence, let the leftmost points have velocity v_x and the rightmost points have velocity $v_x + \Delta v_x$, both at time $t + \Delta t$, which allows a lengthening. Because a distance can be computed via a velocity multiplied by time, we have the length $\Delta x + \Delta v_x \Delta t$ at time $t + \Delta t$. Hence, the rate of lengthening, normalized by the length at time t, is

$$D_{xx} = \lim_{\substack{\Delta t \to 0 \\ \Delta x \to 0}} \frac{1}{\Delta t}\left(\frac{(\Delta x + \Delta v_x \Delta t) - \Delta x}{\Delta x}\right), \tag{7.45}$$

where $\Delta v_x = (\partial v_x/\partial x)\Delta x$ from a Taylor's series expansion: $v_x(x + \Delta x) = v_x(x) + (\partial v_x/\partial x)\Delta x + $ H.O.T. Thus, the rate of extension becomes

$$D_{xx} = \lim_{\substack{\Delta t \to 0 \\ \Delta x \to 0}} \frac{1}{\Delta t}\left(\frac{[\Delta x + (\partial v_x/\partial x)\Delta x \Delta t] - \Delta x}{\Delta x}\right) = \frac{\partial v_x}{\partial x}. \tag{7.46}$$

Likewise, it can be shown that

$$D_{yy} = \frac{\partial v_y}{\partial y}, \qquad D_{zz} = \frac{\partial v_z}{\partial z}. \tag{7.47}$$

Because they quantify rates at which lengths change, D_{xx}, D_{yy}, and D_{zz} are sometimes called components of a *stretching* matrix $[D]$. As an aside, it is interesting to note (do it) that

$$\nabla \cdot \boldsymbol{v} = \frac{\partial v_x}{\partial x} + \frac{\partial v_y}{\partial y} + \frac{\partial v_z}{\partial z} \equiv D_{xx} + D_{yy} + D_{zz}. \tag{7.48}$$

This will prove useful later. Indeed, a flow in which $\nabla \cdot \boldsymbol{v} = 0$ is said to be *incompressible* because it says that a fluid element that is lengthening in one direction must be thinning proportionately in other directions in order to conserve its volume.

Angular deformations of a fluid element involve changes in an angle between two initially perpendicular line segments in the fluid. To measure the time rate of change of angles, or shear rates, consider the differential element in Figure 7.7. In particular, imagine a rectangular fluid element that is contained between two solid plates and initially at rest. If the upper plate is moved with respect to a fixed lower plate, we can easily imagine that the fluid elements would be sheared (Fig. 7.7a). Similar to our analysis of the rigid rotation about the z axis [cf. Eq. (7.38)], let us consider the rate at which the internal angles change. For example (Fig. 7.7b), similar to Figure 7.6

$$\tan \Delta \alpha = \frac{\{[v_y + (\partial v_y/\partial x)\Delta x] - v_y\}\Delta t}{\Delta x}, \tag{7.49}$$

whereas in contrast to Figure 7.6,

$$\tan \Delta \beta = \frac{\{[v_x + (\partial v_x/\partial y)\Delta y] - v_x\}\Delta t}{\Delta y}, \tag{7.50}$$

because the enclosed angle is assumed to get smaller over time during a (positive) shearing motion (i.e., we are not looking at rigid-body rotation as before). Because these results are good for a short interval of time Δt, note that $\Delta \alpha$ and $\Delta \beta$ must likewise be small. By the small-angle approximation, therefore, $\tan \Delta \alpha \cong \Delta \alpha$ and $\tan \Delta \beta \cong \Delta \beta$ and we have

a)

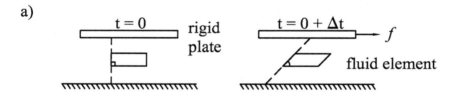

b)

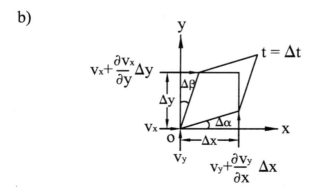

FIGURE 7.7 Panel a: Schema of an experimental system that gives rise to a simple shear of a fluid element; panel b: similar to Figure 7.6 except for a pure shear rather than a rigid-body rotation. In this case, the internal angle decreases a small amount, $(\Delta \alpha + \Delta \beta)$, over an interval of time Δt.

$$\Delta\alpha \approx \frac{\{[v_y + (\partial v_y/\partial x)\Delta x] - v_y\}\Delta t}{\Delta x},$$

$$\Delta\beta \approx \frac{\{[v_x + (\partial v_x/\partial y)\Delta y] - v_x\}\Delta t}{\Delta y}. \tag{7.51}$$

To find the rate at which these angles change, take the limit as $\Delta t \to 0$ of the time-averaged mean of the two changes in angle, namely

$$D_{xy} = \lim_{\Delta t \to 0} \frac{1}{\Delta t}\left(\lim_{\substack{\Delta x \to 0 \\ \Delta y \to 0}} \frac{1}{2}(\Delta\alpha + \Delta\beta)\right) = \lim_{\Delta t \to 0} \frac{1}{\Delta t}\left[\frac{1}{2}\left(\frac{\partial v_y}{\partial x}\Delta t + \frac{\partial v_x}{\partial y}\Delta t\right)\right], \tag{7.52}$$

or

$$D_{xy} = \frac{1}{2}\left(\frac{\partial v_y}{\partial x} + \frac{\partial v_x}{\partial y}\right). \tag{7.53}$$

It is easy to show (do it) that

$$D_{xz} = \frac{1}{2}\left(\frac{\partial v_x}{\partial z} + \frac{\partial v_z}{\partial x}\right), \qquad D_{yz} = \frac{1}{2}\left(\frac{\partial v_y}{\partial z} + \frac{\partial v_z}{\partial y}\right). \tag{7.54}$$

Similar to the situation for strains, which are useful measures of the deformation in solids, it can be shown that, by definition, $D_{xy} = D_{yx}$, $D_{xz} = D_{zx}$ and $D_{yz} = D_{zy}$. These quantities can thus be written in (a symmetric) matrix form as follows:

$$[D] = \begin{bmatrix} D_{xx} & D_{xy} & D_{xz} \\ D_{yx} & D_{yy} & D_{yz} \\ D_{zx} & D_{zy} & D_{zz} \end{bmatrix}, \tag{7.55}$$

where

$$D_{xx} = \frac{\partial v_x}{\partial x}, \qquad D_{xy} = \frac{1}{2}\left(\frac{\partial v_y}{\partial x} + \frac{\partial v_x}{\partial y}\right) = D_{yx},$$

$$D_{yy} = \frac{\partial v_y}{\partial y}, \qquad D_{yz} = \frac{1}{2}\left(\frac{\partial v_y}{\partial z} + \frac{\partial v_z}{\partial y}\right) = D_{zy}, \tag{7.56}$$

$$D_{zz} = \frac{\partial v_z}{\partial z}, \qquad D_{zx} = \frac{1}{2}\left(\frac{\partial v_x}{\partial z} + \frac{\partial v_z}{\partial x}\right) = D_{xz}.$$

Similarly, it can be shown that in cylindricals,

$$[D] = \begin{bmatrix} D_{rr} & D_{r\theta} & D_{rz} \\ D_{\theta r} & D_{\theta\theta} & D_{\theta z} \\ D_{zr} & D_{z\theta} & D_{zz} \end{bmatrix}, \tag{7.57}$$

where

$$D_{rr} = \frac{\partial v_r}{\partial r}, \qquad D_{r\theta} = \frac{1}{2}\left(r\frac{\partial}{\partial r}\left(\frac{v_\theta}{r}\right) + \frac{1}{r}\frac{\partial v_r}{\partial \theta}\right) = D_{\theta r},$$

$$D_{\theta\theta} = \frac{1}{r}\frac{\partial v_\theta}{\partial \theta} + \frac{v_r}{r}, \qquad D_{\theta z} = \frac{1}{2}\left(\frac{1}{r}\frac{\partial v_z}{\partial \theta} + \frac{\partial v_\theta}{\partial z}\right) = D_{z\theta}, \qquad (7.58)$$

$$D_{zz} = \frac{\partial v_z}{\partial z}, \qquad D_{zr} = \frac{1}{2}\left(\frac{\partial v_r}{\partial z} + \frac{\partial v_z}{\partial r}\right) = D_{rz}.$$

Finally, in spherical coordinates, it can be shown that

$$[D] = \begin{bmatrix} D_{rr} & D_{r\theta} & D_{r\phi} \\ D_{\theta r} & D_{\theta\theta} & D_{\theta\phi} \\ D_{\phi r} & D_{\phi\theta} & D_{\phi\phi} \end{bmatrix}, \qquad (7.59)$$

where

$$D_{rr} = \frac{\partial v_r}{\partial r}, \qquad D_{r\theta} = \frac{1}{2}\left(r\frac{\partial}{\partial r}\left(\frac{v_\theta}{r}\right) + \frac{1}{r}\frac{\partial v_r}{\partial \theta}\right) = D_{\theta r},$$

$$D_{\theta\theta} = \frac{1}{r}\frac{\partial v_\theta}{\partial \theta} + \frac{v_r}{r}, \qquad D_{\theta\phi} = \frac{1}{2}\left(\frac{\sin\theta}{r}\frac{\partial}{\partial \theta}\left(\frac{v_\phi}{\sin\theta}\right) + \frac{1}{r\sin\theta}\frac{\partial v_\theta}{\partial \phi}\right) = D_{\phi\theta},$$

$$D_{\phi\phi} = \frac{1}{r\sin\theta}\frac{\partial v_\phi}{\partial \phi} + \frac{v_r}{r} + \frac{v_\theta\cot\theta}{r},$$

$$D_{r\phi} = \frac{1}{2}\left(\frac{1}{r\sin\theta}\frac{\partial v_r}{\partial \phi} + r\frac{\partial}{\partial r}\left(\frac{v_\phi}{r}\right)\right) = D_{\phi r}. \qquad (7.60)$$

Here, let us make a few observations. First, recall from Chapter 2 that although displacements $\boldsymbol{u}$ are intuitive and measured easily, it was discovered that particular combinations of *displacement gradients* (called strains) are more useful in the analysis of the mechanics of solids. So, too, in fluid mechanics, we will see that combinations of *velocity gradients* are more useful in analysis than the intuitive and easily measured velocities $\boldsymbol{v}$. This reminds us that theory and its associated concepts, not simply the ease of making a measurement, must guide experimentation. Second, whereas the diagonal terms of $[D]$ provide information on time rates of change of lengths, or stretching, the off-diagonal terms provide information on time rates of change of internal angles, or shearing. Because of this, some prefer to refer to $[D]$ as a rate of deformation rather than a measure of stretching; indeed, others refer to it as a strain rate, for it is easily seen that the components of $[D]$ could be computed from the components of the strain $[\varepsilon]$ if the displacement gradients were computed with respect to the original not current positions (i.e., with respect to $\boldsymbol{X}$ not $\boldsymbol{x}$). Because $[\varepsilon]$ is based on a linearization of the exact measure of strain $[E]$, however, this correspondence may lead one to the false conclusion that $[D]$ is also a linearization of some measure of the motion—it is not. The small changes in lengths and angles used in deriving the components of $[D]$ are consistent with the small time steps $\Delta t \to 0$; hence, there is no linearization. The take-home message, therefore, is that regardless of its name (stretching, rate of deformation, or strain rate), $[D]$ is an exact, useful measure of the motion of particles in a fluid, just as $[E]$ is an exact, useful measure of the motion of particles in a solid. That said, we shall refer to $[D]$ as a *rate of deformation* and the particular off-diagonal terms as *shear rates*.

Example 7.3 Compute the components of $[D]$ for the velocity field in Example 7.1. Consider only x-y terms.

Solution: Given that $\mathbf{v} = Ax\hat{\boldsymbol{i}} - Ay\hat{\boldsymbol{j}}$ we have

$$D_{xx} = \frac{\partial v_x}{\partial x} = A, \qquad\qquad D_{xy} = \frac{1}{2}\left(\frac{\partial v_y}{\partial x} + \frac{\partial v_x}{\partial y}\right) = 0,$$

$$D_{yx} = \frac{1}{2}\left(\frac{\partial v_y}{\partial x} + \frac{\partial v_x}{\partial y}\right) = 0, \qquad D_{yy} = \frac{\partial v_y}{\partial y} = -A.$$

Relative to x and y, therefore, we would say that this flow is shearless.

Example 7.4 Determine the divergence of the velocity field in Example 7.3.

Solution

$$\nabla \cdot \mathbf{v} = \frac{\partial v_x}{\partial x} + \frac{\partial v_y}{\partial y} + \frac{\partial v_z}{\partial z} = A + (-A) + 0 = 0.$$

Hence, based on the above definition, this flow is incompressible (i.e., volume conserving).

Again, that $\nabla \cdot \mathbf{v} = 0$ implies conservation of mass for an incompressible fluid will be proven in Section 8.1 of Chapter 8.

7.4 Constitutive Behavior

Recall from Chapter 2 that a constitutive relation describes the response of a material to applied loads under conditions of interest, which, of course, depends on the internal constitution of the material. Thus, metals behave differently than polymers, and so too water behaves differently than glycerin, which behaves differently than blood, and so on. Recall, too, that in biosolid mechanics, we sought to relate stress to strain to describe linear and nonlinear elastic behaviors (Chapters 2 and 6). In contrast, experience has proven that it is more useful to relate stress to rates of deformation in order to describe the behavior of most fluids. Nevertheless, as is the case for solids, the formulation of a constitutive relation for a fluid involves five steps (DEICE): delineating general characteristics, establishing an appropriate theoretical framework, identifying a specific form of the relation, calculating best-fit values of the material parameters, and evaluating the predictive capability of the final relation. Let us now consider in detail one class of fluids that we will focus on herein.

7.4.1 Newtonian Behavior

To begin our discussion of the experimental investigation of the constitutive behavior of fluids, let us consider a greatly simplified situation. Assume that we can pour a fluid onto a flat surface such that it does not wet the surface (i.e., it forms a broad "bead" of fluid like water on a newly waxed car). Moreover, assume that we can place a thin rigid solid plate on this thin layer of fluid, giving a situation like that in Figure 7.8. Let us then apply an end load to the plate, which causes it to move at a constant velocity U_0

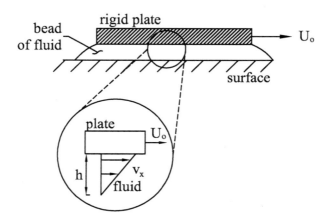

FIGURE 7.8 Couette flow between two rigid parallel plates. The bottom plate is stationary, whereas the top plate moves at velocity U_0 in the positive x direction.

in the x direction, which, in turn, causes the fluid to flow. If *the fluid has the same velocity, at the points of contact, as the solid that it contacts*, and ignoring effects near the ends (e.g., surface tension), the velocity field v in this simple case is $v = v_x(y)\hat{i}$, where

$$v_x = U_0\left(\frac{y}{h}\right), \qquad v_y = 0, \qquad v_z = 0. \qquad (7.61)$$

In particular, note that $v_x(y = 0) = 0$ and $v_x(y = h) = U_0$, the velocities of the bottom surface and the top plate, respectively. Assuming that a fluid has the same velocity as the solid it contacts is called the *no-slip boundary condition*. Albeit an approximation, this assumption tends to be very good in many cases.

Consequently, the only nonzero component of $[D]$, from Eq. (7.56), is

$$D_{xy} = \frac{1}{2}\left(\frac{\partial v_y}{\partial x} + \frac{\partial v_x}{\partial y}\right) = \frac{1}{2}\left(\frac{U_0}{h} + 0\right) = D_{yx}, \qquad (7.62)$$

which we note is the same at all points (x, y, z) in the fluid (if we are away from the end effects). By definition, the shear stress σ_{xy} exerted on the fluid by the plate is equal and opposite the shear stress imposed by the fluid on the plate (by Newton's third law). In this case, $\sigma_{yx} = f/A$, where f is the applied load and A is the surface area of contact between the plate and fluid. It is observed experimentally that the fluid shear stress σ_{yx} is related linearly to its shear rate D_{yx} for many fluids under many conditions (Fig. 7.9). Convention dictates that the slope of this shear stress versus shear-rate relation be denoted as 2μ, where μ is the (absolute) viscosity, much like the shear modulus G for solids. Viscosity is thus an important property of many fluids. Simply put, *viscosity is a measure of the resistance to flow when a fluid is acted upon by a shear stress*; in other words, it is a measure of the "thickness" of the fluid. Molasses, for example, is much more viscous than water (at the same temperature).

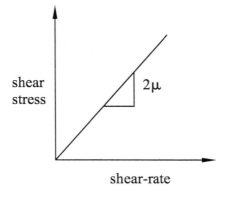

shear
stress

shear-rate

FIGURE 7.9 Shear stress plotted versus shear rate for a Newtonian fluid. Note the linear relationship, the slope of which is denoted by 2μ, where μ is the (absolute) viscosity. The factor 2 is included simply to cancel the $\frac{1}{2}$ that appears in the expression for shear rates [e.g., D_{xy}, as in Eq. (7.56)].

In most fluids, repeating this simple experiment in the y and z directions will result in the same slope 2μ (i.e., the same linear relationship between σ_{yz} and D_{yz} and σ_{xz} and D_{xz}). Hence, the response of many fluids is typically isotropic (i.e., independent of direction). Common experience reveals further that a fluid can also support a pressure p and that if compressible, experiments reveal that the normal stress will vary linearly with $\nabla \cdot v$, a measure of the volume change.

When a fluid exhibits these linear and isotropic characteristics, it is said to be *Newtonian* to commemorate Sir I. Newton's (1642–1727) suggestion that the shear response of fluids was proportional to the shearing force. Newton did not formulate his mechanics for continua, however, nor did he have the concepts of stress, strain, or shear rates. As noted in Chapter 2, these ideas came later, due largely to L. Euler (1707–1783) and A. Cauchy (1789–1857). Hence, formal quantification of this linear behavior of fluids, based on extensive experimentation, was put in mathematical form much later and is given by the Navier–Poisson equations[4]

$$\sigma_{xx} = -p + \lambda(\nabla \cdot v) + 2\mu D_{xx}, \qquad \sigma_{xy} = 2\mu D_{xy} = \sigma_{yx},$$
$$\sigma_{yy} = -p + \lambda(\nabla \cdot v) + 2\mu D_{yy}, \qquad \sigma_{yz} = 2\mu D_{yz} = \sigma_{zy}, \qquad (7.63)$$
$$\sigma_{zz} = -p + \lambda(\nabla \cdot v) + 2\mu D_{zz}, \qquad \sigma_{xz} = 2\mu D_{xz} = \sigma_{zx}.$$

where $\nabla \cdot v = D_{xx} + D_{yy} + D_{zz}$ [Eq. (7.48)] and λ is a second material property/parameter. G. Stokes (1819–1903) hypothesized that $\lambda \sim -2\mu/3$, which is often employed. Regardless, these Navier–Poisson equations are similar for other coordinate systems, such as for cylindricals,

$$\sigma_{rr} = -p + \lambda(\nabla \cdot v) + 2\mu D_{rr}, \qquad \sigma_{r\theta} = 2\mu D_{r\theta} = \sigma_{\theta r},$$
$$\sigma_{\theta\theta} = -p + \lambda(\nabla \cdot v) + 2\mu D_{\theta\theta}, \qquad \sigma_{\theta z} = 2\mu D_{\theta z} = \sigma_{z\theta}, \qquad (7.64)$$
$$\sigma_{zz} = -p + \lambda(\nabla \cdot v) + 2\mu D_{zz}, \qquad \sigma_{rz} = 2\mu D_{rz} = \sigma_{zr}.$$

and sphericals,

$$\sigma_{rr} = -p + \lambda(\nabla \cdot v) + 2\mu D_{rr}, \qquad \sigma_{r\theta} = 2\mu D_{r\theta} = \sigma_{\theta r},$$
$$\sigma_{\theta\theta} = -p + \lambda(\nabla \cdot v) + 2\mu D_{\theta\theta}, \qquad \sigma_{r\phi} = 2\mu D_{r\phi} = \sigma_{\phi r}, \qquad (7.65)$$
$$\sigma_{\phi\phi} = -p + \lambda(\nabla \cdot v) + 2\mu D_{\phi\phi}, \qquad \sigma_{\theta\phi} = 2\mu D_{\theta\phi} = \sigma_{\phi\theta}.$$

In the case of an incompressible behavior, it can be shown (see Chapter 8) that $\nabla \cdot v = 0$ and thus the Stoke's hypothesis for λ becomes a moot point. Of course, when $\nabla \cdot v = 0$, these constitutive equations simplify tremendously. For example, for an incompressible, Newtonian behavior, relative to Cartesian coordinates, we have the *incompressible Navier–Poisson equations* [recalling Eq. (7.56)]

---

[4] L. Navier (1785–1836) and S. D. Poisson (1781–1840).

$$\sigma_{xx} = -p + 2\mu \frac{\partial v_x}{\partial x}, \qquad \sigma_{xy} = \mu\left(\frac{\partial v_y}{\partial x} + \frac{\partial v_x}{\partial y}\right) = \sigma_{yx},$$

$$\sigma_{yy} = -p + 2\mu \frac{\partial v_y}{\partial y}, \qquad \sigma_{xz} = \mu\left(\frac{\partial v_x}{\partial z} + \frac{\partial v_z}{\partial x}\right) = \sigma_{zx}, \qquad (7.66)$$

$$\sigma_{zz} = -p + 2\mu \frac{\partial v_z}{\partial z}, \qquad \sigma_{yz} = \mu\left(\frac{\partial v_y}{\partial z} + \frac{\partial v_z}{\partial y}\right) = \sigma_{zy},$$

which we shall use extensively in Chapters 8 and 9; they are directly analogous to the so-called Hooke's law for solids [Eq. (2.69)], which relates the stresses and strains in a linear fashion.[5] Indeed, Hooke's law was also formulated by Navier and others in the nineteenth century; it contains two independent material parameters in the case of isotropy (E and v), similar to the compressible Navier–Poisson equation (with parameters μ and λ). As noted earlier, however, pressure plays a particularly important role in the constitutive relation here in contrast to that for a Hookean solid.

Example 7.5 Although the p that appears in Eq. (7.66) is actually a Lagrange multiplier that enforces the incompressibility constraint, similar to its role in Eq. (6.76) for nonlinear solids, its value is equivalent to the hydrostatic pressure for an incompressible Newtonian fluid. Prove this.

Solution: Recall from Eq. (7.3) that the hydrostatic pressure is defined as

$$p = -\frac{1}{3}(\sigma_{xx} + \sigma_{yy} + \sigma_{zz});$$

hence, for the incompressible Navier–Poisson relation, we have

$$-\frac{1}{3}(\sigma_{xx} + \sigma_{yy} + \sigma_{zz}) = -\frac{1}{3}(-p + 2\mu D_{xx} - p + 2\mu D_{yy} - p + 2\mu D_{zz})$$

$$= p - \frac{2}{3}\mu(D_{xx} + D_{yy} + D_{zz})$$

$$= p - \frac{2}{3}\mu\left(\frac{\partial v_x}{\partial x} + \frac{\partial v_y}{\partial y} + \frac{\partial v_z}{\partial z}\right)$$

$$= p - \frac{2}{3}\mu(\nabla \cdot v),$$

---

[5] It is interesting that the simple (i.e., linear) descriptions of solid and fluid behavior are called Hookean and Newtonian after the contemporaries/adversaries R. Hooke and Sir I. Newton even though the particular equations were put forth much later.

where we said that $\nabla \cdot \boldsymbol{v}$ will be shown to be zero for incompressible flows. In this case, therefore, the p in Eq. (7.66) is the hydrostatic pressure. This is not true, in general, for incompressible solids.

Example 7.6 In the case of pure *fluid statics*, the fluid is at rest and thus the velocity and acceleration of the fluid are both zero (note: actually, we only need $\boldsymbol{a} = \boldsymbol{0}$ for statics by Newton's second law). Show that in this case, the only possible stress in the fluid is a hydrostatic pressure. Moreover, show that if a cube of a static fluid is oriented with respect to an $(o; x, y, z)$ coordinate system, then the only stresses with respect to an $(o; x', y', z')$ coordinate system are still hydrostatic.

Solution: If $\boldsymbol{v} = \boldsymbol{0}$, then $v_x = v_y = v_z = 0$ and all components of $[D]$ are zero. From Eq. (7.66), therefore,

$$\sigma_{xx} = -p, \qquad \sigma_{yy} = -p, \qquad \sigma_{zz} = -p,$$
$$\sigma_{xy} = 0 = \sigma_{yx}, \qquad \sigma_{xz} = 0 = \sigma_{zx}, \qquad \sigma_{yz} = 0 = \sigma_{zy}$$

and the stress is *hydrostatic* (i.e., the normal stresses at a point equal the negative of the pressure at that point). Moreover, if we consider a rotation about the z axis (i.e., in the x-y plane), then

$$\sigma'_{xx} = \sigma_{xx} \cos^2 \alpha + 2\sigma_{xy} \cos\alpha \sin\alpha + \sigma_{yy} \sin^2 \alpha,$$
$$\sigma'_{xy} = (\sigma_{yy} - \sigma_{xx}) \cos\alpha \sin\alpha + \sigma_{xy} (\cos^2 \alpha - \sin^2 \alpha),$$
$$\sigma'_{yy} = \sigma_{xx} \sin^2 \alpha - 2\sigma_{xy} \cos\alpha \sin\alpha + \sigma_{yy} \cos^2 \alpha$$

from Eqs. (2.13), (2.17), and (2.21). Hence, in our case,

$$\sigma'_{xx} = -p \cos^2 \alpha + 0 - p \sin^2 \alpha = -p,$$
$$\sigma'_{xy} = (-p + p) \cos\alpha \sin\alpha + 0 = 0,$$
$$\sigma'_{yy} = -p \sin^2 \alpha - 0 - p \cos^2 \alpha = -p$$

for all α, and the stress is indeed hydrostatic in pure fluid statics regardless of the coordinate system (Fig. 7.10).

Although discussed in detail in Chapter 8, note that if μ is negligible, then

$$\sigma_{xx} = -p, \qquad \sigma_{yy} = -p, \qquad \sigma_{zz} = -p \qquad\qquad (7.67)$$

and all shear stresses are zero. A fluid that can only support a pressure, not a shear stress, is said to be *inviscid*. Whereas no fluid has zero viscosity, the

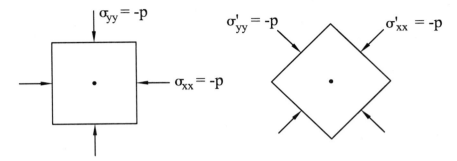

FIGURE 7.10 So-called hydrostatic state of stress (shown in two dimensions simply for convenience) wherein the normal components each equal $-p$, the fluid pressure, and the shear components are all zero—both relative to all coordinate systems.

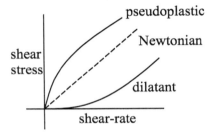

FIGURE 7.11 Two common non-Newtonian (i.e., nonlinear) behaviors exhibited by fluids. A pseudoplastic behavior is characterized by a resistance to flow that decreases with increasing shear rates, whereas a dilatant behavior is characterized by a resistance to flow that increases with increasing shear rates. In both cases, the behavior at high shear rates appears Newtonian; that is, the resistance to flow (or viscosity) is nearly constant at high shear rates. Note, too, that the pseudoplastic and dilatant behaviors are illustrative only; they could appear on either side of the Newtonian curve, the slope of which varies from fluid to fluid and with temperature.

assumption of negligible viscosity (like that of a rigid member in a truss in statics even though no material is truly rigid) has proven useful in many areas of fluid mechanics, particularly in aerospace applications.

7.4.2 Non-Newtonian Behavior

Not all fluids exhibit a Newtonian (i.e., linear) behavior. Figure 7.11 contrasts a Newtonian behavior with two general non-Newtonian behaviors: pseudoplastic and dilatant. *Pseudoplastic behavior* is characterized by a viscosity (i.e., resistance to flow) that is lower at higher shear rates than it is

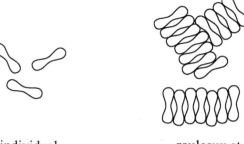

individual
blood cells

rouleaux at
low shear-rates

FIGURE 7.12 Aggregation of red blood cells, called rouleaux, that occurs at low shear rates. Such aggregation tends to increase the resistance of blood to flow at low shear rates and thus is responsible in part for the pseudoplastic type of behavior of blood.

at lower shear rates; for this reason, pseudoplastic behavior is sometimes called *shear thinning*. This can be due to particles within the fluid that aggregate at low shear rates but "break up" at higher shear rates, which lowers the viscosity. Of particular importance in biofluids, whole blood is such a fluid. At low shear rates, the red blood cells tend to aggregate, a phenomenon known as *rouleaux* (Fig. 7.12), which depends on the presence of fibrinogen and the globulins. Of course, in the limit as the shear rate goes to zero, the blood will tend to aggregate further, eventually leading to a process known as clotting, which involves additional mechanisms, including platelet activation and the conversion of fibrinogen to fibrin, an essential component of a clot. As the shear rate increases from low, but nonzero values, however, the rouleaux break up and blood behaves like a Newtonian fluid; the latter is often assumed in large arteries, thus, $\mu \sim$ constant (often cited to be ~ 3.5 cP, or centiPoise) and one can employ the Navier–Poisson equations to describe many flows of blood in large arteries. Plasma (i.e., whole blood minus cells) always behaves as a Newtonian fluid, with a viscosity $\mu \sim 1.2$ cP. It should also be noted, however, that in capillaries, which are ~ 5–$8\,\mu$m in diameter, the red blood cells go through one at a time, with plasma in between. In this case, the blood should be treated as a two-phase flow—a solid and fluid mixture, which is beyond the scope of this book.

Because of its non-Newtonian behavior, whole blood has been modeled using various relations.[6] For example, Fung (1990) advocates the following model for blood:

---

[6] When behavior is linear, the relation is unique; when the behavior is nonlinear, many different constitutive relations can often "fit" the data.

$$\sigma_{xx} = -p + 2\eta(J_2)D_{xx}, \qquad \sigma_{xy} = 2\eta(J_2)D_{xy},$$
$$\sigma_{yy} = -p + 2\eta(J_2)D_{yy}, \qquad \sigma_{yz} = 2\eta(J_2)D_{yz}, \qquad (7.68)$$
$$\sigma_{zz} = -p + 2\eta(J_2)D_{zz}, \qquad \sigma_{xz} = 2\eta(J_2)D_{xz},$$

where

$$\eta(J_2) = \frac{1}{\sqrt{J_2}}\left((\mu^2 J_2)^{1/4} + \frac{1}{\sqrt{2}}\sqrt{\tau_y}\right)^2, \qquad (7.69)$$

$$J_2 = \frac{1}{2}(D_{xx} + D_{yy} + D_{zz} + 2D_{xy}D_{yx} + 2D_{xz}D_{zx} + 2D_{yz}D_{zy}), \qquad (7.70)$$

and μ is a viscosity at a high shear rate and τ_y is a solidlike yield stress at low shear rates. To better appreciate the implications of this complex relation, let us return to the simple experiment in Figure 7.8, which allowed our observation of linear behavior for a Newtonian fluid. In this simple case, D_{xy} is nonzero and all other components of $[D]$ are zero. Hence, Fung's relation reduces to

$$\sigma_{xy} = 2\eta(J_2)D_{xy}, \qquad (7.71)$$

with

$$\eta(J_2) = \frac{1}{\sqrt{D_{xy}^2}}\left((\mu^2 D_{xy}^2)^{1/4} + \frac{1}{\sqrt{2}}\sqrt{\tau_y}\right)^2. \qquad (7.72)$$

When D_{xy} is very small, η tends to become large, its value depending largely on the values of $\tau_y/2D_{xy}$; τ_y is also small, usually on the order of 0.005 Pa. Conversely, when D_{xy} is large, the yield stress becomes negligible and $\eta \cong \mu$, the Newtonian case wherein $\sigma_{xy} = 2\mu D_{xy}$. Hence, Fung's relation accounts for the pseudoplastic character illustrated in Figure 7.11, including Newtonian behavior at high shear rates.

Because the pseudoplastic character of blood is due largely to the red blood cells, note the following. At a low hematocrit H (i.e., percent concentration of red blood cells), such as $H \sim 8.25\%$, $\eta(J_2)$ is nearly constant (i.e., $\eta \sim \mu$) over a range of shear rates from 0.1 to $1000\,\mathrm{s}^{-1}$. For $H \sim 18\%$, however, $\eta(J_2)$ is nearly constant only for shear rates above $600\,\mathrm{s}^{-1}$. Normal shear rates range from 100 to $2000\,\mathrm{s}^{-1}$ in large arteries and from 20 to 200 s^{-1} in large to small veins. Normal values of the hematocrit are about 42% and 47% in women and men, respectively, hence the Newtonian response only at high shear rates. As noted earlier, the value of $\eta(J_2)$ at high shear rates is on the order of 2–3.5 cP for whole blood. Note: 1 Poise = 0.1 kg/ms = 0.1 Ns/m², where cP denotes centiPoise (after J. Poiseuille).

A good example of a *dilatant behavior* is given by a cornstarch solution. As some learn in kindergarten, slowly pulling one's fingers through a cornstarch solution meets little resistance, whereas quickly pulling one's fingers meets with considerable resistance; that is, the apparent viscosity increases with increased shear rate (cf. Fig. 7.11). Dilatant behavior is typical of suspension of solids, in which the solid content is very high (e.g., 80%) so that it forms large masses within the suspension. Increasing shear rates break up such masses.

Finally, note that some fluids exhibit a changing (apparent) viscosity over time when at a constant shear rate; that is, an internal structure may build up or break down over time. A non-Newtonian behavior characterized by a decreasing viscosity over time is called *thixotropic*; one characterized by an increasing viscosity over time of shear is called *rheopectic*. Because of these behaviors, a hysteresis loop will be seen in plots of shear stress versus shear rate if the fluid is sheared at an increasing rate for some time followed by a decreasing rate. Some man-made models of synovial fluid exhibit such non-Newtonian characteristics. Synovial fluid is found in articulating joints and is a remarkable lubricant. It shall be discussed in more detail later.

7.5 Blood Characteristics

Although the flow of air in the lungs and the flow of urine in the renal system are very important problems in biofluid mechanics, both of these fluids can be assumed to exhibit a Newtonian behavior in most cases. Indeed, the flows of physiologic salt solutions in laboratory and clinical settings can likewise be treated as Newtonian. Here, therefore, let us consider blood in more detail, both because of its central role in hemodynamics and as an illustrative non-Newtonian fluid.

7.5.1 Plasma

Blood is a viscous solid–fluid mixture consisting of plasma and cells. Plasma is composed of ~90% water and contains inorganic and organic salts as well as various proteins: albumin, the globulins, and fibrinogen. Collectively, these proteins represent about 7–8% of the plasma by weight. Albumin, the smallest plasma protein, is present in the largest concentration and represents about half of the protein mass; it has a major role in regulating the pH and the colloid osmotic pressure. The alpha and beta globulins, usually 45% of the plasma protein mass, are antibodies that fight infection. Fibrinogen is the largest of the plasma proteins and, through its conversion to long strands of fibrin, has a major role in the process of clotting; it accounts for only about 5% of the plasma protein mass. Serum is simply the fluid that remains after blood is allowed to clot. For the most part, the composition of serum is the same as that of plasma, with the exception that the clotting proteins, primarily fibrinogen, and platelets have been removed. As noted earlier, plasma and, thus, serum exhibit a Newtonian behavior; the viscosity of plasma, for example, is ~1.2 cP at 37°C and its specific gravity is 1.03. For comparison, the viscosity of water is ~1.0 cP at 20°C and ~0.7 cP at 37°C.

7.5.2 Blood Cells

The cellular portion of blood consists of three primary types of cell: erythrocytes, leukocytes, and platelets. The most abundant of these are the erythrocytes, or red blood cells (RBCs), which comprise about 95–97% of the cellular component of blood. This class of cells has a normal life span on average of 120 days, which corresponds to a net turnover rate of ~0.8% per day. They are produced by the bone marrow and removed primarily by the spleen. Erythrocytes consist of a thin, flexible membrane with an interior filled with a hemoglobin solution; whereas the membrane can be modeled as a solid, the hemoglobin solution can be modeled as a fluid with a viscosity of ~6 cP. Clearly, then, there is great advantage to packaging the hemoglobin within deformable membranes rather than transporting it directly through the circulation. The major role played by the RBCs is the transport of oxygen that is bound to the hemoglobin, which constitutes about 95% of their dry weight. Consequently, the density of a red blood cell is higher than that of plasma, and in a quiescent state, the RBCs tend to settle. This settling is used in the laboratory to determine the volume fraction of red blood cells, called the *hematocrit* and often denoted by *H*; its value typically varies between 40% and 50%.

Red blood cells in blood that is not flowing have a unique shape described as a biconcave discoid with a major diameter of approximately 7.6 μm, a maximum thickness of about 2.8 μm, and a minimum thickness of 1.44 μm. The average human red blood cell has a volume of about 98 μm^3 and a surface area of 130 μm^2. There are about 5×10^6 red blood cells in a cubic millimeter of blood; hence, there is tremendous surface area available for gas exchange. When subjected to low shear rates, red blood cells can form face-to-face stacked structures called rouleaux, which, in turn, can clump together to form larger RBC structures called aggregates (Fig. 7.12). Both rouleaux and aggregates break apart under conditions of increased blood flow, or higher shear rates as noted earlier, which contributes to the pseudoplastic character of blood.

Example 7.7 The biconcave shape of the RBC increases its surface area-to-volume ratio. Compute this ratio and compare it to the value that would hold for a spherical cell.

Solution: Given the stated volume *V* and the surface area *A* of 98 μm^3 and 130 μm^2, respectively, we have a ratio of 1.33 μm^{-1} for the biconcave shape. If the RBC were a sphere of volume 98 μm^3, then its radius would be

$$r = \left(3V/4\pi\right)^{1/3} = 2.86 \ \mu\text{m}.$$

The associated surface area would thus be $A = 4\pi r^2 = 103\,\mu m^2$; thus, for a sphere, the ratio of surface area to volume would be $103/98 = 1.05\,\mu m^{-1}$. This value is significantly less than that for the biconcave disk. Of course, in the capillary, the RBC deforms significantly from its baseline shape, which could further increase the ratio of surface area to volume, although some have suggested that the surface area tends to remain constant. A constant surface area can be accounted for constitutively for the membrane as a kinematic constraint (Humphrey, 2002).

The next most abundant cell type in blood is the platelets, which comprise about 4.9% of the cell volume. There are $(2.5–3.0) \times 10^5$ platelets per cubic millimeter of blood, with cell diameters $\sim 2.5\,\mu m$ and thicknesses ~ 0.5 μm. As the name implies, they have a platelike disk shape. The platelets are major players in the coagulation of blood and thus the prevention of blood loss. The remaining 0.1% of the cellular component of blood consists of leukocytes, or white blood cells (WBCs), which form the cellular component of the immune system. There are $(5–8) \times 10^3$ WBCs per cubic millimeter of blood in health. The three primary classes of WBCs are the monocytes ($16–22\,\mu m$ in diameter), granulocytes ($10–12\,\mu m$ in diameter), and lymphocytes ($7\,\mu m$ in diameter). Although much fewer in number than the RBCs, there are $\sim 37 \times 10^9$ (37 billion) WBCs circulating in the blood of a healthy adult. Because the white blood cells and platelets only comprise 5% of the cellular component of blood, their effect on the macroscopic flow characteristics of blood is typically assumed to be negligible; that is, the non-Newtonian character of blood is controlled primarily by the hematocrit and, to a lesser degree, the fibrinogen.

7.5.3 Additional Rheological Considerations

Rheology is a science concerned with the deformation and flow of materials. This name comes from the Greek *rheo*, meaning something that flows. Notwithstanding its non-Newtonian character, which renders blood more difficult to study than Newtonian fluids, the rheological behavior of blood is complicated further by its heterogeneous composition; that is, recall that we mentioned earlier that RBCs tend to go through capillaries in single file, with plasma between them, and, consequently, that such flows are best studied as a solid–fluid mixture. Regardless of the diameter of the vessel through which it flows, blood is always a suspension of blood cells in plasma; thus, its rheological properties depend on the concentration, mechanical properties, and interactions of its constituent parts. In particular, the rheology of blood depends strongly on the deformation of individual cells, especially the erythrocytes. For this reason, cell mechanics is as important in biofluid mechanics as it is in biosolid mechanics (indeed, the study of the inflammatory response due to WBCs depends largely on the mechanics of their adhesion to the endothelium, thus rendering the study of leukocytes likewise important). Actually, because of the ease of isolating RBCs in vitro, they were among the first cells to be studied within the context of mechanics, which began in earnest in the mid- to late-1960s with the birth of modern biomechanics itself.

A marked aggregation of red blood cells at low shear rates is reflected by the yield stress τ_y, which must be exceeded for the material to flow. Many heterogeneous fluids that contain a particulate phase that forms aggregates at low shear rates exhibit a yield stress. The presence of a yield stress alone renders the behavior of a fluid non-Newtonian and, indeed, gives it an initial solidlike behavior. Materials that exhibit a yield stress but thereafter behave as a Newtonian fluid are called *Bingham plastics*—clay suspensions being a prime example. Recall that Fung's proposed constitutive equation for blood [Eq. (7.68)] accounted for both the initial yield stress and the pseudoplastic character. As noted earlier, another characteristic of a non-Newtonian behavior is that the viscosity varies with shear rate. For non-Newtonian fluids, the local slope of the stress versus shear-rate curve at a given value of the shear rate is often called the *apparent viscosity*, sometimes denotes as μ_a. The apparent viscosity of blood is shown in Figure 7.13 as a function of hematocrit H and shear rate at a temperature of 37°C. Consistent with its pseudoplastic character, the apparent viscosity of blood is high at low shear rates due to the presence of rouleaux and aggregates. At shear rates above about $100 s^{-1}$, only individual cells exist, and blood behaves macroscopically as if it were a Newtonian fluid. Actually, blood flow in large arteries and veins is Newtonian only near the vessel wall, where the wall shear rate is significantly higher than $100 s^{-1}$. As one gets

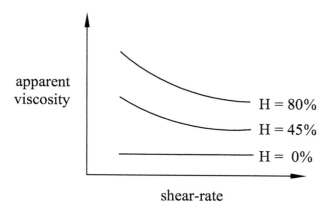

FIGURE 7.13 Viscosity of blood as a function of shear rate and hematocrit H (i.e., percentage of red blood cells by volume). The deformability and the volume fraction of red blood cells both affect the viscosity.

closer to the centerline of the vessel, the shear rate can approach zero (shown in Chapter 9) and blood may exhibit its non-Newtonian character. Many investigators ignore this complexity and model large vessel flows assuming a constant apparent viscosity, yet we must be mindful of the actual physics in cases wherein the heterogeneity of the blood is important.

Whereas the erythrocytes represent the major cell species in determining the flow properties of blood in health, leukocytes, platelets, and blood-borne proteins may play significant roles in abnormal or disease conditions. Recalling that an increase of the hematocrit increases the resistance of blood to flow, note that the same effect is seen when the fibrinogen concentration is increased. Indeed, if the clotting protein fibrinogen is removed, while keeping the hematocrit unchanged, the resulting RBC suspension behaves nearly like a Newtonian fluid for shear rates as low as 0.01 s^{-1}. Fibrinogen and its effect of increasing interactions between RBCs thus appears to play a role in the non-Newtonian behavior at low shear rates; that is, the other plasma proteins, such as albumin and the globulins, do not contribute significantly to the non-Newtonian behavior of blood, although their concentration will affect the viscosity of the plasma.

Finally, it should be noted that the deformability alone (i.e., solid mechanics) of the red blood cells plays a key role in the rheology of blood. Note, therefore, that the solidlike RBC membrane is capable of moving around the fluidlike cell contents much like a tank-tread moves around the wheels of a tank. This movement of the cell membrane appears to aid the RBCs in their adapting to a flow: normal RBCs undergo shearing deformations and their long axes show a preferential alignment with flow. This clearly affects the overall rheology of blood. In diseases such as sickle cell anemia, which alters the shape and properties of the erythrocytes, one sees an associated tremendous change in the rheological properties of the blood. Indeed, the importance of RBC deformability is revealed well by studying suspensions of similarly sized but rigid spheres. The apparent viscosity of a fluid – rigid sphere solution increases nonlinearly with an increase in the volume fraction of the spheres and asymptotes near a 50% concentration, where it ceases to flow. In comparison, blood could flow with even a 98% hematocrit (Fung, 1993). In other words, the viscosity of blood is about one-half that of a similar suspension of hard spheres.

Observation 7.2 Another fluid in the body that exhibits a strong non-Newtonian character is the *synovial fluid* in articulating joints. This fluid is secreted into the joint cavity by the synovial membrane. It is normally clear and colorless, often looking like raw egg white. Indeed, its name comes from *syn* (meaning like) and *ovial* (meaning egg). Synovial fluid is a dialysate of plasma; it contains proteins (e.g., 3.4 g/dL in pigs, including albumin and globulins), hyaluronan (119 mg/dL), and a small amount of phospholipids (19 mg/dL). The hyaluronan, which is commonly called hyaluronic acid, is

Synovial Fluid

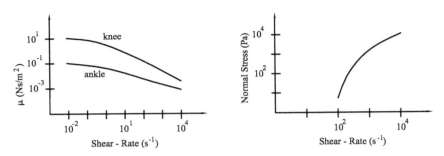

FIGURE 7.14 Shear thinning of synovial fluid, the primary lubricant in articulating joints (left). Another important characteristic of a non-Newtonian fluid is its ability to generate a shear-rate-dependent normal stress (right). Quantification of the 3-D behavior of non-Newtonian fluids is beyond an introductory text, however, and the reader is referred to Tanner (1985) for an introduction.

the simplest glycosaminoglycan; it consists of a sequence of up to 25,000 repeating disaccharide units, but it is, nevertheless, a very large molecule (molecular weight $\sim 8 \times 10^6$). It is the hyaluronic acid that gives synovial fluid its advantageous non-Newtonian characteristics (a shear-thinning viscosity, a normal stress effect, and an elastic effect at high frequencies of loading), which can be lost in diseases such as rheumatoid arthritis. Understanding the mechanical behavior of synovial fluid is thus important in understanding health and disease. Figure 7.14 shows the shear-thinning behavior, which is thought to be due to the increased alignment of the hyaluronic acid molecules at high shear rates.

It is commonly known that synovial fluid plays an important role as a lubricant for the relative motion of cartilage-to-cartilage in a joint such as the knee (Fig. 7.15). It is remarkable, therefore, that this is accomplished with such a small amount of synovial fluid. The human knee joint contains only 0.2 mL of synovial fluid, with a layer thickness that is typically on the order of micrometers to perhaps even nanometers. The associated coefficient of friction ranges from 0.001 to 0.03, which is remarkably small. For more on lubrication within diarthroidal joints, see Mow et al. (1990) and Chapter 5 in *Handbook of Bioengineering* edited by Skalak and Chien (1987).

7.6 Cone-and-Plate Viscometry

Recall from Chapter 3 that we derived governing differential equations for equilibrium (Sections 3.1 and 3.2) that can be solved exactly in certain cases (Section 3.6). Such solutions are very valuable in biomechanics,

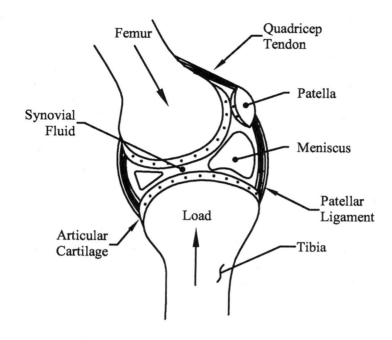

The Human Knee

FIGURE 7.15 Schema of the knee showing the important constituents for geometric modeling and constitutive behavior. Cartilage and joint lubrication are considered briefly in Chapter 11.

particularly in the *design* and interpretation of experiments that are used to determine constitutive relations and likewise in the *analysis* of stress in studies of mechanotransduction. We will derive similar governing differential equations for fluids in Chapter 8 and obtain a number of exact solutions in Chapter 9. Nevertheless, in both biosolid and biofluid mechanics, much simpler approximate solutions are often useful; such as the "strength of materials solutions" for torsion and bending in Chapters 4 and 5. Here, let us consider one such approximate solution of experimental utility in biofluid mechanics: the relationship between the viscosity of a fluid and the applied loads and geometry of a common experimental device.

A device (or meter) that measures viscosity is called a *viscometer*. There are many different types of viscometers, including the capillary viscometer, concentric cylinder viscometer, the parallel-disk viscometer, the falling-sphere viscometer, and the cone-and-plate viscometer. In each case, solu-

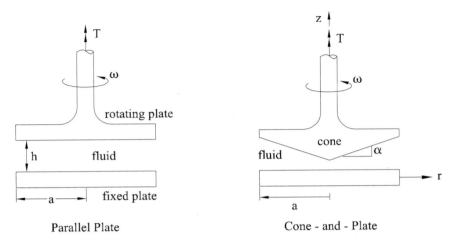

FIGURE 7.16 Schema of two "viscous-meters," or viscometers, that are used to determine the viscosity of a fluid: the parallel-plate (left) and the cone-and-plate (right) devices.

tions are needed for the viscosity μ in terms of experimentally measurable quantities. A concentric cylinder viscometer is discussed in detail in Chapter 9, but here let us consider two approximate solutions. Figure 7.16 shows the so-called parallel-plate and cone-and-plate viscometers. In each case, a fluid is placed between two rigid solids: a fixed-bottom plate and an upper flat plate or a small-angle cone. Obviously, the more viscous the fluid, the more difficult it would be to rotate the upper plate or cone. One thus seeks to relate the viscosity of the fluid to the applied load (torque T) and the geometry of the device (radius a, gap height h, or cone angle α) for a given constant angular velocity ω of the upper plate or cone. Although these may appear to be simple devices based on simple ideas, the generated steady-state flow field (once transients disappear due to starting the device) is at least two dimensional; the velocity will depend on both r and z. Hence, here we derive an approximate (incomplete) rather than exact (full) solution. Let us consider the cone-and-plate device. Obviously, the velocity vector of any point of the cone will have but a circumferential component, which will equal $r\omega$, where ω is the constant angular velocity; that is, the velocity is zero at the centerline and maximum at the outer edge. Let us assume, therefore, that the fluid contacting the cone will have the same velocity as the surface of the cone, namely

$$v_\theta \cong r\omega \rightarrow dv_\theta \cong \omega\,dr, \tag{7.73}$$

where, from geometry, $\tan\alpha = dz/dr$, and thus $dr = \cot\alpha\,dz$. Hence, our velocity gradient at the surface of the cone can be approximated as

$$\frac{\partial v_\theta}{\partial z} \approx \omega \cot \alpha, \tag{7.74}$$

or for small α (i.e., $\alpha \ll 1$ whereby $\cos \alpha \sim 1$ and $\sin \alpha \sim \alpha$), we have a constant shear rate

$$\frac{\partial v_\theta}{\partial z} \approx \frac{\omega}{\alpha}. \tag{7.75}$$

If we now assume a Newtonian behavior, then at the surface of the cone, we have, from Eqs. (7.58) and (7.64),

$$\sigma_{z\theta} = 2\mu \left[\frac{1}{2} \left(\frac{1}{r} \frac{\partial v_z}{\partial \theta} + \frac{\partial v_\theta}{\partial z} \right) \right]. \tag{7.76}$$

Hence, the approximate shear stress in the fluid acting on the surface of the cone is

$$\sigma_{z\theta} \approx \mu \frac{\partial v_\theta}{\partial z} = \mu \frac{\omega}{\alpha}. \tag{7.77}$$

Now, the stress on the solid, called the wall shear stress τ_w, is equal and opposite that of the fluid. It acts over a differential area on the cone, which we approximate as $dA = rd\theta dr$ for $\theta \in [0, 2\pi]$ and $r \in [0, a]$, a reasonable approximation if $\alpha \ll 1$. This wall shear stress, acting at each point over its respective cross-sectional area (i.e., $\tau_w \, dA$), gives rise to a differential force. The sum of all such differential forces acting at a distance r from the centerline of the device gives rise to a differential twisting moment that balances the applied torque T, which is measurable; that is [cf. Eq. (4.16)],

$$T - \int_0^a \int_0^{2\pi} r\tau_w \, dA = 0, \tag{7.78}$$

or

$$T = \int_0^a \int_0^{2\pi} r \left(\frac{\mu\omega}{\alpha} \right) rd\theta \, dr = \frac{2\pi\mu\omega}{\alpha} \int_0^a r^2 dr = \frac{2\pi\mu\omega a^3}{3\alpha}. \tag{7.79}$$

Hence,

$$\mu = \frac{3T\alpha}{2\pi\omega a^3} \tag{7.80}$$

for the cone-and-plate viscometer (Alexandrou, 2001, p. 486). Clearly then, μ is "measurable" given specifications for the device (α and a) as well as measurements of the applied torque T that is required to maintain the cone at a constant angular velocity ω. Note that for a Newtonian fluid, $\mu = $ constant, which is to say, T/ω would likewise be constant. If this ratio is not constant for different values of ω, the fluid is said to be non-Newtonian and μ could be estimated for different values of the shear rate ω/α. Each value of μ so determined would be called an "apparent" viscosity. Because the viscosity of a non-Newtonian fluid varies with shear rate, the cone-and-plate device is useful because the shear rate near the surface of the cone is approximately constant ($= \omega/\alpha$). This solution is approximate (incomplete), of course, because we did not account for the no-slip boundary condition at the bottom fixed plate and we assumed that $\alpha \to 0$.

Using a similar approach, it can be shown that (Alexandrou, 2001, p. 485)

$$\mu \cong \frac{2hT}{\pi \omega a^4} \tag{7.81}$$

for the parallel-plate viscometer. For a description of the design of actual viscometers, see Ferry (1980, pp. 96–102). Just as in biosolid mechanics, quantification of material behavior in biofluid mechanics is fundamental to success in real-life analysis and design. It must remain as a high priority in all R&D.

Appendix 7: Vector Calculus Review

A vector is a mathematical quantity having characteristics of magnitude and direction; a scalar is a quantity having characteristics of magnitude only. Vectors are denoted herein as boldface italics and scalars as italics. Let the vector u be written in terms of its Cartesian components as $u = u_x \hat{i} + u_y \hat{j} + u_z \hat{k}$ and similarly for the vector $v = v_x \hat{i} + v_y \hat{j} + v_z \hat{k}$, where $\hat{i} = (1, 0, 0)$, $\hat{j} = (0, 1, 0)$, and $\hat{k} = (0, 0, 1)$ are orthonormal base vectors (recall that a coordinate system is defined by its origin and a basis); that is, $(\hat{i}, \hat{j}, \hat{k})$ are mutually orthogonal and they are of unit magnitude. We will use the caret to denote a unit magnitude in most cases [e.g., $\hat{w} = w/|w|$, where the magnitude $|w| = (w \cdot w)^{1/2}$.]

Dot Product

The dot product between two vectors yields a scalar quantity. It can be achieved as follows:

$$u \cdot v = v \cdot u = |u||v| \cos \theta, \tag{A7.1}$$

where θ is the angle between u and v. Vectors u and v are orthogonal if the angle between them is $\theta = \pi/2$. Thus, u and v are orthogonal if and only if $u \cdot v = 0$. Alternatively, the dot product can be computed as

$$\begin{aligned}
u \cdot v &= \left(u_x \hat{i} + u_y \hat{j} + u_z \hat{k}\right) \cdot \left(v_x \hat{i} + v_y \hat{j} + v_z \hat{k}\right) = u_x v_x \left(\hat{i} \cdot \hat{i}\right) + u_x v_y \left(\hat{i} \cdot \hat{j}\right) \\
&\quad + u_x v_z \left(\hat{i} \cdot \hat{k}\right) + u_y v_x \left(\hat{j} \cdot \hat{i}\right) + u_y v_y \left(\hat{j} \cdot \hat{j}\right) + u_y v_z \left(\hat{j} \cdot \hat{k}\right) \\
&\quad + u_z v_x \left(\hat{k} \cdot \hat{i}\right) + u_z v_y \left(\hat{k} \cdot \hat{j}\right) + u_z v_z \left(\hat{k} \cdot \hat{k}\right)
\end{aligned} \tag{A7.2}$$

Hence, it is good to remember the dot products between base vectors. For example, $\hat{i} \cdot \hat{i} = |\hat{i}||\hat{i}| \cos \theta = (1)(1) \cos 0 = 1$. In summary,

$$\hat{i} \cdot \hat{i} = 1, \qquad \hat{j} \cdot \hat{i} = 0 \qquad \hat{k} \cdot \hat{i} = 0,$$
$$\hat{i} \cdot \hat{j} = 0, \qquad \hat{j} \cdot \hat{j} = 1, \qquad \hat{k} \cdot \hat{j} = 0,$$
$$\hat{i} \cdot \hat{k} = 0, \qquad \hat{j} \cdot \hat{k} = 0, \qquad \hat{k} \cdot \hat{k} = 1. \qquad \text{(A7.3)}$$

Hence, substituting in Eq. (A7.2), we see that

$$\boldsymbol{u} \cdot \boldsymbol{v} = u_x v_x (1) + u_y v_y (1) + u_z v_z (1). \qquad \text{(A7.4)}$$

The magnitude of a vector, say $\boldsymbol{u}$, can thus be computed as

$$|\boldsymbol{u}| = \sqrt{\boldsymbol{u} \cdot \boldsymbol{u}} = \sqrt{u_x^2 + u_y^2 + u_z^2}. \qquad \text{(A7.5)}$$

Cross Product

The cross product between two vectors yields a vector (i.e., a quantity having both a magnitude and a direction), and specifically a vector that is perpendicular to the plane containing the original two vectors. It can be written as

$$\boldsymbol{u} \times \boldsymbol{v} = -\boldsymbol{v} \times \boldsymbol{u} = |\boldsymbol{u}||\boldsymbol{v}| \sin \theta \hat{e}_\perp, \qquad \text{(A7.6)}$$

where $\hat{e}_\perp$ is a base vector perpendicular to the plane containing $\boldsymbol{u}$ and $\boldsymbol{v}$. As with the dot product, it is useful to know the cross products between the base vectors. For example, $\hat{i} \times \hat{j} = |\hat{i}||\hat{j}| \sin (\pi/2) \hat{k} = \hat{k}$. In summary,

$$\hat{i} \times \hat{i} = \boldsymbol{0}, \qquad \hat{j} \times \hat{i} = -\hat{k}, \qquad \hat{k} \times \hat{i} = \hat{j},$$
$$\hat{i} \times \hat{j} = \hat{k}, \qquad \hat{j} \times \hat{j} = \boldsymbol{0}, \qquad \hat{k} \times \hat{j} = -\hat{i},$$
$$\hat{i} \times \hat{k} = -\hat{j}, \qquad \hat{j} \times \hat{k} = \hat{i}, \qquad \hat{k} \times \hat{k} = \boldsymbol{0} \qquad \text{(A7.7)}$$

Hence, the cross product for the vectors $\boldsymbol{u}$ and $\boldsymbol{v}$ can be computed as

$$\begin{aligned}
\boldsymbol{u} \times \boldsymbol{v} &= \left(u_1 \hat{i} + u_2 \hat{j} + u_3 \hat{k} \right) \times \left(v_1 \hat{i} + v_2 \hat{j} + v_3 \hat{k} \right) \\
&= \left(\hat{i} \times \hat{i} \right)(u_1 v_1) + \left(\hat{i} \times \hat{j} \right)(u_1 v_2) + \left(\hat{i} \times \hat{k} \right)(u_1 v_3) \\
&\quad + \left(\hat{j} \times \hat{i} \right)(u_2 v_1) + \left(\hat{j} \times \hat{j} \right)(u_2 v_2) + \left(\hat{j} \times \hat{k} \right)(u_2 v_3) \\
&\quad + \left(\hat{k} \times \hat{i} \right)(u_3 v_1) + \left(\hat{k} \times \hat{j} \right)(u_3 v_2) + \left(\hat{k} \times \hat{k} \right)(u_3 v_3), \qquad \text{(A7.8)}
\end{aligned}$$

or

$$\boldsymbol{u} \times \boldsymbol{v} = (u_2 v_3 - u_3 v_2)\hat{i} + (u_3 v_1 - u_1 v_3)\hat{j} + (u_1 v_2 - u_2 v_1)\hat{k}. \qquad \text{(A7.9)}$$

There is a special operator that plays a key role in fluid mechanics. Thus, recall the differential operator ∇, also known as the *del operator*, which, relative to a Cartesian coordinate system, can be written as

$$\nabla = \hat{i}\frac{\partial}{\partial x} + \hat{j}\frac{\partial}{\partial y} + \hat{k}\frac{\partial}{\partial z}. \tag{A7.10}$$

This del operator operates on a scalar function ϕ to produce the *gradient* of ϕ, namely

$$\nabla\phi = \left(\hat{i}\frac{\partial}{\partial x} + \hat{j}\frac{\partial}{\partial y} + \hat{k}\frac{\partial}{\partial z}\right)\phi = \hat{i}\frac{\partial}{\partial x}(\phi) + \hat{j}\frac{\partial}{\partial y}(\phi) + \hat{k}\frac{\partial}{\partial z}(\phi)$$

$$= \frac{\partial\phi}{\partial x}\hat{i} + \frac{\partial\phi}{\partial y}\hat{j} + \frac{\partial\phi}{\partial z}\hat{k}. \tag{A7.11}$$

Thus, the gradient of a scalar is a vector; we shall see in Chapter 8 that the "pressure gradient" plays an important role in the governing equations of motion for fluids. The del operator can also form a scalar product with a vector to produce the *divergence* of the vector:

$$\nabla\cdot v = \left(\hat{i}\frac{\partial}{\partial x} + \hat{j}\frac{\partial}{\partial y} + \hat{k}\frac{\partial}{\partial z}\right)\cdot\left(v_x\hat{i} + v_y\hat{j} + v_z\hat{k}\right) = \hat{i}\cdot\frac{\partial}{\partial x}\left(v_x\hat{i} + v_y\hat{j} + v_z\hat{k}\right)$$

$$+ \hat{j}\cdot\frac{\partial}{\partial y}\left(v_x\hat{i} + v_y\hat{j} + v_z\hat{k}\right) + \hat{k}\cdot\frac{\partial}{\partial z}\left(v_x\hat{i} + v_y\hat{j} + v_z\hat{k}\right), \tag{A7.12}$$

which, because $\hat{i}$, $\hat{j}$, and $\hat{k}$ do not vary with position, becomes

$$\nabla\cdot v = \left(\hat{i}\cdot\hat{i}\right)\left(\frac{\partial}{\partial x}v_x\right) + \left(\hat{i}\cdot\hat{j}\right)\left(\frac{\partial}{\partial x}v_y\right) + \left(\hat{i}\cdot\hat{k}\right)\left(\frac{\partial}{\partial x}v_z\right)$$

$$+ \left(\hat{j}\cdot\hat{i}\right)\left(\frac{\partial}{\partial y}v_x\right) + \left(\hat{j}\cdot\hat{j}\right)\left(\frac{\partial}{\partial y}v_y\right) + \left(\hat{j}\cdot\hat{k}\right)\left(\frac{\partial}{\partial y}v_z\right)$$

$$+ \left(\hat{k}\cdot\hat{i}\right)\left(\frac{\partial}{\partial z}v_x\right) + \left(\hat{k}\cdot\hat{j}\right)\left(\frac{\partial}{\partial z}v_y\right) + \left(\hat{k}\cdot\hat{k}\right)\left(\frac{\partial}{\partial z}v_z\right) \tag{A7.13}$$

or, by recalling the above results for dot products between the bases,

$$\nabla\cdot v = \frac{\partial v_x}{\partial x} + \frac{\partial v_y}{\partial y} + \frac{\partial v_z}{\partial z}. \tag{A7.14}$$

Recall that the divergence of the velocity vector provides information on the incompressibility of a fluid flow. The del operator can also form a cross product with a vector to produce the *curl* of the vector:

$$\text{curl } v = \nabla\times v = \left(\hat{i}\frac{\partial}{\partial x} + \hat{j}\frac{\partial}{\partial y} + \hat{k}\frac{\partial}{\partial z}\right)\times\left(v_x\hat{i} + v_y\hat{j} + v_z\hat{k}\right)$$

$$= \hat{i}\times\frac{\partial}{\partial x}\left(v_x\hat{i} + v_y\hat{j} + v_z\hat{k}\right) + \hat{j}\times\frac{\partial}{\partial y}\left(v_x\hat{i} + v_y\hat{j} + v_z\hat{k}\right)$$

$$+ \hat{k}\times\frac{\partial}{\partial z}\left(v_x\hat{i} + v_y\hat{j} + v_z\hat{k}\right), \tag{A7.15}$$

which, again, simplifies considerably because the Cartesian bases do not change with position (x, y, z). Hence,

$$\text{curl } \mathbf{v} = (\hat{\imath} \times \hat{\imath})\left(\frac{\partial}{\partial x}v_x\right) + (\hat{\imath} \times \hat{\jmath})\left(\frac{\partial}{\partial x}v_y\right) + (\hat{\imath} \times \hat{k})\left(\frac{\partial}{\partial x}v_z\right)$$
$$+ (\hat{\jmath} \times \hat{\imath})\left(\frac{\partial}{\partial y}v_x\right) + (\hat{\jmath} \times \hat{\jmath})\left(\frac{\partial}{\partial y}v_y\right) + (\hat{\jmath} \times \hat{k})\left(\frac{\partial}{\partial y}v_z\right)$$
$$+ (\hat{k} \times \hat{\imath})\left(\frac{\partial}{\partial z}v_x\right) + (\hat{k} \times \hat{\jmath})\left(\frac{\partial}{\partial z}v_y\right) + (\hat{k} \times \hat{k})\left(\frac{\partial}{\partial z}v_z\right) \quad \text{(A7.16)}$$

or, by recalling the above results for cross products between the bases,

$$\text{curl } \mathbf{v} = (\mathbf{0})\left(\frac{\partial}{\partial x}v_x\right) + (\hat{k})\left(\frac{\partial}{\partial x}v_y\right) + (-\hat{\jmath})\left(\frac{\partial}{\partial x}v_z\right)$$
$$+ (-\hat{k})\left(\frac{\partial}{\partial y}v_x\right) + (\mathbf{0})\left(\frac{\partial}{\partial y}v_y\right) + (\hat{\imath})\left(\frac{\partial}{\partial y}v_z\right)$$
$$+ (\hat{\jmath})\left(\frac{\partial}{\partial z}v_x\right) + (-\hat{\imath})\left(\frac{\partial}{\partial z}v_y\right) + (\mathbf{0})\left(\frac{\partial}{\partial z}v_z\right), \quad \text{(A7.17)}$$

or, finally,

$$\text{curl } \mathbf{v} = \left(\frac{\partial v_z}{\partial y} - \frac{\partial v_y}{\partial z}\right)\hat{\imath} + \left(\frac{\partial v_x}{\partial z} - \frac{\partial v_z}{\partial x}\right)\hat{\jmath} + \left(\frac{\partial v_y}{\partial x} - \frac{\partial v_x}{\partial y}\right)\hat{k}. \quad \text{(A7.18)}$$

Recall that the curl of the velocity is a measure of the so-called vorticity in a fluid, which provides information on the rotation of fluid elements.

Cylindricals

Because arteries, airways, ureters, medical tubing, and so forth are all cylindrical tubes, we need to be familiar with cylindrical polar coordinates. Although there are different ways to accomplish this, here we recall the relationships between cylindricals and Cartesians. Recall, therefore, that

$$x = r\cos\theta, \qquad y = r\sin\theta$$

or, $\qquad\qquad\qquad\qquad\qquad\qquad\qquad\qquad\qquad\qquad\qquad$ (A7.19)

$$r = \sqrt{x^2 + y^2}, \qquad \theta = \tan^{-1}\left(\frac{y}{x}\right)$$

and, of course, $z = z$. Moreover,

$$\hat{e}_r = \cos\theta\hat{\imath} + \sin\theta\hat{\jmath}, \qquad \hat{e}_\theta = -\sin\theta\hat{\imath} + \cos\theta\hat{\jmath}, \quad \text{(A7.20)}$$

or

$$\hat{\imath} = \cos\theta\hat{e}_r - \sin\theta\hat{e}_\theta, \qquad \hat{\jmath} = \sin\theta\hat{e}_r + \cos\theta\hat{e}_\theta. \quad \text{(A7.21)}$$

The latter results for $\hat{\imath}$ and $\hat{\jmath}$ can be determined from those for $\hat{e}_r$ and $\hat{e}_\theta$ given the two equations for the two "unknowns." Note, therefore, that

$$\frac{\partial \hat{e}_r}{\partial \theta} = -\sin \theta \hat{i} + \cos \theta \hat{j} = \hat{e}_\theta. \tag{A7.22}$$

and, similarly,

$$\frac{\partial \hat{e}_\theta}{\partial \theta} = -\cos \theta \hat{i} - \sin \theta \hat{j} = -\hat{e}_r. \tag{A7.23}$$

Conversely, $\hat{e}_r$ and $\hat{e}_\theta$ do not vary with r.

To determine the del operator ∇ cylindrical coordinates, recall that

$$\nabla = \hat{i} \frac{\partial (\ )}{\partial x} + \hat{j} \frac{\partial (\ )}{\partial y} + \hat{k} \frac{\partial (\ )}{\partial z}. \tag{A7.24}$$

This relation can thus be written (using the transformations for the bases and the chain rule) as

$$\nabla = (\cos \theta \hat{e}_r - \sin \theta \hat{e}_\theta) \left(\frac{\partial (\ )}{\partial r} \frac{\partial r}{\partial x} + \frac{\partial (\ )}{\partial \theta} \frac{\partial \theta}{\partial x} \right)$$
$$+ (\sin \theta \hat{e}_r + \cos \theta \hat{e}_\theta) \left(\frac{\partial (\ )}{\partial r} \frac{\partial r}{\partial y} + \frac{\partial (\ )}{\partial \theta} \frac{\partial \theta}{\partial y} \right) + \hat{e}_z \frac{\partial (\ )}{\partial z}. \tag{A7.25}$$

From Eq. (A7.19), we note that

$$\frac{\partial r}{\partial x} = \frac{1}{2} (x^2 + y^2)^{-1/2} 2x = \frac{x}{\sqrt{x^2 + y^2}} = \frac{r \cos \theta}{r} = \cos \theta \tag{A7.26}$$

and, similarly (show it),

$$\frac{\partial r}{\partial y} = \sin \theta. \tag{A7.27}$$

Likewise, from calculus, we recall the derivative of the arctangent; hence,

$$\frac{\partial \theta}{\partial x} = \frac{1}{1 + (y/x)^2} \left(-\frac{y}{x^2} \right) = -\frac{y}{x^2 + y^2} = -\frac{r \sin \theta}{r^2} = -\frac{\sin \theta}{r} \tag{A7.28}$$

and, similarly (show it),

$$\frac{\partial \theta}{\partial y} = \frac{\cos \theta}{r}. \tag{A7.29}$$

Hence, Eq. (A7.25) can be shown to become (noting that $\cos^2 \theta + \sin^2 \theta = 1$)

$$\nabla = \hat{e}_r \frac{\partial (\ )}{\partial r} + \hat{e}_\theta \frac{1}{r} \frac{\partial (\ )}{\partial \theta} + \hat{e}_z \frac{\partial (\ )}{\partial z}. \tag{A7.30}$$

The Divergence Theorem

Let us now consider an important theorem in solid and fluid mechanics. The *divergence theorem* states that integration over an area of the dot product between a vector B and an outward unit normal $\hat{n}$ is equal to an integration of the divergence of B over a volume:

$$\iint_{Area} B \cdot \hat{n}\, da = \iiint_{Volume} \nabla \cdot B\, d\nu \qquad (A7.31)$$

wherein we use the notation ν for volume in fluids to distinguish it from velocity, even though volume is a scalar and velocity a vector.

Example A7.1 Show numerically that the divergence theorem holds for the vector $B = 4xz\,\hat{i} - y^2\,\hat{j} + yz\,\hat{k}$ over a domain defined by a unit cube.

Solution: Let the six faces of the unit cube be denoted as a, b, c, d, e, and f, where a is the positive x face, b is the positive y face, c is the positive z face, d is the negative x face, e is the negative y face, and f is the negative z face. Hence, note that

$$\iiint \nabla \cdot B\, d\nu = \int_0^1 \int_0^1 \int_0^1 \left(\frac{\partial B_x}{\partial x} + \frac{\partial B_y}{\partial y} + \frac{\partial B_z}{\partial z} \right) dx\, dy\, dz$$

$$= \int_0^1 \int_0^1 \int_0^1 (4z - 2y + y)\, dx\, dy\, dz$$

$$= \int_0^1 \int_0^1 \left[(4z - y) \int_0^1 dx \right] dy\, dz$$

$$= \int_0^1 \left[4zy - \frac{y^2}{2} \right]_0^1 dz = \left[\frac{4z^2}{2} - \frac{1}{2}z \right]_0^1 = \frac{3}{2}.$$

Now, by the divergence theorem, this value must equal that determined by the sum of the surface integrals:

$$\iint B \cdot \hat{n}\, dA = \int_a + \int_b + \int_c + \int_d + \int_e + \int_f,$$

where these integrals represent values over each of the six faces of the unit cube. Note, therefore, that

$(a)\ \hat{n} = \hat{i},$	$dA = dy\, dz,$	at $x = 1,$
$(b)\ \hat{n} = \hat{j},$	$dA = dx\, dz,$	at $y = 1,$
$(c)\ \hat{n} = \hat{k},$	$dA = dx\, dy,$	at $z = 1,$
$(d)\ \hat{n} = -\hat{i},$	$dA = dy\, dz,$	at $x = 0,$
$(e)\ \hat{n} = -\hat{j},$	$dA = dx\, dz,$	at $y = 0,$
$(f)\ \hat{n} = -\hat{k},$	$dA = dx\, dy,$	at $z = 0;$

hence, for surface a,

$$\int_0^1 \int_0^1 \boldsymbol{B} \cdot \hat{\boldsymbol{i}}\, dA = \int_0^1 \int_0^1 4xz\big|_{x=1}\, dy\, dz = 4(y\big|_0^1)\left(\frac{1}{2} z^2 \Big|_0^1\right) = 2.$$

Similarly, show that

$$\int_b = -1, \qquad \int_c = \frac{1}{2}, \qquad \int_d = 0, \qquad \int_e = 0, \qquad \int_f = 0,$$

thus

$$\iint \boldsymbol{B} \cdot \hat{\boldsymbol{n}}\, dA = 2 - 1 + \frac{1}{2} + 0 + 0 + 0 = \frac{3}{2},$$

consistent with the volume integral and thus the divergence theorem as stated.

Exercises

7.1 Generate a list of 20 clinically relevant problems that demand a biofluid mechanical design or analysis.

7.2 Give short, concise definitions of fluid, steady flow, Eulerian approach, viscosity, Newtonian fluid, no-slip boundary condition, and convective acceleration.

7.3 Give short, concise definitions of fully developed flow, vorticity, laminar flow, pseudoplastic behavior, shear-rate, uniform flow, and 1-D flow.

7.4 Show that $\boldsymbol{v} \cdot \nabla \neq \nabla \cdot \boldsymbol{v}$ in Cartesian coordinates.

7.5 Show that

$$\nabla \cdot \boldsymbol{v} = \frac{1}{r}\frac{\partial}{\partial r}(rv_r) + \frac{1}{r}\frac{\partial v_\theta}{\partial \theta} + \frac{\partial v_z}{\partial z},$$

where

$$\nabla = \hat{\boldsymbol{e}}_r \frac{\partial}{\partial r} + \hat{\boldsymbol{e}}_\theta \frac{1}{r}\frac{\partial}{\partial \theta} + \hat{\boldsymbol{e}}_z \frac{\partial}{\partial z}$$

in cylindrical coordinates. Likewise, show that

$$\boldsymbol{v} \cdot \nabla = v_r \frac{\partial}{\partial r} + \frac{v_\theta}{r}\frac{\partial}{\partial \theta} + v_z \frac{\partial}{\partial z}.$$

Hint: Recall that

$$\frac{\partial}{\partial \theta}(\hat{\boldsymbol{e}}_r) = \hat{\boldsymbol{e}}_\theta \quad \text{and} \quad \frac{\partial}{\partial \theta}(\hat{\boldsymbol{e}}_\theta) = -\hat{\boldsymbol{e}}_r.$$

7.6   Show that

$$\nabla \cdot \boldsymbol{v} = \frac{1}{r^2} \frac{\partial}{\partial r} (r^2 v_r) + \frac{1}{r \sin \theta} \frac{\partial}{\partial \theta} (v_\theta \sin \theta) + \frac{1}{r \sin \theta} \frac{\partial v_\phi}{\partial \phi},$$

where

$$\nabla = \hat{\boldsymbol{e}}_r \frac{\partial}{\partial r} + \hat{\boldsymbol{e}}_\theta \frac{1}{r} \frac{\partial}{\partial \theta} + \hat{\boldsymbol{e}}_\phi \frac{1}{r \sin \theta} \frac{\partial}{\partial \phi}$$

in spherical coordinates. Hint: Recall that

$$\frac{\partial}{\partial \theta} (\hat{\boldsymbol{e}}_r) = \hat{\boldsymbol{e}}_\theta, \quad \frac{\partial}{\partial \theta} (\hat{\boldsymbol{e}}_\theta) = -\hat{\boldsymbol{e}}_r, \quad \frac{\partial}{\partial \phi} (\hat{\boldsymbol{e}}_r) = \sin \theta \hat{\boldsymbol{e}}_\phi,$$

$$\frac{\partial}{\partial \phi} (\hat{\boldsymbol{e}}_\theta) = \cos \theta \hat{\boldsymbol{e}}_\phi, \quad \text{and} \quad \frac{\partial}{\partial \phi} (\hat{\boldsymbol{e}}_\phi) = -\sin \theta \hat{\boldsymbol{e}}_r - \cos \theta \hat{\boldsymbol{e}}_\theta.$$

7.7   Consider a velocity vector $\boldsymbol{v} = (xt + 2y)\hat{\boldsymbol{i}} + (xt^2 - yt)\hat{\boldsymbol{j}}$. (a) Is this a steady flow, and why? (b) Is this a possible incompressible flow, and why? (c) Calculate the acceleration.

7.8   Compute the acceleration (in an Eulerian sense) given the following components of the velocity vector: $v_x = xt^2$, $v_y = xyt^2 + y^2$, and $v_z = 0$.

7.9   Let $\boldsymbol{v} = axy\hat{\boldsymbol{i}} - byzt\hat{\boldsymbol{j}}$; a and b are known scalar constants. (a) Calculate the acceleration vector using an Eulerian approach and (b) determine if $\nabla \cdot \boldsymbol{v} = 0$.

7.10   Show that $\nabla \cdot (\nabla \times \boldsymbol{v}) = 0$.

7.11   Given the velocity field $\boldsymbol{v} = (x + y)\hat{\boldsymbol{i}} + (x - y) \hat{\boldsymbol{j}} + 0\hat{\boldsymbol{k}}$, (a) determine if the flow is incompressible (i.e., $\nabla \cdot \boldsymbol{v} = 0$) and (b) determine if it is irrotational (i.e., $\nabla \times \boldsymbol{v} = \boldsymbol{0}$).

7.12   Given the following velocity field $\boldsymbol{v} = az^2\hat{\boldsymbol{i}} + bz\hat{\boldsymbol{k}}$. (a) Is this a possible incompressible flow (i.e., $\nabla \cdot \boldsymbol{v} = 0$)? (b) Is this a possible irrotational flow (i.e., $\nabla \times \boldsymbol{v} = \boldsymbol{0}$)?

7.13   Show that

$$\nabla \times \boldsymbol{v} = \left(\frac{1}{r} \frac{\partial v_z}{\partial \theta} - \frac{\partial v_\theta}{\partial z} \right) \hat{\boldsymbol{e}}_r + \left(\frac{\partial v_r}{\partial z} - \frac{\partial v_z}{\partial r} \right) \hat{\boldsymbol{e}}_\theta + \left(\frac{1}{r} \frac{\partial (r v_\theta)}{\partial r} - \frac{1}{r} \frac{\partial v_r}{\partial \theta} \right) \hat{\boldsymbol{e}}_z$$

in cylindrical coordinates. Hint: Note that

$$\frac{\partial v_\theta}{\partial r} + \frac{v_\theta}{r} \equiv \frac{1}{r} \frac{\partial}{\partial r} (r v_\theta).$$

7.14   Show that $(\boldsymbol{v} \cdot \nabla)\boldsymbol{v} = \frac{1}{2} \nabla (\boldsymbol{v} \cdot \boldsymbol{v}) - \boldsymbol{v} \times (\nabla \times \boldsymbol{v})$.

7.15   Derive the expressions for D_{yy} and D_{zz}.

7.16   Derive the expressions for D_{xz} and D_{yz}.

7.17 If $\mathbf{v} = v_z(r)\hat{\mathbf{e}}_z$, where

$$v_z(r) = c\left(1 - \frac{r^2}{a^2}\right)$$

and c and a are constants, determine (a) if this is a possible incom-pressible flow and (b) if this is a possible irrotational flow. Note that this velocity field will be shown in Chapter 9 to correspond to a steady flow in a rigid cylinder of inner radius a.

7.18 For the velocity field given in the previous exercise, compute D_{rr} and D_{rz}.

7.19 Compute $\nabla^2 \equiv \nabla \cdot \nabla$ in Cartesian coordinates.

7.20 Show that $\nabla^2 \mathbf{v}$ can be written in Cartesians as

$$\nabla^2 \mathbf{v} = \left(\frac{\partial^2 v_x}{\partial x^2} + \frac{\partial^2 v_x}{\partial y^2} + \frac{\partial^2 v_x}{\partial z^2}\right)\hat{\mathbf{i}} + \left(\frac{\partial^2 v_y}{\partial x^2} + \frac{\partial^2 v_y}{\partial y^2} + \frac{\partial^2 v_y}{\partial z^2}\right)\hat{\mathbf{j}}$$
$$+ \left(\frac{\partial^2 v_z}{\partial x^2} + \frac{\partial^2 v_z}{\partial y^2} + \frac{\partial^2 v_z}{\partial z^2}\right)\hat{\mathbf{k}}.$$

7.21 Show that in cylindricals

$$\nabla^2 = \frac{1}{r}\frac{\partial}{\partial r}\left(r\frac{\partial}{\partial r}\right) + \frac{1}{r^2}\frac{\partial^2}{\partial \theta^2} + \frac{\partial^2}{\partial z^2}$$

where ∇^2 is called the Laplacian and therefore

$$\nabla^2 \Phi = \frac{1}{r}\frac{\partial}{\partial r}\left(r\frac{\partial \Phi}{\partial r}\right) + \frac{1}{r^2}\frac{\partial^2 \Phi}{\partial \theta^2} + \frac{\partial^2 \Phi}{\partial z^2}$$

for any scalar Φ. Hint: Remember that the base vectors in cylindrical coordinates may change with direction and that

$$\nabla = \hat{\mathbf{e}}_r\frac{\partial}{\partial r} + \hat{\mathbf{e}}_\theta\frac{1}{r}\frac{\partial}{\partial \theta} + \hat{\mathbf{e}}_z\frac{\partial}{\partial z}.$$

7.22 Sketch the change in the apparent viscosity μ_a as a function of shear rate for pseudoplastic, Newtonian, and dilatant behaviors.

7.23 There are about 5×10^6 RBCs/mm^3 for an average human having 5 L of blood. Compute the number of RBCs circulating in the body, and if the net turnover rate is 0.8% per day, how many cells are produced and removed per day?

7.24 Data from a viscometer suggest the following:

σ_{rz} (Pa)	1.7	2.7	4.8	6.5
D_{rz} (s^{-1})	200	300	470	600

Is this a Newtonian behavior? If not, then what type of behavior might it be?

7.25 Given the data

Shear stress (Pa)	6.5	4.8	2.7	1.7
Shear rate (s^{-1})	600	470	300	200

plot the data and classify the fluid (pseudoplastic, Newtonian, or dilatant).

7.26 It was shown in Section 7.6 that the value of the viscosity can be estimated via a cone-and-plate viscometer, namely

$$\mu = \frac{3T\alpha}{2\pi\omega a^3},$$

where T is the applied torque, α is the cone-angle, ω is the angular velocity (units of s^{-1}), and a is the maximum radius of the cone. If T_m is the maximum torque applied and $B \equiv T/T_m$, show that

$$\mu \approx 6\frac{B}{N}$$

if $\alpha = 1.565°$, $a = 2.409$ cm, and $T_m = 673.7$ dyn cm, where N is the number of revolutions per minute (rpm). Note that ω (rad/s) $= (2\pi$ rad/rev)$(N$ rpm$)(1$ min/60 s$)$.

7.27 Given the results in Exercise 7.26 for a particular cone-and-plate viscometer, plot μ (cP) versus shear rate (s^{-1}) based on the following data:

N (rpm)	3	6	12	30	60
B	2.87	5.77	11.20	21.85	35.50

Classify the fluid (pseudoplastic, Newtonian, or dilatant) based on this plot. (Data from lecture notes by Professor D.J. Schneck, Virginia Tech.)

7.28 We will discover in Chapter 9 that the volumetric flow rate Q for a steady, incompressible flow in a rigid circular tube of radius a is given by

$$Q = \frac{\pi a^4}{8\mu}k,$$

where k is the pressure drop per unit length [i.e., $k = (P_i - P_o)/L$, where P_i and P_o are inlet and outlet pressures, respectively along the tube]. If $a \sim 1.25$ cm and $Q \sim 5$ L/min, compute the requisite percent increase in P_i (assuming P_o does not change) if $\mu = 6$ cP (hemoglobin) rather than $\mu = 3.5$ cP (whole blood). What implications would this have on the human heart with a cardiac output of 5 L/min?

7.29 Derive Eq. (7.81).

7.30 Find the gap height h for the parallel-plate viscometer in Figure 7.16 if the torque T, the angular velocity ω, dimension a, and viscosity μ have values similar to those in Exercise 7.26.

7.31 Similar to Example A7.1, show that the divergence theorem holds for a unit cube if $\boldsymbol{B} = 2xy\hat{\boldsymbol{i}} - 2xy\hat{\boldsymbol{j}}$.

8
Fundamental Balance Relations

Recall from Chapter 1 that one of the best-known equations in science is $F = ma$, which is called Newton's second law of motion. This equation asserts that in an inertial frame of reference, the time rate of change of the linear momentum mv for a mass particle m must *balance* the forces F that are applied to the particle. For this reason, this "law of motion" is also called the principle (actually postulate) of the balance of linear momentum. Whereas Sir I. Newton considered only individual mass points (like the Moon or an apple), L. Euler showed that many bodies can be treated as a continuous collection of mass points (i.e., a continuum), each particle of which obeys Newton's second law. Indeed, as it turns out, there are three basic postulates for any continua,

Balance of mass
Balance of linear momentum
Balance of energy (i.e., first law of thermodynamics)

to which we often add the postulates of the balance of angular momentum and the entropy inequality (i.e., second law of thermodynamics), both of which can provide restrictions on the allowable constitutive relations. For example, recall from Chapter 2 that the balance of angular momentum requires that the Cauchy stress $[\sigma]$ be symmetric, which restricts possible constitutive relations that are formulated in terms of σ. Each of these five postulates can be stated as either differential equations (for systems) or integral equations (for control volumes). In this chapter, we focus on the *governing differential equations* for mass and linear momentum balance; the differential equation for energy balance is useful in bioheat transfer, which is not addressed herein. Chapter 10 addresses the control volume formulation for mass, linear momentum, and energy. Because these postulates are good for all continua, they apply equally well to biosolids and biofluids; we will see, however, that some of these equations specialize for individual material behaviors, which facilitates the formulation and solution of particular problems.

8.1 Balance of Mass

We shall require that the identifiable differential mass Δm be conserved for all time t; that is, in the limit as $\Delta m \to dm$,

$$\frac{d}{dt}(dm) = 0 \to \frac{d}{dt}(\rho\,d\nu) = 0, \tag{8.1}$$

where ρ is the mass density (having units of mass per volume) and $d\nu$ is a differential volume at any time t. Hence, we have by the product rule,

$$\frac{d}{dt}(\rho\,d\nu) = \frac{d\rho}{dt}d\nu + \rho\frac{d}{dt}(d\nu) = 0. \tag{8.2}$$

For simplicity, let us assume that the differential mass of interest is in the shape of a cube both at time $t = 0$ and a particular time t sometime during its history. (Of course, if the original system of interest is cuboidal, it would be expected to assume many different shapes when flowing. We simply assume that, "remarkably," it is again a cube at some time t, which is the instant on which we will focus.) Hence, let the differential volume at time t be $d\nu = dx\,dy\,dz$, which was originally a (possibly) different cube having volume $d\mathcal{V} = dX\,dY\,dZ$. For a cube to deform into another cube, there can be length changes at most (i.e., no shear). Consequently, whereas the position x of a particle in the cube at time t could be a function of X, Y, and Z, in general, and thus by the chain rule

$$dx = \frac{\partial x}{\partial X}dX + \frac{\partial x}{\partial Y}dY + \frac{\partial x}{\partial Z}dZ \tag{8.3}$$

(and similarly for dy and dz), for a cube to deform into a cube, we must have at each time t, only

$$x = x(X) \to dx = \frac{\partial x}{\partial X}dX,$$

$$y = y(Y) \to dy = \frac{\partial y}{\partial Y}dY, \tag{8.4}$$

$$z = z(Z) \to dz = \frac{\partial z}{\partial Z}dZ.$$

Hence, $d\nu$ at time t is

$$d\nu = dx\,dy\,dz = \frac{\partial x}{\partial X}dX\frac{\partial y}{\partial Y}dY\frac{\partial z}{\partial Z}dZ = \frac{\partial x}{\partial X}\frac{\partial y}{\partial Y}\frac{\partial z}{\partial Z}d\mathcal{V}, \tag{8.5}$$

where $d\mathcal{V}$ does not change in time because it is defined at time $t = 0$. The time rate of change of $d\nu$ is thus

$$\frac{d}{dt}(d\mathcal{V}) = \frac{d}{dt}\left(\frac{\partial x}{\partial X}\frac{\partial y}{\partial Y}\frac{\partial z}{\partial Z}\right)d\mathcal{V}. \tag{8.6}$$

Employing the product rule,

$$\frac{d}{dt}(d\mathcal{V}) = \left[\frac{d}{dt}\left(\frac{\partial x}{\partial X}\right)\frac{\partial y}{\partial Y}\frac{\partial z}{\partial Z} + \frac{\partial x}{\partial X}\frac{d}{dt}\left(\frac{\partial y}{\partial Y}\right)\frac{\partial z}{\partial Z}\right.$$
$$\left. + \frac{\partial x}{\partial X}\frac{\partial y}{\partial Y}\frac{d}{dt}\left(\frac{\partial z}{\partial Z}\right)\right]d\mathcal{V}. \tag{8.7}$$

Because the original positions (X, Y, Z) are independent of time, we can interchange the order of the temporal and spatial differentiations. Using Eq. (7.7) and the chain rule, we have

$$\frac{d}{dt}\left(\frac{\partial x}{\partial X}\right) = \frac{\partial}{\partial X}\left(\frac{dx}{dt}\right) = \frac{\partial}{\partial X}(v_x) = \frac{\partial v_x}{\partial x}\frac{\partial x}{\partial X},$$

$$\frac{d}{dt}\left(\frac{\partial y}{\partial Y}\right) = \frac{\partial}{\partial Y}\left(\frac{dy}{dt}\right) = \frac{\partial}{\partial Y}(v_y) = \frac{\partial v_y}{\partial y}\frac{\partial y}{\partial Y}, \tag{8.8}$$

$$\frac{d}{dt}\left(\frac{\partial z}{\partial Z}\right) = \frac{\partial}{\partial Z}\left(\frac{dz}{dt}\right) = \frac{\partial}{\partial Z}(v_z) = \frac{\partial v_z}{\partial z}\frac{\partial z}{\partial Z}.$$

Substituting these results into Eq. (8.7) and then Eq. (8.2), we have

$$\frac{d}{dt}(\rho d\mathcal{V}) = \left[\frac{d\rho}{dt} + \rho\left(\frac{\partial v_x}{\partial x} + \frac{\partial v_y}{\partial y} + \frac{\partial v_z}{\partial z}\right)\right]\frac{\partial x}{\partial X}\frac{\partial y}{\partial Y}\frac{\partial z}{\partial Z}d\mathcal{V} \tag{8.9}$$

or, by recalling Eq. (7.48),

$$\frac{d}{dt}(\rho d\mathcal{V}) = \left[\frac{d\rho}{dt} + \rho(\nabla \cdot \mathbf{v})\right]d\mathcal{V} = 0. \tag{8.10}$$

Because this equation must hold for any $d\mathcal{V}$, not all zero, this implies that

$$\frac{d\rho}{dt} + \rho(\nabla \cdot \mathbf{v}) = 0, \tag{8.11}$$

which is our (local) statement of the balance of mass. Because the mass density could differ at different points (x, y, z) or at different times, then

$$\frac{d\rho}{dt} = \frac{\partial\rho}{\partial t}\frac{dt}{dt} + \frac{\partial\rho}{\partial x}v_x + \frac{\partial\rho}{\partial y}v_y + \frac{\partial\rho}{\partial z}v_z, \tag{8.12}$$

similar to Eq. (7.15) (i.e., the Eulerian description of acceleration). Hence, Eq. (8.11) could also be written as $\partial\rho/\partial t + \nabla \cdot (\rho\mathbf{v}) = 0$. Regardless, if ρ is a constant, $d\rho/dt = 0$, and the balance of mass requires only that

$$\nabla \cdot \mathbf{v} = 0. \tag{8.13}$$

This is the mass balance relation for an incompressible flow, as alluded to in Chapter 7.

Finally, it should be noted that because our final expression for mass balance can be written in vector form, it is independent of coordinate system and therefore completely general; that is, the derivation based on deforming a cube into a cube was simply used for convenience; it is not a restricted case. Using mathematics beyond that typically available to the beginning undergraduate, this derivation can be repeated exactly for an arbitrarily shaped Δm (Humphrey, 2002). Herein, however, we shall simply focus on its use, not its general derivation. Given Eq. (8.13) and the definition of the del operator for various coordinate systems (Appendix 7 of Chapter 7), one can show that mass balance for an incompressible flow requires

$$\nabla \cdot v = \frac{\partial v_x}{\partial x} + \frac{\partial v_y}{\partial y} + \frac{\partial v_z}{\partial z} = 0 \tag{8.14}$$

in Cartesians,

$$\nabla \cdot v = \frac{1}{r}\frac{\partial}{\partial r}(rv_r) + \frac{1}{r}\frac{\partial v_\theta}{\partial \theta} + \frac{\partial v_z}{\partial z} = 0 \tag{8.15}$$

in cylindricals, and

$$\nabla \cdot v = \frac{1}{r^2}\frac{\partial}{\partial r}(r^2 v_r) + \frac{1}{r\sin\theta}\frac{\partial}{\partial \theta}(v_\theta \sin\theta) + \frac{1}{r\sin\theta}\frac{\partial v_\phi}{\partial \phi} = 0 \tag{8.16}$$

in sphericals. The latter two result from Exercises 7.5 and 7.6.

Observation 8.1 Note that $\nabla \cdot v = \text{tr}[D]$; that is, the divergence of the velocity equals the sum of the diagonals of the rate of deformation when it is written in matrix form as $[D]$. Recall Eq. (7.55)–(7.60). In Cartesians, therefore, incompressibility requires that

$$\nabla \cdot v = D_{xx} + D_{yy} + D_{zz} = 0,$$

where D_{xx}, D_{yy}, and D_{zz} are measures of the rates at which line elements change length in the x, y, and z directions. Our intuition is thus supported by this equation: for volume to be conserved, lengthening in at least one direction must be accompanied by shortening in at least one direction.

Example 8.1 Is the following velocity field a possible incompressible flow?

$$v(x, y, z, t) = \frac{\rho g \sin\theta}{\mu}\left(yh - \frac{y^2}{2}\right)\hat{i},$$

where ρ is the mass density of the fluid, g $(= 9.81\,\text{m/s}^2)$ is the gravitational constant, μ is the viscosity of the fluid, θ is some fixed angle (number) relative to a horizontal datum, and h is some depth of a fluid film.

Solution

$$\nabla \cdot v = \frac{\partial}{\partial x}\left[\frac{\rho g \sin \theta}{\mu}\left(yh - \frac{y^2}{2}\right)\right] + \frac{\partial}{\partial y}(0) + \frac{\partial}{\partial z}(0) = 0;$$

so yes, this is a possible incompressible flow field. This velocity field will be determined formally in Example 9.3 via the solution of the equation of motion for a particular boundary value problem.

8.2 Balance of Linear Momentum

As noted earlier, Euler showed that Newton's statement of the balance of linear momentum for a mass particle (i.e., $F = ma$) can be generalized for a continuum (i.e., infinite collection of particles). Hence, let us apply Newton's second law to our differential mass Δm, which we shall again take to be a differential cube having volume $\Delta x \Delta y \Delta z$ and mass density ρ (i.e., $\Delta m = \rho \Delta x \Delta y \Delta z$). Two types of forces of importance in continuum mechanics are those that act on every particle in the continuum, called *body forces*, and those that act on the body only through its surface, the *surface forces*. Let the body force that the fluid element experiences be defined per unit mass and denoted $g = g_x \hat{i} + g_y \hat{j} + g_z \hat{k}$. The most common example of a body force is gravity. Moreover, let the forces acting on the surface of the cube be computed via the appropriate Cauchy stress ($\sigma_{(\text{face})(\text{direction})}$ relative to a prescribed coordinate system) multiplied by its respective surface area. Common surface forces are hydrostatic pressure and frictional forces between fluid particles moving relative to each other. Desiring to let the cube shrink to a point (i.e., in the limit as $\Delta x, \Delta y, \Delta z \to 0$), let the components of the stress at the center of Δm be $\sigma_{xx}, \sigma_{xy}, \sigma_{xz}, \ldots, \sigma_{zz}$. Next, assume that the stress may vary from point to point[1]; thus, the stresses on each of the faces of Δm must differ from those in the center (although the stresses also vary over each face, we shall represent the stresses on a given face by their mean value, which will be appropriate as we shrink to a point). This difference from face to face is expected to be small, however, because the distance from the center, located at (x, y, z), to each face is small (e.g., $\Delta x/2$, $\Delta y/2, \Delta z/2$). Hence, we consider a Taylor's series expansion about the center. For example, for the normal stress on the positive x face, we have

$$\sigma_{xx} + \frac{\partial \sigma_{xx}}{\partial x}\left(\frac{\Delta x}{2}\right) + \text{H.O.T.}, \tag{8.17}$$

---

[1] This is similar to that done in solids, for example, letting the moment in the beam element $M(x)$ be $M(x) + \Delta M(x)$ at $x + \Delta x$ and so on. See also Section 3.1 of Chapter 3.

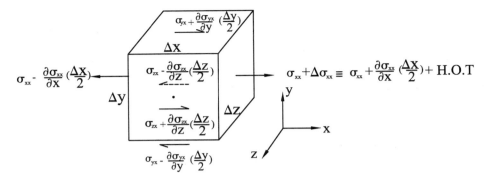

FIGURE 8.1 Force balance for a fluid element of cuboidal shape that is accelerating and subjected to body forces. For simplicity, x-direction contributions only are given.

(where *H.O.T.* stands for higher-order terms such as Δx^2 and so forth, which are negligible with respect to Δx and so forth, as shown in Section 3.1 of Chapter 3) and similarly for each component and each face. Remembering that we are summing forces (i.e., stresses acting over oriented areas), we have (Fig. 8.1)

$$\sum F_x = ma_x \rightarrow \left\{\left[\sigma_{xx} + \frac{\partial \sigma_{xx}}{\partial x}\left(\frac{\Delta x}{2}\right)\right] - \left[\sigma_{xx} - \frac{\partial \sigma_{xx}}{\partial x}\left(\frac{\Delta x}{2}\right)\right]\right\}\Delta y \Delta z$$

$$+ \left\{\left[\sigma_{yx} + \frac{\partial \sigma_{yx}}{\partial y}\left(\frac{\Delta y}{2}\right)\right] - \left[\sigma_{yx} - \frac{\partial \sigma_{yx}}{\partial y}\left(\frac{\Delta y}{2}\right)\right]\right\}\Delta x \Delta z$$

$$+ \left\{\left[\sigma_{zx} + \frac{\partial \sigma_{zx}}{\partial z}\left(\frac{\Delta z}{2}\right)\right] - \left[\sigma_{zx} - \frac{\partial \sigma_{zx}}{\partial z}\left(\frac{\Delta z}{2}\right)\right]\right\}\Delta x \Delta y$$

$$+ \rho g_x \Delta x \Delta y \Delta z = \rho \Delta x \Delta y \Delta z a_x. \tag{8.18}$$

Simplifying and taking the limit, we have

$$\lim_{\substack{\Delta x \to 0 \\ \Delta y \to 0 \\ \Delta z \to 0}} \frac{1}{\Delta x \Delta y \Delta z}\left(\left[\frac{\partial \sigma_{xx}}{\partial x} + \frac{\partial \sigma_{yx}}{\partial y} + \frac{\partial \sigma_{zx}}{\partial z}\right]\Delta x \Delta y \Delta z + \rho g_x \Delta x \Delta y \Delta z - \rho a_x \Delta x \Delta y \Delta z\right)$$

$$= 0, \tag{8.19}$$

or, as our final result in the x direction,

$$\frac{\partial \sigma_{xx}}{\partial x} + \frac{\partial \sigma_{yx}}{\partial y} + \frac{\partial \sigma_{zx}}{\partial z} + \rho g_x = \rho a_x. \tag{8.20}$$

Note that the first subscript on the stress denotes the face on which the force acts, whereas the second subscript denotes the direction of the force—each σ in this equation appropriately has x for the second subscript because

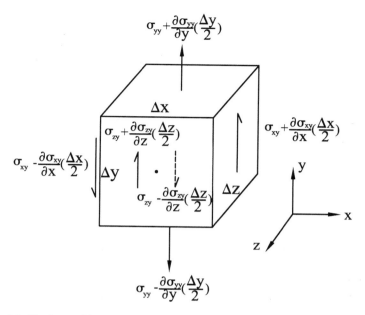

FIGURE 8.2 Similar to Figure 8.1 except for the y-direction.

this is an x-direction force balance. Balance in the y-direction (Fig. 8.2) similarly yields

$$\sum F_y = ma_y \rightarrow \left\{ \left[\sigma_{yy} + \frac{\partial \sigma_{yy}}{\partial y} \left(\frac{\Delta y}{2} \right) \right] - \left[\sigma_{yy} - \frac{\partial \sigma_{yy}}{\partial y} \left(\frac{\Delta y}{2} \right) \right] \right\} \Delta x \Delta z$$

$$+ \left\{ \left[\sigma_{xy} + \frac{\partial \sigma_{xy}}{\partial x} \left(\frac{\Delta x}{2} \right) \right] - \left[\sigma_{xy} - \frac{\partial \sigma_{xy}}{\partial x} \left(\frac{\Delta x}{2} \right) \right] \right\} \Delta y \Delta z$$

$$+ \left\{ \left[\sigma_{zy} + \frac{\partial \sigma_{zy}}{\partial z} \left(\frac{\Delta z}{2} \right) \right] - \left[\sigma_{zy} - \frac{\partial \sigma_{zy}}{\partial z} \left(\frac{\Delta z}{2} \right) \right] \right\} \Delta x \Delta y$$

$$+ \rho g_y \Delta x \Delta y \Delta z = \rho \Delta x \Delta y \Delta z a_y. \tag{8.21}$$

Again, simplifying and taking the limit, we have

$$\frac{\partial \sigma_{xy}}{\partial x} + \frac{\partial \sigma_{yy}}{\partial y} + \frac{\partial \sigma_{zy}}{\partial z} + \rho g_y = \rho a_y. \tag{8.22}$$

Similarly, in the z direction, we find (do it)

$$\frac{\partial \sigma_{xz}}{\partial x} + \frac{\partial \sigma_{yz}}{\partial y} + \frac{\partial \sigma_{zz}}{\partial z} + \rho g_z = \rho a_z. \tag{8.23}$$

Equations (8.20), (8.22), and (8.23) are the general equations of motion relative to a Cartesian coordinate system. Because we did not specify any particular material behavior (i.e., constitutive relation) in this derivation, these equations are true for *all* materials that can be regarded as continua. Indeed, in the case of statics (i.e., no accelerations), we recover Eqs. (3.8)–(3.10), which were derived for solids but likewise are good for all continua.

Similar equations can be found for other coordinate systems. For example, in cylindricals, we have

$$\frac{\partial \sigma_{rr}}{\partial r} + \frac{1}{r}\frac{\partial \sigma_{\theta r}}{\partial \theta} + \frac{\partial \sigma_{zr}}{\partial z} + \frac{\sigma_{rr} - \sigma_{\theta\theta}}{r} + \rho g_r = \rho a_r, \tag{8.24}$$

$$\frac{\partial \sigma_{r\theta}}{\partial r} + \frac{1}{r}\frac{\partial \sigma_{\theta\theta}}{\partial \theta} + \frac{\partial \sigma_{z\theta}}{\partial z} + \frac{2\sigma_{r\theta}}{r} + \rho g_\theta = \rho a_\theta, \tag{8.25}$$

$$\frac{\partial \sigma_{rz}}{\partial r} + \frac{1}{r}\frac{\partial \sigma_{\theta z}}{\partial \theta} + \frac{\partial \sigma_{zz}}{\partial z} + \frac{\sigma_{rz}}{r} + \rho g_z = \rho a_z, \tag{8.26}$$

and in sphericals,

$$\frac{\partial \sigma_{rr}}{\partial r} + \frac{1}{r}\frac{\partial \sigma_{\theta r}}{\partial \theta} + \frac{1}{r\sin\theta}\frac{\partial \sigma_{\phi r}}{\partial \phi} + \frac{1}{r}(2\sigma_{rr} - \sigma_{\theta\theta} - \sigma_{\phi\phi} + \sigma_{\theta r}\cot\theta)$$
$$+ \rho g_r = \rho a_r, \tag{8.27}$$

$$\frac{\partial \sigma_{r\theta}}{\partial r} + \frac{1}{r}\frac{\partial \sigma_{\theta\theta}}{\partial \theta} + \frac{1}{r\sin\theta}\frac{\partial \sigma_{\phi\theta}}{\partial \phi} + \frac{1}{r}[2\sigma_{r\theta} + \sigma_{\theta r} + (\sigma_{\theta\theta} - \sigma_{\phi\phi})\cot\theta]$$
$$+ \rho g_\theta = \rho a_\theta, \tag{8.28}$$

$$\frac{\partial \sigma_{r\phi}}{\partial r} + \frac{1}{r}\frac{\partial \sigma_{\theta\phi}}{\partial \theta} + \frac{1}{r\sin\theta}\frac{\partial \sigma_{\phi\phi}}{\partial \phi} + \frac{1}{r}[2\sigma_{r\phi} + \sigma_{\phi r} + (\sigma_{\phi\theta} + \sigma_{\theta\phi})\cot\theta]$$
$$+ \rho g_\phi = \rho a_\phi. \tag{8.29}$$

8.3 Navier–Stokes Equations

To specialize the equations of motion for an incompressible Newtonian behavior, the incompressible $(\nabla \cdot v = 0)$ Navier–Poisson equation [Eq. (7.66)] can be substituted into the equations of motion [Eqs. (8.20), (8.22), and (8.23)]. For example, for Cartesian coordinates, the x-direction equation

$$\frac{\partial \sigma_{xx}}{\partial x} + \frac{\partial \sigma_{yx}}{\partial y} + \frac{\partial \sigma_{zx}}{\partial z} + \rho g_x = \rho a_x \tag{8.30}$$

becomes

$$\frac{\partial}{\partial x}\left(-p + 2\mu\frac{\partial v_x}{\partial x}\right) + \frac{\partial}{\partial y}\left[2\mu\left(\frac{1}{2}\right)\left(\frac{\partial v_x}{\partial y} + \frac{\partial v_y}{\partial x}\right)\right]$$

$$+ \frac{\partial}{\partial z}\left[2\mu\left(\frac{1}{2}\right)\left(\frac{\partial v_x}{\partial z} + \frac{\partial v_z}{\partial x}\right)\right] + \rho g_x = \rho a_x, \tag{8.31}$$

or

$$-\frac{\partial p}{\partial x} + 2\mu\frac{\partial^2 v_x}{\partial x^2} + \mu\frac{\partial^2 v_x}{\partial y^2} + \mu\frac{\partial^2 v_y}{\partial y \partial x} + \mu\frac{\partial^2 v_x}{\partial z^2} + \mu\frac{\partial^2 v_z}{\partial z \partial x} + \rho g_x = \rho a_x. \tag{8.32}$$

Now, if we let

$$2\mu\frac{\partial^2 v_x}{\partial x^2} = \mu\frac{\partial^2 v_x}{\partial x^2} + \mu\frac{\partial^2 v_x}{\partial x^2} \tag{8.33}$$

and if we interchange the order of mixed derivatives $\partial^2/\partial y\,\partial x$ to $\partial^2/\partial x\,\partial y$ and so forth, then Eq. (8.32) can be written as

$$-\frac{\partial p}{\partial x} + \mu\left(\frac{\partial^2 v_x}{\partial x^2} + \frac{\partial^2 v_x}{\partial y^2} + \frac{\partial^2 v_x}{\partial z^2}\right) + \rho g_x + \mu\frac{\partial}{\partial x}\left(\frac{\partial v_x}{\partial x} + \frac{\partial v_y}{\partial y} + \frac{\partial v_z}{\partial z}\right) = \rho a_x, \tag{8.34}$$

or

$$-\frac{\partial p}{\partial x} + \mu\nabla^2 v_x + \rho g_x + \mu\frac{\partial}{\partial x}(\nabla \cdot \boldsymbol{v}) = \rho a_x. \tag{8.35}$$

Note: The Laplacian $\nabla^2 \equiv \nabla \cdot \nabla$, which is computed easily. Consistent with the above incompressibility assumption, $\nabla \cdot \boldsymbol{v} = 0$; thus, our final relation in the x direction is

$$-\frac{\partial p}{\partial x} + \mu\nabla^2 v_x + \rho g_x = \rho a_x. \tag{8.36}$$

Similarly, the y-direction equation

$$\frac{\partial \sigma_{xy}}{\partial x} + \frac{\partial \sigma_{yy}}{\partial y} + \frac{\partial \sigma_{zy}}{\partial z} + \rho g_y = \rho a_y \tag{8.37}$$

becomes

$$\frac{\partial}{\partial x}\left[2\mu\left(\frac{1}{2}\right)\left(\frac{\partial v_x}{\partial y} + \frac{\partial v_y}{\partial x}\right)\right] + \frac{\partial}{\partial y}\left(-p + 2\mu\frac{\partial v_y}{\partial y}\right)$$

$$+ \frac{\partial}{\partial z}\left[2\mu\left(\frac{1}{2}\right)\left(\frac{\partial v_y}{\partial z} + \frac{\partial v_z}{\partial y}\right)\right] + \rho g_y = \rho a_y, \tag{8.38}$$

or

$$-\frac{\partial p}{\partial y} + \mu\nabla^2 v_y + \rho g_y + \mu\frac{\partial}{\partial y}(\nabla \cdot \boldsymbol{v}) = \rho a_y. \tag{8.39}$$

Incompressibility thus yields

$$-\frac{\partial p}{\partial y} + \mu\nabla^2 v_y + \rho g_y = \rho a_y. \tag{8.40}$$

Finally,

$$\frac{\partial \sigma_{xz}}{\partial x} + \frac{\partial \sigma_{yz}}{\partial y} + \frac{\partial \sigma_{zz}}{\partial z} + \rho g_z = \rho a_z \tag{8.41}$$

can be shown (do it) to reduce to:

$$-\frac{\partial p}{\partial z} + \mu\nabla^2 v_z + \rho g_z = \rho a_z. \tag{8.42}$$

Considering the three component equations, we see that the incompressible Navier–Stokes equations (due to Navier (1785–1836) and Stokes (1819–1903)) can be written more generally in vector notation as

$$-\nabla p + \mu\nabla^2 \boldsymbol{v} + \rho \boldsymbol{g} = \rho \boldsymbol{a}, \tag{8.43}$$

which is good for any coordinate system. Finally, for an Eulerian approach, recall from Eq. (7.21) that the acceleration has two contributions: local and convective. Writing these explicitly yields our final form for the *incompressible Navier–Stokes equation*:

$$-\nabla p + \mu\nabla^2 \boldsymbol{v} + \rho \boldsymbol{g} = \rho\left(\frac{\partial \boldsymbol{v}}{\partial t} + (\boldsymbol{v}\cdot\nabla)\boldsymbol{v}\right). \tag{8.44}$$

Hence, this system of equations consisting of the equation of motion (8.44) and the incompressible mass balance equation [see Eq. (8.13)],

$$\nabla\cdot\boldsymbol{v} = 0, \tag{8.45}$$

represent our four governing differential equations (three scalar momentum equations and one scalar mass equation) for an incompressible Newtonian fluid in terms of our four unknowns (pressure and three components of velocity). Because of the convective acceleration terms, these are nonlinear coupled partial differential equations, which are difficult to solve in general; one must often resort to numerical methods. We shall see in Chapters 9 and 11, however, that a number of useful solutions can be found analytically in Cartesian, cylindrical, and spherical coordinates. In cylindrical coordinates, for example, the incompressible Navier–Stokes equations are

$$-\frac{\partial p}{\partial r} + \mu\left[\frac{\partial}{\partial r}\left(\frac{1}{r}\frac{\partial(rv_r)}{\partial r}\right) + \frac{1}{r^2}\frac{\partial^2 v_r}{\partial\theta^2} - \frac{2}{r^2}\frac{\partial v_\theta}{\partial\theta} + \frac{\partial^2 v_r}{\partial z^2}\right] + \rho g_r$$

$$= \rho\left(\frac{\partial v_r}{\partial t} + v_r\frac{\partial v_r}{\partial r} + \frac{v_\theta}{r}\frac{\partial v_r}{\partial\theta} - \frac{v_\theta^2}{r} + v_z\frac{\partial v_r}{\partial z}\right), \tag{8.46}$$

$$-\frac{1}{r}\frac{\partial p}{\partial \theta}+\mu\left[\frac{\partial}{\partial r}\left(\frac{1}{r}\frac{\partial(rv_\theta)}{\partial r}\right)+\frac{1}{r^2}\frac{\partial^2 v_\theta}{\partial \theta^2}+\frac{2}{r^2}\frac{\partial v_r}{\partial \theta}+\frac{\partial^2 v_\theta}{\partial z^2}\right]+\rho g_\theta$$

$$=\rho\left(\frac{\partial v_\theta}{\partial t}+v_r\frac{\partial v_\theta}{\partial r}+\frac{v_\theta}{r}\frac{\partial v_\theta}{\partial \theta}+\frac{v_r v_\theta}{r}+v_z\frac{\partial v_\theta}{\partial z}\right), \tag{8.47}$$

$$-\frac{\partial p}{\partial z}+\mu\left[\frac{1}{r}\frac{\partial}{\partial r}\left(r\frac{\partial v_z}{\partial r}\right)+\frac{1}{r^2}\frac{\partial^2 v_z}{\partial \theta^2}+\frac{\partial^2 v_z}{\partial z^2}\right]+\rho g_z$$

$$=\rho\left(\frac{\partial v_z}{\partial t}+v_r\frac{\partial v_z}{\partial r}+\frac{v_\theta}{r}\frac{\partial v_z}{\partial \theta}+v_z\frac{\partial v_z}{\partial z}\right), \tag{8.48}$$

which clearly appear as formidable coupled equations (each contains all four unknowns). Because blood vessels, airways, ureters, medical tubing, and so forth are cylindrical in cross section, these equations (combined with mass balance) are perhaps the most important in biofluid mechanics; they are the focus of much of Chapter 9, in which we will find exact solutions for a few important classes of problems.

Here, however, let us note that in certain problems, the Navier–Stokes equations simplify considerably. For example, G. Stokes suggested that it would be useful to consider flows in which the viscous effects are much greater than the inertial (i.e., convective acceleration) effects; that is, in slow (or *creeping*) flows, the Navier–Stokes equation reduces to

$$-\nabla p+\mu\nabla^2 \boldsymbol{v}+\rho \boldsymbol{g}=\rho\frac{\partial \boldsymbol{v}}{\partial t}, \tag{8.49}$$

which is a linear second-order differential equation.

Conversely, another simplification can be made if we assume that the viscous effects are small. Although all fluids resist deformation to some degree, as noted earlier there are problems wherein the viscosity of the fluid is negligible. In this case, the fluid is called *inviscid* and the Navier–Stokes equation reduces to the so-called *Euler equation*:

$$-\nabla p+\rho \boldsymbol{g}=\rho \boldsymbol{a}, \tag{8.50}$$

where the acceleration includes both local and convective parts in general. The Euler equation is thus a nonlinear first-order differential equation. A fluid that experiences only incompressible and inviscid flows is called an *ideal fluid*.

Finally, if the fluid is truly static, then $\boldsymbol{v}=\boldsymbol{0}$ and $\boldsymbol{a}=\boldsymbol{0}$, and the Navier–Stokes equation becomes

$$-\nabla p+\rho \boldsymbol{g}=\boldsymbol{0}, \tag{8.51}$$

which is a linear first-order differential equation. Although this equation is often derived in courses on Engineering Statics, the derivation is typically much different. Regardless, let us examine the following simple example.

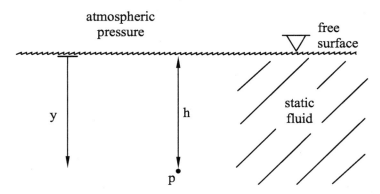

FIGURE 8.3 Determination of the pressure as a function of depth in a static fluid.

Example 8.2 Consider a container of fluid at rest with a Cartesian coordinate defined as positive downward and the origin located at the surface of the fluid (Fig. 8.3). Find the hydrostatic pressure p at the depth h.

Solution: From Eqs. (8.36), (8.40), and (8.42), with $\mathbf{g} = +\rho g \hat{\mathbf{j}}$, given the downward oriented coordinate direction, we have

$$-\frac{\partial p}{\partial x} + 0 = 0, \quad -\frac{\partial p}{\partial y} + \rho g = 0, \quad -\frac{\partial p}{\partial z} + 0 = 0.$$

From the first and third equations, $p = p(y)$ at most, and the partial derivative becomes an ordinary derivative. Solving by integration,

$$\frac{dp}{dy} = \rho g \rightarrow \int \frac{d}{dy}(p)dy = \int \rho g\, dy$$

and, consequently, we have

$$p(y) = \rho g y + c.$$

The integration constant c is found from the boundary conditions. Here, note that the so-called *gauge pressure* is defined as the *absolute pressure* minus atmospheric pressure. If we assume an atmospheric pressure at the surface, then $p(y = 0) = 0$(gauge) and $c = 0$. Thus, $p(y) = \rho g y$. At $y = h$, therefore, we obtain the well-known result that $p = \rho g h$ at depth h (that pressure increases with depth is easily appreciated as we swim deeper in a pool). In a sense, then, this is a solution of the Navier–Stokes equation. Because of the importance and utility of the Navier–Stokes equation, much of Chapter 9 is devoted to its solution.

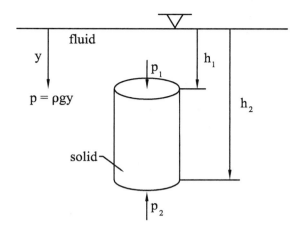

FIGURE 8.4 Schema illustrating Archimedes' principle. Assuming that the submerged solid is a cylinder merely simplifies the analysis, the consequence of which is very general.

Observation 8.2 In approximately 220 B.C., the Greek mathematician Archimedes derived a very important relation in fluid statics that relates the amount of fluid displaced by an immersed solid to the force exerted on that solid by the fluid (the so-called buoyant force). Although we could derive this result by considering an arbitrarily shaped solid, for convenience let us consider a solid cylinder, as shown in Figure 8.4. The weight of the cylinder is $W = \rho_s g(\pi a^2)(h_2 - h_1)$, where ρ_s is the mass density of the solid and a is its radius. Whereas this force tends to cause the solid to "sink," the difference in pressures on the bottom and top surfaces tends to push upward on the solid. This *buoyant force* $F_B = (p_2 - p_1)\pi a^2 = (\rho_f g h_2 - \rho_f g h_1)\pi a^2$; see Example 8.2. If

$W > F_B$, then the soild will sink;
$W = F_B$, then the solid is neutrally buoyant;
$W < F_B$, then the solid will float.

In particular, any body that remains fully submerged at a fixed depth (i.e., where it is placed) is said to be neutrally buoyant. In this case,

$$\rho_s g(\pi a^2)(h_2 - h_1) = \rho_f g(\pi a^2)(h_2 - h_1),$$

or $\rho_s = \rho_f$. Regardless, we see that the buoyant force

$$F_B = \rho_f g(\pi a^2)(h_2 - h_1) = \rho_f g \Psi_s,$$

where Ψ_s is the volume of the solid that is in the fluid, which is equal to the volume of the displaced fluid. Archimedes' principle states, therefore, that the net buoyant force exerted on a solid by a fluid equals the force of gravity on the liquid that is displaced by the solid.

Archimedes' principle is often used in mechanical tests on soft tissues. Because soft tissues tend to have a slightly higher mass density ($\rho \sim 1050$ kg/m^3) than the physiologic solution in which they are placed, they tend to sink, especially when mounting fixtures are affixed to them. To render the tissue neutrally buoyant, therefore, a volume-occupying low-density material (e.g., Styrofoam) can be attached to the fixtures so that the weight of the total volume of fluid displaced by the specimen and fixture equals the tissue–fixture weight. Hence, the only loads on the tissue will be those imposed by the materials testing unit.

[Note to student/instructor: It may be advisable to proceed to Chapter 9 at this time and return to the following sections on inviscid fluids and methods of measurements only if desired.]

8.4 The Euler Equation

Comparison of the incompressible Navier–Stokes equation [Eq. (8.44)] to the Euler equation [Eq. (8.50)] reveals that the former is a system of coupled second-order partial differential equations (PDEs), whereas the latter is a system of coupled first-order PDEs. First-order equations tend to be easier to solve, but because of the convective part of the acceleration [i.e., $(v \cdot \nabla)v$], both equations are nonlinear, and it is often this nonlinearity that poses the greatest difficulty in solution. For this reason, it can be shown that a judicious choice of a coordinate system can be helpful in trying to solve even the Euler equation.

Toward this end, let us define two new terms. Let a *pathline* be defined as the locus of points through which a material particle passes in a flow field. An example would be the path taken by a leaf as it flows down a river. Let a *streamline* be defined as a locus of points where the velocity is everywhere tangent. This mathematical definition is less intuitive than that for the pathline. In cases of steady flows, however, the two lines coincide. Hindsight reveals that it can be convenient, particularly in steady flows, to define a coordinate system such that one of the coordinate axes coincides locally with a streamline. Hence, let us consider the following.

Question: How do we write Euler's equation in terms of locally orthogonal streamline coordinates (s, n, x)? We could use coordinate transformations to get from equations in terms of (x, y, z) to those in terms of (s, n, x), or we could directly rederive Euler's equation in terms of s, n, and x. Let us adopt the second approach, which is straightforward and will reinforce our earlier derivation.

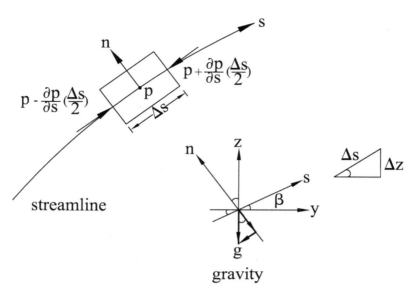

FIGURE 8.5 Differential fluid element in two dimensions relative to streamline coordinates. Remember, therefore, that a streamline is drawn tangent to the velocity vector at every point in a flow field. Note, too, the relation of the streamline coordinates with respect to the Cartesian coordinates, where z is now taken to be vertical, similar to most derivations of this equation in the literature; x is out of the paper.

First, recall, that the constitutive equations for an incompressible, Newtonian fluid are $\sigma_{xx} = -p + 2\mu D_{xx}$, $\sigma_{yy} = -p + 2\mu D_{yy}$, $\sigma_{zz} = -p + 2\mu D_{zz}$, $\sigma_{xy} = 2\mu D_{xy}$, and so on. For inviscid fluids, therefore, $\mu = 0$ and $\sigma_{xx} = \sigma_{yy} = \sigma_{zz} = -p$, which is a hydrostatic state of stress (i.e., an inviscid fluid cannot support a shear stress). Second, recall from Eqs. (7.3) and (7.4) that a hydrostatic state of stress at a point relative to one coordinate system is also hydrostatic relative to any coordinate system at that point. Now, to rederive Euler's equation in terms of s, n, and x, consider the differential fluid element taken along a streamline in Figure 8.5. Let us assume further that the normal stresses are $\sigma_{ss} = \sigma_{nn} = \sigma_{xx} = -p$ at the center of this element and that the pressure can vary from point to point and possibly with time [i.e., $p = p(s, n, x, t)$]. Nonetheless, we shall focus only on possible changes in the streamline direction. Thus, at the positive and negative s faces (i.e., at $s \pm \Delta s/2$), we allow the pressure to be slightly larger or smaller than the value p at the center: On these faces, we let the pressure be $p \pm \Delta p$. Moreover, for the flow to be in the positive s direction, the pressure must be higher at the negative s face. Hence, using a typical Taylor series expansion, at any fixed time t,

$$p\left(s + \frac{\Delta s}{2}\right) = p(s) + \frac{\partial p}{\partial s}\left(\frac{\Delta s}{2}\right),$$

$$p\left(s - \frac{\Delta s}{2}\right) = p(s) - \frac{\partial p}{\partial s}\left(\frac{\Delta s}{2}\right),$$

(8.52)

and we see that the pressure gradient $\partial p/\partial s < 0$ for flow in the positive s direction. Again, higher-order terms in the Taylor's series have been neglected for, in hindsight, they would be negligible. Linear momentum balance in s thus requires that

$$\sum F_s = ma_s \rightarrow \left(p - \frac{\partial p}{\partial s}\frac{\Delta s}{2}\right)\Delta n\Delta x - \left(p + \frac{\partial p}{\partial s}\frac{\Delta s}{2}\right)\Delta n\Delta x$$
$$- \rho(g\sin\beta)\Delta s\Delta n\Delta x = \rho a_s\Delta s\Delta n\Delta x.$$

(8.53)

Simplifying, we have

$$-\frac{\partial p}{\partial s}\Delta s\Delta n\Delta x - \rho g(\sin\beta)\Delta s\Delta n\Delta x = \rho a_s\Delta s\Delta n\Delta x.$$

(8.54)

Dividing this equation by the differential volume and taking the limit, we obtain

$$\lim_{\substack{\Delta s \to 0 \\ \Delta n \to 0 \\ \Delta x \to 0}} \frac{1}{\Delta s\Delta n\Delta x}\left(-\frac{\partial p}{\partial s}\Delta s\Delta n\Delta x - \rho g(\sin\beta)\Delta s\Delta n\Delta x - \rho a_s\Delta s\Delta n\Delta x\right) = 0,$$

(8.55)

or

$$-\frac{\partial p}{\partial s} - \rho g\sin\beta = \rho a_s.$$

(8.56)

Because $\sin\beta = \partial z/\partial s$, we have

$$-\frac{\partial p}{\partial s} - \rho g\frac{\partial z}{\partial s} = \rho a_s = \rho\left(\frac{\partial v_s}{\partial t} + v_s\frac{\partial v_s}{\partial s} + v_n\frac{\partial v_s}{\partial n} + v_x\frac{\partial v_s}{\partial x}\right).$$

(8.57)

Let us now exploit our choice of a streamline coordinate system. Because the velocity vector is everywhere tangent to a streamline, the only component of the velocity is v_s; that is, $\boldsymbol{v} = v_s(s, n, x, t)\hat{\boldsymbol{e}}_s$ relative to streamline coordinates, whereas $\boldsymbol{v} = v_x(x, y, z, t)\hat{\boldsymbol{e}}_x + v_y(x, y, z, t)\hat{\boldsymbol{e}}_y + v_z(x, y, z, t)\hat{\boldsymbol{e}}_z$ relative to a usual Cartesian system. Both represent possible unsteady 3-D flows, but the simplification is clear for the streamline system. Hence, with $v_n = v_x = 0$, the s-direction Euler equation becomes

$$-\frac{\partial p}{\partial s} - \rho g\frac{\partial z}{\partial s} = \rho\left(\frac{\partial v_s}{\partial t} + v_s\frac{\partial v_s}{\partial s}\right).$$

(8.58)

Enforcing linear momentum balance via a summation of the forces in the n direction similarly yields

$$\sum F_n = ma_n \rightarrow \left(p - \frac{\partial p}{\partial n}\frac{\Delta n}{2}\right)\Delta s\Delta x - \left(p + \frac{\partial p}{\partial n}\frac{\Delta n}{2}\right)\Delta s\Delta x$$
$$- \rho(g\cos\beta)\Delta s\Delta n\Delta x = \rho a_n \Delta s\Delta n\Delta x. \qquad (8.59)$$

Simplifying,

$$-\frac{\partial p}{\partial n}\Delta s\Delta n\Delta x - \rho g(\cos\beta)\Delta s\Delta n\Delta x = \rho a_n \Delta s\Delta n\Delta x. \qquad (8.60)$$

Dividing this by $\Delta\nu$ and taking the limit, we obtain

$$\lim_{\substack{\Delta s\to 0 \\ \Delta n\to 0 \\ \Delta x\to 0}} \frac{1}{\Delta s\Delta n\Delta x}\left(-\frac{\partial p}{\partial n}\Delta s\Delta n\Delta x - \rho g(\cos\beta)\Delta s\Delta n\Delta x - \rho a_n \Delta s\Delta n\Delta x\right) = 0, \quad (8.61)$$

or

$$-\frac{\partial p}{\partial n} - \rho g\cos\beta = \rho a_n. \qquad (8.62)$$

Because $\cos\beta = \partial z/\partial n$, we get

$$-\frac{\partial p}{\partial n} - \rho g\frac{\partial z}{\partial n} = \rho a_n. \qquad (8.63)$$

For a centripetal acceleration, $a_n = \partial v_n/\partial t - v_s^2/R$, where $v_n \equiv 0$ and R is the radius of curvature for the streamline. Thus,

$$-\frac{\partial p}{\partial n} - \rho g\frac{\partial z}{\partial n} = \rho\left(-\frac{v_s^2}{R}\right). \qquad (8.64)$$

In summary, for flow in the s-n plane, the Euler equation relative to streamline coordinates reduces to two equations [Eqs. (8.58) and (8.64)] in terms of two unknowns: v_s and p. Clearly, these equations should be easier to solve than the more general equations in terms of four unknowns (pressure and three components of the velocity).

Observation 8.3 It can be shown that the Laplacian of the velocity

$$\nabla^2 v = \nabla(\nabla\cdot v) - \nabla\times(\nabla\times v).$$

Hence, for an incompressible flow,

$$\nabla^2 v = -\nabla\times\zeta,$$

where ζ is the vorticity vector. In this case, the incompressible Navier–Stokes equations can be written as

$$-\nabla p + \mu\nabla^2 v + \rho g = \rho a \rightarrow -\nabla p - \mu(\nabla\times\zeta) + \rho g = \rho a.$$

Note, therefore, that the Navier–Stokes equation reduces to the Euler equation ($\mu = 0$) when the flow is irrotational ($\zeta = \mathbf{0}$) regardless of the viscosity. In other words, any incompressible, irrotational flow that satisfies the Euler equation will likewise satisfy the full Navier–Stokes equations, as we will see in Chapter 11. One must be careful, however, because Euler solutions will not satisfy viscous boundary conditions, such as those on shear stress.

8.5 The Bernoulli Equation

The so-called Bernoulli equation is one of the most used, yet probably *most misused*, equations in fluid mechanics. As we shall see, Bernoulli's equation is an algebraic equation that is much easier to solve than the differential equations of Navier–Stokes or Euler. This simplification does not come without a price, however, for there are five important restrictions that must be respected for the Bernoulli equation to apply. The best way to appreciate restrictions is to derive carefully the equation of interest—let us so begin.

8.5.1 Bernoulli Equation for Flow Along a Streamline

Let us first derive Bernoulli's equation from the Euler equation, relative to a streamline coordinate system (s, n, x). Hence, the first two restrictions are those for Euler's equation: incompressible flow and negligible viscosity. Recall that such a fluid is said to be ideal. Next, let us restrict our attention to a steady flow wherein $\partial v/\partial t = \mathbf{0}$. This provides our third restriction. Hence, Eq. (8.58) reduces to

$$\frac{\partial p}{\partial s} + \rho g \frac{\partial z}{\partial s} + \rho v_s \frac{\partial v_s}{\partial s} = 0. \tag{8.65}$$

Next, note that, in general, the pressure and velocity can each vary from point to point: that is, $p = p(s, n, x)$ and $v_s = (s, n, x)$. Consequently,

$$dp = \frac{\partial p}{\partial s} ds + \frac{\partial p}{\partial n} dn + \frac{\partial p}{\partial x} dx,$$

$$dv_s = \frac{\partial v_s}{\partial s} ds + \frac{\partial v_s}{\partial n} dn + \frac{\partial v_s}{\partial x} dx. \tag{8.66}$$

Yet, if we restrict our attention to flow *along a streamline s*, whereby $dn = dx = 0$, then

$$dp = \frac{\partial p}{\partial s} ds, \quad dv_s = \frac{\partial v_s}{\partial s} ds, \quad dz = \frac{\partial z}{\partial s} ds. \tag{8.67}$$

This suggests that if we integrate Eq. (8.65) *along* a streamline, we obtain

$$\int \frac{\partial p}{\partial s} ds + \int \rho g \frac{\partial z}{\partial s} ds + \int \rho v_s \frac{\partial v_s}{\partial s} ds = \int 0 \, ds, \quad (8.68)$$

or

$$\int dp + \int \rho g \, dz + \int \rho v_s \, dv_s = c. \quad (8.69)$$

Assuming further that the mass density and gravitational constant do not vary with position in the z direction, our final relation is

$$p + \rho g z + \frac{1}{2} \rho v_s^2 = c, \quad (8.70)$$

or, as it is most often written,

$$\frac{p}{\rho} + g z + \frac{v_s^2}{2} = C, \quad (8.71)$$

where $C = c/\rho$. Again, however, we emphasize that this—Bernoulli's—equation can be used only if all of the following restrictions are met:

1. Incompressible flow
2. Inviscid fluid
3. Steady flow
4. Flow along a streamline
5. Constant gravitational forces

Before illustrating some solutions to the Bernoulli equation, let us consider a few additional interesting findings.

8.5.2 Bernoulli Equation for Irrotational Flow

In this subsection, we show that Bernoulli's equation holds at all points in a flow field, not just along a streamline, if the flow is irrotational and the other four restrictions are still satisfied. Hence, recall that an irrotational flow is one in which fluid elements moving in the flow field do not undergo any rigid rotation. Moreover, the vorticity vanishes if the flow is irrotational (i.e., $\zeta = 0 = \nabla \times v$). Recall, too, that for an incompressible fluid, mass balance requires that $\nabla \cdot v = 0$, and for an inviscid fluid, $\mu = 0$; thus, the linear momentum equation for an ideal fluid reduces to Euler's equation, $-\nabla p + \rho g = \rho a$, where

$$a = \frac{\partial v}{\partial t} + (v \cdot \nabla)v \quad (8.72)$$

in an Eulerian formulation. Substituting this equation into Euler's equation for a steady flow, we obtain

$$-\frac{1}{\rho}\nabla p + \boldsymbol{g} = (\boldsymbol{v} \cdot \nabla)\boldsymbol{v}. \tag{8.73}$$

Now, from vector calculus, it can be shown that (see Exercise 7.14)

$$(\boldsymbol{v} \cdot \nabla)\boldsymbol{v} = \frac{1}{2}\nabla(\boldsymbol{v} \cdot \boldsymbol{v}) - \boldsymbol{v} \times (\nabla \times \boldsymbol{v}). \tag{8.74}$$

For an irrotational flow, however, $\nabla \times \boldsymbol{v} = \boldsymbol{0}$; thus, Euler's equation for steady, irrotational flow can be written

$$-\frac{1}{\rho}\nabla p + \boldsymbol{g} = \frac{1}{2}\nabla(\boldsymbol{v} \cdot \boldsymbol{v}), \tag{8.75}$$

where $\boldsymbol{v} \cdot \boldsymbol{v} = |\boldsymbol{v}||\boldsymbol{v}|\cos 0 = v^2$, with v^2 a scalar. Hence, Euler's equation becomes

$$-\frac{1}{\rho}\nabla p + \boldsymbol{g} = \frac{1}{2}\nabla v^2. \tag{8.76}$$

At this point, it is important to note that we have not yet specified a coordinate system and, in particular, we have not specified streamline coordinates. Thus, consider a generic displacement of a particle in the flow field from position $\boldsymbol{r}$ to position $\boldsymbol{r} + d\boldsymbol{r}$ (Fig. 8.6). The displacement vector $d\boldsymbol{r}$ is an arbitrary infinitesimal displacement in any direction. If the only body force is the force due to gravity, then $\boldsymbol{g} = -g\hat{\boldsymbol{k}}$, with z a vertical direction, as in most applications of Bernoulli's equation. Taking the dot product of $d\boldsymbol{r} = dx\,\hat{\boldsymbol{i}} + dy\,\hat{\boldsymbol{j}} + dz\,\hat{\boldsymbol{k}}$ with each of the terms in Eq. (8.76), we have

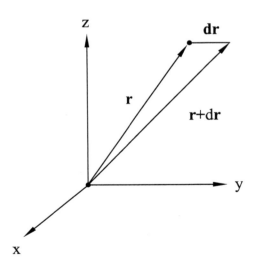

FIGURE 8.6 Position vector $\boldsymbol{r}$ and a small change therefrom.

$$-\frac{1}{\rho}\left(\hat{\boldsymbol{i}}\,\frac{\partial p}{\partial x}+\hat{\boldsymbol{j}}\,\frac{\partial p}{\partial y}+\hat{\boldsymbol{k}}\,\frac{\partial p}{\partial z}\right)\cdot\left(dx\,\hat{\boldsymbol{i}}+dy\,\hat{\boldsymbol{j}}+dz\,\hat{\boldsymbol{k}}\right)+\left(-g\,\hat{\boldsymbol{k}}\right)\cdot\left(dx\,\hat{\boldsymbol{i}}+dy\,\hat{\boldsymbol{j}}+dz\,\hat{\boldsymbol{k}}\right)$$

$$=\frac{1}{2}\left(\hat{\boldsymbol{i}}\,\frac{\partial}{\partial x}(v^2)+\hat{\boldsymbol{j}}\,\frac{\partial}{\partial y}(v^2)+\hat{\boldsymbol{k}}\,\frac{\partial}{\partial z}(v^2)\right)\cdot\left(dx\,\hat{\boldsymbol{i}}+dy\,\hat{\boldsymbol{j}}+dz\,\hat{\boldsymbol{k}}\right), \qquad (8.77)$$

or

$$-\frac{1}{\rho}\left(\frac{\partial p}{\partial x}\,dx+\frac{\partial p}{\partial y}\,dy+\frac{\partial p}{\partial z}\,dz\right)-g\,dz$$

$$=\frac{1}{2}\left(\frac{\partial(v^2)}{\partial x}\,dx+\frac{\partial(v^2)}{\partial y}\,dy+\frac{\partial(v^2)}{\partial z}\,dz\right), \qquad (8.78)$$

which can be written more compactly as

$$-\frac{1}{\rho}\,dp-g\,dz=\frac{1}{2}\,d(v^2). \qquad (8.79)$$

Finally, integration yields the final result, namely

$$\int\frac{1}{\rho}\,dp+\int g\,dz+\int\frac{1}{2}\,d(v^2)=C\;\rightarrow\;\frac{p}{\rho}+gz+\frac{v^2}{2}=C, \qquad (8.80)$$

which is the same equation that we obtained in Eq. (8.71) by focusing our attention along a streamline. We see, therefore, that Bernoulli's equation is valid between *any* two points in the field if the flow is irrotational; if the flow is not irrotational, the Bernoulli equation is still valid at any two points along a streamline. Summarizing then, our five basic restrictions for using the Bernoulli equation are (1) incompressible, (2) inviscid, (3) steady, (4) along a streamline *or* in an irrotational flow, and (5) constant gravitational forces.

For any two appropriate points, say 1 and 2, the Bernoulli equation thus becomes

$$\frac{p_1}{\rho}+gz_1+\frac{1}{2}v_1^2=\frac{p_2}{\rho}+gz_2+\frac{1}{2}v_2^2, \qquad (8.81)$$

which reveals its simple algebraic character and, consequently, why many are tempted to (mis)use it. We will consider a few simple examples later to illustrate how we might use this simple equation.

First, however, note the following. Because it came from Euler's equation, Bernoulli's equation is also a statement of the balance of linear momentum in an inertial reference frame. Being a single algebraic equation, it can be solved for only one unknown. Of course, regardless of the formulation—Navier–Stokes, Euler, or Bernoulli—one must always simultaneously satisfy both the balance of mass and the balance of linear momentum, with mass balance providing one additional equation and thus the ability to solve for one additional unknown. Although we have derived a differential equation for mass balance, let us consider a special case here. For flow into and out of a rigid, impermeable pipe or nozzle, the net *volumetric flow* in, Q_{in}, must equal the net volumetric flow out, Q_{out}. These flows are defined by

$$Q_{in} = \int v \cdot \hat{n} \, dA_{in} = \int v \cdot \hat{n} \, dA_{out} = Q_{out}, \qquad (8.82)$$

where $\hat{n}$ is an outward unit vector normal to the cross-sectional area A of interest. If v is taken to be uniform across the differential area of interest and in the $\pm$ direction of the outward unit normal vector, then

$$Q_{in} = Q_{out} \rightarrow \bar{v}_1 A_1 = \bar{v}_2 A_2, \qquad (8.83)$$

where 1 and 2 denote the inlet and outlet, respectively, and the overbar denotes a mean value. This simple form of mass balance is often used in problems using the Bernoulli equation, as we will now see. In combination with Bernoulli, it allows us to solve for two unknowns between two appropriate points 1 and 2.

Example 8.3 It can be shown experimentally that Bernoulli's equation can be used in computations for flows through constrictions but *not* for flows through expansions. The reason for this is that in the latter case, adverse pressure gradients can disturb the flow such that there is a reversal and thus significant viscous losses (Fig. 8.10). Bernoulli assumes no viscous effects and therefore does not apply. We shall see in Chapter 10 that the flow in an expansion can be handled easily using the energy equation. For a constriction, such as a nozzle or needle, find the injection pressure needed to achieve an exit flow of v_o if the flow exits into a fluid of pressure $P_o = P_{atm}$.

Solution: Assuming that we know the cross-sectional area within the inlet to the needle A_i and its exit area A_o, mass balance requires that $v_i A_i = v_o A_o$,

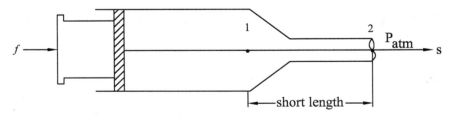

FIGURE 8.7 Flow through a nozzle (e.g., syringe and needle) over a short length. The importance of length on the viscous effects will be demonstrated in Chapter 10.

which allows us to compute v_i given the value of v_o. If we assume that the needle is short, we would expect negligible viscous losses. Indeed, if we further select a centerline streamline, where the velocity gradient $\partial v/\partial r$ should be zero due to the symmetry of v, viscous losses should be small and, thus, we can use Bernoulli. Assuming a horizontal situation,

$$\frac{P_i}{\rho} + \frac{1}{2}\left(\frac{v_o A_o}{A_i}\right)^2 = \frac{P_o}{\rho} + \frac{1}{2}(v_o)^2 \rightarrow P_i = P_o + \frac{1}{2}\rho v_o^2\left[1 - \left(\frac{A_o}{A_i}\right)^2\right].$$

Example 8.4 Water flows steadily up a short, vertical, 2.54-cm-diameter pipe and discharges to atmospheric pressure (Fig. 8.8). If a pressure of 16 kPa drives the fluid at a volumetric flow rate Q of 5 L/min, what height does the fluid reach?

Solution:

Given:

$A_1 = \pi(1.27)^2 = 5.07\,\text{cm}^2$ $z_1 = 0$

$Q = 5\dfrac{\text{L}}{\text{min}}$ $z_2 = h\,\text{m}$

$p_1 = 16\,\text{kPa}$ $g = 9.81\dfrac{\text{m}}{\text{s}^2}$

$p_2 = 0\,\text{(gauge)}$ $\rho = 1000\dfrac{\text{kg}}{\text{m}^3}.$

Assume:

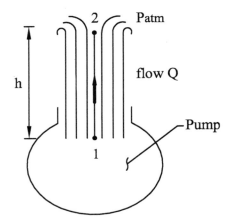

FIGURE 8.8 Flow from a vertical tube/pump that discharges to atmosphere. Because of the influence of gravity, fluid particles will rise to a particular height and then fall.

1. Incompressible
2. Inviscid
3. Steady flow (given)
4. Along a streamline (given)
5. Constant gravitational forces

Moreover, let us assume that the velocity $v_2 = 0$ at the maximum height of the fluid column. Hence, from mass balance

$$Q = v_1 A_1 \rightarrow v_1 = \frac{Q}{A_1} = \frac{5\,L/min}{5.07\,cm^2} \left(\frac{1000\,cm^3}{1\,L} \right) = 986 \frac{cm}{min}$$

or $v_1 = 0.164\,m/s$. Hence, from Bernoulli,

$$\frac{p_1}{\rho} + gz_1 + \frac{v_1^2}{2} = \frac{p_2}{\rho} + gz_2 + \frac{v_2^2}{2} \rightarrow \frac{p_1}{\rho} + \frac{1}{2}v_1^2 = gh,$$

or

$$h = \frac{p_1}{\rho g} + \frac{1}{2g}v_1^2 = \frac{16000\,N/m^2}{(1000\,kg/m^3)(9.81\,m/s^2)} + \frac{(0.164\,m/s)^2}{2(9.81\,m/s^2)} = 1.64\,m.$$

Note that ρg is sometimes called the *specific weight* and denoted by γ, not to be confused with the *specific gravity* SG $= \rho/\rho_{H_2O}$ at 4°C. Given that 1 kPa $= 7.5\,mm\,Hg$, what might this suggest with regard to how far blood might travel if an open needle (having a different diameter) were placed in the heart?

Example 8.5 Note that Bernoulli and mass balance provide two equations:

$$\frac{p_1}{\rho} + gz_1 + \frac{1}{2}v_1^2 = \frac{p_2}{\rho} + gz_2 + \frac{1}{2}v_2^2, \quad v_1 A_1 = v_2 A_2,$$

which can be used to solve for the two velocities, v_1 and v_2, along a straight horizontal streamline s in a steady, converging, ideal flow, with A_1 and A_2 known. To do so, however, we must independently compute or measure the pressures p_1 and p_2. Assuming a negligible gravitational field, determine if the pressure gauges in Figure 8.9 can be used to determine the pressures along the center streamline.

Solution: Because we do not know $\boldsymbol{v}$ as a function of (x, y, z) or (r, θ, z), we cannot determine if Bernoulli holds across the streamline (i.e., if $\nabla \times \boldsymbol{v} = \mathbf{0}$, then Bernoulli may hold for any two points). Hence, let us recall the original Euler equations for a steady ideal flow:

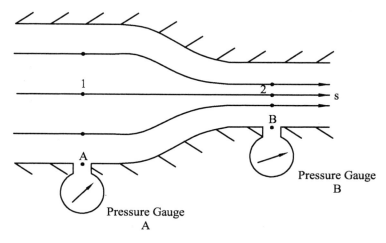

FIGURE 8.9 A simple internal flow that converges from a larger to a smaller diameter tube. Assume that the pressure gauges are connected flush to the wall of the tubing and that they are filled with an incompressible fluid.

$$-\frac{\partial p}{\partial s} - \rho g \frac{\partial z}{\partial s} = \rho v_s \frac{\partial v_s}{\partial s}, \quad -\frac{\partial p}{\partial n} - \rho g \frac{\partial z}{\partial n} = -\frac{\rho v_s^2}{R}$$

from Eqs. (8.58) and (8.64). In particular, from the n-direction equation with $g \sim 0$,

$$\frac{\partial p}{\partial n} = \frac{\rho v_s^2}{R},$$

where R is the radius of curvature of the streamline. Noting that $R \to \infty$ for the locally parallel horizontal streamlines within the regions associated with gauges A and B, then at each gauge, $\partial p / \partial n = 0$, which states that p does not vary in the normal direction when the streamlines are locally parallel. Hence, the pressure measured by these gauges, at the wall, equals the pressures at 1 and 2, and Bernoulli and mass balance can determine v_1 and v_2 in terms of measured p_1, p_2, A_1, and A_2.

8.5.3 Further Restrictions for the Bernoulli Equation

We have suggested that Bernoulli's equation is perhaps the most used and misused equation in fluid mechanics. The latter observation should cause us to respect the noted restrictions: the flow must be incompressible, inviscid, steady, irrotational or along a streamline, and within a constant gravitational field. The last restriction is seldom a concern in the research laboratory or clinical environment; hence, let us focus on the first four restrictions. If we know the velocity field, it is obviously easy to check the incompressible ($\nabla \cdot \mathbf{v} = 0$), steady ($\partial \mathbf{v}/\partial t = \mathbf{0}$), and irrotational ($\nabla \times \mathbf{v} = \mathbf{0}$) restrictions. This would be the case wherein we *measure* $\mathbf{v}(x, y, z, t)$ and seek to use Bernoulli to *calculate* the pressure field. In many cases, however, we may only know the velocity at a few select points, not everywhere; hence, rigorously checking these restrictions is not always so easy. With regard to the inviscid ($\mu = 0$) restriction, we know that all fluids resist flowing to some degree and, thus, have a nonzero μ. The key question then is whether the viscous effects (losses) are negligible with respect to other factors in the problem. This can often be answered only via experience or by comparing solutions of the problem with and without viscosity, which defeats the purpose of seeking an easier approach. Hence, let us record some well-established observations based on others' experiences.

It is well known that viscous (frictional) effects become more and more important over longer lengths of tubes. The Bernoulli equation should thus be restricted to short lengths (e.g., in a needle). In cases of long lengths (e.g., IV tubing from the bag to the patient), one must solve the full differential equations of motion or employ the semi-empirical methods of Chapter 10. Note, too, that flow from a syringe into a needle is an example of a converging flow. Experience reveals that Bernoulli holds in many converging flows for which the flow field is not *turbulent* (i.e., fluctuating randomly). In contrast, Bernoulli should not be used to compute flows in diverging geometries or sudden expansions. Adverse pressure gradients can disturb the flow within such geometries, resulting in separation of the flow from the wall and the formation of recirculation zones (e.g., eddies; Fig. 8.10). Note, therefore, that stenoses in the vasculature can be considered as

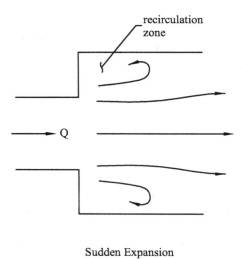

Sudden Expansion

FIGURE 8.10 Formation of recirculation zones (sometimes referred to as eddies) downstream (i.e., distal) of a sudden expansion. Such eddies can dissipate considerable energy.

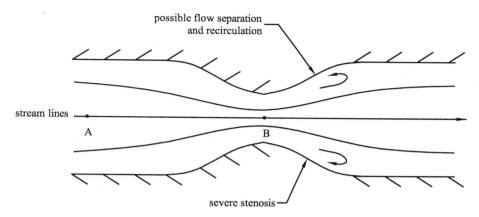

FIGURE 8.11 Similar to that in Figure 8.10 except for flow through a stenosis.

a converging geometry upstream (proximal) but a diverging geometry downstream (distal). Hence, there is a possibility of complex flows, particularly *flow separation* and recirculation zones just distal to the stenosis (Fig. 8.11). Therefore, Bernoulli's equation should not be used across a severe stenosis, although it may be used to estimate the maximum velocity in the stenosis, given proximal data. Bernoulli may sometimes be used in cases of gentle bends, although complex *secondary flows* can develop in curved tubes, which disallow the use of Bernoulli (Fig. 8.12). Likewise, Bernoulli may be used for internal flows entering a rounded entrance (converging flow), but a sharp entrance may disturb the flow and disallow its use. Finally, Bernoulli cannot be applied across a pump or propeller. Hence, Bernoulli could not be used to compare inlet and outlet velocities for an intravascular ventricular-assist device (IVAD). In this case, the semi-empirical methods of Chapter 10 would be an appropriate first approximation.

Example 8.6 Determine the time t_f it takes for a cylindrical container with a small central hole to drain.

Solution: Referring to Fig. 8.13, let us consider a streamline from the free surface, at 1, to the drain, at 2. Assuming an atmospheric pressure at 1 and 2, Bernoulli's equation reduces to

$$gh + \frac{1}{2}v_1^2 = \frac{1}{2}v_2^2,$$

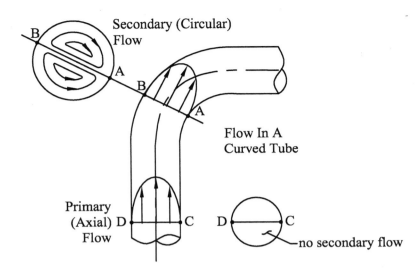

FIGURE 8.12 Secondary flows develop in curved tubes and are characterized by components of the velocity in the circumferential as well as the axial direction.

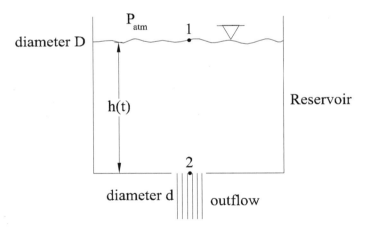

FIGURE 8.13 Fluid draining from a reservoir through a centrally located bottom hole.

or

$$v_2^2 - v_1^2 = 2gh(t),$$

where we emphasize that h varies with time t. Mass balance gives $v_1 A_1 = v_2 A_2$; thus,

$$v_2 = v_1 \frac{\pi D^2/4}{\pi d^2/4} = v_1 \frac{D^2}{d^2}$$

and, therefore,

$$v_1^2\left(\frac{D^4}{d^4} - 1\right) = 2gh(t) \rightarrow v_1 = \sqrt{\frac{d^4 2gh(t)}{D^4 - d^4}}.$$

Now, we recognize that $v_1 = -dh/dt$ and, therefore,

$$\frac{1}{\sqrt{h(t)}}\left(-\frac{dh}{dt}\right) = \frac{d^2\sqrt{2g}}{\sqrt{D^4 - d^4}}.$$

Integrating with respect to time,

$$\int_H^0 \frac{1}{\sqrt{h}}\frac{dh}{dt}\,dt = \int_0^{t_f} \frac{-d^2\sqrt{2g}}{\sqrt{D^4 - d^4}}\,dt,$$

or

$$-2\sqrt{H} = \frac{-d^2\sqrt{2g}}{\sqrt{D^4 - d^4}}t_f;$$

thus,

$$t_f = 2\sqrt{\frac{H(D^4 - d^4)}{2gd^4}} = \sqrt{\frac{2H(D^4 - d^4)}{gd^4}}.$$

Example 8.7 Evaluate the pressure difference between points A and B in Fig. 8.11. Assume aortic values such that $\bar{v}$ at A is 0.15 m/s, that the diameter at A is 0.03 m, and that the diameter at B is 0.01 m. Assume that $\rho = 1060\,\text{kg/m}^3$.

Solution: Although Bernoulli should not be used across a sudden expansion, it can be used along a central streamline between sections at A and B. Bernoulli becomes

$$\frac{p_1}{\rho} + \frac{1}{2}v_1^2 = \frac{p_2}{\rho} + \frac{1}{2}v_2^2,$$

where $v_1 A_1 = v_2 A_2$. Hence, the pressure difference is

$$p_1 - p_2 = \frac{1}{2}\rho(v_2^2 - v_1^2) = \frac{1}{2}\rho v_1^2\left[\left(\frac{A_1}{A_2}\right)^2 - 1\right] = \frac{1}{2}\rho v_1^2\left[\left(\frac{\pi d_1^2}{\pi d_2^2}\right)^2 - 1\right],$$

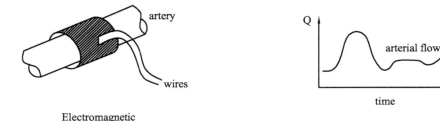

FIGURE 8.14 Schema of the time-varying volumetric flow rate Q measured in vivo using an electromagnetic flowmeter.

or

$$p_1 - p_2 = \frac{1}{2}\left(1060\,\frac{kg}{m^3}\right)\left(0.15\,\frac{m}{s}\right)^2\left[\left(\frac{0.03\,m}{0.01m}\right)^4 - 1\right]$$

$$= 954\,\frac{kg}{ms^2} = 954\left(kg\,\frac{m}{s^2}\right)\Big/m^2 = 954\,Pa,$$

where $7.5\,mm\,Hg = 1\,kPa$; hence, $p_1 - p_2 = 7.155\,mm\,Hg$. Note that pressures can be measured chronically in animals using indwelling catheters whereas flows are often measured with implanted flowmeters (e.g., Fig. 8.14).

Example 8.8 Under what conditions can you compute the pressure in the system in Figure 8.15? Recall that streamlines must be parallel and straight, where the radius of curvature is infinity, in order for $\partial p/\partial n = 0$. Consider multiple possibilities.

Solution 1: For flow along a streamline between points 2 and 4,

$$\frac{p_2}{\rho} + gz_2 + \frac{1}{2}v_2^2 = \frac{p_4}{\rho} + gz_4 + \frac{1}{2}v_4^2.$$

From overall mass balance, $v_2A_2 = v_4A_4$. For the pipe from point 2 to point 4, $A_2 = A_4$; therefore, $v_2 = v_4$. With $v_2 = v_4$ and $z_2 = z_4$, Bernoulli's equation suggests that the pressures p_2 and p_4 are equal. Because we discharge to atmospheric pressure at an assumed subsonic velocity, $p_4 = 0$ (gauge) and the pressure at point 2 is also predicted to be zero. Would we expect this to be the case particularly given that we are driving the flow only via a pressure gradient? Recall that Bernoulli should not be used over long distances.

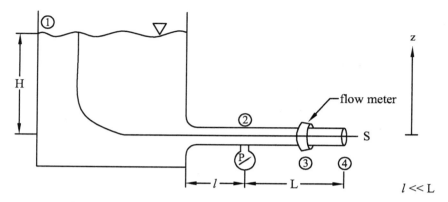

FIGURE 8.15 Flow from a reservoir through a segment of rigid tubing.

Solution 2: For flow along a streamline between points 1 and 2, assuming a rounded entrance at the chamber–tube interface,

$$\frac{p_1}{\rho} + gz_1 + \frac{1}{2}v_1^2 = \frac{p_2}{\rho} + gz_2 + \frac{1}{2}v_2^2,$$

where the pressure at point 1 is zero (gauge) and $v_1 \ll v_2$ if $A_1 \gg A_2$; thus,

$$gz_1 = \frac{p_2}{\rho} + gz_2 + \frac{1}{2}v_2^2 \rightarrow p_2 = \rho gH - \frac{1}{2}\rho v_2^2,$$

where v_2 is nonzero and equal to v_3 (which is measured via the flowmeter) by mass balance. We observe, therefore, that one application of Bernoulli suggests that $p_2 = 0$, whereas another yields $p_2 = \rho gH - \frac{1}{2}\rho v_2^2$. Bernoulli is applicable across contractions and short distances.

Example 8.9 A forced vortex flow is given by $v = r\omega_o \hat{e}_\theta$, where ω_o is constant (Fig. 8.16). Determine if Bernoulli's equation can be used to determine the pressure difference between two radial locations. Ignore gravity.

Solution: For $v = v_r \hat{e}_r + v_\theta \hat{e}_\theta + v_z \hat{e}_z$, we have

$$v_r = 0, \quad v_\theta = r\omega_o, \quad v_z = 0.$$

To use the Bernoulli equation, several assumptions must be met:

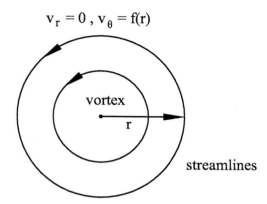

$$v_r = 0, v_\theta = f(r)$$

FIGURE 8.16 Schema of a vortex flow. In a forced vortex, $v_\theta = r\omega$, whereas in a free vortex, $v_\theta = c/r$, where c is a constant. What is the vorticity for each?

1. The flow must be steady. Checking in each direction, we get

$$\frac{\partial}{\partial t}(v_r) = 0, \quad \frac{\partial}{\partial t}(r\omega_o) = 0, \quad \frac{\partial}{\partial t}(v_z) = 0;$$

thus, this restriction is satisfied.
2. The fluid must be incompressible. For cylindrical coordinates,

$$\nabla \cdot v = \frac{1}{r}\frac{\partial}{\partial r}(rv_r) + \frac{1}{r}\frac{\partial}{\partial \theta}(v_\theta) + \frac{\partial}{\partial z}(v_z).$$

Checking $\nabla \cdot v = 0$, we get

$$\frac{1}{r}\frac{\partial}{\partial r}(r(0)) + \frac{1}{r}\frac{\partial}{\partial \theta}(r\omega_o) + \frac{\partial}{\partial z}(0) = 0;$$

thus, this restriction is satisfied.
3. Under certain situations, we can assume that the fluid is inviscid. The validity of this assumption must be established via experience.
4. The flow must be along a streamline or it must be irrotational, where $\nabla \times v = \mathbf{0}$. For cylindrical coordinates, $\nabla \times v = \mathbf{0}$ is given by

$$\left(\frac{1}{r}\frac{\partial v_z}{\partial \theta} - \frac{\partial v_\theta}{\partial z}\right)\hat{e}_r + \left(\frac{\partial v_r}{\partial z} - \frac{\partial v_z}{\partial r}\right)\hat{e}_\theta + \left(\frac{1}{r}\frac{\partial(rv_\theta)}{\partial r} - \frac{1}{r}\frac{\partial v_r}{\partial \theta}\right)\hat{e}_z = \mathbf{0}.$$

With $v_r = v_z = 0$, we have

$$\nabla \times v = \left(-\frac{\partial v_\theta}{\partial z}\right)\hat{e}_r + \left(\frac{1}{r}\frac{\partial(rv_\theta)}{\partial r}\right)\hat{e}_z = \frac{1}{r}\frac{\partial(r^2\omega_o)}{\partial r}\hat{e}_z$$

or, in the z direction,

$$\frac{1}{r}(2r\omega_o) = 2\omega_o.$$

Therefore, the flow is not irrotational, and Bernoulli's equation cannot be used unless along a streamline.

8.6 Measurement of Pressure and Flow

One of the most important advances in the development of the modern method of scientific investigation was the realization (by Galileo and others) that theory and experiment must go hand in hand. Theory is needed to design and interpret experiments, which, in turn, are needed to test theories. Experimentation often involves the identification of specific functional relationships between the dependent and independent variables that theory establishes to be important, as well as the calculation of the numerical values of the associated material parameters. Recall from Figure 1.9 of Chapter 1 that theories, like hypotheses, are motivated by basic observations. Both observation and experimentation require measurements.

Measurement implies that we assign a numerical value to a quantity, often via a comparison to some standard. For example, if we desire to measure the length of an object, we may choose to quantify the length in terms of meters, where 1 m = 1,650,763.73 wavelengths of the orange–red radiation of krypton-86 in a vacuum. Standard weights and measures are kept by governmental agencies such as the National Institute for Standards and Technology (NIST) in the United States and the International Bureau of Weights and Measures in Sèvres, France.

In the modern laboratory, most systems for measurement consist of three components: a transducer, a signal conditioner, and a recorder. A *transducer* is simply any device that converts a physical quantity of interest into another quantity that is more easily measured. Perhaps the simplest transducer is the mercury thermometer, which "converts" temperature (thermal energy) into the displacement of a column of mercury that is easily measured against a ruled background. Most modern transducers convert physical quantities into electrical outputs, either a voltage or a current. A signal conditioner often consists of a combination of amplifiers and filters. Amplifiers modify the range of a signal, whereas filters remove unwanted portions of a signal. A recorder may be any device that archives the measurement; it may take various forms, including a still camera or video-cassette recorder (VCR), but most often includes an analog-to-digital (A/D) converter and digital memory. Amplifiers, conditioners, and recorders are discussed in detail in courses on instrumentation. Here, let us focus on a few basic transducers.

8.6.1 Pressure

A pressure is a net force per unit area that acts normal to and into a surface area. Stephen Hales was apparently the first, in 1733, to measure the pressure in an artery under "normal" conditions. Specifically, Hales inserted a small-diameter vertical tube into the carotid artery of a horse and recorded the height to which the blood rose in the tube. A simple free-body diagram of such an experiment reveals that the blood was acted upon by atmospheric pressure p_0 from above and arterial pressure p_a from below. With $p_a > p_0$, the net vertical force due to these pressures, $(p_a - p_0)\pi a^2$, where a is the inner radius of the tube, balanced the weight of the column of blood, $W = \rho g(\pi a^2 h)$ where h is the height. Hence, this simple transducer (tube) allowed the gauge pressure $p = p_a - p_0$ to be inferred simply in terms of the height that the blood rose ($p = \rho g h$), as we know from fluid statics. Albeit the first method of measurement, this clearly is not the easiest. (Note: The blood could easily reach a value of 2 m in an excited animal.)

A major advancement in the measurement of blood pressure, therefore, was the use of a U-shaped mercury manometer by Poiseuille in 1828. The principle of operation of a U-tube manometer is very simple (Fig. 8.17). Mercury proved convenient because of its high density: $SG_{Hg} = \rho_{Hg}/\rho_{H_2O}(4°C) = 13.55$, where $\rho_{H_2O}(4°C) = 1000 \, \text{kg/m}^3$. Why? Of course, blood pressure (e.g., 120/80) continues to be measured by physicians using the units mm Hg (where 7.5 mm Hg = 1 kPa).

Electrical-based resistance strain gauges were first used in the physiologic measurement of blood pressure in 1947. Briefly, the fluid pressure elastically deformed a thin metal diaphragm within the transducer, the deformation of which was measured by a strain gauge and calibrated. Hence, an

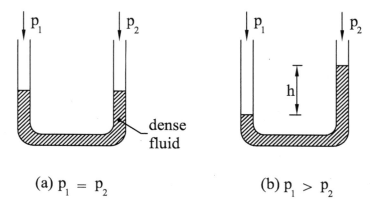

(a) $p_1 = p_2$ (b) $p_1 > p_2$

FIGURE 8.17 A U-tube manometer. Relative differences in height between the fluid in the two tubes indicates a difference in pressure acting on each column of fluid.

analysis similar to that of LEHI beam bending (Chapter 5) allowed the design of such transducers. Although strain gauge transducers are still used, the ability to use miniature piezoelectric crystals or fiber optics in catheters has revolutionized in vivo measurements (see, e.g., the website for Millar Instruments in Houston, TX). A piezoelectric material is one that generates an electrical output in direct response to an applied load. For more on physiologic measurements and, in particular, the need for adequate frequency responses, see Chapter 11 of Milnor (1989).

8.6.2 Flow

The history of measuring physiological flows dates back to at least 1628 and Harvey, but recent advances in technology have revolutionized the field. Nevertheless, let us consider a simple, theoretically motivated method. Noting from the previous subsection that static pressures are easy to measure, let us exploit Bernoulli's equation (i.e., a theory) to design a device to measure (i.e., perform an experiment or make an observation) the velocity of a flowing fluid. If we consider a horizontal streamline, then Bernoulli's equation (8.71) becomes

$$\frac{p_1}{\rho} + \frac{1}{2}v_1^2 = \frac{p_2}{\rho} + \frac{1}{2}v_2^2. \tag{8.84}$$

Now, let us define the so-called stagnation pressure p_0. A *stagnation pressure* is that value of pressure at a point in a flow field where the fluid is decelerated to zero velocity due to nonviscous effects. Hence, from Bernoulli, we see that if point 2 is a stagnation point, then

$$\frac{p_1}{\rho} + \frac{1}{2}v_1^2 = \frac{p_0}{\rho} \rightarrow v_1 = \sqrt{\frac{2(p_0 - p_1)}{\rho}}. \tag{8.85}$$

In other words, we can infer the velocity by measuring a pressure difference along a streamline for a fluid of known, constant mass density ρ. A possible experimental setup to exploit this theoretical result is shown in Figure 8.18a. Recall that $\partial p/\partial n = 0$ if the streamlines are parallel [Eq. (8.64)] in the absence of gravity and with $R \rightarrow \infty$; hence, a wall tap can measure the pressure at point 1, whereas a tube filled with an incompressible fluid/gel will stop the flow at point 2 and thus create a stagnation point. The difference in pressures $p_0 - p_1$ can thus be measured simply by a U-shaped manometer and we see again that theory guides the design of many transducers. Shown in Figure 8.18b is a pitot-static tube (pronounced pea-toe), which is designed based on a similar idea and assuming that the thin tube (~0.0625 in. diameter) does not disturb the flow significantly. Indeed, another method of measuring a flow velocity is to use a heated wire. The rate of cooling of the wire can be related to the velocity of the flow; actually, one measures the current supplied to the wire to maintain it at

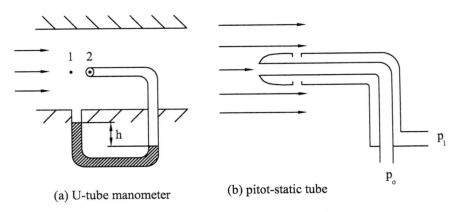

(a) U-tube manometer   (b) pitot-static tube

FIGURE 8.18 A pitot-static tube, which allows velocity to be inferred via the simpler measurement of pressure. The motivation for the simple device lies in the theory (Bernoulli equation) and reminds us that theory should always guide the design and interpretation of experiments.

a constant temperature. Such *hot-wire* anemometers are commercially available as small as 0.02 mm in diameter and 0.1 mm long, with a 50-kHz frequency response.

Whereas the pitot-static tube and hot-wire devices measure the velocity at a point, many physiological and clinical situations necessitate that one simply measure the volumetric flow rate Q. An important advance in this regard was the *electromagnetic flowmeter*, developed between 1968 and 1974. Briefly, these devices are based on the fundamental discoveries of Michael Faraday (around 1832) that the motion of an electrically conductive material within a magnetic field generates an electromotive force [see Milnor (1989) for further details]. These flowmeters must be surgically placed around the vessel directly (Fig. 8.14), and with calibration, the output signal is related to the mean flow. Blood is electrically conductive, of course, because of the many ions within.

In some cases, of course, one may wish to simply know qualitative characteristics about the flow field rather than quantitative information. Visualizing flows can be as simple as placing floats on the surface and watching their motion or similarly seeding a flow field with neutrally buoyant fluorescent markers and imaging their motions. Another method is to inject a dye into the flow field (e.g., Fig. 8.19); indeed the common diagnostic tool of angiography uses X-rays to image the motion of a radio-opaque contrast agent that is injected into the bloodstream. Angiography remains the primary method for diagnosing coarctations, aneurysms, and obstructive atherosclerotic lesions (Fig. 8.20).

Advances in technology have led to many additional, sophisticated methods for quantifying velocities and flows. Ultrasonic (1–8 MHz) trans-

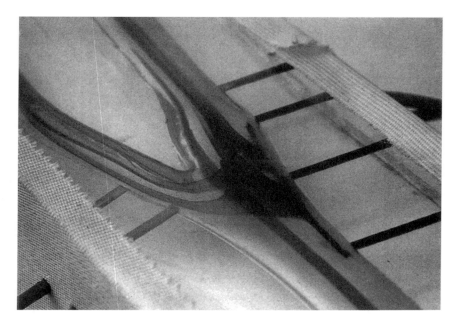

FIGURE 8.19 Visualization of the flow through a model carotid bifurcation. Colored dye is introduced into the flow stream, which allows pathlines to be visualized. In steady flows, pathlines and streamlines coincide. Albeit not quantitative, flow visualization can provide important clues into important aspects of a flow, which, in turn, allow us to focus theoretically or computationally on that which is important. With permission from Lippincott Williams & Wilkins.

ducers and laser Doppler anemometers (LDA) both rely on the Doppler shift (i.e., the frequency shift experienced by waves when the distance between the generator and receiver changes). In the LDA, for example, one focuses a laser beam on a point (i.e., small volume) in the flow field, which scatters when it hits indigenous or seeded particles in the flow. A frequency shift between the scattered and reference light is proportional to the velocity of the scatterer. LDA is widely used in the laboratory to study the complex flow fields within tapering, branching models of the vasculature or airways. Clinically, Doppler ultrasound and magnetic resonance angiography are powerful tools for noninvasively measuring local flows. The interested student is encouraged to research these modalities further.

8.7 Navier–Stokes Worksheets

One quickly discovers that the solution of the Navier–Stokes and Euler equations for many different problems and different coordinate systems follow the same steps. Consequently, we have found it useful to use "work-

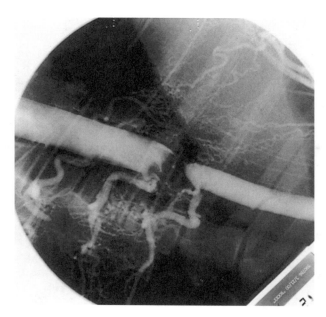

FIGURE 8.20 An aortogram (X-ray) from an experimental model of aortic coarctation, which is induced by inflating an occluding balloon around the aorta. Coarctation models are used to induce and then study hypertension proximal to the occlusion. Note the extensive development of collateral vessels to shunt blood around the obstruction and thereby respond to the insult. The development of new blood vessels is called angiogenesis, which is an important area of current research, as it relates to cancer (tumors develop vessels to supply nutrients and oxygen), tissue engineering (tissues which need to be fed in vivo if they are to survive post-implantation), and recovery from severe injuries such as a myocardial infarction. Angiogenesis research requires input from both biosolid and biofluid mechanics. (Courtesy Dr. M. Miller, Texas A&M University).

sheets" to formulate such problems in a consistent way. These worksheets guide us through the identification of the physical problem (e.g., via free-body diagrams) and the identification of appropriate assumptions such as steady flow ($\partial \mathbf{v}/\partial t = \mathbf{0}$), axisymmetric flow ($\partial \mathbf{v}/\partial \theta = \mathbf{0}$), no body forces ($\mathbf{g} = \mathbf{0}$), and so forth. Listing such assumptions and then identifying the terms within the mass balance and Navier–Stokes (or Euler) equations that drop out accordingly allow us to find the reduced differential equations that require solution, subject to appropriate initial-boundary conditions. We encourage the student to make multiple photocopies of these worksheets, which can then be used to solve the problems of interest.

Navier–Stokes Worksheet (Cartesians)

Problem Statement:

Assumptions:
1. Newtonian 3. 5. 7. 9.
2. Incompressible 4. 6. 8. 10.

Mass Balance: $\nabla \cdot \mathbf{v} = 0$:

$$\frac{\partial v_x}{\partial x} + \frac{\partial v_y}{\partial y} + \frac{\partial v_z}{\partial z} = 0$$

Linear Momentum: $-\nabla p + \mu \nabla^2 \mathbf{v} + \rho \mathbf{g} = \rho \mathbf{a}$:

$$-\frac{\partial p}{\partial x} + \mu\left(\frac{\partial^2 v_x}{\partial x^2} + \frac{\partial^2 v_x}{\partial y^2} + \frac{\partial^2 v_x}{\partial z^2}\right) + \rho g_x = \rho\left(\frac{\partial v_x}{\partial t} + v_x \frac{\partial v_x}{\partial x} + v_y \frac{\partial v_x}{\partial y} + v_z \frac{\partial v_x}{\partial z}\right),$$

$$-\frac{\partial p}{\partial y} + \mu\left(\frac{\partial^2 v_y}{\partial x^2} + \frac{\partial^2 v_y}{\partial y^2} + \frac{\partial^2 v_y}{\partial z^2}\right) + \rho g_y = \rho\left(\frac{\partial v_y}{\partial t} + v_x \frac{\partial v_y}{\partial x} + v_y \frac{\partial v_y}{\partial y} + v_z \frac{\partial v_y}{\partial z}\right),$$

$$-\frac{\partial p}{\partial z} + \mu\left(\frac{\partial^2 v_z}{\partial x^2} + \frac{\partial^2 v_z}{\partial y^2} + \frac{\partial^2 v_z}{\partial z^2}\right) + \rho g_z = \rho\left(\frac{\partial v_z}{\partial t} + v_x \frac{\partial v_z}{\partial x} + v_y \frac{\partial v_z}{\partial y} + v_z \frac{\partial v_z}{\partial z}\right)$$

Reduced Governing Differential Equations:

Boundary/Initial Conditions:

Navier–Stokes Worksheet (Cylindricals)

Problem Statement:

Assumptions:
1. Newtonian   3.   5.   7.   9.
2. Incompressible   4.   6.   8.   10.

Mass Balance: $\nabla \cdot \mathbf{v} = 0$:

$$\frac{1}{r}\frac{\partial}{\partial r}(rv_r) + \frac{1}{r}\frac{\partial v_\theta}{\partial \theta} + \frac{\partial v_z}{\partial z} = 0$$

Linear Momentum: $-\nabla p + \mu\nabla^2\mathbf{v} + \rho\mathbf{g} = \rho\mathbf{a}$:

$$-\frac{\partial p}{\partial r} + \mu\left[\frac{\partial}{\partial r}\left(\frac{1}{r}\frac{\partial(rv_r)}{\partial r}\right) + \frac{1}{r^2}\frac{\partial^2 v_r}{\partial \theta^2} - \frac{2}{r^2}\frac{\partial v_\theta}{\partial \theta} + \frac{\partial^2 v_r}{\partial z^2}\right] + \rho g_r$$

$$= \rho\left(\frac{\partial v_r}{\partial t} + v_r\frac{\partial v_r}{\partial r} + \frac{v_\theta}{r}\frac{\partial v_r}{\partial \theta} - \frac{v_\theta^2}{r} + v_z\frac{\partial v_r}{\partial z}\right),$$

$$-\frac{1}{r}\frac{\partial p}{\partial \theta} + \mu\left[\frac{\partial}{\partial r}\left(\frac{1}{r}\frac{\partial(rv_\theta)}{\partial r}\right) + \frac{1}{r^2}\frac{\partial^2 v_\theta}{\partial \theta^2} + \frac{2}{r^2}\frac{\partial v_r}{\partial \theta} + \frac{\partial^2 v_\theta}{\partial z^2}\right] + \rho g_\theta$$

$$= \rho\left(\frac{\partial v_\theta}{\partial t} + v_r\frac{\partial v_\theta}{\partial r} + \frac{v_\theta}{r}\frac{\partial v_\theta}{\partial \theta} + \frac{v_r v_\theta}{r} + v_z\frac{\partial v_\theta}{\partial z}\right)$$

$$-\frac{\partial p}{\partial z} + \mu\left[\frac{1}{r}\frac{\partial}{\partial r}\left(r\frac{\partial v_z}{\partial r}\right) + \frac{1}{r^2}\frac{\partial^2 v_z}{\partial \theta^2} + \frac{\partial^2 v_z}{\partial z^2}\right] + \rho g_z$$

$$= \rho\left(\frac{\partial v_z}{\partial t} + v_r\frac{\partial v_z}{\partial r} + \frac{v_\theta}{r}\frac{\partial v_z}{\partial \theta} + v_z\frac{\partial v_z}{\partial z}\right)$$

Reduced Governing Differential Equations:

Boundary/Initial Conditions:

Appendix 8: Differential Equations

Albeit not without controversy and debate (Boyer, 1949; Bell, 1986), it is generally agreed that Sir I. Newton invented the basic ideas of calculus and that he was so motivated largely by problems of mechanics. The two basic areas of this subject are, of course, the differential and the integral calculus.

Differential equations allow us to determine how quantities of interest (dependent variables) vary in space and time (independent variables). Such equations can depend on but one independent variable (yielding an ordinary differential equation) or they can simultaneously depend on multiple independent variables (thus yielding partial differential equations); they can appear singly or as systems of equations that must be solved simultaneously; and they can be linear or nonlinear. There is, therefore, great motivation for the biomechanicist to be well versed in methods of solving differential equations and the student is well advised to complete multiple courses in this important area.

Although we see that the Navier–Stokes and Euler equations of motion, in combination with mass balance, represent coupled nonlinear partial differential equations (PDEs), we will consider only simple cases herein and thereby focus primarily on linear ordinary differential equations (ODEs). For example, consider a simple ODE of the form

$$\frac{d^n}{dx^n}(f(x)) = g(x). \tag{A8.1}$$

Such equations arise frequently, as in Chapters 5 and 9, particularly when $n = 2$ or 4. The best way to solve such equations is directly via integration. Note, therefore, that Eq. (A8.1) can be written as

$$\frac{d}{dx}\left(\frac{d^{n-1}}{dx^{n-1}}(f(x))\right) = g(x) \rightarrow \frac{d}{dx}(\text{something}) = g(x), \tag{A8.2}$$

whereby we can integrate with respect to x to obtain

$$\int \frac{d}{dx}(\text{something})dx = \int g(x)dx \rightarrow \text{something} = \int g(x)dx + c. \tag{A8.3}$$

The constant of integration c requires additional information for solution. If the integration is with respect to a spatial variable, we say that we need a *boundary condition* to find c; if the integration is with respect to time, we say that we need an *initial condition* (i.e., a condition at the time of initiation of the process, usually at time $t = 0$ or perhaps $t = -\infty$).

Note from Eq. (A8.3) that the form d/dx (*something*) dx permits a simple integration; hence, we should try to put ODEs in this form whenever possible. For example, note that

$$\frac{df}{dx} + \frac{1}{x} f(x) \equiv \frac{1}{x} \frac{d}{dx}(x f(x)); \qquad (A8.4)$$

hence, if we have

$$\frac{df}{dx} + \frac{1}{x} f(x) = g(x) \rightarrow \frac{1}{x} \frac{d}{dx}(x f(x)) = g(x) \qquad (A8.5)$$

and multiplication by x permits a simple solution for $f(x)$,

$$\int \frac{d}{dx}(xf(x))dx \equiv xf(x) = \int xg(x)dx + c. \qquad (A8.6)$$

This form [Eq. (A8.4)] occurs frequently in cylindrical coordinates, with x replaced by the radial coordinate r. A similar situation arises if

$$\frac{df}{dx} + \frac{2}{x} f(x) \equiv \frac{1}{x^2} \frac{d}{dx}(x^2 f(x)). \qquad (A8.7)$$

Multiplication by x^2 thus yields d/dx (*something*), which is easily integrated. Regardless of the form, the direct integration of an ODE reduces the problem to one of integral calculus, and integration tables for $\int g(x)dx$, $\int xg(x)dx$, and so forth become very useful, as do methods such as integration by parts:

$$\int_a^b udv = uv\Big|_a^b - \int_a^b vdu. \qquad (A8.8)$$

Of course, not all ODEs can be put in a simple form to allow direct integration. Another commonly encountered form in mathematics is the linear, second-order ODE with constant coefficients:

$$\frac{d^2 f}{dx^2} + a_1 \frac{df}{dx} + a_2 f = 0. \qquad (A8.9)$$

Experience reveals that such equations admit exponential solutions of the form $f(x) \propto e^{\lambda x}$. Because the equation is linear, we know that its solution is unique. Hence, if we can find any solution (e.g., by trial and error, by guessing, etc.), then we will have found THE solution. If we guess that $f(x) \propto e^{\lambda x}$, then Eq. (A8.9) becomes

$$\lambda^2 e^{\lambda x} + a_1 \lambda e^{\lambda x} + a_2 e^{\lambda x} = (\lambda^2 + a_1\lambda + a_2)e^{\lambda x} = 0 \quad \forall x. \qquad (A8.10)$$

Clearly, this equation is satisfied for all x by ensuring that $\lambda^2 + a_1\lambda + a_2 = 0$, which is a (simple) quadratic equation in λ. Hence, whereas the method of Eq. (A8.3) reduces the differential equation to a problem of integral calculus, here we have reduced it to one of algebra, noting that the solution of our quadratic equation is

$$\lambda^2 + a_1\lambda + a_2 = 0 \rightarrow \lambda_{1,2} = \frac{-a_1 \pm \sqrt{a_1^2 - 4a_2}}{2}. \qquad \text{(A8.11)}$$

Clearly, there are three possible types of solution. (1) The values of λ_1 and λ_2 can be real and distinct, whereby

$$f(x) = c_1 e^{\lambda_1 x} + c_2 e^{\lambda_2 x}, \qquad \text{(A8.12)}$$

where c_1 and c_2 are arbitrary constants (like the constant of integration in equation A8.6) that must be determined via boundary or initial conditions. (2) The values of λ_1 and λ_2 can be real and equal, whereby

$$f(x) = c_1 e^{\lambda x} + c_2 x e^{\lambda x} \quad (\lambda \equiv \lambda_1 = \lambda_2). \qquad \text{(A8.13)}$$

Finally, (3) the values of λ_1 and λ_2 can be complex conjugates (with a_1 and a_2 real), whereby

$$f(x) = d_1 e^{(a+ib)x} + d_2 e^{(a-ib)x}. \qquad \text{(A8.14)}$$

Using Euler's relations, however,

$$e^{ibx} = \cos bx + i \sin bx, \qquad e^{-ibx} = \cos bx - i \sin bx, \qquad \text{(A8.15)}$$

we can alternatively write the solution as

$$f(x) = d_1(\cos bx + i \sin bx)e^{ax} + d_2(\cos bx - i \sin bx)e^{ax} \qquad \text{(A8.16)}$$

or by defining $c_1 = d_1 + d_2$ and $c_2 = (d_1 - d_2)i$, we have

$$f(x) = e^{ax}(c_1 \cos bx + c_2 \sin bx). \qquad \text{(A8.17)}$$

In summary, it cannot be overemphasized that the student of biomechanics must be well versed in the methods of applied mathematics, including differential equations, which requires undergraduate and graduate courses beyond the basic 2-year sequence required of all students in engineering. This appendix merely addressed two of the simpler cases encountered in introductory problems.

Exercises

8.1 Derive the equation of motion [Eq. (8.23)] for the z direction.

8.2 Derive the incompressible Navier–Stokes equation (8.42) for the z direction taking into account the contributions from the local and convective acceleration.

8.3 The general incompressible Navier–Stokes equation can be written as $-\nabla p + \mu \nabla^2 v + \rho g = \rho a$. (a) Write the equation that governs fluid statics and (b) use it to find the relationship among the pressure p, the density ρ, and gravity g in a beaker of water.

8.4 Given the vectorial form of the incompressible Navier–Stokes equation, find the r-, θ-, and z-direction equations in cylindrical coordinates. Hint: Remember that the base vectors change with direction in cylindrical coordinates (see Exercise 7.5).

8.5 Repeat Exercise 8.4 for spherical coordinates (see Exercise 7.6). This is nontrivial.

8.6 An incompressible fluid flows through the device shown in Figure 8.21. If the pressure at the two gauges is the same, then find the value of the diameter at section 2 given that the diameter at section 1 is 2 cm, the velocity at section 1 is 10 cm/s, and the height h is 10 cm.

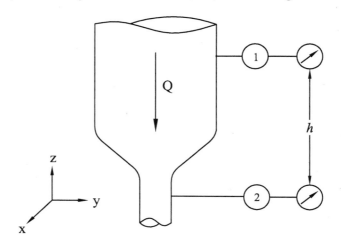

FIGURE 8.21

8.7 Derive the streamline direction Euler equation for the case where the streamlines are horizontal and all parallel; assume an arbitrary body force vector g, which may include gravity and electromagnetic effects (see figure).

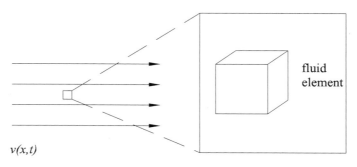

FIGURE 8.22

8.8 Explain why the pressure is constant across straight-parallel stream-lines if the fluid is ideal and the flow is steady. Assume no gravity.

8.9 Specific gravity SG is defined as

$$SG = \frac{\rho}{\rho_{H_2O} \text{ at } 4°C},$$

where ρ_{H_2O} at $4°C$ is $1000\,kg/m^3$. Compute the mass density for the following fluids at room temperature ($20°C$) if SG = 1.26 (glycerin), SG = 13.55 (mercury), SG = 1.025 (seawater), and SG = 0.998 (water).

8.10 Glycerin exits a pipe at a mean velocity of $v_o = 1\,m/s$ and rises to a height $h = 2\,m$ as in Figure 8.8. Find the value of the exit pressure assuming gravity $g = 9.81\,m/s^2$ $(-\hat{j})$ acts down and the pipe is vertical.

8.11 Water flows through a 90° elbow of a water slide that is open on the top to the atmosphere (Fig. 8.23). Assume steady, ideal, irrotational flow and that $v_s = c/r$ in the bend, where c is constant. For $b > a$, and including gravity, determine whether the flowing fluid will be deeper on the inside (i.e., at $r = a$) or outside (at $r = b$) in the bend.

Inlet Flow

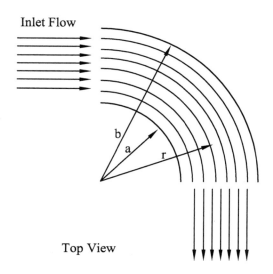

Top View

FIGURE 8.23

8.12 Can Bernoulli's equation be used to relate the pressures and veloci-ties at any two points in a flow field given $v_x = x + y$, $v_y = x - y$, $v_z = 0$ plus negligible viscosity and negligible body forces? Why?

8.13 A dentist uses a device similar to that in Figure 8.24. If the supply volumetric flow rate is Q, and the nozzle tip is d in diameter, what is the exit velocity.

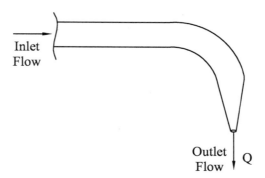

FIGURE 8.24

8.14 Rederive Archimedes' principle if the immersed solid is a cube of length a.

8.15 Rederive Archimedes' principle if the immersed solid is of arbitrary shape.

8.16 Compare the buoyant force on a ring of artery versus a ring of silicone rubber. Assume that each is 4 mm in internal diameter, 0.8 mm in thickness, and 3 mm in length. Compute the volume of Styrofoam that would need to be glued to each ring to render it neutrally buoyant. Assume that the density of an artery is 1050 kg/m^3, that of silicone is 1500 kg/m^3, and that of Styrofoam is 40 kg/m^3.

8.17 If the density of blood is ~1060 kg/m^3 and the blood rose 2 m in Hales' experiment on blood pressure in a horse, what was the arterial blood pressure (gauge) in kPa and mm Hg, where 7.5 mm Hg = 1 kPa and 1 Pa = 1 N/m^2 (1 N = 1 kg m/s^2).

8.18 If blood pressure is 120 mm Hg as measured using a mercury manometer, what would the value be using a water manometer (i.e., ×mm H$_2$O)? Recall that $SG_{Hg} = 13.55$.

8.19 The *venturi meter* is a commonly used device in engineering to measure internal flows. Research this device and rederive the requisite equations that guide its design and use.

8.20 Show that

$$\nabla^2 v = \nabla(\nabla \cdot v) - \nabla \times (\nabla \times v).$$

8.21 Given that

$$\nabla \cdot v = \mathrm{tr}[D]$$

is true regardless of coordinates, show that it is true for sphericals see [Eq. (7.60)].

8.22 Take two sheets of paper and hold them closely together while blowing air between them. What happens? Does the forced flow between the paper cause the sheets to move farther apart or closer together? Use Bernoulli's equation to explain your observation.

8.23 Bernoulli's equation is often used to explain *lift* of an airfoil (i.e., an airplane wing). In particular, airfoils are designed such that the distance from the leading edge to the trailing edge is longer on the top than on the bottom surface. Consequently, the air must move faster over the top surface (to reach the trailing edge at the same time as the air traveling along the bottom surface). Hence, from Bernoulli, we predict a lift (i.e., a pressure greater on the bottom than on the top surface).

Some have similarly argued that Bernoulli can be used to "explain" why red blood cells tend to move to the central region of an artery. Note, therefore, that it will be shown in Chapter 9 that the velocity field for a steady, incompressible flow in a circular tube is

$$v = c\left(1 - \frac{r^2}{a^2}\right)\hat{e}_z,$$

where c is constant, a is the inner radius of the tube, and r is a cylindrical-polar coordinate $r \in [o, a]$. This velocity field suggests that, similar to the airfoil, the velocity is higher on the side of the cell closest to the centerline. Is it reasonable to use Bernoulli's equation in this case and why? See Figure 8.25.

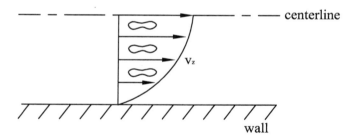

FIGURE 8.25

9
Some Exact Solutions

The Navier–Stokes equations are the most famous and perhaps the most important equations in fluid mechanics. In biofluid mechanics, these equations can be used to compute the flow of air within the airways, the flow of blood in large vessels at sufficiently high shear rates, the flow of urine from the bladder, the flow of crystalloid perfusates in in vitro experiments, and so on. Because few analytical solutions are available, one must often resort to numerical methods to solve these important equations. Nonetheless, in this chapter, we will consider five important analytical solutions to the Navier–Stokes equations.

Recall from Chapter 8 that the two governing differential equations for an incompressible, Newtonian flow are

$$\nabla \cdot \boldsymbol{v} = 0 \tag{9.1}$$

which enforces the balance of mass, and

$$-\nabla p + \mu \nabla^2 \boldsymbol{v} + \rho \boldsymbol{g} = \rho \left[\frac{\partial \boldsymbol{v}}{\partial t} + (\boldsymbol{v} \cdot \nabla)\boldsymbol{v} \right], \tag{9.2}$$

which enforces the balance of linear momentum. Again, $\boldsymbol{v}$ is the velocity, p is the fluid pressure, μ is the viscosity, ρ is the mass density, and $\boldsymbol{g}$ is the body force vector per unit mass; Eq. (9.2) is the so-called incompressible Navier–Stokes equation.

Here, we will consider three classes of incompressible flows: (1) in vivo flows, (2) in vitro flows in experiments that are useful in cell biology, and (3) in vitro flows that can be used to quantify the viscous behavior of particular Newtonian fluids. With regard to the first class of flows, note that several complicating conditions exist in large arteries and large airways, such as the pulsatility (i.e., unsteadiness) of the pressures as well as the distensibility of the tubes. These perhaps seemingly simple characterisitics add tremendous complexity to the formulation and solution of the associated initial-boundary value problems. In particular, computing flows within distensible tubes necessitates the solution of coupled solid–fluid mechanics problems. Such solutions are the topic of current research and generally beyond the scope of an introductory text; we address them only briefly in Chapter 11. Fortunately, there are still many in vivo situations for which it may be reasonable to assume a steady flow within a "rigid" tube. For example, some left ventricular-assist devices output a steady flow to the aorta, and aortic stiffness allows only small changes in radius in this situation. Likewise, venous flows are nearly steady and there is little change in the radius of the vessel at a given location. If we integrate over time (i.e., over the cardiac cycle), we also find that the time-averaged shear stress on the arterial wall is approximated reasonably well in some situations by the steady-flow solution. Hence, we will carefully consider a number of solutions for steady flow. Finally, we must remember that blood only exhibits a Newtonian response at high shear rates. There is a need, therefore, to consider non-Newtonian behavior as well, which we do briefly in the last section of this chapter.

9.1 Flow Between Parallel Flat Plates

9.1.1 Biological Motivation

Cellular function is so complex that it was realized many years ago that it can be very useful to study the response of isolated cells to well-controlled stimuli. For example, the endothelial cells that line the inner (luminal) surface of the vasculature are very sensitive to fluid-induced shear stresses. Large arteries appear to vasoconstrict or vasodilate, via smooth muscle contraction and relaxation, so as to maintain the wall shear stress $\tau_w \sim 1.5\,Pa$. This vasomotion is controlled, in part, by endothelial production of vasoactive molecules such as the vasodilators nitric oxide (NO) and prostacyclin (PGI_2) and the vasoconstrictors endothelin-1 (ET-1) and thromboxane (TXA_2). In general, as τ_w is increased by an increased flow (e.g., during exercise), the endothelium produces more vasodilators to increase the lumen and thereby restore the wall shear stress to its normal value; the converse occurs when τ_w is decreased by a decreased flow (e.g., due to a sedentary lifestyle). See Figure 9.1. Of course, many conditions can alter blood flow, including changes during pregnancy due to the presence of placental flow, changes in luminal geometry due to atherosclerosis, and the surgical creation of an arterio-venous fistula (or shunt) in kidney dialysis patients. An important research goal is to correlate the endothelial production of these various molecules with alterations in flow-induced shear stresses. To do this, *we need solutions of initial-boundary value problems that represent convenient experimental situations*, one of which is flow between rigid parallel plates.

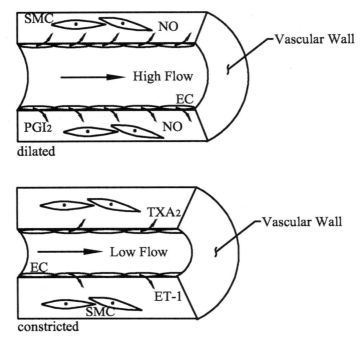

FIGURE 9.1 Schema showing the lumen and cross section of a vascular wall in response to two different altered stimuli: a sustained increase in flow, which increases the lumen, and a sustained decrease in flow, which decreases the lumen. It is not uncommon, for example, for marathon runners to have significantly larger iliac arteries due to the consistent elevation of blood flow because of training. It is thought that such changes in lumen maintain the wall shear stress at its homeostatic value. Albeit not shown, a sustained increase in pressure results in an increase in the thickness of the vascular wall. It is thought that this response maintains the circumferential wall stress at its homeostatic value.

9.1.2 Mathematical Formulation

A useful in vitro experiment is to culture a monolayer of endothelial cells on one of two parallel flat plates and then to subject the system to a pressure-driven flow (Fig. 9.2). Although the monolayer has a cobblestone or undulating appearance on a microscale, the cells being thicker in the region of their nucleus, we shall assume for our purposes that both the cell layer and the opposing plate are flat. This is a reasonable assumption when the variation in cell height is on the order of micrometers and the spacing between the plates is on the order of millimeters or centimeters. Moreover, we will assume that the flow is steady, incompressible, and Newtonian, thus

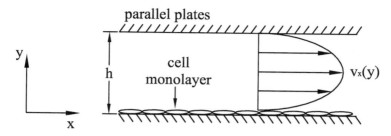

FIGURE 9.2 Steady flow of a Newtonian fluid (e.g., culture media) between two rigid, impermeable, flat plates that allows one to investigate mechanobiological responses of a monolayer of cells to changes in flow. Note the parabolic velocity profile, which is called a Poiseuille flow.

allowing us to formulate the experimental boundary value problem within the context of an exact solution of the Navier–Stokes equation. Indeed, because we are the ones who design the experiment, we can ensure the use of a fluid that exhibits an incompressible Newtonian response as well as a steady flow; that is, we can appropriately choose the culture media, the pump, and the geometry of the plates. Let us also assume that the effects of gravity on the flow field are negligible in comparison to the effects of the applied pressure gradient (i.e., difference between the upstream and downstream pressures), that the flow is unidirectional, and that the flow develops fully before reaching the cells. Hence, mathematically we have the following assumptions/restrictions:

1. Newtonian fluid (μ = constant)
2. Incompressible flow ($\nabla \cdot v = 0$)
3. Steady flow ($\partial v/\partial t = 0$)
4. Unidirectional flow ($v_y = v_z = 0$)
5. Negligible body forces ($g = 0$)
6. Fully developed flow ($\partial v/\partial x = 0$)
7. 1-D flow ($\partial v_x/\partial z = 0$, $\partial v_x/\partial x = 0$)

We recommend at this point that the student use the worksheet from Section 8.7 of Chapter 8 to follow the subsequent derivations.

Note that assumption 6 suggests that there are no "end effects"; that is, the flow is the same along the length of the test section (recall from Chapters 2 and 3 that we similarly avoided end effects in designing uniaxial experiments in biosolid mechanics for this simplified the analysis of the data). The constraint of an incompressible flow, in Cartesian coordinates, requires that mass balance according to [Eq. (8.14)]

$$\frac{\partial v_x}{\partial x} + \frac{\partial v_y}{\partial y} + \frac{\partial v_z}{\partial z} = 0,$$

(9.3)

which is clearly satisfied identically given our assumption that $\mathbf{v} = v_x(y)\hat{\mathbf{i}}$ only. Linear momentum balance, for a Newtonian behavior, requires that we satisfy the Navier–Stokes equation, which in Cartesian components is [Eqs. (8.36), (8.40), and (8.42)]

$$\hat{\mathbf{i}}: \quad -\frac{\partial p}{\partial x} + \mu\left(\frac{\partial^2 v_x}{\partial x^2} + \frac{\partial^2 v_x}{\partial y^2} + \frac{\partial^2 v_x}{\partial z^2}\right) + \rho g_x$$

$$= \rho\left(\frac{\partial v_x}{\partial t} + v_x\frac{\partial v_x}{\partial x} + v_y\frac{\partial v_x}{\partial y} + v_z\frac{\partial v_x}{\partial z}\right), \qquad (9.4)$$

$$\hat{\mathbf{j}}: \quad -\frac{\partial p}{\partial y} + \mu\left(\frac{\partial^2 v_y}{\partial x^2} + \frac{\partial^2 v_y}{\partial y^2} + \frac{\partial^2 v_y}{\partial z^2}\right) + \rho g_y$$

$$= \rho\left(\frac{\partial v_y}{\partial t} + v_x\frac{\partial v_y}{\partial x} + v_y\frac{\partial v_y}{\partial y} + v_z\frac{\partial v_y}{\partial z}\right), \qquad (9.5)$$

$$\hat{\mathbf{k}}: \quad -\frac{\partial p}{\partial z} + \mu\left(\frac{\partial^2 v_z}{\partial x^2} + \frac{\partial^2 v_z}{\partial y^2} + \frac{\partial^2 v_z}{\partial z^2}\right) + \rho g_z$$

$$= \rho\left(\frac{\partial v_z}{\partial t} + v_x\frac{\partial v_z}{\partial x} + v_y\frac{\partial v_z}{\partial y} + v_z\frac{\partial v_z}{\partial z}\right). \qquad (9.6)$$

Canceling out terms consistent with the above assumptions (do it), we are left with

$$-\frac{\partial p}{\partial x} + \mu\frac{\partial^2 v_x}{\partial y^2} = 0, \qquad -\frac{\partial p}{\partial y} = 0, \qquad -\frac{\partial p}{\partial z} = 0. \qquad (9.7)$$

The second and third of these equations show that the pressure is a function of x at most (i.e., it is independent of y and z), thus the final governing differential equation is

$$\frac{dp}{dx} = \mu\frac{d^2 v_x}{dy^2}. \qquad (9.8)$$

Note: The only way for a function of x (at most) to equal a function of y (at most) for all (x, y) is for each function to be a constant. Hence, the pressure gradient is constant and so too for the right-hand side of Eq. (9.8). Therefore, integrating this equation, we have

$$\mu\int \frac{d}{dy}\left(\frac{dv_x}{dy}\right) dy = \int \frac{dp}{dx}\, dy, \qquad (9.9)$$

where, from Appendix 8 of Chapter 8, the integral of $d(something)/dy$ with respect to y yields that *something*, thus,

$$\mu\frac{dv_x}{dy} = \frac{dp}{dx}y + c_1. \qquad (9.10)$$

Integrating again, we have

$$\mu \int \frac{d}{dy}(v_x)\, dy = \int \left(\frac{dp}{dx} y + c_1 \right) dy, \qquad (9.11)$$

or

$$\mu v_x = \frac{dp}{dx} \frac{y^2}{2} + c_1 y + c_2. \qquad (9.12)$$

As expected, we need two boundary conditions to find the two constants of integration because we began with a second-order differential equation. Enforcing the no-slip condition (that a fluid velocity equals that of a solid it contacts) at the bottom plate, $v_x(y = 0) = 0$, and likewise at the top plate, $v_x(y = h) = 0$, allows us to find the constants

$$0 = \frac{dp}{dx} \frac{(0)^2}{2} + c_1(0) + c_2 \rightarrow c_2 = 0 \qquad (9.13)$$

and

$$0 = \frac{dp}{dx} \frac{(h)^2}{2} + c_1(h) + 0 \rightarrow c_1 = -\frac{dp}{dx} \frac{h}{2}. \qquad (9.14)$$

Thus, the velocity field $\mathbf{v} = v_x(y)\hat{\mathbf{i}}$ is described fully by the x-direction component,

$$v_x(y) = \frac{1}{2\mu} \frac{dp}{dx}(y^2 - hy). \qquad (9.15)$$

Note that the velocity distribution is parabolic (Fig. 9.2); it is called a *Poiseuille flow* in honor of the French physician J. Poiseuille (1799–1869), who studied pressure–flow relations for blood flow. After checking that the boundary conditions are indeed satisfied, we should then calculate the maximum velocity of the flow. To do this, we must first determine that value of y at which the maximum occurs. This value can be calculated by taking the derivative of the velocity profile with respect to y and setting it equal to zero as follows:

$$\frac{dv_x}{dy} = \frac{1}{2\mu} \frac{dp}{dx}(2y - h) = 0 \rightarrow y = \frac{h}{2}. \qquad (9.16)$$

Occurring at $y = h/2$, the maximum velocity is

$$v_x)_{max} = -\frac{h^2}{8\mu} \frac{dp}{dx}. \qquad (9.17)$$

Question: Does it make sense that this expression for the maximum velocity has a minus sign? The answer is yes, of course, because the pressure gradient must be negative to drive the flow in the positive x direction. In fact, returning to Eq. (9.8), which reveals that the pressure gradient equals a constant, say c_3, we have

$$\frac{dp}{dx} = c_3 \rightarrow \int \frac{dp}{dx}\, dx = \int c_3\, dx, \tag{9.18}$$

or

$$p = c_3 x + c_4. \tag{9.19}$$

The constants c_3 and c_4 can be found from upstream (proximal) and downstream (distal) pressures. For example, if $p = p_1$ at $x = x_1$ and $p = p_2$ at $x = x_2$, with $p_1 > p_2$ for $x_1 < x_2$ in order to drive the flow in the positive x direction, then

$$c_3 = \frac{p_1 - p_2}{x_1 - x_2}, \qquad c_4 = -\left(\frac{p_1 x_2 - p_2 x_1}{x_1 - x_2}\right), \tag{9.20}$$

where c_3 is clearly negative and so too is the pressure gradient. Hence, the pressure field is

$$p = \left(\frac{p_1 - p_2}{x_1 - x_2}\right)x - \left(\frac{p_1 x_2 - p_2 x_1}{x_1 - x_2}\right). \tag{9.21}$$

To calculate the volumetric flow rate Q in the x direction, where $Q = \int_A (\mathbf{v} \cdot \mathbf{n})\, dA$, and $\mathbf{n} \equiv \hat{\mathbf{i}}$ in this problem, we have

$$Q = \int_A v_x\, dA = \int_0^h \int_0^w \frac{1}{2\mu}\frac{dp}{dx}(y^2 - hy)\, dz\, dy = \frac{w}{2\mu}\frac{dp}{dx}\int_0^h (y^2 - hy)\, dy, \tag{9.22}$$

or

$$Q = -\frac{h^3 w}{12\mu}\left(\frac{dp}{dx}\right), \tag{9.23}$$

where w is the width of the plates, which is assumed to be large compared to h so that edge effects are also negligible on each side. Again, the minus sign appears because the pressure gradient is negative and the value of Q is positive, as it should be. The average velocity (speed) of the flow is given by

$$\bar{v} = \frac{Q}{A} = -\frac{h^2}{12\mu}\frac{dp}{dx}. \tag{9.24}$$

We see, therefore, that $v_x)_{\max} = 3\bar{v}/2$.

Before finishing this problem by computing the fluid shear stress and associated wall shear stress, let us formalize two ideas that have been alluded to. First, we see from Eq. (9.15) that the computed velocity field $\mathbf{v}$

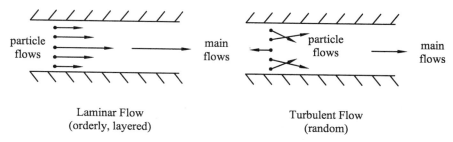

Laminar Flow
(orderly, layered)

Turbulent Flow
(random)

FIGURE 9.3 Schematic comparison of laminar versus turbulent flow. The former (left) is characterized by an orderly flow in which layers of fluid move relative to one another, whereas the latter (right) is characterized by a random flow superimposed on a net mean flow. A laminar flow can transition to a turbulent flow if the velocity increases sufficiently or if the surface of the constraining solid is sufficiently rough.

is a smooth function of position (x, y, z). Indeed, as we pick increasingly larger values of $y \in [0, h]$ the velocity changes in an ordered way, one in which the particles appear to travel in layers that slide relative to each other (Fig. 9.3). Such an ordered flow is called *laminar* for obvious reasons. In contrast, there are cases in which in fluid particles tend to have a random motion superimposed on an overall mean flow. Such a flow is termed *turbulent* (Fig. 9.3), which is mathematically very challenging to describe. Turbulence could occur in the flow between parallel plates if the velocity is very large or if the plates are very rough. This issue is not considered until Chapter 10, however. Second, we see from Eq. (9.15) that v does not depend on x; that is, the flow is assumed to be the same, within the region of interest, anywhere along the direction of flow. We would expect, of course, that the flow would not necessarily be the same at the point that it enters the parallel plates. Indeed, experiments reveal that the flow may develop from a uniform profile, to a blunted profile, to the final parabolic profile represented by Eq. (9.15). Illustrated in Figure 9.4, such a region over which a flow develops is called an *entrance length*. Because the velocity field depends on multiple coordinates when developing, it is much more difficult to describe mathematically. For both computational and experimental convenience, such boundary value problems are often formulated so that the region of interest is within the *fully developed* region. In other cases, of course, the entrance length may play a key role in the flow and thus must be considered fully. This often requires numerical methods.

To calculate the shear stress on the cells, which is essential to enable the experimentalist to correlate changes in the production of various molecules by the endothelial cells with changes in flow, we need to calculate σ_{yz} at $y = 0$. Thus, recall the (Navier–Poisson) constitutive relation for Newtonian fluids [Eq. (7.66)], the yx equation of which is

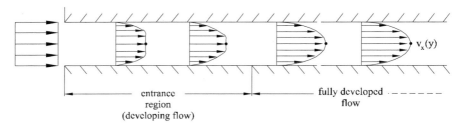

FIGURE 9.4. Because of the no-slip boundary condition for a viscous fluid, fluid particles slow (or stop) when they encounter a stationary solid. Because of the friction between layers of a viscous fluid, these slowed particles will, in turn, tend to slow neighboring fluid particles. Regions wherein such solid–fluid interactions are strong are called *boundary layers*. The boundary layer in a tube develops until the entire flow is affected. The entrance length is the region in which in the boundary layer is developing; thereafter the flow is called *fully developed*.

$$\sigma_{yx} = 2\mu \left[\frac{1}{2} \left(\frac{\partial v_x}{\partial y} + \frac{\partial v_y}{\partial x} \right) \right]. \tag{9.25}$$

Given that $v_y \equiv 0$, we have from Eq. (9.15)

$$\sigma_{yx} = \mu \frac{\partial v_x}{\partial y} = \mu \left(\frac{1}{2\mu} \frac{dp}{dx} (2y - h) \right), \tag{9.26}$$

which we emphasize is the shear stress *in the fluid* at any point y (Fig. 9.5). By Newton's third law (for every action there is an equal and opposite reaction), *the wall shear stress* τ_w is equal and opposite the shear stress in the fluid at $y = 0$ and $y = h$. Note, therefore, that at $y = 0$,

$$\sigma_{yx} = -\frac{h}{2} \left(\frac{dp}{dx} \right), \tag{9.27}$$

whereas at $y = h$,

$$\sigma_{yx} = \frac{h}{2} \left(\frac{dp}{dx} \right). \tag{9.28}$$

Because $dp/dx < 0$, σ_{yx} is positive at $y = 0$ and negative at $y = h$. Just as in statics (cf. Example A1.6 in Chapter 1), when the sign direction is taken positive but the value is found to be negative, we simply switch the direction of the force or stress. Hence, the shear stresses in the fluid act in the direction opposite the flow, whereas the fluid-induced wall shear stresses act in the positive x direction, as expected because the fluid tends to "push" on the solid, whereas the solid tends to "push back."

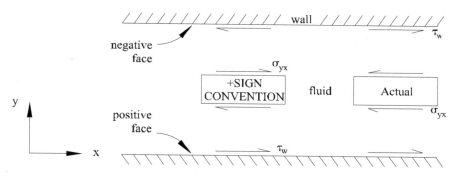

FIGURE 9.5 Positive sign convention for fluid stresses $\sigma_{(face)\,(direction)}$ relative to a Cartesian coordinate system and the associated actual shear stresses in a fluid element σ_{xy} and on the wall τ_w in a parallel-plate experiment.

At this juncture, it is useful to note that endothelial cells not only produce various molecules in response to changes in the magnitude of the applied shear stress, they also tend to align themselves and elongate in the direction of the applied shear. Indeed, for this reason, the in vivo orientation and elongation of an endothelial cell provides indirect information on the value of the wall shear stress that exists at that point. Experimentally, then, it is useful to correlate changes in morphology and the production of vasoactive molecules with the magnitude of the applied shear stress, which in the parallel-plate device is now seen to be

$$\tau_w = \left| \frac{h}{2} \frac{dp}{dx} \right|, \tag{9.29}$$

or in terms of the volumetric flow rate Q, from Eq. (9.23),

$$\tau_w = \frac{6\mu Q}{wh^2}. \tag{9.30}$$

Not all parallel-plate experiments are designed the same, however. In some cases, investigators leave a small gap of air (actually 95% air and 5% CO_2) between the upper plate and the fluid, with the cells cultured on the bottom plate. In this case, it can be shown that (see Exercise 9.3)

$$\tau_w = \frac{3\mu Q}{wh^2}. \tag{9.31}$$

Clearly, therefore, theory is essential in the design and interpretation of experiments.

Observation 9.1 Many investigators have used parallel-plate devices to subject cultured endothelial or epithelial cells to well-controlled steady or pulsatile shear stresses and to monitor their responses. Here, however, let us consider but a few of the early findings on endothelial cells as recorded in Chapter 6 in Frangos (1993), studies which began in 1974 but began to attract heightened attention around 1981.

Endothelial cells are often taken from either the bovine aorta (BAEC) or the human umbilical vein (HUVEC). Regardless, the cells are typically cultured on glass slides coated with synthetic (e.g., polyester or Mylar) or biologic (elastin or fibronectin) substrates and exposed to a 37°C culture media (often with 2–20% fetal bovine serum and antibiotics) and a 95% air/5% CO_2 gas mixture. After reaching confluence on the substrate, the cells are typically subjected to laminar flows (often with the quantity $\rho \bar{v} h / \mu$ ≤ 100) for periods of hours to days. Flows between parallel plates tend to become turbulent only after $\rho \bar{v} h / \mu > 1400$; it will be shown later that this nondimensional combination of terms is called the Reynolds' number, an important parameter in classifying many flows.

Among other results, it has been shown that endothelial cells tend to have a polygonal shape under static conditions, but in response to a constant shear stress ~1–10 Pa, the cells tend to elongate and then align with the direction of flow within 1–4 h of exposure to the shear (Fig. 9.6). Associated

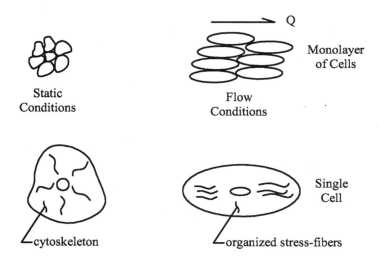

ENDOTHELIAL RESPONSE TO UNI-DIRECTIONAL FLOW

FIGURE 9.6 Schema of the effects of fluid flow on the shape and constitution of endothelial cells. Whereas the cells tend to be polygonal in shape under no-flow conditions, they tend to elongate and reorient with the direction of flow. Such shape changes are accompanied by changes in the cytoskeleton that include the production and orienting of stress fibers (cf. Fig. 1.5 of Chapter 1) in the direction of flow.

with these morphological changes are changes in the organization of the cytoskeleton. For example, the F-actin microfilaments tend to form dense peripheral bands under static conditions, bands that in the presence of shear stress tend to give way, over time, to the appearance of more centrally located stress fibers that are oriented in the direction of flow. Hence, the formation and orientation of the stress fibers coincide with gross changes in cell shape and orientation. That these two observations are coupled is revealed by tests in which cells are exposed to cytochalasin B, which disrupts actin assembly. Cells so treated do not elongate significantly in response to increases in shear stress. The intermediate filament vinculin, which participates in focal adhesion complexes, shows similar changes. Whereas vinculin may form nearly uniformly around the periphery of a cell under static conditions, it appears to localize at the upstream edge of cells subjected to flow. Hence, cells change their adhesion character- istics in response to flow. Indeed, studies have also shown that endothe- lial cells produce and organize fibronectin, an important extracellular adhesion molecule, which aids alignment in the direction of flow. Interest- ingly, this production tends to be diminished (with respect to static con- trols) early on during the exposure to flow, when the cell needs to be mobile and realign, but to be increased after alignment. Understanding cell–matrix interactions is obviously critical to understanding overall vascular biology.

In summary, parallel-plate experiments have revealed tremendous insights into correlations between shear stress and changes in cell mor- phology, cytoskeletal organization, the production of a host of molecules (vasoactive, growth regulatory, inflammatory, degradatory, and adhesive), and the production of extracellular matrix. The primary caveat, however, is that because cells are so sensitive to their environment, one must be cau- tious when trying to extrapolate results in culture to in vivo settings. Cell response likely depends primarily on its recent history (including an initial growth under static conditions), the particular substrate and its possible deformation, the specific culture media (i.e., chemical milieu, including antibiotics and growth factors), the flow characteristics (steady, pulsatile, laminar, turbulent), and cell–cell and cell–matrix interactions. Much remains to be learned.

Example 9.1 It appears that Rosen et al. (1974) first showed in vitro that endothelial cells alter their production of a specific molecule (histamine) in response to altered shear stresses. They accomplished this using cultured cells placed within a parallel-plate device. Their device was $1.3 \times 1.3 \times 23.5\,cm$ in dimension, with the cells placed in the fully developed region (15 cm from the entrance). Because the flow chamber was not much wider than it was deep, however, the equations to compute the wall shear stress differed from Eq. (9.30). Ensuring that $h \ll w$ allows Eq. (9.30) to be used and facilitates the easy design and interpretation of the experiment. Toward this end, let us consider the work by Levesque and Nerem (1985). Their flow chamber was $0.025 \times 1.3 \times 5$ cm in dimension; hence, $h \ll w$ and our equations hold. They plotted the morphological measures for the cells versus wall shear stress τ_w, which they computed as

$$\tau_w = \frac{6\mu^2}{\rho h^2} \, \text{Re},$$

where Re is the Reynolds' number. Show that this relation is correct.

Solution: As noted earlier, let the Reynolds' number be given by Re = $\rho \bar{v} h / \mu$. Recall, therefore, that [from Eq. (9.24)]

$$\bar{v} = -\frac{h^2}{12\mu} \frac{dp}{dx} \rightarrow \frac{dp}{dx} = -\frac{12\mu\bar{v}}{h^2};$$

hence, from Eq. (9.29),

$$\tau_w = \left| \frac{h}{2} \frac{dp}{dx} \right| = \left| \frac{h}{2} \left(-\frac{12\mu\bar{v}}{h^2} \right) \right| = \left| -\frac{6\mu\bar{v}}{h} \right|.$$

Now, simply multiply by "one," namely

$$\tau_w = \left| -\frac{6\mu\bar{v}}{h} \left(\frac{h}{h} \frac{\rho}{\rho} \frac{\mu}{\mu} \right) \right| = \left| -\frac{6\mu^2}{h^2\rho} \left(\frac{\rho\bar{v}h}{\mu} \right) \right| = \frac{6\mu^2}{h^2\rho} \text{Re},$$

with $\rho > 0$ and Re > 0 by definition, thus yielding our desired result.

Note, too, that Levesque and Nerem stated that Re < 2000 ensured a laminar flow, and they subjected the cells to steady shear stresses of 1.0, 3.0, and 8.5 Pa for up to 24 h. Given that the viscosity and density where assumed to equal those of water and that $h = 250\,\mu$m, find the exact values of Re for their reported values of τ_w to check if Re < 2000. Finally, note a few of their findings: "After 24 hours of exposure at shear stresses of 30 and 85 dyn/cm^2, there was a significant reduction in cell surface area, an increase in cell perimeter and length, and a decrease in cell width ... the more elongated cells have a higher degree of alignment with the flow axis. This effect becomes accentuated with increasing shear stress." Note that 1 dyn/cm^2 = 0.1 Pa.

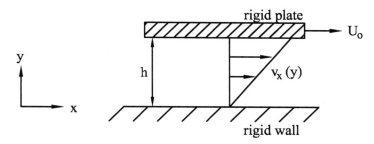

FIGURE 9.7 Couette flow induced by the relative motion of an upper plate, at constant velocity U_0, relative to a fixed rigid plate at the bottom. The associated velocity profile in the fluid is linear and, consequently, the fluid shear stress is uniform.

Example 9.2 The flow of water at room temperature ($\mu = 1.0 \times 10^{-3} \, \text{N s/m}^2$) between parallel plates need not be due to a pressure gradient; one can also generate a flow by moving the plates relative to each other while maintaining the gap distance at a constant value. Thus, consider the flow in Figure 9.7, where the top plate is moving at a constant velocity $U_0 = 0.1131$ m/s with no pressure gradient in the x direction. The fluid layer is 2 mm thick and the plate is 1 m wide. Use the Navier–Stokes equation for Newtonian flows to find (a) the velocity field, (b) the volumetric flow rate, and (c) the shear stress field.

Solution: Given

$$U_0 = 0.1131 \frac{\text{m}}{\text{s}}, \quad h = 2 \, \text{mm}, \quad w = 1 \, \text{m}, \quad \frac{dp}{dx} = 0 \frac{\text{Pa}}{\text{m}}.$$

Assume:

1. Newtonian fluid (μ = constant)
2. Incompressible flow ($\nabla \cdot \mathbf{v} = 0$)
3. Steady flow ($\partial \mathbf{v}/\partial t = \mathbf{0}$)
4. Unidirectional laminar flow ($v_y = v_z = 0$)
5. Negligible body forces ($\mathbf{g} = \mathbf{0}$)
6. Fully developed flow ($\partial \mathbf{v}/\partial x = \mathbf{0}$)
7. 1-D flow ($\partial v_x/\partial z = 0$, $\partial v_x/\partial x = 0$)
8. No pressure gradient in x ($\partial p/\partial x = 0$)

Mass balance is given by Eq. (9.3); it is again satisfied identically because $\mathbf{v} = v_x(y)\hat{\mathbf{i}}$ only. The balance of linear momentum in Cartesian coordinates is given by Eqs. (9.4)–(9.6). Eliminate the terms that disappear given the above assumptions and show that we have

$$\mu \frac{\partial^2 v_x}{\partial y^2} = 0, \qquad -\frac{\partial p}{\partial y} = 0, \qquad -\frac{\partial p}{\partial z} = 0.$$

The second and third equations, together with assumption 8, reveal that p = constant. Hence, in the absence of a pressure-driven flow, the governing differential equation of motion is

$$\mu \frac{d^2 v_x}{dy^2} = 0.$$

Integrating twice with respect to y, we obtain

$$\mu v_x = c_1 y + c_2.$$

Invoking the no-slip condition at the bottom plate, $v_x(y = 0) = 0$, and the no-slip condition at the top plate, $v_x(y = h) = U_0$, we find that

$$0 = c_1(0) + c_2 \rightarrow c_2 = 0,$$

$$\mu U_0 = c_1(h) + 0 \rightarrow c_1 = \frac{\mu U_0}{h}.$$

Thus, the velocity profile is

$$v_x(y) = \frac{U_0}{h} y,$$

which is called a *Couette flow*. The volumetric flow rate Q, with $\boldsymbol{n} = \hat{\boldsymbol{i}}$, is thus given by

$$Q = \int_A v_x \, dA = \int_0^h \int_0^w \frac{U_0}{h} y \, dz \, dy \rightarrow Q = \frac{U_0 w h}{2}.$$

To calculate the shear stress, recall again that for Newtonian fluids,

$$\sigma_{yx} = 2\mu \left[\frac{1}{2} \left(\frac{\partial v_x}{\partial y} + \frac{\partial v_y}{\partial x} \right) \right],$$

where, consistent with the assumptions, $v_y \equiv 0$. Hence,

$$\sigma_{yx} = \mu \frac{\partial v_x}{\partial y} \rightarrow \sigma_{yx} = \mu \frac{U_0}{h},$$

or in terms of the volumetric flow rate Q,

$$\sigma_{yx} = \frac{2\mu Q}{w h^2}.$$

In contrast to the pressure-driven flow wherein σ_{yx} varied with position y and indeed went to zero at $y = h/2$, we see that σ_{yx} is constant in this Couette flow. Moreover, because the computed value of σ_{yx} is everywhere positive, each fluid element experiences a *simple shear* (cf. Fig. 7.7). The wall shear stress τ_w is equal and opposite the fluid shear stress at $y = 0$ and h.

As we emphasize throughout, although the problem statement requires a specific computation, it is always better to first solve the problem generally. Now that we have the general relations, we can substitute the numerical values given in the problem statement into our equations for the volumetric flow rate and the shear stress. They are respectively

$$Q = \frac{1}{2}\left(0.1131\ \frac{m}{s}\right)(1\ m)(0.002\ m) = 0.000113\ \frac{m^3}{s} = 113\ \frac{mL}{s},$$

$$|\tau_w| = \frac{\mu U_0}{h} = \frac{(1.0 \times 10^{-3}\ N\ s/m^2)(0.1131\ m/s)}{0.002\ m} \approx 0.0566\ \frac{N}{m^2} = 0.0566\ Pa.$$

We will discover in Section 9.3.3 that this simple (general) solution has important implications in various real-world problems, including determination of the viscosity of a fluid.

Example 9.3 Fluid flow down an inclined plane is influenced by the force of gravity. Consider the flow in Figure 9.8, where a fluid film is subjected to the effects of gravity alone. Assume that the height of the fluid layer, h,

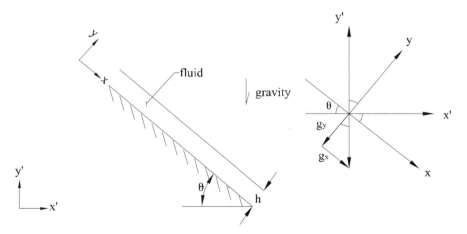

FIGURE 9.8 Uniform thickness flow of a Newtonian fluid down an inclined surface. The upper surface of the fluid is exposed to quiescent air at atmospheric pressure (i.e., zero pressure gauge), which is thus called a free surface. The only body force is gravity.

remains constant and that there is no pressure gradient in the x direction. Use the Navier–Stokes equation for steady, incompressible Newtonian flows, given the coordinate system in the figure, to find (a) the pressure distribution in the y direction and (b) the velocity field.

Solution

Assume:

1. Newtonian fluid (μ = constant)
2. Incompressible flow ($\nabla \cdot \mathbf{v} = 0$)
3. Steady flow ($\partial \mathbf{v}/\partial t = \mathbf{0}$)
4. Unidirectional laminar flow ($v_y = v_z = 0$)
5. Fully developed flow ($\partial \mathbf{v}/\partial x = \mathbf{0}$)
6. 1-D flow ($\partial v_x/\partial z = 0$, $\partial v_x/\partial x = 0$)
7. No pressure gradient in x ($\partial p/\partial x = 0$)
8. Shear stress due to airflow over the surface of the film is negligible ($\sigma_{yx(\text{air})} \approx 0$)

The balance of mass, given by Eq. (9.3), is again satisfied identically because $\mathbf{v} = v_x(y)\hat{\mathbf{i}}$ only. The balance of linear momentum, in Cartesian coordinates, is given by Eqs. (9.4)–(9.6). Given the above assumptions, we again eliminate the appropriate terms and find that we are left with

$$\mu \frac{\partial^2 v_x}{\partial y^2} + \rho g_x = 0, \qquad -\frac{\partial p}{\partial y} + \rho g_y = 0, \qquad -\frac{\partial p}{\partial z} = 0,$$

wherein, due to gravity $\mathbf{g} = -g\hat{\mathbf{j}}'$, we include the body force acting on the fluid. This force can be resolved into x and y components using the given coordinate systems (remember that coordinate systems should be picked for convenience). Doing so, we see that $g_x = g\sin\theta$ and $g_y = -g\cos\theta$. Thus, we have

$$\mu \frac{\partial^2 v_x}{\partial y^2} + \rho g \sin\theta = 0, \qquad -\frac{\partial p}{\partial y} - \rho g \cos\theta = 0, \qquad -\frac{\partial p}{\partial z} = 0.$$

Clearly, the pressure is a function of y alone, which can be determined via

$$\frac{dp}{dy} = -\rho g \cos\theta,$$

which, upon integration, yields

$$p(y) = -\rho g(\cos\theta)y + c.$$

Now, we need a boundary condition to solve for the constant c. Knowing that the surface of the fluid film, at $y = h$, is subjected to atmospheric pressure conditions or P_{atm}, we get

$$P_{\text{atm}} = -\rho g (\cos \theta) h + c \rightarrow c = P_{\text{atm}} + \rho g (\cos \theta) h.$$

Thus, the (absolute) pressure distribution in the y direction is

$$p(y) = \rho g (h - y) \cos \theta + P_{\text{atm}}.$$

The gauge pressure is the absolute pressure minus atmospheric pressure. We also see from the x-direction equation of motion that the velocity is a function of y alone, thus the final governing differential equation of motion is

$$\frac{d^2 v_x}{dy^2} = -\frac{\rho g \sin \theta}{\mu}.$$

Integrating twice, we obtain

$$v_x(y) = -\frac{\rho g \sin \theta}{2\mu} y^2 + c_1 y + c_2.$$

Invoking the no-slip condition at the face of the inclined plane, $v_x(y = 0) = 0$, we find that

$$0 = 0 + c_1(0) + c_2 \rightarrow c_2 = 0.$$

Now, we need an appropriate boundary condition for the top surface of the fluid film. Recall that when we formulated the problem, we assumed that the shear stress due to airflow over the fluid film was negligible. Hence,

$$\sigma_{yx(\text{air})} \approx 0 \rightarrow \sigma_{yx(\text{fluid})}\big|_{y=h} = \mu \frac{\partial v_x}{\partial y}\bigg|_{y=h} = 0,$$

where $\sigma_{yx(\text{air})}\big|_{y=h} = -\sigma_{yx(\text{fluid})}\big|_{y=h}$. Thus,

$$\mu \frac{\partial v_x}{\partial y}\bigg|_{y=h} = 0 = \mu\left(-\frac{\rho g h \sin \theta}{\mu} + c_1\right),$$

or

$$c_1 = \frac{\rho g h \sin \theta}{\mu}.$$

Thus, the velocity profile is

$$v_x(y) = \frac{\rho g \sin \theta}{\mu}\left(yh - \frac{y^2}{2}\right).$$

Once we are finished (i.e., when we have found the velocity and pressure fields), we should always examine special cases, the correctness of which gives us added confidence in our formulation and solution. Note, therefore, that $v_x = 0$ when $\theta = 0$, as expected, because there is no pressure gradient or moving solid to drive the flow. In conclusion, then, having computed the velocity and pressure fields, we can now calculate any quantity of interest, such as the shear stress, acceleration, or vorticity. This is left as an exercise, however.

9.2 Steady Flow in Circular Tubes

9.2.1 Biological Motivation

Most conduits for the flow of fluids within the body are cylindrical or nearly so, such as arteries, veins, airways, and ureters. Again, because of the sensitivity of endothelial and epithelial cells to applied shear stresses, it is important to have full solutions for these flows. In this way, we can understand better both physiology and pathology, and perhaps, most importantly, we can design better strategies for diagnosis and treatment. As it is well known, atherosclerosis is one of the leading causes of morbidity and mortality in the Western world. Briefly, atherosclerosis is a disease of the innermost layer of the arterial wall, the intima; it generally begins as a localized accumulation of lipids, sometimes called "fatty streaks," that form in preferential sites within the vascular tree. Over time, these lesions become more complex due to the accumulation of proliferating smooth muscle cells, excess matrix proteins (e.g., collagen) synthesized by the smooth muscle, and, in later stages, calcium and necrotic debris (Fig. 9.9). As a result, these lesions begin to compromise the lumen, the region of the obstruction being called a *stenosis*.

The three primary methods of treating atherosclerosis all rely heavily on biomechanics, or at least they should. Surgery typically involves the implantation of a graft that either replaces or bypasses the diseased region; angioplasty involves the dilatation of a balloon-tipped catheter within the stenosis to expand the lumen; and stenting involves the deployment of a metallic device to "hold open" the diseased region (see Figure 9.10a–9.10c). Currently, vascular grafts consist of two basic classes: *Synthetic grafts* are fabricated from man-made materials such as Dacron (see Fig. 10.5), whereas *natural grafts* include the use of arteries (e.g., internal mammary) or veins (saphenous) from other vascular beds within the patient. Finally, note that an exciting frontier is that of *tissue engineering* wherein one seeks to grow replacement tissues from the patient's cells. Herein, let us consider briefly the case of vein grafts.

Perhaps the first research into the potential use of veins as arterial grafts was that of Carrel and Guthrie (1906). Briefly, they transplanted canine jugular veins into the position of the carotid artery. Whereas the carotid

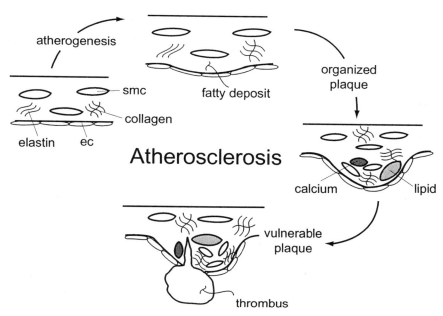

FIGURE 9.9 Schema of the atherosclerotic process, potentially leading from a fatty streak (defined by the accumulation of lipids in the subintimal space), to an organized plaque consisting of excess cells and matrix as well as lipids and calcium, to a possibly vulnerable plaque that may rupture and thrombose. Vulnerable plaques appear to be characterized by a thin collagenous cap that covers a softer core containing significant amounts of necrotic debris and lipids. Understanding the rupture of plaque thus requires knowledge of the solid mechanics (properties of and stresses in the plaque) and the fluid mechanics (the fluid-induced loads on the plaque, which serve as boundary conditions in the solid mechanics problem). ec = endothelial cell and smc = smooth muscle cell. [From Humphrey (2002), with permission.]

artery typically experiences pulsatile flows with luminal pressures changing from ~80 to 120 mm Hg, the jugular vein typically experiences nearly steady flows and a luminal pressure ~10 mm Hg. Because *structure closely follows function in the body*, the normal microstructure and gross morphology of the jugular vein is (as expected) much different from that of the carotid artery. In humans in particular, the typical luminal diameter is ~10.1 mm in the jugular vein and ~6.4 mm in the nearby carotid artery (Talbot et al, 1990).

As expected, Carrel and Guthrie found that veins transplanted to the arterial system thicken dramatically in response to increased pressure (Fig. 9.11). In order to understand, and ultimately to control, the response of vein grafts to their new environment (i.e., their growth and remodeling processes), we must understand well the normal properties and environ-

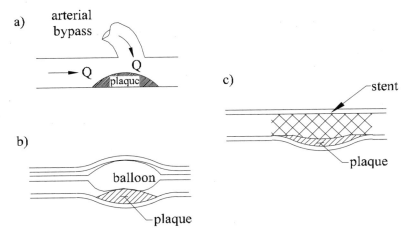

FIGURE 9.10 Three methods by which obstructive atherosclerotic lesions are treated: bypass grafts seek to restore flow to distal tissue by bypassing the obstruction, balloon angioplasty seeks to modify the plaque and weaken the wall so that it can distend more under normal physiologic pressures, and intravascular stents seek to hold open the lumen. Some stents are made of shape-memory alloys, and thus require advanced theories of material behavior.

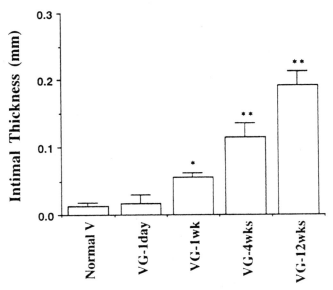

FIGURE 9.11 Intimal thickening over time of a vein used as a graft in the arterial system; note that the asterisks denote statistical significance; VG = Vein graft. [From Han et al. (1998), with permission from ASME.]

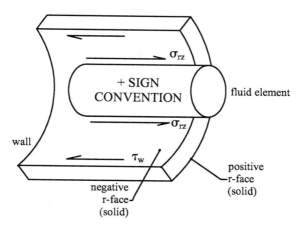

FIGURE 9.12 Sign convention for fluid shear stresses σ_{rz} for flow in a rigid circular tube.

ment of the vein itself. Indeed, as we just saw in Section 9.1, the activity of the endothelium is tightly controlled by the wall shear stress τ_w, and as we saw in Chapter 6, the activity of the intramural cells (smooth muscle and fibroblasts) is tightly controlled by the pressure-induced circumferential stress. Here, let us begin such a study by considering a simple case—the steady, incompressible flow of a Newtonian fluid in a nondistending circular tube—in order to determine the wall shear stress in terms of clinically measurable quantities. In this case, the Navier–Stokes equations admit another exact solution. Thus, consider the tube in Figure 9.12, of inner radius a, wherein the flow is one dimensional (e.g., no lateral diffusion or secondary flows).

9.2.2 Mathematical Formulation

Because of the circular geometry, it will prove convenient to employ cylindrical coordinates. Let us assume further the following:

1. Newtonian fluid (μ = constant)
2. Incompressible flow ($\nabla \cdot \boldsymbol{v} = 0$)
3. Steady flow ($\partial \boldsymbol{v}/\partial t = \boldsymbol{0}$)
4. Axial flow only ($v_r = v_\theta = 0$)
5. Fully developed flow ($\partial \boldsymbol{v}/\partial z = \boldsymbol{0}$)
6. Axisymmetric flow ($\partial \boldsymbol{v}/\partial \theta = \boldsymbol{0}$)
7. Negligible body forces ($\boldsymbol{g} = \boldsymbol{0}$)
8. Laminar flow

Note that the flow tends to be fully developed when its distance from the entrance of the tube is greater than

$$L_e \sim 0.06\left(\frac{\rho \bar{v} D}{\mu}\right)D,$$ (9.32)

where $\bar{v}$ is the mean forward velocity and D is the diameter. This length L_e is called the *entrance length*. Moreover, the flow tends to remain laminar if

$$\mathrm{Re} \equiv \frac{\rho \bar{v} D}{\mu} < 2100$$ (9.33)

where this combination of terms is the aforementioned Reynolds' number.
Mass balance in cylindrical coordinates [Eq. (8.15)] requires that

$$\frac{1}{r}\frac{\partial (rv_r)}{\partial r} + \frac{1}{r}\frac{\partial v_\theta}{\partial \theta} + \frac{\partial v_z}{\partial z} = 0,$$ (9.34)

which again is satisfied identically given this set of assumptions (i.e., v_z is a function of r alone). In cylindrical coordinates, linear momentum balance (i.e., the Navier–Stokes equation, $-\nabla p + \mu \nabla^2 v + \rho g = \rho a$) requires [Eqs. (8.46), (8.47), and (8.48)] the following:

$$\hat{e}_r: \quad -\frac{\partial p}{\partial r} + \mu\left[\frac{\partial}{\partial r}\left(\frac{1}{r}\frac{\partial (rv_r)}{\partial r}\right) + \frac{1}{r^2}\frac{\partial^2 v_r}{\partial \theta^2} - \frac{2}{r^2}\frac{\partial v_\theta}{\partial \theta} + \frac{\partial^2 v_r}{\partial z^2}\right] + \rho g_r$$

$$= \rho\left(\frac{\partial v_r}{\partial t} + v_r\frac{\partial v_r}{\partial r} + \frac{v_\theta}{r}\frac{\partial v_r}{\partial \theta} - \frac{v_\theta^2}{r} + v_z\frac{\partial v_r}{\partial z}\right),$$ (9.35)

$$\hat{e}_\theta: \quad -\frac{1}{r}\frac{\partial p}{\partial \theta} + \mu\left[\frac{\partial}{\partial r}\left(\frac{1}{r}\frac{\partial (rv_\theta)}{\partial r}\right) + \frac{1}{r^2}\frac{\partial^2 v_\theta}{\partial \theta^2} + \frac{2}{r^2}\frac{\partial v_r}{\partial \theta} + \frac{\partial^2 v_\theta}{\partial z^2}\right] + \rho g_\theta$$

$$= \rho\left(\frac{\partial v_\theta}{\partial t} + v_r\frac{\partial v_\theta}{\partial r} + \frac{v_\theta}{r}\frac{\partial v_\theta}{\partial \theta} + \frac{v_r v_\theta}{r} + v_z\frac{\partial v_\theta}{\partial z}\right),$$ (9.36)

$$\hat{e}_z: \quad -\frac{\partial p}{\partial z} + \mu\left[\frac{1}{r}\frac{\partial}{\partial r}\left(r\frac{\partial v_z}{\partial r}\right) + \frac{1}{r^2}\frac{\partial^2 v_z}{\partial \theta^2} + \frac{\partial^2 v_z}{\partial z^2}\right] + \rho g_z$$

$$= \rho\left(\frac{\partial v_z}{\partial t} + v_r\frac{\partial v_z}{\partial r} + \frac{v_\theta}{r}\frac{\partial v_z}{\partial \theta} + v_z\frac{\partial v_z}{\partial z}\right).$$ (9.37)

After eliminating terms consistent with the above assumptions (do it using the worksheet in Section 8.7), we are left with

$$-\frac{\partial p}{\partial r} = 0, \qquad -\frac{\partial p}{\partial \theta} = 0, \qquad -\frac{\partial p}{\partial z} + \mu\left[\frac{1}{r}\frac{\partial}{\partial r}\left(r\frac{\partial v_z}{\partial r}\right)\right] = 0.$$ (9.38)

The first two equations show that the pressure is a function of z at most. Noting that the velocity is a function of r alone, the only way that a function of z (the pressure gradient) can equal a function of r (the viscous term) for all (r, z) is for both functions to be a constant. Thus, the pressure gradient is a constant, and so too for the viscous term; hence,

$$\frac{dp}{dz} = \mu \left[\frac{1}{r} \frac{\partial}{\partial r} \left(r \frac{\partial v_z}{\partial r} \right) \right] \tag{9.39}$$

is the governing differential equation. Multiplying through by r and integrating this equation with respect to r, we obtain

$$\int \frac{\partial}{\partial r} \left(r \frac{\partial v_z}{\partial r} \right) dr = \int \frac{1}{\mu} \frac{dp}{dz} r \, dr, \tag{9.40}$$

or

$$r \frac{\partial v_z}{\partial r} = \frac{1}{2\mu} \frac{dp}{dz} r^2 + c_1. \tag{9.41}$$

Dividing through by r and integrating again, we have

$$\int \frac{\partial}{\partial r} (v_z) \, dr = \int \left(\frac{1}{2\mu} \frac{dp}{dz} r + \frac{c_1}{r} \right) dr, \tag{9.42}$$

or

$$v_z(r) = \frac{1}{4\mu} \frac{dp}{dz} r^2 + c_1 \ln r + c_2. \tag{9.43}$$

In order for $v_z(r)$ to be finite at all r, including the centerline at $r = 0$, c_1 must be zero (because the natural logarithm is not finite at $r = 0$). Note that this condition is not a boundary condition; rather, it is an extra condition that requires that the solution be physically reasonable. Similar "additional conditions" were used in earlier chapters on biosolid mechanics, as, for example, the requirement that the deflection be the same in a bone and prosthesis or in the left and right halves of a transversely loaded beam. The identification of such conditions comes primarily from intuition or experience. Next, applying the no-slip boundary condition at the wall of the cylinder, $v_z(r = a) = 0$, we find that

$$0 = \frac{1}{4\mu} \frac{dp}{dz} a^2 + c_2 \rightarrow c_2 = -\frac{1}{4\mu} \frac{dp}{dz} a^2. \tag{9.44}$$

Thus, the (fully developed) velocity field is $\mathbf{v} = v_z(r) \hat{e}_z$, where

$$v_z(r) = \frac{1}{4\mu} \frac{dp}{dz} (r^2 - a^2) = \frac{-a^2}{4\mu} \frac{dp}{dz} \left(1 - \frac{r^2}{a^2} \right). \tag{9.45}$$

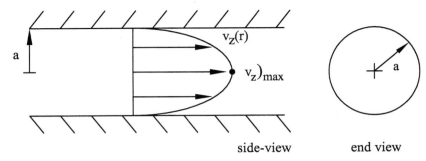

side-view   end view

FIGURE 9.13 Velocity profile for the steady, laminar, fully developed, incompressible flow of a Newtonian fluid in a rigid circular tube. Like the flow between parallel plates, this is also called a Poiseuille flow.

Now, to calculate the maximum velocity in the flow field (Fig. 9.13), we must first determine the value of r at which the maximum occurs. This value can be calculated by taking the derivative of the velocity profile with respect to r and setting it equal to zero as follows:

$$\frac{dv_z}{dr} = \frac{1}{2\mu}\frac{dp}{dz}r = 0 \rightarrow r = 0. \tag{9.46}$$

The maximum velocity, at $r = 0$, is thus

$$v_z)_{max} = v_z(r = 0) = -\frac{a^2}{4\mu}\left(\frac{dp}{dz}\right), \tag{9.47}$$

again realizing that the pressure gradient is negative in order to drive the fluid in the positive z direction; thus, the value of the maximum velocity is positive as it should be. To calculate the volumetric flow rate Q, we have (with $\mathbf{n} = \hat{\mathbf{e}}_z$ and $\mathbf{v} = v_z\hat{\mathbf{e}}_z$)

$$Q = \int_A v_z\,dA = \int_0^{2\pi}\int_0^a \left[\frac{1}{4\mu}\frac{dp}{dz}(r^2 - a^2)r\right]dr\,d\theta, \tag{9.48}$$

where $dA = rd\theta dr$ or

$$Q = \frac{2\pi}{4\mu}\frac{dp}{dz}\left(\frac{r^4}{4} - \frac{a^2 r^2}{2}\right)\Bigg|_0^a = -\frac{\pi a^4}{8\mu}\frac{dp}{dz}. \tag{9.49}$$

The average velocity (speed) of the flow is given by

$$\bar{v} = \frac{Q}{A} = \left(-\frac{\pi a^4}{8\mu}\frac{dp}{dz}\right)\left(\frac{1}{\pi a^2}\right) = -\frac{a^2}{8\mu}\frac{dp}{dz}. \tag{9.50}$$

To calculate the shear stress on the vessel wall τ_w, we need to calculate the shear stress in the fluid σ_{rz} at all r and then at $r = a$. Recall that for Newtonian fluids [Eqs. (7.64) and (7.58) of Chapter 7],

$$\sigma_{rz} = 2\mu\left[\frac{1}{2}\left(\frac{\partial v_z}{\partial r} + \frac{\partial v_r}{\partial z}\right)\right],$$

(9.51)

but given that $v_r \equiv 0$, we have

$$\sigma_{rz} = \mu\frac{\partial v_z}{\partial r} = \frac{dp}{dz}\frac{r}{2},$$

(9.52)

again noting that dp/dz is negative. Hence, the direction of σ_{rz} is opposite its positive sign convention (Fig. 9.12) and the wall shear stress is

$$\tau_w = \left|\frac{dp}{dz}\frac{a}{2}\right|$$

(9.53)

in the positive z direction. From Eqs. (9.49) and (9.50), however, the pressure gradient is

$$\frac{dp}{dz} = -\frac{8\mu Q}{\pi a^4} \quad \text{or} \quad \frac{dp}{dz} = -\frac{8\mu\bar{v}}{a^2};$$

(9.54)

hence the wall shear stress can be written as

$$\tau_w = \frac{4\mu Q}{\pi a^3} \quad \text{or} \quad \tau_w = \frac{4\mu\bar{v}}{a}.$$

(9.55)

The former result is one of the most often cited equations in vascular biology related to endothelial mechanotransduction [cf. Eq. (9.30)]. Nonetheless, we must remember all of the assumptions embodied in its derivation, including the assumptions of a rigid wall and steady flow.

Example 9.4 The so-called *skin friction coefficient* c_f is defined as the wall shear stress divided by the mean dynamic pressure. Find a formula for c_f for a steady flow in a rigid tube.

Solution: The *dynamic pressure* is defined as $\rho\bar{v}^2/2$, where $\bar{v}$ is a scalar measure of the mean velocity [i.e., the speed; see Bernoulli's equation (8.80)]. For the case of a tube flow, therefore, the mean dynamic pressure is

$$P_{dyn} = \frac{1}{2}\rho\bar{v}^2 = \frac{1}{2}\rho\left(-\frac{a^2}{8\mu}\frac{dp}{dz}\right)^2,$$

where the mean velocity is given by Eq. (9.50). Hence, in this case (Fung, 1993),

$$c_f = -\frac{a}{2}\left(\frac{dp}{dz}\right)\left\{\frac{\rho}{2}\left[\frac{a^4}{64\mu^2}\left(\frac{dp}{dz}\right)^2\right]\right\}^{-1} = -64\mu^2\left[\rho a^3\left(\frac{dp}{dz}\right)\right]^{-1},$$

which, via Eq. (9.54), can be written as

$$c_f = \frac{-64\mu^2}{\rho a^3\left(-8\mu\bar{v}/a^2\right)} = \frac{16}{\rho(2a)\bar{v}/\mu} = \frac{16}{\text{Re}},$$

where the Reynolds' number is

$$\text{Re} = \frac{\rho\bar{v}d}{\mu},$$

with $d = 2a$ the diameter of the tube. Re is a very important nondimensional parameter in fluid mechanics; it will be discussed in greater depth in Chapter 10. In summary, the wall shear stress in this tube flow can also be written as

$$\tau_w = \frac{1}{2}\rho\bar{v}^2 c_f = \frac{1}{2}\rho\bar{v}^2\left(\frac{16}{\text{Re}}\right).$$

For example, in the human aorta,

$$\bar{v} = 0.15 \text{ m/s}, \qquad d = 0.03 \text{ m}$$
$$\rho = 1060 \text{ kg/m}^3, \qquad \mu = 3.3\times10^{-3} \text{ N s/m}^2;$$

hence,

$$\text{Re} = \frac{(1060 \text{ kg/m}^3)(0.15 \text{ m/s})(0.03 \text{ m})}{3.3\times10^{-3} \text{ N s/m}^2} = 1445,$$

where $1 \text{ kg m/s}^2 = 1 \text{ N}$. This value is less than that expected for turbulent flow ($\text{Re} > 2100$); hence, our laminar assumption applies. Note, too, that the associated value of $\tau_w = 0.13 \text{ kg/m s}^2 = 0.13 \text{ Pa}$, which is within the reported range albeit lower than that which is usually reported (1.5 Pa).

9.3 Circumferential Flow Between Concentric Cylinders

Let us next consider a general case that has two important applications (Fig. 9.14): The steady, incompressible flow of a Newtonian fluid between two rigid, impermeable, circular cylinders, each of which can rotate at a differ-

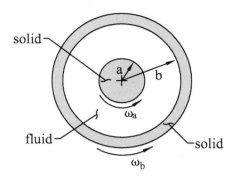

FIGURE 9.14 Steady flow of an in-compressible Newtonian fluid between two concentric, rigid, circular cylinders which may rotate at different angular velocities ω_a and ω_b. In practice, the gap distance $h = b - a$ would be small in comparison to a; thus, the gap is exaggerated for illustrative purposes only.

ent angular velocity. For example, such a problem describes the concentric cylinder viscometer wherein one seeks to measure the viscosity of a fluid by measuring the twisting moment (torque) T that is required to rotate the inner cylinder at angular velocity ω while the outer cylinder is fixed. Another application of this problem is the NASA bioreactor (Wolf and Schwarz, 1991).

9.3.1 Bioreactor Application

Atmospheric pressure near sea level is ~14.7 psi. If we estimate the surface area of our skin via a cylinder ~60 in. high and 12 in. in diameter, then the surface area is ~$2\pi(6)(60) = 2,260\,in^2$. At 14.7 psi, therefore, our bodies resist an amazing ~33,250 lbs of total force. That we do not feel the "burden" of this load is an example of adaptation; that is, it is as if our skin senses gauge pressures, not absolute pressures, where $P_{abs} = P_{atm} + P_{gauge}$.

Likewise, our bodies are well accommodated to the normal gravitational pull of 1 g. It should not be surprising, therefore, that when astronauts experience microgravity in space, the cells in their bodies sense and seek to respond to this change. Because we plan to extend a human's duration in space via the International Space Station or a voyage to Mars, we must understand better the responses of cells to a microgravity environment. Indeed, recall from Chapter 1 that the *Apollo* program in the 1960s provided an important motivation for the development of biomechanics.

Space-based experimentation is clearly the most natural approach to study the effects of microgravity, but it is also the most expensive. NASA scientists and engineers have thus sought Earth-based experiments to simulate the effects of microgravity; such tests should be both cost-effective and revealing. One such experiment is the so-called hindlimb unloading of a rat. Note, therefore, that any human activity on Earth that involves an upright posture (e.g., standing, walking, running) induces a normal gradient on the hydrostatic blood pressure from head to foot (cf. Example 8.2 of Chapter 8). Consequently, blood vessels in the legs may constrict to prevent blood from pooling in the lower extremities. In a microgravity environment, however, there is a loss of this normal head-to-foot gradient in pressure and there is an associated shift in fluids from the lower to the upper portions of the body. Such changes, if sustained over long periods, trigger adaptations in the cardiovascular system that can cause problems when the astronaut returns to Earth. For example, astronauts may become dizzy or faint when standing upright soon after their return to Earth. One way to examine the effects of such *orthostatic intolerance* is to induce similar head-to-foot fluid shifts in laboratory animals and then to study changes in vascular structure and function. In the hindlimb unloading experiment, a rat is suspended by its tail so that the hindlimbs are elevated and not weight bearing. As noted by Delp et al. (2000), this animal model "induces the cephalic fluid shift and postural muscle unloading that occur in microgravity. Additionally, the hindlimb unloaded animals manifest many of the adaptations that are characteristic of exposure to microgravity, including postural muscle atrophy, hypovolemia, a diminished capacity to elevate vascular resistance, orthostatic hypotension, and a reduced aerobic capacity." Of course, to understand fully such experiments, one must understand the associated biosolid and biofluid mechanics.

Another experimental setup to study the effects of microgravity is the so-called NASA *bioreactor*. Briefly, living cells are either allowed to float freely within a culture media contained between two concentric rotating cylinders or the cells are cultured on the walls of one of the cylinders. The basic idea is to confuse the cells as to which "direction is up," which is to say, to subject them to a changing gravitational vector and thereby simulate the microgravity via the absence of a consistent gravitational field. More on these specific applications later. Here, let us formulate and solve the general problem (Fig. 9.14) independent of the specific application.

9.3.2 Mathematical Formulation

Let the fluid flow be in the θ and possibly r directions, with no-slip boundary conditions at $r = a$ and $r = b$ requiring v_θ to vary with radius. Hence, let $v = v_r(r)\hat{e}_r + v_\theta(r)\hat{e}_\theta$ consistent with the following assumptions:

1. Newtonian fluid (μ = constant)
2. Incompressible flow ($\nabla \cdot v = 0$)
3. Steady flow ($\partial v/\partial t = 0$)
4. $v_z = 0$
5. No change in the z direction ($\partial v/\partial z = 0$)
6. Axisymmetric flow ($\partial v/\partial \theta = 0$)
7. Negligible body forces ($g = 0$)
8. Laminar flow

Recall that mass balance, in cylindrical coordinates, requires [Eq. (8.15)]

$$\frac{1}{r}\frac{\partial(rv_r)}{\partial r} + \frac{1}{r}\frac{\partial v_\theta}{\partial \theta} + \frac{\partial v_z}{\partial z} = 0, \tag{9.56}$$

which, upon invoking the above assumptions, reduces to

$$\frac{1}{r}\frac{\partial(rv_r)}{\partial r} = 0. \tag{9.57}$$

Multiplying through by r and integrating, we obtain

$$\int \frac{\partial(rv_r)}{\partial r}\, dr = \int 0\, dr \rightarrow rv_r = c_1 \quad \text{or} \quad v_r = \frac{c_1}{r}. \tag{9.58}$$

Applying the no-slip condition at the inner $(r = a)$ or outer $(r = b)$ walls of the cylinders, $v_r(r = a) = 0 = v_r(r = b)$, we get $c_1 = 0$. Thus, the velocity is zero in the r direction (in this idealized case), and we again have a unidirectional 1-D flow: $\mathbf{v} = v_\theta(r)\hat{\mathbf{e}}_\theta$ only. Recall, too, that the Navier–Stokes equation, $-\nabla p + \mu\nabla^2\mathbf{v} + \rho\mathbf{g} = \rho\mathbf{a}$), in cylindrical coordinates is [Eqs. (8.46)–(8.48)]

$$\hat{\mathbf{e}}_r: \quad -\frac{\partial p}{\partial r} + \mu\left[\frac{\partial}{\partial r}\left(\frac{1}{r}\frac{\partial(rv_r)}{\partial r}\right) + \frac{1}{r^2}\frac{\partial^2 v_r}{\partial \theta^2} - \frac{2}{r^2}\frac{\partial v_\theta}{\partial \theta} + \frac{\partial^2 v_r}{\partial z^2}\right] + \rho g_r$$

$$= \rho\left(\frac{\partial v_r}{\partial t} + v_r\frac{\partial v_r}{\partial r} + \frac{v_\theta}{r}\frac{\partial v_r}{\partial \theta} - \frac{v_\theta^2}{r} + v_z\frac{\partial v_r}{\partial z}\right), \tag{9.59}$$

$$\hat{\mathbf{e}}_\theta: \quad -\frac{1}{r}\frac{\partial p}{\partial \theta} + \mu\left[\frac{\partial}{\partial r}\left(\frac{1}{r}\frac{\partial(rv_\theta)}{\partial r}\right) + \frac{1}{r^2}\frac{\partial^2 v_\theta}{\partial \theta^2} + \frac{2}{r^2}\frac{\partial v_r}{\partial \theta} + \frac{\partial^2 v_\theta}{\partial z^2}\right] + \rho g_\theta$$

$$= \rho\left(\frac{\partial v_\theta}{\partial t} + v_r\frac{\partial v_\theta}{\partial r} + \frac{v_\theta}{r}\frac{\partial v_\theta}{\partial \theta} + \frac{v_r v_\theta}{r} + v_z\frac{\partial v_\theta}{\partial z}\right), \tag{9.60}$$

$$\hat{\mathbf{e}}_z: \quad -\frac{\partial p}{\partial z} + \mu\left[\frac{1}{r}\frac{\partial}{\partial r}\left(r\frac{\partial v_z}{\partial r}\right) + \frac{1}{r^2}\frac{\partial^2 v_z}{\partial \theta^2} + \frac{\partial^2 v_z}{\partial z^2}\right] + \rho g_z$$

$$= \rho\left(\frac{\partial v_z}{\partial t} + v_r\frac{\partial v_z}{\partial r} + \frac{v_\theta}{r}\frac{\partial v_z}{\partial \theta} + v_z\frac{\partial v_z}{\partial z}\right). \tag{9.61}$$

Canceling out terms using the above assumptions (do it, using the worksheet from Section 8.7 of Chapter 8), we are left with

$$-\frac{\partial p}{\partial r} = -\rho\frac{v_\theta^2}{r} \quad \text{and} \quad \mu\frac{\partial}{\partial r}\left(\frac{1}{r}\frac{\partial(rv_\theta)}{\partial r}\right) = 0. \tag{9.62}$$

These two governing differential equations of motion are decoupled (i.e., we can solve v_θ from the second equation and then p from the first equation rather than having to solve simultaneously two equations for two unknowns); hence, let us solve them sequentially. Integrating the second equation and putting it in the form of $d(\text{something})/dr$ as suggested in Appendix 8 of Chapter 8, we obtain

$$\int \frac{\partial}{\partial r}\left(\frac{1}{r}\frac{\partial}{\partial r}(rv_\theta)\right)dr = \int 0\, dr \to \frac{1}{r}\frac{\partial}{\partial r}(rv_\theta) = c_2. \tag{9.63}$$

Multiplying through by r and integrating again, we have

$$\int \frac{\partial}{\partial r}(rv_\theta)\, dr = \int rc_2\, dr \to rv_\theta = \frac{c_2}{2}r^2 + c_3, \tag{9.64}$$

or

$$v_\theta(r) = \frac{c_2}{2}r + \frac{c_3}{r}. \tag{9.65}$$

The no-slip condition at the inner cylinder, which may rotate at angular velocity ω_a, is $v_\theta(r = a) = a\omega_a$ (remember, ω has units of inverse time); likewise, the no-slip condition at the outer cylinder, which may rotate at angular velocity ω_b, is $v_\theta(r = b) = b\omega_b$. Hence, we have

$$a\omega_a = \frac{c_2}{2}a + \frac{c_3}{a}, \qquad b\omega_b = \frac{c_2}{2}b + \frac{c_3}{b}, \tag{9.66}$$

which are simply two algebraic equations in terms of two unknowns. Solving these two equations simultaneously, we find that

$$c_2 = \frac{2(b^2\omega_b - a^2\omega_a)}{b^2 - a^2}, \qquad c_3 = \frac{a^2b^2(\omega_a - \omega_b)}{b^2 - a^2}. \tag{9.67}$$

Thus, the velocity field is $\mathbf{v} = v_\theta(r)\hat{\mathbf{e}}_\theta$, where

$$v_\theta(r) = \left(\frac{b^2\omega_b - a^2\omega_a}{b^2 - a^2}\right)r + \left(\frac{a^2b^2(\omega_a - \omega_b)}{b^2 - a^2}\right)\frac{1}{r}. \tag{9.68}$$

Again, check that the boundary conditions are enforced at $r = a$ and $r = b$. Given this "general" solution, it is useful to consider numerous special cases. For example, if we rotate the cylinders at the same angular velocity (i.e., let $\omega_a = \omega_b = \omega$), then

$$v_\theta(r) = r\omega; \tag{9.69}$$

that is, when the angular velocities are the same, in magnitude and direction, the fluid moves like a rigid body. Question: What is the associated vorticity? Recalling Eq. (7.43), $\boldsymbol{\zeta} = \nabla \times \mathbf{v} = 2\omega\hat{\mathbf{e}}_z$ as we would expect. Hence, this flow is not irrotational.

Recalling that many cells are very sensitive to imposed shear stresses, note that NASA sought to minimize flow-induced shear in their afore-mentioned bioreactor so that the effects of the simulated microgravity could be isolated and studied separately. Recall, therefore, that for Newtonian fluids [cf. Eqs. (7.64) and (7.58)],

$$\sigma_{r\theta} = \mu \left[r \frac{\partial}{\partial r} \left(\frac{v_\theta}{r} \right) + \frac{1}{r} \frac{\partial v_r}{\partial \theta} \right], \tag{9.70}$$

where $v_r \equiv 0$ here and, thus,

$$\sigma_{r\theta} = \mu r \frac{\partial}{\partial r} \left(\frac{v_\theta}{r} \right). \tag{9.71}$$

To find the shear stress at any point in the fluid, including at the walls, we know from Eq. (9.69) that $v_\theta = r\omega$ if the cylinders rotate at the same angular velocity. In this case, then, the shear stress is

$$\sigma_{r\theta} = \mu r \frac{\partial}{\partial r} (\omega) = 0 \tag{9.72}$$

because ω is not a function of r. Hence, rotating the fluid as a rigid-body results in a shearless flow as desired by NASA. Clearly, the design of both the device and the experimental protocol is possible only through a fluid mechanical analysis.

Observation 9.2 Although our analysis of the NASA bioreactor gives us a good "feel" for the fluid mechanics, actual research has been based on a more general mathematical analysis. For example, Tsao et al. (1994) considered a bioreactor having dimensions $a = 2.86$ cm, $b = 4.0$ cm, and length (height) $h = 11$ cm. For a Newtonian fluid of mass density $\rho = 1020$ kg/m^3 and viscosity $\mu = 0.97$ cP $= 0.97 \times 10^{-3}$ N s/m^2, they showed that two secondary flows (cf. Fig. 8.12) occur in addition to the primary circumferential flow; that is, assuming only a steady, axisymmetric flow, the numerical solution of the mass balance and Navier–Stokes equations reveals a counterclockwise circulation and an opposing clockwise circulation pattern in the r-z plane; that is, both v_r and v_z are nonzero, in general, in the actual bioreactor wherein $h = 11$ cm is much larger than the gap $b - a = 1.14$ cm. They suggest that these countercirculation patterns facilitate good mixing, which enables oxygen and nutrient transport to free-floating cells. The numerical details are beyond the present scope, however, and the interested reader is referred to the original paper.

9.3.3 Viscometer Application

Recall that an incompressible, Newtonian fluid is characterized by a single material parameter, the viscosity μ (Fig. 7.11 of Chapter 7). Fundamental to the solution of Newtonian flows, therefore, is determination of the numerical value of μ for the fluid of interest and under the conditions of interest. For example, the viscosity of many Newtonian fluids varies strongly with temperature (e.g., motor oil is much more viscous at low temperatures); indeed, this relationship is often approximated via $\log \mu = \log A + 0.434 B/T$, where A and B are material parameters and T is temperature (i.e., $\mu = A e^{B/T}$).

There are numerous ways to determine the value of μ for a Newtonian fluid. Devices designed specifically for such experimental determinations are called viscous meters or *viscometers*. Common designs of viscometers are the capillary-tube, the falling-sphere, the cone-and-plate, and the concentric-cylinder viscometer. In a falling-sphere viscometer, one measures the time of descent of a solid sphere in a column of fluid of interest. This situation can be modeled as the flow of a fluid around a stationary sphere. Under conditions wherein the viscous effects are much greater than the inertial effects, the Navier–Stokes equations simplify and one obtains a solution (Slattery, 1981) that allows the viscosity to be inferred from the mass and diameter of the (smooth) sphere and the descent time over a prescribed length of travel (Fig. 9.15). This is discussed further in Chapter 10 using a nondimensional approach.

In the cone-and-plate viscometer, one measures the torque T that is necessary to rotate a small-angle cone at a specified angular velocity within a fluid of interest (Fig. 7.16 of Chapter 7; see, e.g., Slattery, 1981). Such devices have been used commonly to quantify the viscosity of blood as well as to

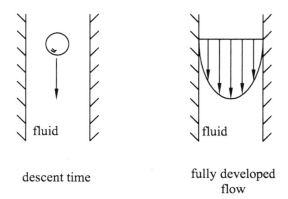

descent time   fully developed
                                      flow

FIGURE 9.15 Schema of two additional setups that are used as viscometers (cf. Fig. 7.16 of Chapter 7).

subject cultured monolayers of cells to known shear stresses. An approximate relation among the torque, viscosity, and device parameters is given in Section 7.6 of Chapter 7. Let us now consider an exact solution for the concentric-cylinder device based on the general solution found in the previous section.

In the concentric-cylinder viscometer, one likewise measures the torque needed to rotate the inner cylinder at a constant angular velocity. Specifically, if we fix the outer cylinder ($\omega_b = 0$) and only rotate the inner cylinder at ($\omega_a \equiv \omega$) then Eq. (9.68) reduces to

$$v_\theta(r) = \frac{a^2\omega}{b^2 - a^2}\left(\frac{b^2}{r} - r\right).$$

(9.73)

Now, the torque (or twisting moment) T is, of course, a force acting at a distance. In particular, for the inner cylinder, we have

$$\sum M_z)_0 = 0 \rightarrow T - \int_A a\tau_w\, dA = 0,$$

(9.74)

where $\tau_w\, dA$ is the differential force acting at distance a from the centerline (Fig. 9.16), the torque being balanced by all such differential torques. Clearly, we need to compute the wall shear stress τ_w on the inner cylinder, which is equal and opposite $\sigma_{r\theta}$ in the fluid at $r = a$. From Eq. (9.73) and the constitutive relation for the fluid [Eq. (7.64)], we have

$$\sigma_{r\theta} = \mu r \frac{\partial}{\partial r}\left(\frac{v_\theta}{r}\right) = \mu r \frac{\partial}{\partial r}\left[\frac{a^2\omega}{b^2 - a^2}\left(\frac{b^2}{r^2} - 1\right)\right].$$

(9.75)

Thus, the shear stress at any point r is

$$\sigma_{r\theta} = \mu r\left[\frac{a^2\omega}{b^2 - a^2}\left(-\frac{2b^2}{r^3}\right)\right] \rightarrow \sigma_{r\theta} = -\frac{2\mu b^2 a^2\omega}{b^2 - a^2}\left(\frac{1}{r^2}\right);$$

(9.76)

the negative sign reveals a direction opposite that for a positive sign convention. To find the shear stress on the wall, we need to calculate $\sigma_{r\theta}$ at $r = a$, which is

$$\sigma_{r\theta}(r = a) = -\frac{2\mu b^2\omega}{b^2 - a^2}.$$

(9.77)

Because $b > a$, the shear stress $\sigma_{r\theta}$ is negative on an inner (negative) face of the fluid, and the free-body diagram for the inner cylinder in terms of τ_w is as shown in Fig. 9.16. The shear stress on the wall is thus

$$\tau_w = |\sigma_{r\theta}(r = a)| = \frac{2\mu b^2\omega}{b^2 - a^2},$$

(9.78)

in the direction shown.

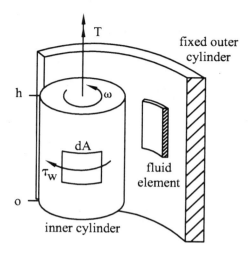

concentric cylinder
viscometer

FIGURE 9.16 Detail of the wall shear stresses τ_w that act on the inner cylinder of the concentric-cylinder viscometer. The applied torque T must balance these stresses in a steady-state situation.

With the differential area $dA = dz\,ad\theta$ (where $r = a$), Eq. (9.74) becomes

$$T = \int_0^h \int_0^{2\pi} a^2 \tau_w \, d\theta dz = \frac{2\mu b^2 \omega}{b^2 - a^2}(a^2)\int_0^h \int_0^{2\pi} d\theta dz, \qquad (9.79)$$

or

$$T = \frac{4\pi\mu a^2 b^2 \omega h}{b^2 - a^2}. \qquad (9.80)$$

Hence, the viscosity μ can be "measured" (actually inferred) by measuring the geometry (a, b, h), angular speed (ω), and applied torque T, namely

$$\mu = \frac{T(b^2 - a^2)}{4\pi a^2 b^2 h\omega}. \qquad (9.81)$$

Again, therefore, we see that analysis allows one to design an experiment (i.e., to determine what to measure, why, and to what resolution).

It should be noted that many concentric cylinder viscometers have a small gap distance $(b - a)$ relative to the inner radius a. Although Eq. (9.81) should be used to compute μ, it is interesting to note that when $b - a \ll a$, we can think of the problem locally as a "flat" (inner) plate moving relative to an "outer" stationary one. In this case, we can exploit the solution in Example 9.2 for the Couette flow. In that case, $U_0 = a\omega$ and $h = b - a$, and the associated shear stress is

$$\sigma_{xy} = \frac{\mu a \omega}{b - a} \rightarrow \tau_w = \left| \frac{\mu a \omega}{b - a} \right|. \tag{9.82}$$

The associated torque is thus

$$T = \int \int \tau_w a\, dA = \frac{\mu a^2 \omega}{b - a}(2\pi a h) \rightarrow \mu = \frac{T(b - a)}{2\pi a^3 \omega h}. \tag{9.83}$$

Calling this solution for the viscosity μ_{approx} and Eq. (9.81) μ_{exact}, note that the error due to the flat plate assumption is

$$\text{error} = \frac{\mu_{approx} - \mu_{exact}}{\mu_{exact}} = \frac{2b^2 - (ab + a^2)}{ab + a^2}. \tag{9.84}$$

Note: Viscosity has units of $N\,s/m^2$ in the SI system, but values are sometimes reported as Poise (P) or centiPoise (cP). The conversion is $1\,P = 0.1\,N\,s/m^2$ or $1\,cP = 1 \times 10^{-3}\,N\,s/m^2$; alternatively, $1\,P = 1\,dyn\,s/cm^2 = 1\,g/cm\,s$.

Example 9.5 In an experiment, one has

$$a = 9\,\text{mm}, \qquad b = 9.2\,\text{mm}, \qquad \omega = 1200\,\text{rpm},$$
$$h = 60\,\text{mm}, \qquad T = 0.0036\,\text{N m}.$$

Compute the error in "measuring" μ based on the flat plate assumption.

Solution: Based on the exact solution

$$\mu_{exact} = \frac{(0.0036)(0.0092^2 - 0.009^2)}{4\pi(0.009)^2(0.0092)^2(0.06)(125.66)} = 0.02017\frac{N\,s}{m^2},$$

where $\omega = (1200\,\text{rpm})(1\,\text{min}/60\,\text{s})(2\pi\text{rad/rev}) = 125.66\,\text{rad/s}$, whereas, based on the approximate solution,

$$\mu_{approx} = \frac{T(b - a)}{2\pi a^3 \omega h} = \frac{0.0036(0.0002)}{2\pi(0.009)^3(125.66)(0.06)} = 0.02085\frac{N\,s}{m^2}.$$

Hence, our error is only

$$\text{error} = \frac{\mu_{approx} - \mu_{exact}}{\mu_{exact}} = \frac{0.02085 - 0.02017}{0.02017} = 3.37\%.$$

Indeed, from our general formula [Eq. (9.84)],

$$\text{error} = \frac{2(0.0092)^2 - \left((0.009)(0.0092) + (0.009)^2\right)}{(0.009)(0.0092) + (0.009)^2} = 3.35\%,$$

the difference being due to numerical round-off errors.

9.4 Steady Flow in an Elliptical Cross Section

9.4.1 Biological Motivation

Many blood vessels are embedded within a particular soft tissue. Examples include the arteries within muscular organs such as the diaphragm, heart, uterus, and skeletal muscle. It is easy to imagine, therefore, that as the surrounding tissue deforms, the cross section of the embedded vessel can likewise change (Fig. 9.17). For example, blood vessels in the heart are compressed by the contracting muscle; indeed, vessels in the left heart are compressed closed during systole, which is why the left heart is perfused only during diastole. There is a need, therefore, to consider flows in noncircular geometries. Here, let us consider a steady flow within a tube of elliptical cross section, defined by major and minor radii a and b, respectively.

9.4.2 Mathematical Formulation

Let the z axis go through the center of the elliptical cross section with a boundary given by $x^2/a^2 + y^2/b^2 = 1$. Moreover, similar to our analysis of flow through the cylindrical tube, let us assume the following:

1. Newtonian fluid ($\mu = $ constant)
2. Incompressible flow ($\nabla \cdot \mathbf{v} = 0$)
3. Steady flow ($\partial \mathbf{v}/\partial t = \mathbf{0}$)
4. Axial flow ($v_x = v_y = 0$)
5. Negligible body forces ($\mathbf{g} = \mathbf{0}$)
6. Fully developed flow ($\partial \mathbf{v}/\partial z = \mathbf{0}$)
7. Laminar flow

To solve this problem using the Navier–Stokes equation, we have two choices: use elliptical coordinates or use Cartesian coordinates. We choose the latter here. Because of the no-slip boundary condition, the velocity $\mathbf{v} = v_z(x, y)\hat{\mathbf{e}}_z$ must be zero at all x and y around the inner surface of the ellipse. This will play a key role in our solution. First, however, recall that mass balance requires

$$\frac{\partial v_x}{\partial x} + \frac{\partial v_y}{\partial y} + \frac{\partial v_z}{\partial z} = 0, \tag{9.85}$$

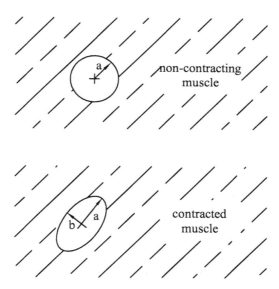

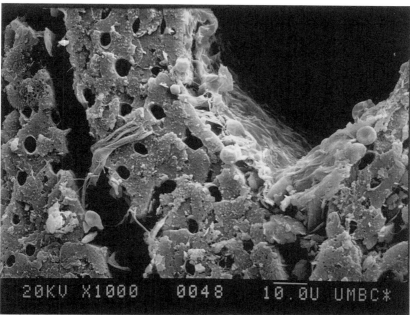

FIGURE 9.17 Illustration (upper) that many blood vessels are contained within surrounding soft tissue and that the deformation of the tissue can alter the cross-sectional shape of the vessel. Shown, too, is a scanning electron micrograph of heart tissue (lower): The many small holes are capillaries and the large hole is an artery. Note the contiguous endothelial cell layer inside the artery and the remnant red blood cells (biconcave disks about $8\,\mu$m in diameter) around the opening of the artery. It is easy to imagine that the cross sections of each of these vessels can be altered significantly by the finite strains experienced by the wall of the heart (recall Fig. 2.19 of Chapter 2).

which is satisfied identically given our assumptions. Linear momentum balance (i.e., the Navier–Stokes equation) requires [cf. Eqs. (8.36), (8.40), and (8.42)]

$$\hat{i}: \quad -\frac{\partial p}{\partial x} + \mu\left(\frac{\partial^2 v_x}{\partial x^2} + \frac{\partial^2 v_x}{\partial y^2} + \frac{\partial^2 v_x}{\partial z^2}\right) + \rho g_x$$

$$= \rho\left(\frac{\partial v_x}{\partial t} + v_x\frac{\partial v_x}{\partial x} + v_y\frac{\partial v_x}{\partial y} + v_z\frac{\partial v_x}{\partial z}\right), \quad (9.86)$$

$$\hat{j}: \quad -\frac{\partial p}{\partial y} + \mu\left(\frac{\partial^2 v_y}{\partial x^2} + \frac{\partial^2 v_y}{\partial y^2} + \frac{\partial^2 v_y}{\partial z^2}\right) + \rho g_y$$

$$= \rho\left(\frac{\partial v_y}{\partial t} + v_x\frac{\partial v_y}{\partial x} + v_y\frac{\partial v_y}{\partial y} + v_z\frac{\partial v_y}{\partial z}\right), \quad (9.87)$$

$$\hat{k}: \quad -\frac{\partial p}{\partial z} + \mu\left(\frac{\partial^2 v_z}{\partial x^2} + \frac{\partial^2 v_z}{\partial y^2} + \frac{\partial^2 v_z}{\partial z^2}\right) + \rho g_z$$

$$= \rho\left(\frac{\partial v_z}{\partial t} + v_x\frac{\partial v_z}{\partial x} + v_y\frac{\partial v_z}{\partial y} + v_z\frac{\partial v_z}{\partial z}\right). \quad (9.88)$$

After canceling out terms consistent with the above assumptions (do it using the worksheets in Section 8.7), we are left with

$$-\frac{\partial p}{\partial x} = 0, \quad -\frac{\partial p}{\partial y} = 0, \quad -\frac{\partial p}{\partial z} + \mu\left(\frac{\partial^2 v_z}{\partial x^2} + \frac{\partial^2 v_z}{\partial y^2}\right) = 0. \quad (9.89)$$

The first and second of these equations show that the pressure is a function of z at most, similar to the solution for a cylindrical tube. Because the velocity is a function of x and y at most, both the pressure gradient and the viscous term must equal a constant. Thus, our governing differential equation is

$$\frac{1}{\mu}\frac{dp}{dz} = \frac{\partial^2 v_z}{\partial x^2} + \frac{\partial^2 v_z}{\partial y^2}. \quad (9.90)$$

To solve this problem, one could first seek a solution to the homogenous differential equation of the form[1]

$$\frac{\partial^2 v_z}{\partial x^2} + \frac{\partial^2 v_z}{\partial y^2} = 0. \quad (9.91)$$

Although there are many different approaches to solve this linear partial differential equation, here we shall consider a very simple, yet powerful

---

[1] One may recognize that this is a 2-D Laplace equation, written as $\nabla^2 v_z = 0$, which appears widely in physics.

approach to solve the full nonhomogeneous equation. Note that the full solution v_z must satisfy the no-slip boundary condition around the inner perimeter. Consequently, let us consider a function $g(x, y)$ that is zero over the entire boundary of the flow, which for the elliptical boundary is $x^2/a^2 + y^2/b^2 - 1 = 0$. Hence, as a trial solution (i.e., guess), let

$$v_z(x, y) = cg(x, y) = c\left(\frac{x^2}{a^2} + \frac{y^2}{b^2} - 1\right), \tag{9.92}$$

where c is a yet unknown parameter. Taking the partial derivatives with respect to x, we get

$$\frac{\partial v_z}{\partial x} = c\frac{2x}{a^2} \quad \text{and} \quad \frac{\partial^2 v_z}{\partial x^2} = c\frac{2}{a^2}, \tag{9.93}$$

and similarly taking the partial derivatives with respect to y, we get

$$\frac{\partial v_z}{\partial y} = c\frac{2y}{b^2} \quad \text{and} \quad \frac{\partial^2 v_z}{\partial y^2} = c\frac{2}{b^2}. \tag{9.94}$$

Substituting these relations into the governing differential equation and solving for c, we obtain

$$c\left(\frac{2}{a^2}\right) + c\left(\frac{2}{b^2}\right) = \frac{1}{\mu}\frac{dp}{dz} \rightarrow c = \frac{1}{2\mu}\frac{dp}{dz}\left(\frac{a^2b^2}{a^2 + b^2}\right). \tag{9.95}$$

Substituting this expression into Eq. (9.92), we get the following solution for the velocity field:

$$v_z(x, y) = \frac{1}{2\mu}\frac{dp}{dz}\left(\frac{a^2b^2}{a^2 + b^2}\right)\left(\frac{x^2}{a^2} + \frac{y^2}{b^2} - 1\right), \tag{9.96}$$

which satisfies both the differential equation and the boundary conditions. Because the governing equations are linear, this trial solution is THE solution (i.e., mathematicians have proved uniqueness theorems for such linear differential equations). Note, too, that if $b = a$, with $x^2 + y^2 = r^2$, we recover the solution for the circular tube [cf. Eq. (9.45)], as we should. Such checks provide added confidence in the formulation and solution of the problem. Finally, given solutions for the pressure and velocity fields, other quantities of interest are calculated easily as in prior sections. This is left as an exercise.

9.5 Pulsatile Flow

Of the assumptions invoked in Section 9.2, the most suspect for many biological problems is that of steady flow. Here, therefore, let us consider an analysis of pulsatile flows based on a solution by J. R. Womersley in the 1950s. Because of the additional complexity due to the pulsatility, this shall return us to the Navier–Stokes solution for flow in a cylindrical rigid tube.

9.5.1 Some Biological Motivation

The cardiac cycle consists of four primary phases: diastolic filling, isovolumetric contraction, ejection, and isovolumetric relaxation (Fig. 9.18, top), which corresponds to the primary electrical events in the heart as revealed by an electrocardiogram (Fig. 9.18, bottom). As a result, the heart is a "pulsatile pump" and the associated flow within the arterial tree is pulsatile. Figure 9.19 shows, for example, the pressure P history in a typical artery (see Fig. 8.14 for data on the flow); each can be described well by a Fourier series (Milnor, 1989). For example, the pressure waveform for an aortic flow can be described by

$$p(t) = p_m + \sum_{n=1}^{N} (A_n \cos n\omega t + B_n \sin n\omega t), \qquad (9.97)$$

where p_m is the mean pressure, ω is the fundamental (circular) frequency, and A_n and B_n are the Fourier coefficients for N harmonics. Table 9.1 lists typical values.

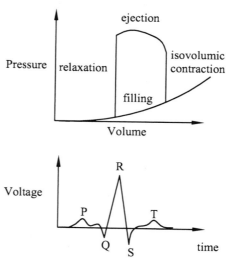

FIGURE 9.18 Schematic illustration of the four phases of the cardiac cycle: diastolic filling, isovolumic contraction (which builds up the ventricular pressure), ejection, and isovolumic relaxation. These mechanical phases are controlled by the electrical activity of the heart, which is monitored easily with an electrocardiogram (or EKG). This reminds us that coupled effects such as electromechanics are very important, as noted in Section 11.6 of Chapter 11.

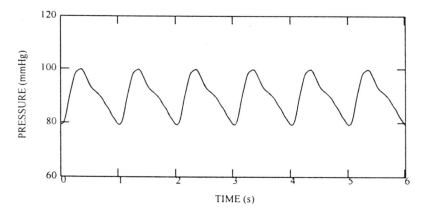

FIGURE 9.19 Typical variations in arterial pressures over the cardiac cycle, which can be described well via a Fourier series representation.

TABLE 9.1. Values of the Fourier coefficients for a Fourier series representation of an aortic pressure.

	$n = 1$	$n = 2$	$n = 3$	$n = 4$	$n = 5$	$n = 6$	$n = 7$
Pressure C_n	18.6	8.6	5.1	2.9	1.3	1.4	1.2
Pressure Φ_n	−1.67	−2.25	−2.61	−3.12	−2.91	−2.81	2.93
Flow C_n	202	157	103	62	47	42	31
Flow Φ_n	−0.78	−1.50	−2.11	−2.46	−2.59	−2.91	2.92

Note: The modulus $C_n = \sqrt{A_n^2 + B_n^2}$, in mm Hg or mL/s, whereas the phase $\Phi_n = \tan^{-1}(B_n/A_n)$, in radians, for each harmonic n. The mean values are $P_m = 85$ mm Hg and $Q_m = 110$ mL/s, with a fundamental frequency of 1.25 Hz.
Source: From Milnor (1989).

9.5.2 Mathematical Formulation

Womersley suggested that the pulsatile axial flow of an incompressible Newtonian fluid in a rigid tube could be studied by assuming that the pressure gradient $\partial p/\partial z$ could likewise be described by multiple harmonics. Consequently, he suggested that we let

$$\frac{\partial p}{\partial z} = \phi_0 + \sum_{n=1}^{N} (\phi_n \cos n\omega t + \psi_n \sin n\omega t), \tag{9.98}$$

where ϕ_0 is the mean (steady) portion of the pressure gradient and ϕ_n and ψ_n are Fourier coefficients for the nth harmonic. For analytical expediency, however, note that if $\Psi = \phi - i\psi$, where $i = \sqrt{-1}$, then

$$\Psi e^{i\omega t} = (\phi - i\psi)(\cos\omega t + i\sin\omega t) = \phi\cos\omega t + \psi\sin\omega t + i(\phi\sin\omega t - \psi\cos\omega t).$$
(9.99)

Hence, if we take the real part

$$\text{Re}(\Psi e^{i\omega t}) = \phi\cos\omega t + \psi\sin\omega t,$$
(9.100)

we obtain terms that appear in the Fourier representation of Eq. (9.98). (Note: Do not confuse the real part of a complex function Re( ) with the Reynolds' number Re). In particular, we can now let

$$\frac{\partial p}{\partial z} = \phi_0 + \sum_{n=1}^{N} \text{Re}(\Psi_n e^{in\omega t}),$$
(9.101)

which is simply a compact way to represent the assumed variation in the pressure gradient.

Recall from Eq. (9.38), therefore, that for steady flow in a circular tube, the Navier–Stokes equation reduces to

$$\frac{dp}{dz} = \mu\left[\frac{1}{r}\frac{\partial}{\partial r}\left(r\frac{\partial v_z}{\partial r}\right)\right].$$
(9.102)

The assumptions under which this flow takes place are essentially the same for the pulsatile flow that we are considering, with the exception that the assumption of steady flow does not apply. Again, therefore, mass balance [Eq. (9.34)] is satisfied identically for a $v = v_z(r, t)\hat{e}_z$, and the Navier–Stokes equations [Eqs. (9.35)–(9.37)] reduce to

$$-\frac{\partial p}{\partial r} = 0, \qquad -\frac{\partial p}{\partial \theta} = 0, \qquad -\frac{\partial p}{\partial z} + \mu\left[\frac{1}{r}\frac{\partial}{\partial r}\left(r\frac{\partial v_z}{\partial r}\right)\right] = \rho\frac{\partial v_z}{\partial t}. \quad (9.103)$$

The first two equations show that the pressure must be a function of z and time at most: $p = p(z, t)$. Hence, the single governing differential equation is

$$\rho\frac{\partial v_z(r, t)}{\partial t} + \frac{\partial p(z, t)}{\partial z} = \mu\left[\frac{1}{r}\frac{\partial}{\partial r}\left(r\frac{\partial v_z(r, t)}{\partial r}\right)\right].$$
(9.104)

Note that Eq. (9.104) is linear in both the pressure $p(z, t)$ and velocity $v_z(r, t)$ and, therefore, solutions of this linear differential equation can be superimposed. Let us deal with the steady and unsteady parts of the flow independently. This is a very important observation because the steady part of the flow has already been solved in Section 9.2. Thus, here we simply need to consider the unsteady part.

To look at the steady and unsteady parts of the flow separately, let the subscripts s and u denote steady flow and unsteady flow, respectively. Thus, we can write the unknown pressure and velocity fields as

$$p(z, t) = p_s(z) + p_u(z, t) \quad \text{and} \quad v_z(r, t) = v_s(r) + v_u(r, t). \tag{9.105}$$

Substituting these into Eq. (9.104), we obtain

$$\left\{ \frac{\partial p_s(z)}{\partial z} - \mu \left[\frac{1}{r} \frac{\partial}{\partial r} \left(r \frac{\partial v_s(r)}{\partial r} \right) \right] \right\} + \left\{ \rho \frac{\partial v_u(r, t)}{\partial t} + \frac{\partial p_u(z, t)}{\partial z} \right.$$
$$\left. - \mu \left[\frac{1}{r} \frac{\partial}{\partial r} \left(r \frac{\partial v_u(r, t)}{\partial r} \right) \right] \right\} = 0 \tag{9.106}$$

wherein we grouped terms that depend on time separate from those that do not. Because of the differences between the two groups of terms, each group must equal zero separately. The former is simply that which was solved in Section 9.2; hence, v_s is known. Let us focus our attention on the governing equation for the unsteady part of the flow, namely

$$\rho \frac{\partial v_u}{\partial t} + \frac{\partial p_u}{\partial z} - \mu \left[\frac{1}{r} \frac{\partial}{\partial r} \left(r \frac{\partial v_u}{\partial r} \right) \right] = 0. \tag{9.107}$$

Similar to the steady-flow solution, let us assume that the pressure gradient does not depend on z in a fully developed flow. Hence, the pulsatile pressure gradient depends on time t only, as given by Eq. 9.101.

Moreover, because of the linearity of the governing equation, let us solve Eq. (9.107) separately for each harmonic n ($= 1, 2, \ldots, N$). Our governing equation for unsteady flow thus becomes

$$\mu \left[\frac{1}{r} \frac{\partial}{\partial r} \left(r \frac{\partial v_u}{\partial r} \right) \right] - \rho \frac{\partial v_u}{\partial t} = \Psi_n e^{in\omega t} \tag{9.108}$$

for *each* n. Using separation of variables, we can separate the equation for $v_u(r, t)$ into one part that depends on r only and another that depends on t only; that is, let

$$v_u(r, t) = V_n(r) e^{in\omega t} \tag{9.109}$$

for each n. Substituting this equation into Eq. (9.108), we obtain

$$\mu \left[\frac{1}{r} \frac{\partial}{\partial r} \left(r \frac{\partial}{\partial r} [V_n(r) e^{in\omega t}] \right) \right] - \rho \frac{\partial}{\partial t} [V_n(r) e^{in\omega t}] = \Psi_n e^{in\omega t} \tag{9.110}$$

The common term $e^{in\omega t}$ cancels throughout, leaving the ordinary differential equation

$$\frac{d^2 V_n}{dr^2} + \frac{1}{r} \frac{dV_n}{dr} - \frac{\rho}{\mu} (in\omega) V_n = \frac{1}{\mu} \Psi_n \tag{9.111}$$

for each harmonic n. Thus, our governing ordinary differential equation for each n (now suppressed notationally) becomes

$$\frac{d^2V}{dr^2} + \frac{1}{r}\frac{dV}{dr} + \lambda^2 V = \frac{\Psi}{\mu}, \tag{9.112}$$

where $\lambda^2 \equiv i^3 n\omega\rho/\mu$. This governing differential equation has the form of a standard Bessel's equation

$$\frac{d^2y}{dx^2} + \frac{1}{x}\frac{dy}{dx} + y = 0, \tag{9.113}$$

which has a solution of the form

$$y(x) = c_1 J_0(x) + c_2 Y_0(x), \tag{9.114}$$

where $J_0(x)$ and $Y_0(x)$ are Bessel functions of order zero and the first and second kinds, respectively. For example,

$$J_0(x) = 1 - \frac{x^2}{2^2} + \frac{x^4}{2^2 4^2} - \frac{x^6}{2^2 4^2 6^2} + \cdots, \tag{9.115}$$

$$Y_0(x) = J_0(x)\log(x) + \frac{x^2}{4} - \frac{3x^4}{128} + \cdots. \tag{9.116}$$

Now, if we consider a change of variables, $x \equiv \lambda z$, then

$$\frac{d}{dx} = \left(\frac{d}{dz}\right)\left(\frac{dz}{dx}\right) = \frac{1}{\lambda}\frac{d}{dz} \quad \text{and} \quad \frac{d^2}{dx^2} = \frac{1}{\lambda}\frac{d}{dz}\left(\frac{1}{\lambda}\frac{d}{dz}\right) = \frac{1}{\lambda^2}\frac{d^2}{dz^2}, \tag{9.117}$$

thus, an equation of the form

$$\frac{d^2y}{dz^2} + \frac{1}{z}\frac{dy}{dz} + \lambda^2 z = 0 \tag{9.118}$$

admits a solution of the form

$$y(z) = c_1 J_0(\lambda z) + c_2 Y_0(\lambda z). \tag{9.119}$$

The homogeneous solution of our governing equation (9.112) is thus

$$V_H(r) = c_1 J_0(\lambda r) + c_2 Y_0(\lambda r), \tag{9.120}$$

where c_2 must be zero to maintain $V(r)$ finite at the centerline $r = 0$ (a similar restriction was used in the steady-flow solution). For the particular solution, we let $V_p(r) = c_3$ and find that $c_3 = \Psi/\mu\lambda^2$ for each harmonic. Hence,

$$V(r) = V_H(r) + V_p(r) = c_1 J_0(\lambda r) + \frac{\Psi}{\mu\lambda^2} \tag{9.121}$$

for each harmonic n. Now, this solution must satisfy the no-slip boundary condition $v_z(r = a, t) = 0$ for all time and, thus, $V(r = a) = 0$. Hence,

$$c_1 J_0(\lambda a) = -\frac{\Psi}{\mu \lambda^2} \rightarrow c_1 = -\frac{\Psi}{\mu \lambda^2}\left(\frac{1}{J_0(\lambda a)}\right) \qquad (9.122)$$

and

$$V(r) = \frac{\Psi}{\mu \lambda^2}\left(1 - \frac{J_0(\lambda r)}{J_0(\lambda a)}\right) \qquad (9.123)$$

for each harmonic n. Our final solution, therefore, for the assumed pressure gradient

$$\frac{\partial p}{\partial z}(t) = \frac{\partial p}{\partial z}(\text{steady}) + \sum_{n=1}^{N} \text{Re}(\Psi_n e^{in\omega t}) \qquad (9.124)$$

is

$$v(r,t) = v_s(r) + \text{Re}\left\{\sum_{n=1}^{N}\left[\frac{\Psi_n}{\mu \lambda_n^2}\left(1 - \frac{J_0(\lambda_n r)}{J_0(\lambda_n a)}\right)e^{in\omega t}\right]\right\}, \qquad (9.125)$$

where the number of harmonics $n = 1,2,\ldots,N$ is dictated by the Fourier series fit to the pressure gradient data [cf. Eq. (9.98)]. The wall shear stress can thus be computed in the normal way, where

$$\sigma_{rz}(r,t) = \mu \frac{\partial v_z}{\partial r}. \qquad (9.126)$$

Note, therefore, that

$$\frac{d}{dx}[J_0(kx)] = -kJ_1(kx), \qquad (9.127)$$

where J_1 is a first-order Bessel function of the first kind.

In particular, for each harmonic n, the unsteady contribution is

$$(\tau_w)_u = \text{Re}\left\{\frac{|\Psi_n|}{\lambda_n}\left(\frac{J_1(\lambda_n a)}{J_0(\lambda_n a)}\right)e^{in\omega t}\right\}. \qquad (9.128)$$

Likewise, the unsteady contribution to the volumetric flow rate is

$$Q_u(t) = \text{Re}\left\{\frac{\pi a^4 \Psi_n e^{in\omega t}}{\mu(\lambda_n a)^2}\left(1 - \frac{2J_1(\lambda_n a)}{\lambda_n a J_0(\lambda_n a)}\right)\right\}. \qquad (9.129)$$

Observation 9.3 Computations based on Womersley's results are clearly complex, and the interested reader is referred to Zamir (2000) for more details. Nonetheless, Figure 9.20 shows velocity profiles (fully developed) at five different times in the cardiac cycle. In particular, note the near-

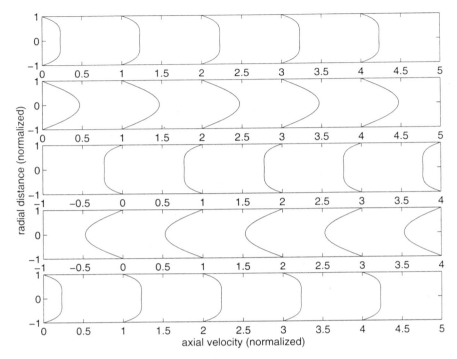

FIGURE 9.20 Pulsatile velocity profiles computed for low frequencies (1 Hz). Results are shown for different phase angles ωt, values being $0°$ at the top and increasing by $90°$ for subsequent panels. Note that the flow "develops" over time at a single location similar to its development with distance in an entrance length. [From Zamir (2000), with permission from Springer-Verlag.]

parabolic profiles in the second and fourth panels, but the more blunted profiles in the first, middle, and fifth panels (the first and fifth are the same because the pressure gradient is periodic). Note, too, that we see a flow reversal in the third and fourth panels. Because wall shear stress is proportional to the slope of the velocity profile at the wall (i.e., the velocity gradient), we see that the wall shear stress is oscillatory. There has been considerable attention in the literature on delineating the effects of the oscillatory versus mean wall shear stress on atherogenesis and other pathologies. The interested reader should research this.

Indeed, because of the potential importance of the unsteadiness, a nondimensional parameter called the *Womersley number* α is defined as

$$\alpha = \sqrt{\frac{a\omega\rho}{\mu}}.$$

Typical values in man are $\alpha = 22.2$ in the aorta and $\alpha = 4.0$ in the femoral artery; in comparison, $\alpha = 4.3$ in the rat aorta and $\alpha = 1.5$ in the rat femoral artery. Localization of disease is also correlated with α in some works (see Milnor, 1989).

Finally, pulsatility raises important issues with regard to the generation and reflection of waves in distensible tubes. Again, however, the reader is referred to Fung (1984) or Zamir (2000) for more on this advanced topic.

9.6 Non-Newtonian Flow in a Circular Tube

Whereas the flow of air in the airways, the flow of urine in the ureters, and the flow of blood in large arteries at sufficiently high shear rates can all be modeled assuming a Newtonian response, non-Newtonian behavior can be important in the vasculature. Hence, let us consider a brief introduction to an analysis of a relevant non-Newtonian flow.

9.6.1 Motivation

Figure 7.11 of Chapter 7 shows that blood, among other biological fluids, exhibits a non-Newtonian (pseudoplastic) behavior under certain circumstances (e.g., low shear rates). Whereas quantification of linear (e.g., Newtonian) material behavior is simplified by the uniqueness of linear relations, quantification of nonlinear behavior remains an area of active research. The interested reader is reminded of Eq. (7.68), but referred to texts on nonlinear rheology (e.g., Tanner, 1985). Here, we shall restrict our attention to the simplest nonlinear behavior—a 1-D power-law model, which is not without mathematical limitations, but does serve to illustrate some nonlinear effects; that is, whereas a 1-D constitutive relation for a Newtonian fluid can be written as

$$\sigma_{rz} = 2\mu D_{rz} = \mu \frac{\partial v_z}{\partial r} \tag{9.130}$$

for an axial flow in a circular tube characterized by $v = v_z(r)\hat{e}_z$ [cf. Eq. (9.45)], a generalization of this relation has been proposed of the form

$$\sigma_{rz} = \underbrace{k \left| \frac{dv_z}{dr} \right|^{n-1} \frac{dv_z}{dr}}_{\mu_a} = k \left(\frac{dv_z}{dr} \right)^n, \tag{9.131}$$

where μ_a is an "apparent" viscosity, k an empirical parameter, and n a nonintegral material parameter. Of course, when $\mu_a \equiv \mu$ and $n = 1$, we recover the Newtonian result. When $n > 1$, we have a dilatant behavior, and when $n < 1$, we have a pseudoplastic behavior (cf. Fig. 7.11 of Chapter 7).

In the special case that $n = 0$, we have $\sigma_{rz} = $ constant, independent of deformation. Such a model is called perfectly plastic in solid mechanics, hence the name pseudoplastic for n approaching zero.

9.6.2 Mathematical Formulation

Consider a differential annulus as shown in Figure 9.21. The sum of the z components of force acting on the full annulus must be zero in the case of a steady flow (i.e., no local acceleration) and in the absence of a convective acceleration. Enforcing equilibrium, we have

$$p2\pi r\Delta r - \left(p + \frac{\partial p}{\partial z}\Delta z\right)2\pi r\Delta r + \left(\sigma_{rz} + \frac{\partial \sigma_{rz}}{\partial r}\Delta r\right)2\pi(r + \Delta r)\Delta z - \sigma_{rz}2\pi r\Delta z = 0.$$

(9.132)

Simplifying, we have

$$-\frac{\partial p}{\partial z}r\Delta r\Delta z + \frac{\partial \sigma_{rz}}{\partial r}r\Delta r\Delta z + \sigma_{rz}\Delta r\Delta z + \frac{\partial \sigma_{rz}}{\partial r}\Delta r^2\Delta z = 0.$$ (9.133)

Dividing this equation by $r\Delta r\Delta z$ and letting $\Delta r \to 0$ and $\Delta z \to 0$, the pressure gradient is found to be [cf. Eq. (8.26)]

$$\frac{\partial p}{\partial z} = \frac{\partial \sigma_{rz}}{\partial r} + \frac{\sigma_{rz}}{r} = \frac{1}{r}\frac{\partial}{\partial r}(r\sigma_{rz}).$$ (9.134)

Assuming p varies with z alone and integrating with respect to r yields

$$\int \frac{d}{dr}(r\sigma_{rz})\,dr = \int \frac{dp}{dz}r\,dr,$$ (9.135)

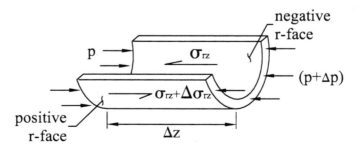

FIGURE 9.21 Free-body diagram of half a differential annulus of fluid and associated shear stresses for purposes of deriving a general equation of motion. Compare to Figure 8.1 of Chapter 8 for a cuboidal fluid element.

or

$$r\sigma_{rz} = \frac{dp}{dz}\frac{r^2}{2} + c_1 \rightarrow \sigma_{rz} = \frac{dp}{dz}\frac{r}{2} + \frac{c_1}{r}. \tag{9.136}$$

Now, we can apply a constitutive equation for a Newtonian fluid or a non-Newtonian fluid (because this derivation thus far has been independent of the material). For example, substituting the expression for a power-law fluid [Eq. (9.131)] into Eq. (9.136), we obtain

$$k\left(\frac{dv_z}{dr}\right)^n = \frac{dp}{dz}\frac{r}{2} + \frac{c_1}{r}. \tag{9.137}$$

Applying the symmetry condition that $dv_z/dr = 0$ at the centerline ($r = 0$), $c_1 = 0$. Therefore,

$$\left(\frac{dv_z}{dr}\right)^n = \frac{1}{2k}\frac{dp}{dz}r \rightarrow \frac{dv_z}{dr} = \left(\frac{1}{2k}\frac{dp}{dz}\right)^{1/n} r^{1/n}. \tag{9.138}$$

Integrating with respect to r again,

$$\int \frac{d}{dr}(v_z)\,dr = \left(\frac{1}{2k}\frac{dp}{dz}\right)^{1/n} \int r^{1/n}\,dr, \tag{9.139}$$

we obtain

$$v_z(r) = \left(\frac{1}{2k}\frac{dp}{dz}\right)^{1/n} \frac{r^{1+1/n}}{1+1/n} + c_2 \rightarrow v_z(r) = \left(\frac{1}{2k}\frac{dp}{dz}\right)^{1/n} \frac{n}{1+n} r^{(1+n)/n} + c_2. \tag{9.140}$$

Applying the no-slip boundary condition at the wall, $v_z(r = a) = 0$, we find that

$$0 = \left(\frac{1}{2k}\frac{dp}{dz}\right)^{1/n} \frac{n}{1+n} a^{(1+n)/n} + c_2 \rightarrow c_2 = -\left(\frac{1}{2k}\frac{dp}{dz}\right)^{1/n} \frac{n}{1+n} a^{(1+n)/n}. \tag{9.141}$$

Therefore,

$$v_z(r) = \left(\frac{1}{2k}\frac{dp}{dz}\right)^{1/n} \frac{n}{1+n} (r^{(1+n)/n} - a^{(1+n)/n}). \tag{9.142}$$

As $n = 1$, we recover the result for Newtonian flows [Eq. (9.45)], as we should. Clearly, the volumetric flow rate Q can be computed and, hence, so too the wall shear stress in terms of Q.

For example,

$$Q = \frac{\pi n}{1+3n}\left[\frac{a^{1+3n}}{2k}\left(-\frac{dp}{dz}\right)\right]^{1/n}.$$ (9.143)

Noting that Eq. (9.142) can be written as

$$v_z(r) = -\frac{n}{1+n}\left(\frac{a^{1+n}}{2k}\frac{dp}{dz}\right)^{1/n}\left(1-\left(\frac{r}{a}\right)^{(1+n)/n}\right),$$ (9.144)

the maximum velocity is

$$v_z)_{max} = v_z(r=0) = -\frac{n}{1+n}\left(\frac{a^{1+n}}{2k}\frac{dp}{dz}\right)^{1/n};$$ (9.145)

hence, we can write

$$\frac{v_z(r)}{v_z)_{max}} = 1-\left(\frac{r}{a}\right)^{(1+n)/n},$$ (9.146)

which facilitates plotting the velocity profile. Figure 9.22 shows, for example, a profile for $n < 1$, which differs from the parabolic profile for $n = 1$. In particular, note the blunted profile for the pseudoplastic response, remembering from Chapter 7 that blood exhibits a pseudoplastic character.

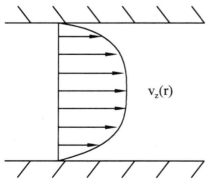

$v_z(r)$

Non-Newtonian Fluid

FIGURE 9.22 Non-Newtonian velocity profiles; note the more blunted profile than for the Newtonian case (cf. Fig. 9.13).

Appendix 9: Biological Parameters

Fundamental to computations in mechanics is knowledge of geometry, material properties, and applied loads for the system of interest. Here, we list information of importance to hemodynamics and airflow mechanics.

TABLE A9.1. Values for blood pressure in humans in health and hypertension.

	Diastolic	Systolic
Normotensive	<85	<130
High	85–89	130–139
Hypertensive		
Stage 1	90–99	149–159
Stage 2	100–109	169–179
Stage 3	110–119	180–209
Stage 4	>120	>210

Source: From J. H. Laragh and B. M. Brener (1995) *Hypertension, Pathophysiology, Diagnosis, and Management.* Raven Press, New York.

TABLE A9.2. Mean blood vessel characteristics (man).

Vessel	Lumen radius	Wall thickness	Pressure (mm Hg)	CSA (cm^2)
Aorta	1.25 cm	2 mm	120/80	4.5
Artery	0.4 cm	1 mm	112/79	20
Arteriole	15 mm	20 μm	45/35	400
Capillary	6 μm	1 μm	30	4500
Venule	10 μm	2 μm	20	4000
Vein	0.25 cm	0.5 mm	15	40
Vena cava	1.5 cm	1.5 mm	10	18

Note: Pressure is given as systolic/diastolic if pulsatile; CSA is the accumulative cross-sectional area. Vena cava pressures fluctuate with pulmonary inspiration and expiration but are given as representative.
Source: From Johnson (1991).

TABLE A9.3. Hemodynamic characteristics (man).

	α	$\bar{v}$ (cm/s)	Mean Re	Max Re
Ascending aorta	21	18	1500	9400
Abdominal aorta	12	14	640	3600
Renal artery	4	40	700	1300
Femoral artery	4	12	200	860
Inferior vena cava	17	21	1400	3000

Source: From Milnor (1989), p. 148.

TABLE A9.4. Mean airway characteristics (man).

	Generation	Number	Diameter (mm)	Length (mm)	CSA (cm^2)	$\bar{v}$ (cm/s)	Re
Trachea	0	1	18	120	2.6	393	4350
Main bronchus	1	2	12.2	47.6	2.3	427	3210
Lobar bronchus	2	4	8.3	19.0	2.2	462	2390
Lobar bronchus	3	8	5.6	7.6	2.0	507	1720
Segmental Bronchus	4	16	4.5	12.7	2.6	392	1110
Terminal Bronchus	11	2050	1.09	3.9	19	52.3	34
Alveoli	Last	300×10^6	0.28	0.28	—	—	—

Note: CSA is the accumulative cross-sectional area.
Source: From Weibel (1963).

TABLE A9.5. Ventilation–perfusion ratios.

	% Lung volume	Alveolar Q (cm^3/s)	Perfusion Q (cm^3/s)	Ventilation/ perfusion ratio
Top	7	4	1.2	3.3
	8	5.5	3.2	1.8
	10	7.0	5.5	1.3
	11	8.7	8.3	1.0
	12	9.8	11.0	0.9
	13	11.2	13.8	0.8
	13	12.0	16.3	0.73
	13	13.0	19.2	0.68
Bottom	13	13.7	21.5	0.63
Total	100%			

Source: From Johnson (1991), p. 178.

TABLE A9.6. Density and viscosity for common fluids.

Material	Density (kg/m^3)		Viscosity (Ns/m^2)	
	20°C	37°C	20°C	37°C
Air	1.208	1.142	1.8×10^{-5}	2.0×10^{-5}
Water	998	995	1.0×10^{-3}	7.5×10^{-4}
Plasma			1.9×10^{-3}	1.2×10^{-3}
Glycerin	1260		1.6	4.5×10^{-1}

Notes: The density of water @ 4°C is 1000 kg/m^3, which is used to compute specific gravities: SG = ρ/ρ_{H_2O} at 4°C. 1 P = 0.1 N s/m^2 and thus 1 cP = 1×10^{-3} N s/m^2. Finally, the viscosity tends to vary as $\mu \sim Ae^{B/T}$ for fluids, where A and B are material parameters, and T is temperature.

Exercises

9.1 Design (sketch) an experimental setup that ensures a constant steady flow within a parallel-plate device. Discuss various options with regard to how to generate the requisite constant pressure gradient.

9.2 The velocity profile in Eq. (9.15) represents a flow between parallel plates relative to an $(o; x, y)$ coordinate system with the origin at the bottom plate. Show that the solution can alternatively be written as

$$v_x(y) = -\frac{1}{2\mu}\left(\frac{dp}{dx}\right)\left(\frac{h^2}{4} - y^2\right)$$

if the origin of the coordinate system is at the centerline (i.e., $y \in [-h/2, h/2]$). Show, too, that regardless of the particular coordinate system used, the values of Q and τ_w are the same for this Poiseuille flow.

9.3 To facilitate access to the cells, some researchers use a parallel flow setup in an incubator wherein the bottom plate is stationary but the top surface of the fluid is exposed to an air/CO_2 environment (i.e., a free surface). Assuming a steady, incompressible, fully developed, 1-D flow, show that

$$Q = -\frac{h^3 w}{3\mu}\left(\frac{dp}{dx}\right) \quad \text{and} \quad \tau_w = \frac{3\mu Q}{wh^2}.$$

9.4 A constant-pressure gradient $dp/dx = 0.2\,\text{kPa/m}$ is used to drive *glycerin* through a parallel-plate device with gap $h \sim 0.2\,\text{m}$. Find the maximum velocity and volumetric flow rate per unit width. Compare these values to those for water. Assume that the temperature is ~20°C. Values for the viscosities are in Appendix 9.

9.5 Is the flow field in Example 9.2 irrotational?

9.6 Assuming a steady flow with no gravity, find the velocity in the x direction for the problem in Figure 9.23, where U is constant and the fluid is at a constant pressure.

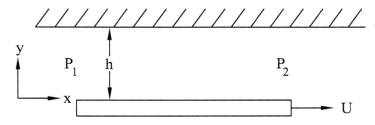

FIGURE 9.23

9.7   Solve for the flow between parallel plates (cf. Example 9.2) with both $dp/dx \neq 0$ (i.e., $p_1 > p_2$) and $U_0 \neq 0$. Calculate Q and $v_x)_{max}$, as well as σ_{xy} and τ_w in terms of Q.

9.8   For the problem in Example 9.3, show that

$$Q = \frac{\rho g \sin \theta h^3}{3\mu}.$$

Furthermore, plot the velocity distribution $v_x/v_x)_{max}$ on the abscissa versus the depth of the fluid (y/h) on the ordinate. Note that both variables are nondimensional and that they will vary from 0 to 1 regardless of the specific values in the problem. Finally, plot a normalized shear stress $\sigma_{xy}/\sigma_{xy})_{max}$ versus y/h and discuss based on the shape of the velocity distribution curve.

9.9   A biomedical device is thrombogenic and thus must be coated with a thin biocompatible film as in Figure 9.24. Assume that the fluid adheres to the device (no slip) as the device is pulled through it. Assume a constant film thickness h, and that the fluid behaves as Newtonian and incompressible. By solving Navier–Stokes, show that

$$h = \sqrt{\frac{2\mu U}{\rho g_x}}.$$

Hint: Assume $v_x(y = h) = 0$ in addition to the free-surface boundary condition $\partial v_x/\partial y \ (y = h) = 0$. What is implied by the latter condition?

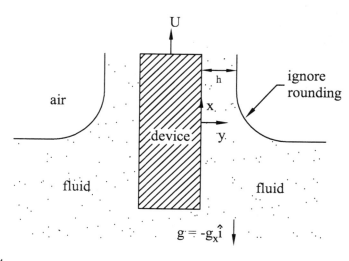

FIGURE 9.24

9.10 A plate with area A and mass M is sliding down an incline covered by a fluid film of constant thickness h (Fig. 9.25). (a) Determine the velocity profile in the fluid. (b) Determine the velocity of the plate U_0. Hint: Assume that the pressure gradient in the x direction is zero. Draw a free-body diagram of the plate and sum the forces to find an expression for the shear stress acting on the plate in terms of M, A, and g.

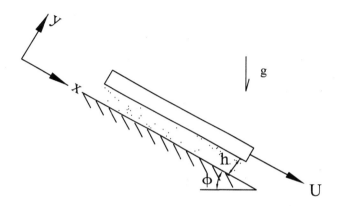

FIGURE 9.25

9.11 A rigid membrane with negligible thickness is located between two belts and is free to move (Fig. 9.26). The top belt is moving to the right with velocity $2U$. The bottom belt is moving to the left with velocity U. The fluid in section 1 (lower-half) has viscosity μ_1 and the fluid in section 2 (upper-half) has viscosity μ_2, with $\mu_1 = 3\,\mu_2$. (a) Determine the velocity field in sections 1 and 2. (b) Determine the velocity of the membrane using the given coordinate system. Hint: Draw a free-body diagram of the membrane to find the boundary condition at $y = h$.

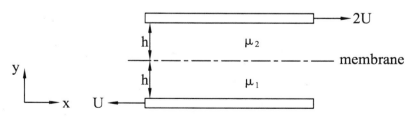

FIGURE 9.26

9.12 A fluid film of constant thickness h is sliding down a vertical wall (Fig. 9.27). Show that the velocity field is given by

$$v_y = \frac{\rho g h^2}{\mu} \left[\frac{x}{h} - \frac{1}{2} \left(\frac{x}{h} \right)^2 \right]$$

for $x \in [0, h]$. Hint: Assume no-slip at the wall ($x = 0$) and assume that the shear stress due to the airflow over the fluid is negligible (i.e., $\partial v_y / \partial x \big|_{x=h} = 0$). Show that the velocity field is equivalent to that obtained in Example 9.3 if $\theta = 90°$.

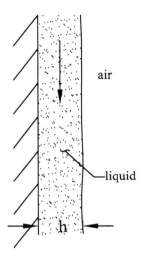

air

—liquid

h

FIGURE 9.27

9.13 Show that the governing differential equation [Eq. (9.39)] for a steady flow of a Newtonian fluid in a rigid circular tube can be written as

$$\frac{1}{\mu} \frac{dp}{dz} = \frac{d^2 v_z}{dr^2} + \frac{1}{r} \frac{dv_z}{dr}.$$

Consequently, the equation can be solved by assuming a solution of the form $v_z \propto r^n$. Show that the solution is the same as that in Eq. (9.45).

9.14 Ex vivo perfusion systems are becoming more common in vascular research. Assuming a fully developed laminar flow, the wall shear stress in the perfused vessel is estimated by

$$\tau_w = \frac{4\mu Q}{\pi r_i^3}.$$

If $r_i \sim 2.2\,\text{mm}$ (porcine carotid) and $\mu = 4\,\text{cP}$ (Han and Ku, 2001) find the value of Q necessary to produce a physiologic wall shear stress $\sim 1.5\,\text{Pa}$.

9.15 Estimate the value of the Reynolds' number in the vena cava in a human. For comparison, note that for the vena cava of a dog, the diameter is ~1.25 cm, the mean velocity ~33 cm/s, the wall shear rate ~211 s^{-1}, the viscosity ~3 cP, and, thus, τ_w ~0.63 Pa. What is the associated value of the skin friction coefficient c_f?

9.16 Both in vivo and in vitro experiments show that the erythrocytes in blood vessels do not distribute themselves evenly across the cross section of a large blood vessel. Instead, they tend to accumulate along the centerline, thereby allowing, in a statistical sense, a thin cell-free layer to form along the wall of the vessel called the plasma layer (Fig. 9.28). Let the central core region containing cells have a viscosity μ_c and the cell-free plasma layer have a viscosity μ_p and thickness δ. In each region, assume that the flow is Newtonian. Use the Navier–Stokes equation to find (a) the velocity profile in the core region $v_z^c(r)$, (b) the velocity profile in the plasma layer $v_z^p(r)$, (c) the core volumetric flow rate Q_c, and (d) the plasma layer volumetric flow rate Q_p. Hint: Assume steady, unidirectional flow with a pressure gradient in the z direction to drive the flow. The velocity and shear stress in each region must be the same at the interface $r = a - \delta$.

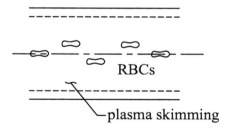

FIGURE 9.28

9.17 For a pressure-driven axial flow between long concentric cylinders (similar to that in Fig. 9.14 but for an axial flow; see Fig. 9.29), find the expression for the velocity profile in the z direction if the inner cylinder is of radius b and outer cylinder is of radius a. This problem relates to flow in an airway or blood vessel in which a central catheter has been placed. In particular, show that (Stattery, 1981)

$$v_z(r) = \frac{-a^2}{4\mu}\left(\frac{dp}{dz}\right)\left(1 - \frac{r^2}{a^2} + \frac{1-\beta^2}{\ln(1/\beta)}\ln\frac{r}{a}\right),$$

where $b = \beta a$ and $\beta < 1$. In addition, find an expression for the volumetric flow rate Q. Note that $b = 0$ recovers Eq. 9.45.

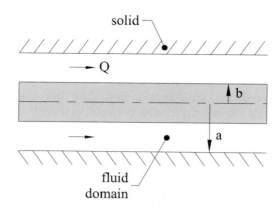

FIGURE 9.29

9.18 The viscosity in a concentric cylinder viscometer was shown to be calculated via

$$\mu = \frac{T(b^2 - a^2)}{4\pi\omega ha^2 b^2},$$

where T is the applied torque, a and b are the inner and outer radii, respectively, h is the height, and ω is the angular velocity. In contrast, a so-called capillary viscometer allows one to measure viscosity according to

$$\mu = \frac{-\pi a^4}{8Q}\left(\frac{dp}{dz}\right),$$

where dp/dz is the applied pressure gradient, a is the radius of the capillary (i.e., straight tube), and Q is the volumetric flow rate. Show that this equation is correct and state the associated restrictions that govern the experimental setup.

9.19 We have seen that there are many different types of viscometers, including the concentric-cylinder (Section 9.3) and the cone-and-plate (Section 7.6) devices. Explain how the results of Section 9.2 can be used to design a "capillary viscometer." Also discuss why, in contrast to the cone-and-plate viscometer, the capillary viscometer would not be useful for non-Newtonian fluids.

9.20 Similar to the previous exercise, explain how the results of Section 9.1 can be used to design a parallel-plate *flowmeter* to measure Q.

9.21 Synthecon, Inc. is a company that produces rotary cell culture systems, or *bioreactors*. According to literature on their website, a bioreactor is any device that monitors and controls the environment of a population of cells so as to promote normal metabolic and other activities. They write further that "the fluid-filled rotating wall vessel (RWV) bioreactor is a recently developed cell culture device that is able to successfully integrate cell–cell and cell–matrix co-localization and three-dimensional interaction with excellent low-shear mass-transfer of nutrients and wastes, without sacrificing one parameter for the other. Designed by Ray Schwarz, David Wolf, and Tinh Trinh at the Johnson Manned Spaceflight Center, the RWV bioreactor consists of a cylindrical growth chamber that contains an inner co-rotating cylinder with a gas exchange membrane." Write a three-page summary and critique of the NASA RWV bioreactor with particular emphasis on the fluid mechanics.

9.22 The velocity field for the NASA bioreactor (rotating cylinders) was assumed to be $v = v_\theta(r)\hat{e}_\theta$, with $v_\theta = r\omega$. This flow was shown to be "shearless" but not irrotational. How would the situation change if the angular velocity was a function of time [i.e., $\omega = \omega(t)$, with both cylinders still moving together]? Why? Is it possible to construct a Wormersley-type solution for an oscillating case?

9.23 Confirm the result in Eq. (9.84) for the error in the inferred torques based on the exact versus flat plate approximations.

9.24 Let $b = a + h$ in Eq. (9.84). Show numerically how the error increases as the gap h increases. Hint: Consider different values of h as a percentage of a and plot the error versus h/a.

9.25 In some cases, the *pressure gradient* in a tube flow is computed as $-\partial p/\partial z = \Delta p/L$, where Δp is called the *pressure drop* ($\Delta p = p_i - p_o$, where subscripts i and o denote inlet and outlet) and L is the distance between the locations at which p_i and p_o are measured. If the straight tube is angled upward at angle α, show that the results from Section 9.2 hold provided that the pressure gradient is taken to be (Slattery, 1981, p. 73)

$$-\frac{\partial p}{\partial z} = \frac{p_i - p_o - \rho g L \sin\alpha}{L}.$$

9.26 A catheter is to be coated by a nonthrombogenic film. A schema of the manufacturing setup is shown in Fig. 9.30. Assume that the flow is steady, laminar, and fully developed in the section L and that the coating fluid is Newtonian. Find the volumetric flow rate Q of the fluid through this section. Assuming the coating (film) thickness δ is uniform, find an expression for δ.

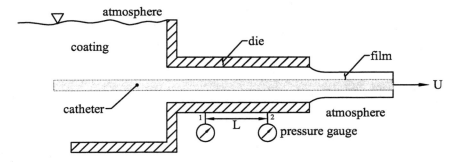

FIGURE 9.30

9.27 In Section 9.4, we found the solution to the flow in an elliptical cross section by assuming a form of $v_z(x, y)$ that satisfies the no-slip boundary condition exactly. Rederive the solution for steady, incompressible, fully developed, Newtonian flow in a rigid, straight, circular tube using this same idea. Hint: Let $v_z(r) = c(x^2 + y^2 - a^2) = c(r^2 - a^2)$.

9.28 We found that the velocity in an elliptical cross section is given by

$$v_z(x, y) = -\frac{1}{2\mu}\left(\frac{dp}{dz}\right)\frac{a^2 b^2}{a^2 + b^2}\left(1 - \frac{x^2}{a^2} - \frac{y^2}{b^2}\right).$$

Show that, as reported by Zamir (2000), the volumetric flow rate and wall shear stress are given by

$$Q = -\frac{\pi a^3 b^3}{4\mu(a^2 + b^2)}\left(\frac{dp}{dz}\right)$$

and

$$\tau_w = \frac{dp}{dz}\left(\frac{a^2 b^2}{a^2 + b^2}\right)\sqrt{\frac{x_1^2}{a^4} + \frac{y_1^2}{b^4}},$$

where (x_1, y_1) are points on the elliptic boundary.

9.29 The Fourier series representation for pressure in Eq. (9.97) can be written alternatively as

$$p(t) = p_m + \sum_{n=1}^{N} C_n \cos\left(\frac{2\pi n t}{T} + \Phi_n\right),$$

where T is the period (not temperature) and Φ_n is a phase shift. Prove that this is the case, and in so doing, relate the Fourier coefficients A_n and B_n to the amplitude C_n and the phase Φ_n. See Table 9.1.

9.30 Plot the function

$$f(t) = \sum_{n=1}^{N} C_n \cos\left(\frac{2\pi n t}{T} + \Phi_n\right)$$

given

C_n	7.5803	5.4124	1.5210	0.5217	0.8311
Φ_n	−173.9200	88.9220	−21.7046	−33.5370	−126.8100

Pick a constant value of $2\pi/T$ for a typical cardiac cycle.

9.31 Nagel et al. (J Clin Invest 94: 885–891, 1994) used a cone-and-plate device to subject cultured endothelial cells to various shear stresses. They showed that cultured human umbilical vein endothelial cells (HUVEC) exhibit time-dependent changes in the production of adhesion molecules for wall shear stresses τ_w from 0.25 to 4.6 Pa. They report that τ_w was given by

$$\tau_w = \frac{\mu\omega}{\alpha}\left[1 - 0.4743\left(\frac{r^2\omega\alpha^2\rho}{12\mu}\right)^2\right],$$

where μ is the viscosity of the culture media, ρ is its mass density, ω is the angular velocity of the cone, α is its inclination angle, and r is a radial distance from the symmetry axis. Show that the formula is approximately "correct."

9.32 Usami et al. (Ann Biomed Eng 21: 77–83, 1993) reported that a multidirectional steady-flow solution between parallel flat plates is given by

$$v_x = \frac{6}{h^2} z(h - z)\bar{v}_x(x, y), \qquad v_y = \frac{6}{h^2} z(h - z)\bar{v}_y(x, y),$$

where h is the gap distance, z is the out-of-plane direction, and $\bar{v}_x$ and $\bar{v}_y$ are mean values over the gap at any (x, y). Show that in the case of $\bar{v}_x = c$, a constant, and $\bar{v}_y = 0$, one recovers the simple Poiseuille flow. Find the "value" of c. Show, too, that these results for $v_x(x, y, z)$ and $v_y(x, y, z)$ are solutions to the incompressible Navier–Stokes equation and that

$$\tau_w\mathbf{n} = \tau_{13}\hat{\mathbf{e}}_1 + \tau_{23}\hat{\mathbf{e}}_2 = \frac{6\mu}{h}\bar{\mathbf{v}}.$$

Finally, note that this formulation allowed the design of a unique flow chamber, with particular advantages over the standard parallel-plate device.

9.33 Repeat the analysis in Section 9.6 for the flow of a power-law fluid between parallel plates. Let the gap distance be h and the width w and assume a fully developed, steady flow. If $y \in [-h/2, h/2]$, find the velocity distribution $v_x(y)$.

9.34 Plot the velocity distribution $v_x/v_x)_{max}$ in Exercise 9.33 versus depths y/h (ordinate) for $n = 0.8$, $n = 1$, and $n = 1.5$.

9.35 Plot Q(non-Newtonian)/Q(Newtonian) on the ordinate versus the power-law exponent n on the abscissa for n from 0.5 to 1.5 for the parallel-plate flow.

9.36 Given the following data for water

Temperature (°C)	Density (kg/m³)	Viscosity (Ns/m²)
4	1000.00	1.568×10^{-3}
15	999.13	1.145×10^{-3}
20	998.00	1.009×10^{-3}
30	996.00	0.800×10^{-3}
40	992.00	0.653×10^{-3}

use interpolation methods to find precisely $\rho(37°C)$ and $\mu(37°C)$. Given that $\mu = Ae^{B/T}$ (where T is the absolute temperature), find A and B for water.

9.37 A power-law (Ostwald-deWalle) model is given in Eq. (9.131) as

$$\sigma_{rz} = \kappa\left(\frac{\partial v_z}{\partial r}\right)^n,$$

where n is a parameter ($n > 1$ for dilatant and $n < 1$ for pseudoplastic) and κ is related to the "apparent" viscosity. Actually, it was proposed that the viscosity varied with the shear rate, namely

$$\sigma_{rz} = \mu(\text{shear rate})\frac{\partial v_z}{\partial r},$$

where the viscosity was found from experiments to be given by

$$\mu(\text{shear rate}) \sim \mu_a\left(\frac{\partial v_z}{\partial r}\right)^{n-1},$$

which recovers a Newtonian behavior if $n = 1$. A similar but different model (Hermes and Fredrickson) is of the form (Slattery, 1981, p. 53)

$$\mu \sim \frac{m\mu_0}{m + \mu_0(\partial v_z/\partial r)^{1-n}},$$

where m, n, and μ_0 are parameters. Repeat the analysis in Section 9.6 using this model.

9.38 If we denote an arbitrary shear rate (e.g., $\partial v_z/\partial r$ or $\partial v_\theta/\partial r$) via the symbol γ, then the viscosity for the power-law model is

$$\mu = \mu_a\gamma^{n-1}.$$

Hence,

$$\ln \mu = \ln \mu_a + (n-1)\ln \gamma$$

can be interpreted as a straight line $y = b + mx$. Use a linear regression method to compute μ_a and n for the data in Exercise 7.27 of Chapter 7.

9.39 The cross-sectional area A of an artery tends to taper exponentially as a function of distance from the heart. In particular, it has been shown in canines that

$$A = \pi a_0 e^{(-Bx/a_0)}$$

where a_0 is the radius at an upstream site, x is a distance along the aorta from that site, and $B \sim 0.02$–0.05. Plot the change in cross-sectional area for different values of B over a length of 20 cm.

9.40 For the solution of a flow down an inclined surface (Example 9.3), show that the maximum velocity occurs at the free surface, where

$$\left(v_x\right)_{max} = \frac{\rho g \sin \theta h^2}{2\mu}$$

and the volumetric flow rate is

$$Q = \frac{\rho g \sin \theta h^3 w}{3\mu}.$$

Also find the wall shear stress τ_w (at $y = 0$).

9.41 We have used the no-slip boundary condition at all solid–fluid interfaces (i.e., the fluid has the same velocity as the solid it contacts). This is clearly an approximation. Consider therefore, a *slip* boundary condition whereby

$$v_{surface} = \gamma \tau_w,$$

where γ is an empirical coefficient ($\gamma = 0$ for no-slip and a stationary surface). In the case of parallel-plate flow, with the coordinate system at the centerline,

$$v_{surface} \equiv v_x\left(y = \pm \frac{h}{2}\right) = \gamma\left[-\sigma_{yx}\left(\frac{h}{2}\right)\right] = -\gamma \frac{\partial v_x}{\partial y}\left(\frac{h}{2}\right).$$

Show that, in this case,

$$v_x(y) = c\left(1 - \frac{y^2}{h^2/4 - \gamma\mu h}\right),$$

where c can be written in terms of the volumetric flow rate or the pressure gradient. Hint: Use the condition that the velocity field is symmetric, $\partial v_x/\partial y = 0$ at $y = 0$. Note that if $\gamma = 0$, then we should recover our previous answer.

9.42 Equation (9.49) provides a relationship between the volumetric flow rate Q and the pressure gradient dp/dz for a rigid circular tube. The ratio of $|dp/dz|$ to Q provides a measure of the resistance to flow, which for the circular tube is $8\mu/\pi a^4$, which is to say, the resistance decreases as the luminal radius increases (to the fourth power). Similarly, find the ratio of $|dp/dz|$ to Q for flow in an elliptical tube [cf. Eq. (9.96)]. Compare the resistance to flow for these two geometries given the same cross-sectional area A.

9.43 Show that the velocity profile for a power-law fluid flowing between two parallel flat plates is given by (with the coordinate system centered between the plates with no-slip boundary conditions at $y = \pm h/2$)

$$v_x(y) = \frac{n}{1+n}\left(\frac{1}{k}\frac{dp}{dx}\right)^{1/n}\left(y^{1+1/n} - \left(\frac{h}{2}\right)^{1+1/n}\right)$$

and, consequently, that

$$Q = \left(-\frac{2nw}{2n+1}\right)\left(\frac{1}{k}\frac{dp}{dx}\right)^{1/n}\left(\frac{h}{2}\right)^{2+1/n}.$$

9.44 For Exercise 9.43, show that

$$v_x(y) = v_x)_{max}\left[1-\left(\frac{2y}{h}\right)^{1+1/n}\right].$$

Plot the velocity profile, normalized by $v_x)_{max}$, for various values of n.

10
Control Volume and Semi-empirical Methods

Chapters 7–9 focus on the formulation and solution of governing *differential equations* (one for mass balance and three for linear momentum balance) to determine the pressure and velocity fields in fluids under various conditions of interest. Solutions of these differential equations allow us to compute values of many important quantities of interest at each point in the flow field at each time. Such detail is often necessary, for example, to compute wall shear stress from velocity gradients, which, in turn, can be correlated with mechanosensitive responses of various cells.

Nevertheless, we do not always need such detail to analyze a problem of clinical or industrial importance or to design a revealing experiment. In some cases, *average* rather than *pointwise* information is sufficient. Toward this end, let us consider a fundamentally different approach based on the concept of a control volume.

10.1 Fundamental Equations

Before deriving the governing control volume equations, however, let us consider an organ system wherein such equations are very useful clinically. Indeed, as we will see, there are many situations in vivo and ex vivo for which average or global information suffices.

Observation 10.1 The primary function of the lungs is to facilitate gas exchange—oxygen from the atmosphere to the blood and carbon dioxide from the blood to the atmosphere. In addition, however, the lungs also filter some toxic materials from the blood and metabolize others. The human has a right lung, consisting of three lobes and 55% of the total lung volume, and a left lung, consisting of two lobes and 45% of the volume (Fig. 10.1). Obviously, the left lung is a little smaller to accommodate the heart, which lies below and slightly to the left of the sternum.

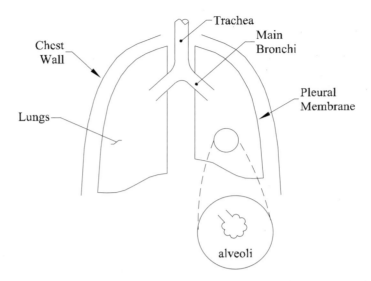

FIGURE 10.1 Schema of the lungs, which consists of two lungs (right and left), each of which are subdivided into lobes: three on the right and two on the left. The lungs are covered by a thin collagenous membrane called the visceral pleura; the inner, functional, tissue is a spongelike material consisting of air sacs called alveoli.

Air communicates with the lungs via a complex system of branching airways that become shorter and smaller in diameter as they penetrate deeper into the lungs (Table A9.4 of Chapter 9). The largest airway, which is about 12 cm long and 2 cm in diameter, is called the trachea. It is a flexible tube composed primarily of cartilage, smooth muscle, and elastic fibers, its inner surface being lined with mucosa and a monolayer of ciliated epithelial cells. The mucosa helps to trap inhaled particulate matter, which the cilia then transport to the throat. Consistent exposure to irritants such as cigarette smoke leads to an increase in the submucous glands.

From a biosolid mechanics perspective, the distinguishing feature of the trachea is its supporting framework of 15–20 C-shaped cartilaginous rings, which are separated vertically by fibromuscular tissue and bounded on the posterior surface by smooth muscle. This structure protects the trachea from collapsing while affording considerable flexibility.

The trachea divides prior to entering the lungs, thus forming the two primary bronchi that "enter" the right and left lungs at the hilum. These bronchi then divide into five lobar bronchi (~0.9 cm in diameter), one per lobe. The larger bronchi also contain plates of cartilage, which reduce in size and number until they disappear in bronchi of ~1 mm in diameter. Smooth muscle thus becomes more prominent in the smaller bronchi. The continuously branching bronchi transition to the bronchioles, which are

~0.3–0.5 mm in diameter; there are no mucosal glands or cartilage in the bronchioles and the smooth muscle forms in discrete bundles rather than in a continuous circumferential layer.

Collectively, the trachea, bronchi, and bronchioles represent a system of 16 generations of over 75,000 branching tubes; this is the so-called *conducting portion* of the airways, which is to say, the main function of this network is to conduct air from (to) the atmosphere to (from) the parenchymal tissue of the lung, where gas exchange occurs via diffusion. The remaining millions of *respiratory airways* terminate in over 300 million polyhedral-shaped air sacs called alveoli, which have a collective surface area of ~140 m^2 for gas exchange. Question: How does this value compare to the surface area of our skin? Scanning electron micrographs of the lung parenchyma reveal a spongelike appearance due to the alveoli (Fig. 10.2), which have a diameter on the order of 300 μm and a wall thickness of only ~11 μm. More importantly, however, Fick's law of diffusion states that the transport of a substance (gas) across a sheet of material (tissue) is proportional to the area of the sheet and inversely proportional to the thickness. The distance from the air to the red blood cells in the alveolar capillaries is only ~0.2–0.5 μm, which is due to a pulmonary epithelium (if present), a basal layer, interstitial space, basal lamina of the capillary, and its endothelium (Fig. 10.3). The presence of elastin (very compliant) and surfactant

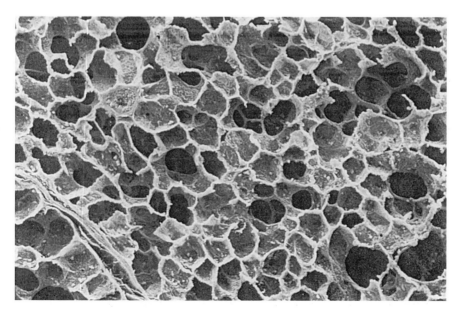

FIGURE 10.2 Scanning electron micrograph of the parenchyma (air sacs) of the lungs. [From Fawcett (1986), with permission].

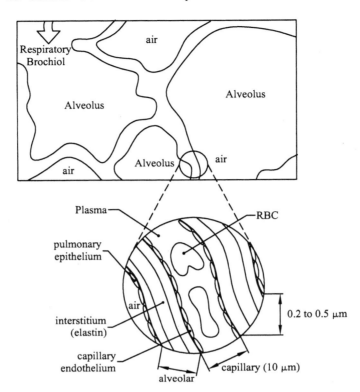

FIGURE 10.3 Schema of the alveolar walls in the lung. Note that the flow of blood in the alveolar capillaries appears to occur in sheets rather than in tubes. This increases the surface area available for diffusion of O_2 and CO_2 into and out of the blood. The thin-walled alveoli also minimize the diffusion pathway, with the red blood cells being only ~0.2–0.5 μm from the alveolar air.

(which reduces the surface tension) enables the lung to be distended easily. For example, a normal breath of ~500 mL of air is accomplished via a distending pressure of only 0.3 kPa (West, 1979). Total alveolar air volume is ~3 L.

Of primary importance in gas exchange is the so-called ventilation–perfusion ratio (i.e., the ratio of fresh air that is brought into the lungs per unit time versus the deoxygenated blood that perfuses the pulmonary capillaries per unit time). Some typical numbers for a healthy person are Q_{lung} ~7.5 L/min (at a frequency of 15 breaths per minute or 0.25 Hz), but $Q_{alveoli}$ ~5.25 L/min and Q_{blood} ~5 L/min. This yields a ventilation–perfusion ratio of about 1. Mismatching of the air and blood flow is responsible for most of the poor gas exchange in many pulmonary diseases. We see, therefore, that this aspect of pulmonary mechanics requires only average, not pointwise, descriptors of the flow field; this is also true for many other issues of pulmonary physiology (West, 1979).

10.1.1 Theoretical Framework

Recall from Chapter 8 that the five fundamental "postulates" of mechanics are the balance of mass, linear momentum, and energy as well as the balance of angular momentum and the entropy inequality. These postulates are defined most naturally in terms of a *system* (i.e., a fixed identifiable mass). Examples of systems are differential masses Δm, mass particles, and continuum bodies (i.e., collections of particles). In contrast to a fixed identifiable mass, one can also define a fixed identifiable volume, called a *control volume*. A control volume is, therefore, a fictitious volume in space that is convenient for study; mass can enter or leave a control volume, taking momentum and energy with it.

The first goal of this chapter is to derive equations that enforce the balance of mass, linear momentum, and energy in a control volume. Although these equations can be derived in different ways, it can be shown that this is best accomplished by exploiting known results for a system, which is to say to identify consequences of prior results for a system in terms of a control volume. Hence, note that the three balance relations of interest here can be written as

$$
\left. \frac{dM}{dt} \right|_{sys} = 0, \qquad \left. \frac{dP}{dt} \right|_{sys} = \Sigma F, \qquad \left. \frac{dE}{dt} \right|_{sys} = \dot{Q} + \dot{W}, \tag{10.1}
$$

where M, P, and E represent respectively the total mass, linear momentum, and energy in the system, whereas F are the applied forces, $\dot{Q}$ is the rate at which heat is added *to* the system, and $\dot{W}$ is the rate at which work is done *on* the system.[1] It will prove convenient later to let N represent the total (extensive) quantity of interest (mass, linear momentum, energy) in a system and, likewise, to let η represent these quantities (intensive) defined per unit mass. Below, we will let N represent M, P, or E and similarly η represent 1, v, or e for mass, linear momentum, and energy, respectively; hence, dM/dt, dP/dt, and dE/dt for a system can each be represented simply as dN/dt.

Now, to relate what happens in a system to what happens in a control volume, consider the following. First, recall the differential mass Δm that we defined and used in Chapter 8. We assumed that Δm occupied a cube in space at time $t = 0$, with $\Delta m = \rho \Delta X \Delta Y \Delta Z$ and that it deformed into various shapes at other times $t > 0$; we assumed, however, that at some time t, this

---

[1] In some texts, the rate at which heat is taken from the system or the rate at which work is done by the system is of interest. This simply affects the signs of the right-hand sides of the equations, of which one must be mindful.

Δm was once again shaped as a cube—this is a useful simplification, for it says that current and original positions are related by extensions alone,

$$x = x(X), \qquad y = y(Y), \qquad z = z(Z), \qquad (10.2)$$

rather than a more general relation $x = x(X,Y,Z)$ and so forth which describes extensions and shears. Thus, for a cube deforming into another cube,

$$dx = \frac{\partial x}{\partial X} dX, \qquad dy = \frac{\partial y}{\partial Y} dY, \qquad dz = \frac{\partial z}{\partial Z} dZ \qquad (10.3)$$

rather than the more general expression in Eq. (8.3) of Chapter 8. Here, we will further assume that at time t, the differential mass Δm occupies fully and precisely a control volume $\Delta\forall$ of interest; that is, at time t, $\Delta m = \rho\Delta x\Delta y\Delta z$ occupies fully the control volume $\Delta\forall = \Delta x\Delta y\Delta z$. Hence, at time t, the system results and control volume results shall correspond exactly.

Recalling from Chapter 8 that when a cube deforms into another cube [cf. Eq. (8.5)],

$$dxdydz = \frac{\partial x}{\partial X} \frac{\partial y}{\partial Y} \frac{\partial z}{\partial Z} dXdYdZ, \qquad (10.4)$$

then for any extensive quantity N (where N can represent M, P, or E),

$$\frac{dN}{dt}\bigg|_{sys} \equiv \frac{d}{dt}\int_{sys} \eta\rho d\forall = \frac{d}{dt}\int_{sys} \rho\eta \frac{\partial x}{\partial X} \frac{\partial y}{\partial Y} \frac{\partial z}{\partial Z} d\forall. \qquad (10.5)$$

Because $d\forall$ is the original volume, and thus independent of time t, we can now interchange the order of the differentiation and integration. By the product rule, we have

$$\begin{aligned}
\frac{dN}{dt}\bigg|_{sys} &= \int_{sys} \frac{d}{dt}\left(\rho\eta \frac{\partial x}{\partial X} \frac{\partial y}{\partial Y} \frac{\partial z}{\partial Z}\right)d\forall \\
&= \int_{sys}\left[\frac{d}{dt}(\rho\eta) \frac{\partial x}{\partial X} \frac{\partial y}{\partial Y} \frac{\partial z}{\partial Z} + \rho\eta \frac{d}{dt}\left(\frac{\partial x}{\partial X} \frac{\partial y}{\partial Y} \frac{\partial z}{\partial Z}\right)\right]d\forall \\
&= \int_{sys}\left[\frac{d}{dt}(\rho\eta) \frac{\partial x}{\partial X} \frac{\partial y}{\partial Y} \frac{\partial z}{\partial Z} + \rho\eta \frac{d}{dt}\left(\frac{\partial x}{\partial X}\right) \frac{\partial y}{\partial Y} \frac{\partial z}{\partial Z}\right. \\
&\quad \left. + \rho\eta \frac{\partial x}{\partial X} \frac{d}{dt}\left(\frac{\partial y}{\partial Y}\right) \frac{\partial z}{\partial Z} + \rho\eta \frac{\partial x}{\partial X} \frac{\partial y}{\partial Y} \frac{d}{dt}\left(\frac{\partial z}{\partial Z}\right)\right]d\forall, \qquad (10.6)
\end{aligned}$$

where we can also interchange the order of the time and space derivatives because X, Y, and Z relate to the original configuration and thus do not change with time. For example,

$$\frac{d}{dt}\left(\frac{\partial x}{\partial X}\right) = \frac{\partial}{\partial X}\left(\frac{dx}{dt}\right) = \frac{\partial v_x}{\partial X} \qquad (10.7)$$

and so forth, from which we recognize the velocity component v_x [cf. Eq. (7.7) of Chapter 7]. Using the chain rule as we did in Chapter 8,

$$\frac{\partial v_x}{\partial X} = \frac{\partial v_x}{\partial x}\frac{\partial x}{\partial X} \tag{10.8}$$

and so forth, thus it can be shown (do it) that

$$\rho\eta\left(\frac{\partial v_x}{\partial X}\right)\frac{\partial y}{\partial Y}\frac{\partial z}{\partial Z} + \rho\eta\frac{\partial x}{\partial X}\left(\frac{\partial v_y}{\partial Y}\right)\frac{\partial z}{\partial Z} + \rho\eta\frac{\partial x}{\partial X}\frac{\partial y}{\partial Y}\left(\frac{\partial v_z}{\partial Z}\right)$$

$$= \rho\eta(\nabla \cdot v)\frac{\partial x}{\partial X}\frac{\partial y}{\partial Y}\frac{\partial z}{\partial Z}, \tag{10.9}$$

where $\nabla \cdot v$ in Cartesians is given by Eq. (8.14). Substituting this result into Eq. (10.6), we obtain

$$\left.\frac{dN}{dt}\right|_{sys} = \int_{sys}\left(\frac{d}{dt}(\rho\eta) + \rho\eta\nabla \cdot v\right)\frac{\partial x}{\partial X}\frac{\partial y}{\partial Y}\frac{\partial z}{\partial Z}d\mathcal{V} = \int_{sys}\left(\frac{d}{dt}(\rho\eta) + \rho\eta\nabla \cdot v\right)dv, \tag{10.10}$$

with

$$dv = \frac{\partial x}{\partial X}\frac{\partial y}{\partial Y}\frac{\partial z}{\partial Z}d\mathcal{V}.$$

Because $\rho\eta$ can change with time and position (x, y, z), and $v_x = dx/dt$ and so forth, this equation can be written as

$$\left.\frac{dN}{dt}\right|_{sys} \equiv \int_{sys}\left[\frac{\partial}{\partial t}(\rho\eta)\frac{dt}{dt} + \frac{\partial}{\partial x}(\rho\eta)\frac{dx}{dt} + \frac{\partial}{\partial y}(\rho\eta)\frac{dy}{dt} + \frac{\partial}{\partial z}(\rho\eta)\frac{dz}{dt}\right.$$

$$\left. + \rho\eta\left(\frac{\partial v_x}{dx} + \frac{\partial v_y}{dy} + \frac{\partial v_z}{dz}\right)\right]dv, \tag{10.11}$$

or

$$\left.\frac{dN}{dt}\right|_{sys} = \int_{sys}\frac{\partial}{\partial t}(\rho\eta) + \frac{\partial}{\partial x}(\rho\eta)v_x + \frac{\partial}{\partial y}(\rho\eta)v_y + \frac{\partial}{\partial z}(\rho\eta)v_z$$

$$ + \rho\eta\left(\frac{\partial v_x}{\partial x} + \frac{\partial v_y}{\partial y} + \frac{\partial v_z}{\partial z}\right)\right]dv \tag{10.12}$$

and, therefore, if we use the product rule and split the integral into two parts, we have

$$\left.\frac{dN}{dt}\right|_{sys} = \int_{sys}\left(\frac{\partial}{\partial t}(\rho\eta) + \nabla \cdot (\rho\eta v)\right)dv = \int_{sys}\frac{\partial}{\partial t}(\rho\eta)dv + \int_{sys}\nabla \cdot (\rho\eta v)dv. \tag{10.13}$$

Recalling the divergence theorem [Eq. (A7.31)],

$$\iiint_{\mathcal{V}}\nabla \cdot B dv = \iint_{Area} B \cdot \hat{n}da,$$

where, in this case, the arbitrary vector $B = \rho\eta v$, we obtain our final result:

$$\left.\frac{dN}{dt}\right|_{sys} = \int_{sys} \frac{\partial}{\partial t}(\rho\eta)d\textbf{v} + \int_{sys} \rho\eta\textbf{v} \cdot \textbf{n}da, \qquad (10.14)$$

which is valid for all time $t \geq 0$ with $\textbf{n} \equiv \hat{\textbf{n}}$. Because this relation is good at that time t when the system and control volume coincide, we have

$$\left.\frac{dN}{dt}\right|_{sys} = \int_{C\textbf{v}} \frac{\partial}{\partial t}(\rho\eta)d\textbf{v} + \int_{CS} \rho\eta\textbf{v} \cdot \textbf{n}da \qquad (10.15)$$

where $C\textbf{v}$ and CS denote "control volume" and "control surfaces," respectively. Note that the term $\left.\frac{dN}{dt}\right|_{sys}$ describes the rate of change of any extensive property N of the system, the term $\int_{C\textbf{v}} \partial(\rho\eta)/\partial t\,d\textbf{v}$ describes the time rate of change of the arbitrary property within the control volume, where η is the intensive property corresponding to N, and the term $\int_{CS} \rho\eta\textbf{v} \cdot \textbf{n}\,da$ describes the net flux of the property through a control surface, where $\textbf{n}$ is an outward unit normal vector to the control surface CS.

Here, it is important to note that because our control volume is fixed, it does not change in time. This allows us to write Eq. (10.15) as

$$\left.\frac{dN}{dt}\right|_{sys} = \frac{\partial}{\partial t}\int_{C\textbf{v}}(\rho\eta)d\textbf{V} + \int_{CS} \rho\eta\textbf{v} \cdot \textbf{n}dA, \qquad (10.16)$$

wherein we use $\textbf{V}$ and A to emphasize that the $C\textbf{v}$ and CS are fixed (i.e., the current control volume and the original control volume coincide). This is the fundamental equation in this chapter; commit it to memory.

10.1.2 Special Cases for Mass and Momentum

It is useful to recognize two potentially useful simplifications. First, for time-independent processes,

$$\frac{\partial}{\partial t}\int (\rho\eta)d\textbf{V} = 0 \qquad (10.17)$$

and, second, for incompressible flow, the density ρ of the fluid is constant and thus it can come out of the integrals. Let us now consider the specific balance relations for the case of a fixed identifiable control volume.

Conservation of mass states that the total mass M of a system is constant; hence with $N \equiv M$ and $\eta \equiv 1$, Eq. (10.16) becomes

$$\left.\frac{dM}{dt}\right|_{sys} = 0 = \frac{\partial}{\partial t}\int_{C\textbf{v}} \rho d\textbf{V} + \int_{CS} \rho(\textbf{v} \cdot \textbf{n})dA. \qquad (10.18)$$

Thus, conservation of mass for a steady, incompressible flow requires

$$\left.\frac{dM}{dt}\right|_{sys} = 0 = \int_{CS} \boldsymbol{v} \cdot \boldsymbol{n} dA, \tag{10.19}$$

which is to say, the flow into the C∀ through its control surfaces CS must balance the flow out. This is certainly not surprising, but it is a useful result to remember that for a steady, incompressible flow into a fixed identifiable volume, $Q_{in} = Q_{out}$, where Q is the volumetric flow rate.

Balance of linear momentum, or Newton's second law of motion for a system, states that the time rate of change of the linear momentum must balance the applied forces. Hence, letting $N \equiv \boldsymbol{P}$ and $\eta \equiv \boldsymbol{v}$, Eq. (10.16) becomes

$$\left.\frac{d\boldsymbol{P}}{dt}\right|_{sys} = \Sigma\boldsymbol{F} = \frac{\partial}{\partial t}\int_{C\forall} \rho\boldsymbol{v} d\forall + \int_{CS} \rho\boldsymbol{v}(\boldsymbol{v} \cdot \boldsymbol{n}) dA. \tag{10.20}$$

Again, considerable simplification occurs for steady and incompressible flows, namely

$$\Sigma\boldsymbol{F} = \rho\int_{CS} \boldsymbol{v}(\boldsymbol{v} \cdot \boldsymbol{n}) dA. \tag{10.21}$$

10.1.3 The Energy Equation

Conservation of energy, otherwise known as the *first law of thermodynamics*, states that the time rate of change of energy in a system must balance the rate at which work is done on the system plus the rate at which heat is added to the system. Letting $N \equiv E$ and $\eta \equiv e$, where e includes internal, potential, and kinetic energy contributions, Eq. (10.16) becomes

$$\left.\frac{dE}{dt}\right|_{sys} = \dot{Q} + \dot{W} = \frac{\partial}{\partial t}\int_{C\forall} \rho e d\forall + \int_{CS} \rho e(\boldsymbol{v} \cdot \boldsymbol{n}) dA. \tag{10.22}$$

Balance of energy is very important, but will not be considered until Section 10.6. In the next three sections, we will focus on isothermal problems and, thus, the balance of mass and linear momentum alone.

In closing, it is important to recognize that the general C∀ (i.e., control volume) formulation was achieved without specifying a particular class of fluids (Newtonian, non-Newtonian, ideal, or otherwise). Hence, the associated statements for the balance of mass, momentum, and energy are general even though they provide only averaged (global) information, not pointwise (local) information that results from differential equations like the Navier–Stokes and Euler equations. As such, control volume analyses are generally mathematically much simpler than the analyses found in Chapter 9. The key things for the student to master, therefore, are (1) to determine when a control volume formulation is sufficient and (2) how to pick a useful C∀. These are best learned via experience; hence, let us consider the following examples.

10.2 Control Volume Analyses in Rigid Conduits

10.2.1 Clinical Motivation

Albeit taken for granted today, open-heart surgery is heralded as one of the most important advances in health care delivery during the period 1945–1975 (Comroe and Dripps, 1977). Whereas we must be quick to acknowledge the advanced skills and self-confidence of the surgeons in making this advance a reality, the procedure would not have been possible without the biomedical engineering development of the so-called heart –lung machine; that is, to enable precise surgical manipulation of the heart, the surgeon needed a quiescent, nonbeating heart on which to work. The design criteria for the assist technology thus included the need to pump and oxygenate the blood, to maintain body temperature at a prescribed level, to minimize clotting, and to do so without damaging the cells within the blood. As we now know, this can be accomplished by shunting blood outside the body via roller pumps and through membranes that allow the requisite gas exchange. Although detailed solutions of the governing differential equations provide important information on the effects of the artificial flow field on the cells and, in particular, on minimizing the hemolysis, a more global analysis of the flow can also provide important information on the requisite design of the pumps needed to maintain a desired volumetric flow rate Q. Toward this end, a control volume analysis can be very helpful. Although we will not consider the complexities of the heart–lung machine, let us now illustrate the use of Eqs. (10.18) and (10.20) via a few simple examples.

10.2.2 Illustrative Examples

Consider a control volume analysis of the bifurcation in Figure 10.4, wherein an atherosclerotic plaque occludes a vessel distal to a bifurcation. We seek to determine the reduction in flow distal to the obstruction as a function of the extent of the disease (i.e., stenosis), given the simplifying assumptions of a steady, incompressible flow within a rigid section. A reasonable control volume in this case is simply one that coincides with the fluid in the fluid domain of interest. Let us also consider steady flow with a fully developed entrance velocity [cf. Eq. (9.45)] $v_1 = c(1 - r^2/a^2)\hat{e}_z$ with *mean* outlet velocities of $v_2 = \bar{v}_2\hat{e}_z'$ and $v_3 = \bar{v}_3\hat{e}_z''$. For large vessels, we can assume that the density ρ is constant, which is to say that we neglect effects like plasma skimming. Thus, ρ can be taken outside the integral and then deleted as we divide each side of the mass balance equation by ρ. We have

$$0 = \rho \int_{CS} (v \cdot n)dA \rightarrow \int_{A_1} (v_1 \cdot n_1)dA + \int_{A_2} (v_2 \cdot n_2)dA$$

$$+ \int_{A_3} (v_3 \cdot n_3)dA + \underbrace{\int_{CS} (v \cdot n)dA}_{\text{all other surfaces}} = 0. \tag{10.23}$$

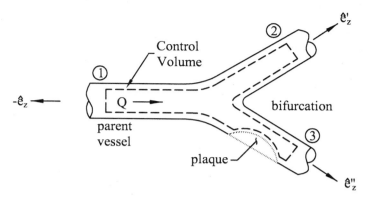

FIGURE 10.4 Possible control volume for the analysis of mean flows in an arterial bifurcation. Control surfaces 1, 2, and 3 represent cross sections; the remaining control surface along the length of the vessel is not shown because there is no flux through this surface.

Note that the velocity is zero through all control surfaces except CS_1, CS_2, and CS_3, which are defined by *outward* unit normal vectors $\boldsymbol{n}_1 = -\hat{\boldsymbol{e}}_z$, $\boldsymbol{n}_2 = \hat{\boldsymbol{e}}_z'$, $\boldsymbol{n}_3 = \hat{\boldsymbol{e}}_z''$. Hence,

$$0 = \int_{A_1} c\left(1 - \frac{r^2}{a^2}\right)(\hat{\boldsymbol{e}}_z \cdot -\hat{\boldsymbol{e}}_z)dA + \int_{A_2} \bar{v}_2(\hat{\boldsymbol{e}}_z' \cdot \hat{\boldsymbol{e}}_z')dA + \int_{A_3} \bar{v}_3(\hat{\boldsymbol{e}}_z'' \cdot \hat{\boldsymbol{e}}_z'')dA. \quad (10.24)$$

Because the mean velocities at CS_2 and CS_3 do not vary over the cross section by definition, we have

$$0 = -c\int_0^a \int_0^{2\pi}\left(1 - \frac{r^2}{a^2}\right)rd\theta dr + \bar{v}_2\int_{A_2} dA + \bar{v}_3\int_{A_3} dA. \quad (10.25)$$

Integrating and simplifying, we get

$$\bar{v}_2 A_2 + \bar{v}_3 A_3 = \frac{\pi a^2 c}{2}, \quad (10.26)$$

which simply states that *the net flow out equals the flow in*. Because we have but one equation, we must know the inlet velocity (i.e., c and a) as well as A_2 and $\bar{v}_2$ to calculate $Q_3 \equiv \bar{v}_3 A_3$ as desired, where c contains information on the proximal pressure gradient that drives the flow.

Because we may need to bypass or replace the diseased vessel with a vascular graft (e.g., Fig. 10.5), we may also like to know the net forces borne by the sutures that hold the graft in place. To solve this problem, which is a solid mechanics problem, we first need to construct a free-body diagram of the graft (Fig. 10.6). The proximal suture forces will arise primarily from two sources: first, the axial load on the graft due to the surgeon's attempt to restore homeostatic axial stresses in the native vessel (recall Section 3.4 of Chapter 3) and, second, the forces due to the flow-induced shear stresses;

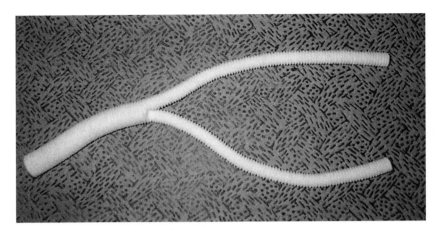

FIGURE 10.5 Photograph of a synthetic vascular graft (artificial artery) produced by Meadox Medicals in New Jersey. This is a large vessel graft (aorta), which has had good clinical success. The current challenge in vascular grafts is to develop robust small-diameter grafts that will not develop a thrombosis, significant neointima hyperplasia, or an immune rejection. Tissue engineering has great promise in this regard.

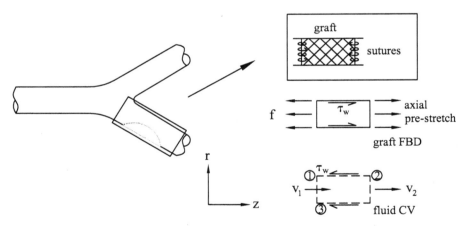

FIGURE 10.6 Free-body diagram and associated control volume for a combined solid mechanics/fluid mechanics analysis of stresses in an in-line arterial graft.

that is, it is well known that arteries experience significant axial loads in vivo. A prime example is the carotid artery in the neck; it will retract ~50% when cut (i.e., when the in vivo force is released). Although it is not clear how these loads develop, they likely do so during normal development. In cases of "adaptation," these axial loads can change. For example, arteries in patients with hypertension retract less when cut; this is thought to be due to the addition of tissue in the axial direction (which unloads the vessel) that arises as a natural consequence of the vessel adding material in the radial direction to better resist the increased distension pressure (recall discussion in Chapter 3). However, let us focus on the flow-induced shear stresses.

To find the wall shear stress τ_w on the wall of the graft, one could find all of the shear stresses σ_{rz} in the fluid, and then evaluate those at the wall. This would require solving for the velocity field $v_z(r, z)$; because the flow may not be fully developed throughout this region (we assume fully developed at CS_2 only), this would likely require a numerical solution of the 3-D Navier–Stokes equations, which would be very time-consuming. Alternatively, one could use the control volume approach to estimate the net effect of the shear stress. Hence, consider the $C\!\!\!\!V$ in Fig. 10.6. Assuming a steady flow and a constant mass density, mass balance requires

$$0 = \int_{CS} \boldsymbol{v} \cdot \boldsymbol{n} dA \rightarrow 0 = \int_{A_1} \boldsymbol{v}_1 \cdot \boldsymbol{n}_1 dA + \int_{A_2} \boldsymbol{v}_2 \cdot \boldsymbol{n}_2 dA + \int_{A_3} \boldsymbol{v}_3 \cdot \boldsymbol{n}_3 dA, \quad (10.27)$$

where for CS_1, $\boldsymbol{v}_1 = \bar{v}_1 \hat{\boldsymbol{e}}_z$ and $\boldsymbol{n}_1 = -\hat{\boldsymbol{e}}_z$, for CS_2, $\boldsymbol{v}_2 = c(1 - r^2/a^2)\hat{\boldsymbol{e}}_z$ and $\boldsymbol{n}_2 = \hat{\boldsymbol{e}}_z$, and for CS_3, $\boldsymbol{v}_3 = \boldsymbol{0}$ (by no slip) and $\boldsymbol{n}_3 = \hat{\boldsymbol{e}}_r$. Hence,

$$0 = \int_{A_1} \bar{v}_1 (\hat{\boldsymbol{e}}_z \cdot -\hat{\boldsymbol{e}}_z) dA + \int_{A_2} c\left(1 - \frac{r^2}{a^2}\right)(\hat{\boldsymbol{e}}_z \cdot \hat{\boldsymbol{e}}_z) dA. \quad (10.28)$$

Considering a mean velocity through CS_1, we have

$$0 = -\bar{v}_1 \int_{A_1} dA + c \int_0^a \int_0^{2\pi} \left(1 - \frac{r^2}{a^2}\right) r d\theta dr. \quad (10.29)$$

Integrating and simplifying, we get

$$0 = -\bar{v}_1 A_1 + c(2\pi)\left(\frac{a^2}{4}\right) \rightarrow \bar{v}_1 = \frac{\pi a^2 c}{2 A_1}. \quad (10.30)$$

To solve for the force that the proximal sutures must withstand, we need to know the net force applied to the graft due to the fluid flow. We use linear momentum balance for steady, incompressible flow [Eq. (10.21)]. Summing forces and expanding for this particular $C\!\!\!\!V$, we obtain

$$p_1 A_1 \hat{\boldsymbol{e}}_z + p_2 A_2 (-\hat{\boldsymbol{e}}_z) + f_\tau \hat{\boldsymbol{e}}_z = \rho \int_{A_1} \boldsymbol{v}_1 (\boldsymbol{v}_1 \cdot \boldsymbol{n}_1) dA + \rho \int_{A_2} \boldsymbol{v}_2 (\boldsymbol{v}_2 \cdot \boldsymbol{n}_2) dA + \boldsymbol{0}, \quad (10.31)$$

where p_1 and p_2 are the mean pressures at CS_1 and CS_2, respectively, and f_τ is the total force due to all the shear stresses that act on the fluid volume (it is equal and opposite that which acts on the wall), or

$$p_1 A_1 \hat{\boldsymbol{e}}_z + p_2 A_2 (-\hat{\boldsymbol{e}}_z) + f_\tau \hat{\boldsymbol{e}}_z = \rho \int_{A_1} v_1 \hat{\boldsymbol{e}}_z [v_1 (\hat{\boldsymbol{e}}_z \cdot -\hat{\boldsymbol{e}}_z)] dA$$

$$+ \rho \int_{A_2} c\left(1 - \frac{r^2}{a^2}\right)\hat{\boldsymbol{e}}_z \left[c\left(1 - \frac{r^2}{a^2}\right)(\hat{\boldsymbol{e}}_z \cdot \hat{\boldsymbol{e}}_z)\right] dA, \quad (10.32)$$

where $dA = r\,d\theta\,dr$. With $\bar{v}_1$ the mean value, integration over the entire CS_1 cross section yields, in the z direction,

$$p_1 A_1 - p_2 A_2 + f_\tau = -\rho \bar{v}_1^2 A_1 + \rho 2\pi c^2 \int_0^a \left(1 - \frac{r^2}{a^2}\right)^2 r\,dr, \qquad (10.33)$$

where

$$\int_0^a \left(1 - \frac{r^2}{a^2}\right)^2 r\,dr = \int_0^a \left(r - 2\frac{r^3}{a^2} + \frac{r^5}{a^4}\right)dr = \frac{a^2}{6}, \qquad (10.34)$$

thus, substituting the value for $\bar{v}_1$ obtained from mass balance into this equation, we get

$$f_\tau = p_2 A_2 - p_1 A_1 - \rho A_1 \left(\frac{\pi a^2 c}{2 A_1}\right)^2 + \frac{\rho \pi c^2 a^2}{3}, \qquad (10.35)$$

which is the total shear force experienced by the fluid due to the graft. That experienced by the graft is equal and opposite this value. This allows the solid mechanics problem to be addressed given the axial forces that are borne by the wall in vivo.

Example 10.1 To illustrate further a CV formulation, consider flow through a nozzle (Fig. 10.7). Examples include needles on syringes and nozzles on

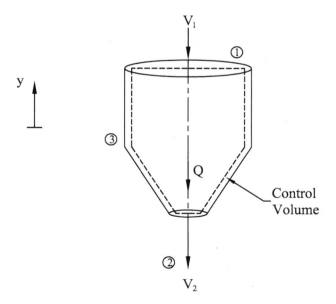

FIGURE 10.7 Possible control volume for the analysis of flow through a vertical nozzle (e.g., needle).

laboratory sinks. Many needles have Luer connections to ensure that the needle does not "fly off" when the fluid is pressurized. Likewise, laboratory nozzles are threaded. With regard to the latter, one may be interested, for example, in ensuring that the threads will be sufficient to withstand the loads on the nozzle, which will result from the solid–fluid interaction as well as the weight of the nozzle. Let us find the net fluid induced shear force on a nozzle through which the fluid flows.

Solution: To find the net shear force exerted on the solid by the flow of the fluid, consider a *CV* to determine the forces that act on the fluid and then invoke Newton's third law. For a steady, incompressible flow within a (fixed) control volume, mass balance again requires that

$$0 = \int_{CS} \boldsymbol{v} \cdot \boldsymbol{n} dA,$$

whereas for linear momentum balance, we have

$$\Sigma \boldsymbol{F} = \rho \int_{CS} \boldsymbol{v}(\boldsymbol{v} \cdot \boldsymbol{n}) dA.$$

For CS_1, $\boldsymbol{v}_1 = v_1(-\hat{\boldsymbol{j}})$ and $\boldsymbol{n}_1 = \hat{\boldsymbol{j}}$, for CS_2, $\boldsymbol{v}_2 = v_2(-\hat{\boldsymbol{j}})$ and $\boldsymbol{n}_2 = -\hat{\boldsymbol{j}}$, and for CS_3, $\boldsymbol{v}_3 = \boldsymbol{0}$ by no slip or $\boldsymbol{v}_3 \perp \boldsymbol{n}_3$. Hence, for mass balance,

$$0 = \int_{A_1} \boldsymbol{v}_1 \cdot \boldsymbol{n}_1 dA + \int_{A_2} \boldsymbol{v}_2 \cdot \boldsymbol{n} dA \rightarrow 0 = \int_{A_1} v_1(-\hat{\boldsymbol{j}} \cdot \hat{\boldsymbol{j}}) dA + \int_{A_2} v_2(-\hat{\boldsymbol{j}} \cdot -\hat{\boldsymbol{j}}) dA.$$

Considering only average values of the velocity,

$$0 = -\bar{v}_1 \int_{A_1} dA + \bar{v}_2 \int_{A_2} dA \rightarrow \bar{v}_2 A_2 = \bar{v}_1 A_1,$$

which is what we expect for mass to be conserved; that is, the flow in must equal the flow out, where $\bar{v}_2 A_2 = \bar{v}_1 A_1 = Q$. For linear momentum balance, summing forces and expanding for this *CV*, we have

$$p_1 A_1 (-\hat{\boldsymbol{j}}) + p_2 A_2 (\hat{\boldsymbol{j}}) + W_{fld} (-\hat{\boldsymbol{j}}) + f_\tau(-\hat{\boldsymbol{j}})$$
$$= \rho \int_{A_1} v_1 (-\hat{\boldsymbol{j}})[v_1(-\hat{\boldsymbol{j}} \cdot \hat{\boldsymbol{j}})] dA + \rho \int_{A_2} v_2 (-\hat{\boldsymbol{j}})[v_1(-\hat{\boldsymbol{j}} \cdot -\hat{\boldsymbol{j}})] dA$$

or, for mean velocities,

$$p_1 A_1 (-\hat{\boldsymbol{j}}) + p_2 A_2 (\hat{\boldsymbol{j}}) + W_{fld}(-\hat{\boldsymbol{j}}) + f_\tau(-\hat{\boldsymbol{j}}) = \rho \bar{v}_1^2 (\hat{\boldsymbol{j}}) \int_{A_1} dA + \rho \bar{v}_2^2 (-\hat{\boldsymbol{j}}) \int_{A_2} dA.$$

Integrating and considering the effects in $\hat{\boldsymbol{j}}$ alone,

$$-p_1 A_1 - W_{fld} - f_\tau = \rho \bar{v}_1^2 A_1 - \rho \bar{v}_2^2 A_2,$$

where the gauge pressure acting on CS_2 is approximately zero if we discharge to atmosphere (i.e., $p_2 = 0$). Thus, summing forces in the y direction yields

$$f_\tau = \rho(\bar{v}_2^2 A_2 - \bar{v}_1^2 A_1) - p_1 A_1 - W_{fld},$$

where $\bar{v}_1$ and W_{fld} are measured easily and p_1 potentially, and the solid mechanics problem for the stresses on the threads can now be formulated and solved.

Example 10.2 We have noted various times throughout this book that the vascular endothelium responds to an altered wall shear stress τ_w by increasing its production of, among other molecules, vasodilators, and vasoconstrictors that induce a dilatation or constriction that restores τ_w to its baseline value. Clearly, however, the endothelium cannot sustain arbitrarily large increases in wall shear stress; there must be a value at which the endothelium becomes damaged. Well before the discovery of endothelial mechanotransduction, D. L. Fry, a scientist at the National Institutes of Health at Bethesda, MD, showed in 1968 that aortic endothelial cells are damaged at values of τ_w of 40 Pa and above (recall that normal values are ~1.5 Pa in arteries). This finding led to additional questions, such as whether a jet flow from a needle may likewise be able to damage, literally erode, the endothelium. (Note: The term used in the literature is *denude*, which means to lay bare.) A logical question, therefore, is: How do we design and interpret an experiment to quantify an "erosion" stress? Toward this end, consider the stress on the wall of the vessel created by the injection of a fluid through a needle. Find the erosion stress σ_{xx} on the endothelium by assuming a steady, incompressible flow with a volumetric flow rate of Q. Neglect gravity.

Solution: Consider a C∀ for the fluid as well as a free-body diagram for the arterial wall (Fig. 10.8). To calculate the erosion stress on the arterial wall, we need to calculate the reaction force $\boldsymbol{R}$ that the artery exerts on the fluid and the oriented area over which $\boldsymbol{R}$ acts. To calculate $\boldsymbol{R}$, we can appeal to Newton's third law and use the C∀ equation for the balance of linear momentum for a steady, incompressible flow [Eq. (10.21)],

$$\sum \boldsymbol{F} = \rho \int_{CS} \boldsymbol{v}(\boldsymbol{v} \cdot \boldsymbol{n}) dA.$$

For CS_1, $\boldsymbol{v}_1 = v_1(\hat{\boldsymbol{i}})$ and $\boldsymbol{n}_1 = -\hat{\boldsymbol{i}}$, for CS_2, $\boldsymbol{v}_2 = v_2(\hat{\boldsymbol{j}})$ and $\boldsymbol{n}_2 = \hat{\boldsymbol{j}}$, for CS_3, $\boldsymbol{v}_3 = v_3(-\hat{\boldsymbol{j}})$ and $\boldsymbol{n}_3 = -\hat{\boldsymbol{j}}$, and for CS_4, $\boldsymbol{v}_4 = \boldsymbol{0}$ due to impenetrability. Summing the forces and expanding for this C∀, we obtain

$$p_1 A_1(\hat{\boldsymbol{i}}) + p_2 A_2(-\hat{\boldsymbol{j}}) + p_3 A_3(\hat{\boldsymbol{j}}) + R(-\hat{\boldsymbol{i}})$$
$$= \rho \int_{A_1} v_1(\hat{\boldsymbol{i}})[v_1(\hat{\boldsymbol{i}} \cdot -\hat{\boldsymbol{i}})] dA + \rho \int_{A_2} v_2(\hat{\boldsymbol{j}})[v_2(\hat{\boldsymbol{j}} \cdot \hat{\boldsymbol{j}})] dA$$
$$+ \rho \int_{A_3} v_3(-\hat{\boldsymbol{j}})[v_3(-\hat{\boldsymbol{j}} \cdot -\hat{\boldsymbol{j}})] dA.$$

Assuming average values for velocities yields

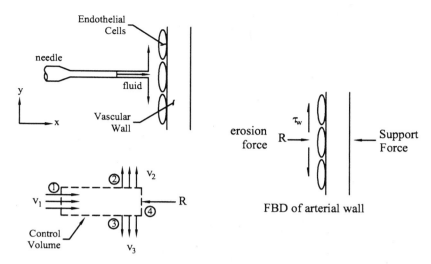

FIGURE 10.8 Experimental setup to determine the erosion stress for endothelial cells. Shown are a possible control volume for the fluid and a free-body diagram for the arterial segment. Although the fluid will exert shear stresses on the cells, we are interested primarily in the normal force R in this experiment.

$$p_1 A_1 (\hat{i}) - p_2 A_2 (\hat{j}) + p_3 A_3 (\hat{j}) + R(-\hat{i})$$
$$= -\rho \bar{v}_1^2 (\hat{i}) \int_{A_1} dA + \rho \bar{v}_2^2 (\hat{j}) \int_{A_2} dA + \rho \bar{v}_3^2 (-\hat{j}) \int_{A_3} dA.$$

By summing the forces in the x direction, assuming that all pressures are gauge pressures, and then integrating, we obtain

$$R(-\hat{i}) = -\rho \bar{v}_1^2 A_1 (\hat{i}) \rightarrow R = \rho \bar{v}_1^2 A_1.$$

Therefore, from the free body-diagram of the artery wall,

$$R = \int_A \sigma_{xx} dA = \rho \bar{v}_1^2 A_1,$$

thus allowing us to address the solid mechanics problem. Again, we see that an analysis helps us to design an experiment for we now know what needs to be measured and why.

Example 10.3 For a steady incompressible flow of water through the reducing elbow in Fig. 10.9, the entrance area A_1 is $30\,\text{cm}^2$ and the exit area A_2 is $\frac{1}{2}A_1$. The mean velocity $\bar{v}_1$ entering the elbow is $5\,\text{m/s}$ with an inlet pressure of $5\,\text{Pa}$ and outlet pressure equal to atmospheric. Find the total force required to hold the bend in place.

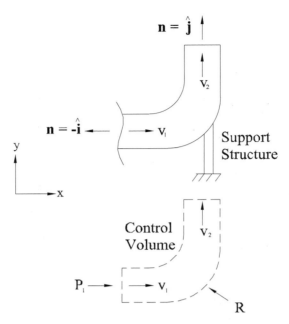

FIGURE 10.9 Academic experiment to determine the support reactions due to a fluid flowing through an elbow. Shown, too, is a possible control volume that ignores possible effects of the fluid-induced shear stresses.

Solution: We are given the following:

$$A_1 = 30 \, \text{cm}^2 = 0.003 \, \text{m}^2,$$

$$A_2 = \frac{1}{2} A_1 = 15 \, \text{cm}^2 = 0.0015 \, \text{m}^2$$

$$\bar{v}_1 = 5 \, \text{m/s},$$

$$P_1 = 5 \, \text{Pa} = 5 \, \text{N/m}^2,$$

$$P_2 = 0 \, \text{(gauge)},$$

$$\rho_{H_2O} = 1000 \, \text{kg/m}^3$$

We first need to solve for the velocity v_2 leaving the elbow. From the mass balance equation for a steady, incompressible flow,

$$0 = \int_{A_1} \boldsymbol{v}_1 \cdot \boldsymbol{n}_1 \, dA + \int_{A_2} \boldsymbol{v}_2 \cdot \boldsymbol{n}_2 \, dA,$$

where we have for CS_1, $\boldsymbol{v}_1 = v_1(\hat{\boldsymbol{i}})$ and $\boldsymbol{n}_1 = -\hat{\boldsymbol{i}}$, and for CS_2, $\boldsymbol{v}_2 = v_2(\hat{\boldsymbol{j}})$ and $\boldsymbol{n}_2 = \hat{\boldsymbol{j}}$. Hence, mass balance requires

$$0 = \int_{A_1} v_1(\hat{\boldsymbol{i}} \cdot -\hat{\boldsymbol{i}}) \, dA + \int_{A_2} v_2(\hat{\boldsymbol{j}} \cdot \hat{\boldsymbol{j}}) \, dA.$$

Using mean velocities, we have

$$0 = -\bar{v}_1 \int_{A_1} dA + \bar{v}_2 \int_{A_2} dA \to \bar{v}_1 A_1 = \bar{v}_2 A_2$$

or, with $A_2 = A_1/2$,

$$\bar{v}_1 A_1 = \bar{v}_2 \left(\frac{1}{2} A_1 \right) \to \bar{v}_2 = 2\bar{v}_1.$$

From Newton's third law we know that for every action, there is an equal and opposite reaction. To determine the total force required to hold the elbow in place, we need to use the balance of linear momentum equation for a steady, incompressible flow. Summing the forces for this $C\!\!\!V$, we obtain

$$p_1 A_1 (\hat{i}) + p_2 A_2 (-\hat{j}) + \boldsymbol{R} = \rho \int_{A_1} v_1 (\hat{i}) [v_1 (\hat{i} \cdot -\hat{i})] dA + \rho \int_{A_2} v_2 (\hat{j}) [v_2 (\hat{j} \cdot \hat{j})] dA,$$

where $\boldsymbol{R}$ is the reaction force, or with $p_2 = P_{\text{atm}} = 0$ (gauge), integration yields

$$p_1 A_1 \hat{i} + \boldsymbol{R} = -\rho \bar{v}_1^2 A_1 \hat{i} + \rho \bar{v}_2^2 A_2 \hat{j}.$$

Collecting terms, we have

$$\boldsymbol{R} = -(p_1 A_1 + \rho \bar{v}_1^2 A_1)\hat{i} + (\rho \bar{v}_2^2 A_2)\hat{j}.$$

By substituting the numerical values into this equation, we obtain

$$\boldsymbol{R} = -\left[\left(\frac{5\,\text{N}}{\text{m}^2} \right)(0.003\,\text{m}^2) + \left(\frac{1000\,\text{kg}}{\text{m}^3} \right)(5\,\text{m/s})^2 (0.003\,\text{m}^2) \right]\hat{i}$$

$$+ \left[\left(\frac{1000\,\text{kg}}{\text{m}^3} \right)(10\,\text{m/s})^2 (0.0015\,\text{m}^2) \right]\hat{j} = +(75\,\text{N})\hat{i} + (150\,\text{N})\hat{j}.$$

This force, which acts on the fluid is equal and opposite that exerted by the flow on the support. We could now solve the solid mechanics problem and determine the reactions at the base of the support and the stresses due to axial and bending loads.

10.3 Control Volume Analyses in Deforming Containers

Whereas it is very natural to employ a fixed identifiable volume in space (i.e., a control volume) in problems wherein the flow goes through a rigid conduit, there are cases wherein we must compute mean values associated with flows in organs or devices that deform. The approach must be the same, but it deserves some special attention.

10.3.1 Clinical Motivation

The manufacture of easily used, disposable, low-profile catheters has given rise to whole new fields in medicine, including interventional cardiology and interventional radiology. Briefly, such catheters can be placed percutaneously into a blood vessel of choice and advanced under fluoroscopic guidance to a target organ or lesion. The catheter can then be used to inject contrast agents for visualization, to deploy intravascular stents or coils, to biopsy tissue, to deliver thermal energy to treat a pathology, to inflate a balloon to dilate an atherosclerotic plaque or temporarily obstruct/prevent blood flow, or to measure local flows or pressures.

Here, let us consider but one simple example: the inflation of a silicone-based balloon to temporarily obstruct blood flow (Fig. 10.10). Such a device could be used in cardiology, for example, to determine how the preload (venous pressure) affects function in the right ventricle. One could block the venous return to the right heart by inflating occluding balloons in the superior and inferior vena cava (i.e., via caval occlusion). In addition to the biomedical engineer designing and manufacturing an appropriate device (including biocompatibility issues in materials selection and understanding the mechanics of the balloon), one may desire to know the rate at which the balloon expands based on the rate at which the inflating syringe is operated.

10.3.2 Mathematical Formulation

Let us consider the syringe and the balloon separately, assuming that the catheter connecting them does not distend or extend due to the driving pressure (Fig. 10.11). Moreover, let us assume that the syringe plunger is

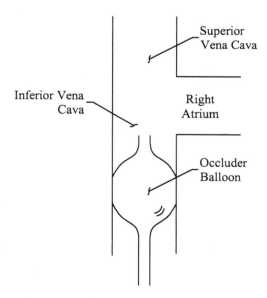

FIGURE 10.10 Experimental situation wherein an occluding balloon is used to reduce caval blood flow to the heart and thereby alter the preload on the heart.

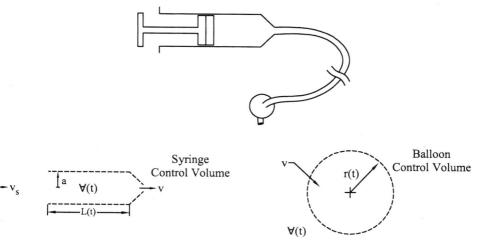

FIGURE 10.11 Inflation of a balloon via a syringe. Shown are separate control volumes for the syringe and the balloon.

advanced steadily, thus giving rise to a steady flow. First, for the balloon: We could define a fixed control volume to be a spherical volume in space. At time t, we could further let the balloon radius $r(t)$ match exactly that of the control volume at that instant. Hence, with a steady flow and fixed control volume, we have

$$\frac{\partial}{\partial t}\int_{CV}\rho d\forall + \int_{CS}\rho(\mathbf{v}\cdot\mathbf{n})dA = 0, \qquad (10.36)$$

which reduces to

$$\int_{CS_1}(\mathbf{v}\cdot\mathbf{n})dA + \int_{CS_2}(\mathbf{v}_r\cdot\mathbf{n}_r)dA = 0, \qquad (10.37)$$

where CS_1 can be taken to be the inlet where the fluid enters the balloon; CS_2, therefore, is the surface area of the fixed control volume, $\sim 4\pi r^2$ at time t. As the balloon distends, the velocity of the fluid at CS_2 will equal that of the expanding balloon. If the displacement of the balloon is $u_r = r(t) - R$, where R is the original radius at time $t = 0$, then $v_r = du_r/dt = dr/dt$. Hence, we have

$$[\bar{v}(-\mathbf{n})\cdot\mathbf{n}]A_1 + \left(\frac{dr}{dt}\mathbf{n}_r\cdot\mathbf{n}_r\right)\int_{CS_2}dA = 0, \qquad (10.38)$$

or

$$-\bar{v}A + \frac{dr}{dt}(4\pi r^2) = 0 \rightarrow \frac{dr}{dt} = \frac{\bar{v}A}{4\pi r^2}. \qquad (10.39)$$

Now, for the syringe: Its mass balance relation

$$\frac{\partial}{\partial t}\int_{CV}\rho d\mathcal{V}+\int_{CS}\rho(\mathbf{v}\cdot\mathbf{n})dA = 0 \qquad (10.40)$$

reduces to

$$\int_{CS_1}(\mathbf{v}\cdot\mathbf{n})dA+\int_{CS_2}(\mathbf{v}\cdot\mathbf{n})dA = 0, \qquad (10.41)$$

wherein we have assumed that there is no flux through the side of the syringe. If the syringe is of radius a, then

$$\bar{v}_s\pi a^2 = \bar{v}A \rightarrow \bar{v} = \frac{\bar{v}_s\pi a^2}{A}, \qquad (10.42)$$

where $\bar{v}A$ is the same quantity that enters the balloon. Hence, from Eqs. (10.39) and (10.42), we have

$$4\pi r^2 \frac{dr}{dt} = \bar{v}_s\pi a^2, \qquad (10.43)$$

where the velocity of the syringe is assumed to be known, as is the radius a of the syringe. Hence, we see that the rate at which the balloon expands depends strongly on its current diameter. The larger the balloon, the more slowly it expands given a constant inflow; that is, there is but one physical answer (assuming uniqueness), which must be obtained regardless of the particular formulation.

Example 10.4 Given the result of Eq. (10.43), find the relation for the balloon radius given a constant $\bar{v}_s$.

Solution: Equation (10.43) can be written as

$$[r(t)]^2 \frac{dr}{dt} = \frac{\bar{v}_s a^2}{4}.$$

Hence, integrating from time $t = 0$ to time t, we have

$$\int_0^t [r(t)]^2 \frac{dr}{dt}\, dt = \int_0^t \frac{\bar{v}_s a^2}{4}\, dt,$$

or

$$\frac{[r(t)]^3}{3} - \frac{[r(0)]^3}{3} = \frac{\bar{v}_s a^2}{4} t$$

and, thus,

$$r(t) = \left([r(0)]^3 + \frac{3\bar{v}_s a^2}{4} t\right)^{1/3}.$$

10.4 Murray's Law and Optimal Design

In 1926, C. D. Murray suggested that the inner radius a of a blood vessel arises as an outcome of a "compromise" between the advantage of increasing the lumen, which decreases the resistance to flow, and the disadvantage of increasing overall blood volume, which increases the metabolic demand to maintain the blood (remember that the red blood cells must be produced continuously by the bone marrow to replace cells as they die after ~120 days). Murray postulated an associated "cost" function C and suggested that the optimal radius is determined by minimizing this cost with respect to the radius. Specifically, he assumed that C consisted of two terms: one representing the mechanical power associated with the flowing blood and another that is proportional to the volume of blood that had to be maintained metabolically.

Mechanical *power* is defined as the rate at which work is done; work on the other hand, can be computed as a force acting through a distance (e.g., fx). If the force is constant, then the associated rate is computed as the force times the time rate of change of the distance, or force times velocity ($f \cdot v$). It can be shown that, in terms of stress, the mechanical power per unit volume is given by the sum of all stresses acting at a point multiplied by their associated rates of deformation, as, for example (Humphrey, 2002),

$$\sigma_{rr}D_{rr} + \sigma_{r\theta}D_{r\theta} + \sigma_{rz}D_{rz} + \cdots + \sigma_{zr}D_{zr} + \sigma_{z\theta}D_{z\theta} + \sigma_{zz}D_{zz}. \quad (10.44)$$

10.4.1 Straight Segment

To investigate one consequence of Murray's postulate, let us consider a steady, fully developed, unidirectional, incompressible, Newtonian flow in a straight rigid circular tube. From Section 9.2 of Chapter 9, we recall that the only nonzero component of stress is σ_{rz}, which equals σ_{zr} by the balance of angular momentum. Specifically,

$$\sigma_{rz} = 2\mu D_{rz} = \mu\left(\frac{\partial v_z}{\partial r}\right) \quad (10.45)$$

where

$$\frac{\partial v_z}{\partial r} = \frac{r}{2\mu}\left(\frac{dp}{dz}\right),$$

as seen in Eq. (9.46). Recall, too, that [Eq. (9.54)]

$$\frac{dp}{dz} = -\frac{8\mu Q}{\pi a^4}; \quad (10.46)$$

hence, the total mechanical power associated with the flowing blood is

$$\int_v (\sigma_{rz}D_{rz} + \sigma_{zr}D_{zr})dv = \int_0^L \int_0^{2\pi} \int_0^a 2\mu\left(\frac{\partial v_z}{\partial r}\right)\left(\frac{1}{2}\frac{\partial v_z}{\partial r}\right)r\,dr\,d\theta\,dz, \quad (10.47)$$

where L is the length of the tube. Hence,

$$2\pi L \int_0^a \mu \left[\frac{r}{2\mu} \left(-\frac{8\mu Q}{\pi a^4} \right) \right]^2 r\, dr = 2\pi L \int_0^a \frac{16\mu Q^2 r^2}{\pi^2 a^8} r\, dr$$
$$= \frac{32\mu L Q^2}{\pi a^8} \left(\frac{a^4}{4} \right) = \frac{8\mu L Q^2}{\pi a^4}. \qquad (10.48)$$

Murray assumed that the metabolic cost of maintaining the blood was proportional to its volume $(\pi a^2 L)$, with γ a material parameter. Hence, he let the overall, or global, cost be

$$C = \frac{8\mu L Q^2}{\pi a^4} + \gamma \pi a^2 L \qquad (10.49)$$

whereby a minimum requires that

$$\frac{dC}{da} = 0, \qquad \frac{d^2 C}{da^2} > 0. \qquad (10.50)$$

Clearly, the first condition requires

$$0 = -4 \left(\frac{8\mu L Q^2}{\pi} \right) a^{-5} + 2\gamma \pi a L \to a^6 = \frac{16\mu Q^2}{\gamma \pi^2}, \qquad (10.51)$$

or

$$Q = \left(\frac{\gamma}{\mu} \right)^{1/2} \frac{\pi a^3}{4} \to (\gamma\mu)^{1/2} \frac{\pi a^3}{4\mu} = Q. \qquad (10.52)$$

It can be shown that the second derivative is positive; thus, Murray's simple analysis suggests that the volumetric flow rate is related to the optimal radius cubed. This conclusion led Zamir to suggest in 1977 that, because the wall shear stress is related inversely to the radius cubed [i.e., $\tau_w = 4\mu Q/\pi a^3$ from Eq. (9.55)], a consequence of Murray's minimum postulate is that the vasculature seeks to maintain τ_w constant because $\sqrt{\gamma\mu}$ is just a constant (see Zamir, 2000). This conclusion has been supported by most data, with the exception that the mean wall shear stress is very different in veins (0.1–0.6 Pa) than it is in most arteries (1.2–1.8 Pa). To explain such differences, Pries et al. (1995) proposed a "pressure-shear" hypothesis stating that "vascular systems grow and adapt in response to hemodynamic conditions so as to maintain local wall shear stress at a set point that is a function of the local transmural pressure." See Humphrey (2002) for a discussion of how this hypothesis can be addressed mathematically via an extension of Murray's law that was put forth by L. A. Taber.

10.4.2 Bifurcation Areas

One of the most conspicuous characteristics of the vasculature is that it divides into smaller and smaller vessels down to the level of capillaries.

Moreover, these divisions occur as *bifurcations* (i.e., a single parent vessel gives rise to a pair of daughter vessels). Within the context of Murray's optimization postulate, it is interesting to ask if such bifurcations are also optimized.

Let the inner radius of the parent vessel be a_0 and that of the daughter vessels be a_1 and a_2. If we use the convention that $a_1 \geq a_2$ then useful indices of a bifurcation are

$$\alpha = \frac{a_2}{a_1}, \qquad \beta = \frac{a_1^2 + a_2^2}{a_0^2}, \tag{10.53}$$

the latter of which is a measure of the ratio of the net cross-sectional area of the daughter branches to that of the parent vessel. Regardless of the value of this ratio, balance of mass requires that $Q_0 = Q_1 + Q_2$. If we assume that the viscosity μ and the metabolic parameter γ are constant throughout the bifurcation, then "Murray's law" $Q = ka^3$ from Eq. (10.52), where k is a constant, requires from mass balance that

$$a_0^3 = a_1^3 + a_2^3 = (1 + \alpha^3)a_1^3, \tag{10.54}$$

or

$$\frac{a_1}{a_0} = \frac{1}{(1 + \alpha^3)^{1/3}}, \qquad \frac{a_2}{a_0} = \frac{\alpha}{(1 + \alpha^3)^{1/3}} \tag{10.55}$$

and, thus,

$$\beta = \frac{1 + \alpha^2}{(1 + \alpha^3)^{2/3}}. \tag{10.56}$$

For a "symmetric bifurcation," therefore, $\alpha = 1$, $a_1/a_0 = a_2/a_0 \sim 0.794$ and $\beta \sim 1.26$; that is, each symmetrical bifurcation that obeys Murray's law would increase the net cross-sectional area by $\sim 26\%$ and thereby decrease the mean velocity by the same (recall, $Q = \bar{v}A$).

Fung (1993) noted that n generations of symmetrical bifurcations would yield $a_1^{(n)} = (0.794)^n a_0^{(1)}$. He suggests, therefore, that if a capillary has a radius of $5\,\mu m$ and the mean radius of the aorta is $\sim 1.5\,cm$, then the number of generations n would be about 30 (actually $n \sim 35$, which can be found by taking logarithms). Fung notes that if each bifurcation multiplies the number of vessels by 2, then the total number of vessels would be $2^n = 2^{30} \sim 1 \times 10^9$ (or, actually $2^{35} \sim 3.4 \times 10^{10}$). Fung emphasizes that these numbers are merely rough estimates for the vasculature, which, unlike the airways, does not branch symmetrically (cf. Fig. 7.5 of Chapter 7). Nevertheless, it is interesting to note that Milnor (1989) reports that a 20-kg dog has 1 aorta + 116,540 arteries + 2.8×10^6 arterioles + 2.7×10^9 capillaries $\sim 2.7 \times 10^9$ vessels. Given Milnor's report that the maximum radius of the aorta is $1.0\,cm$ and the capillary diameter is $8\,\mu m$, Murray's law would predict (for symmetric bifurcations) $n \sim 31$, and, thus $2^{31} \sim 2.15 \times 10^9$ vessels, which is remarkably close (20%) to the reported number given the gross assumption.

10.4.3 Bifurcation Patterns

Whereas we have considered the cross-sectional areas of daughter vessels in a bifurcation relative to that of the parent vessel, let us now consider the bifurcation angles (Fig. 10.12). One way to address the optimal geometry of a bifurcation is to simply ask if it is better to deliver fluid from point A to points C and D (Fig. 10.12) through two tubes of equal diameter, one from A to C and the other from A to D, or if it is better to deliver fluid from A to C and D via a tube that bifurcates. If the distances between B to C and B to D are small in comparison to L_0, this question can be answered approximately by comparing the mechanical power for one vessel of length L_0 and radius a_0 to that for two vessels, one of radius a_1 and one of radius a_2, each of approximate length L_0; that is, from Eq. (10.48), we compare

$$\frac{8\mu L_0 Q_0^2}{\pi a_0^4} \quad \text{to} \quad \frac{8\mu L_0}{\pi}\left(\frac{Q_1^2}{a_1^4} + \frac{Q_2^2}{a_2^4}\right). \tag{10.57}$$

If Murray's (cube) law holds, then $Q_1 = ka_1^3$, $Q_2 = ka_2^3$, and $Q_0 = ka_0^3$ from Eq. (10.52); hence, the fractional difference between the power required for two versus one supply vessel is

$$\left[\frac{8\mu L_0}{\pi}\left(\frac{k^2 a_1^6}{a_1^4} + \frac{k^2 a_2^6}{a_2^4}\right) - \frac{8\mu L_0}{\pi}\left(\frac{k^2 a_0^6}{a_0^4}\right)\right]\left[\frac{8\mu L_0}{\pi}\left(\frac{k^2 a_0^6}{a_0^4}\right)\right]^{-1}$$
$$= \frac{a_1^2 + a_2^2 - a_0^2}{a_0^2} = \beta - 1 \tag{10.58}$$

from Eq. (10.53). Because β is generally greater than 1, the power requirement for two supply vessels is generally higher. For example, for a symmetrical bifurcation, $\beta \sim 1.26$, and thus there is a 26% lower power requirement for one supply vessel over two identical "parallel" vessels. Hence, the vessel wants to bifurcate.

To address the optimal bifurcation angle, first note from Eqs. (10.49) and (10.52) that the minimum cost function with respect to radius is

$$C(\text{min on } a) = \frac{1}{2}\gamma\pi a^2 L + \gamma\pi a^2 L = \frac{3}{2}\gamma\pi a^2 L. \tag{10.59}$$

The total cost for a bifurcation shown in Figure 10.12 with optimal radii is thus

$$C = \frac{3}{2}\gamma\pi(a_0^2 L_0 + a_1^2 L_1 + a_2^2 L_2). \tag{10.60}$$

It can be shown (Fung, 1993; Zamir, 2000) that minimization of this cost function yields the optimal branch angles

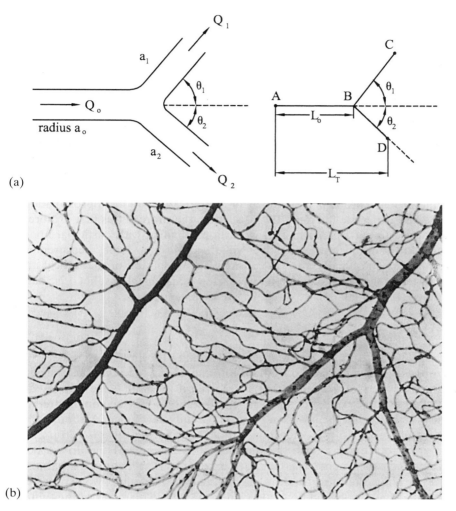

(a)

(b)

FIGURE 10.12 (a) Bifurcation angles and Murray's law. Bifurcations are one of the basic structural units of the vasculature; they are likely dictated by genetic programming but also functional adaptations (recall Fig. 8.20 of Chapter 8). A basic question, therefore, is whether such angles are optimized in each case. (b) Photomicrograph of an actual vascular bed (from the retina), which reveals the many generations of branches and different branch angles. [From Fawcett (1986), with permission].

$$\cos\theta_1 = \frac{a_0^4 + a_1^4 - a_2^4}{2a_0^2 a_1^2}, \qquad \cos\theta_2 = \frac{a_0^4 - a_1^4 + a_2^4}{2a_0^2 a_1^2}. \qquad (10.61)$$

10.5 Buckingham Pi and Experimental Design

Experimental methods are essential to three distinct areas of continuum biomechanics. As noted in Chapters 2 and 7, quantification of constitutive behaviors can be accomplished only via the performance of (theoretically motivated) experiments. Experiments are likewise essential for the evaluation of theoretical or computational findings. Finally, in some cases, we cannot solve a problem analytically or numerically without approximations based on experimental data. In this section, therefore, let us first explore a tool that supports experimental work in mechanics, with particular application to biofluid mechanics. We will then consider some measurement techniques and, finally, we will consider a combination of theoretical and experimental results that enables us to solve many problems that arise daily in the research laboratory or clinical setting.

10.5.1 Motivation

Consider the "simple" problem of determining the pressure gradient (or drop) that is required to move fluid through a long tube that has a significant roughness of the luminal surface. Even if the fluid is Newtonian and the flow is steady and incompressible, the solution may be complicated by the possibility that the flow is not laminar or fully developed. Assuming that one must resort to the determination of the requisite *pressure drop* $\Delta p = p_1 - p_2$ from experiments, the question thus becomes: How do we best design the experiment? The first need, of course, is to identify the parameters on which the pressure drop may depend. For example, we expect that Δp will, for each desired flow rate Q, depend on the diameter of the tubing D, its length L, and the surface roughness e (Fig. 10.13). Likewise, even in an isothermal test, the pressure–flow relation will likely depend on the choice of the infused fluid and, thus, its viscosity μ and density ρ. At the minimum, therefore, we must experimentally determine a functional relation of the form

$$\Delta p = f(D, L, e, \mu, \rho, \bar{v}), \qquad (10.62)$$

where $\bar{v} = Q/A = 4Q/\pi D^2$; thus we can use the mean velocity $\bar{v}$ rather than the volumetric flow rate Q in our desired functional relationship. To determine how Δp varies as a function of the tube diameter D, we must perform multiple tests (say $n = 5$ for repeatability) using tubes of different diameters while all the other parameters are held fixed. Assuming one uses only 5 different diameters, with $n = 5$ each, this requires 25 experiments with L, e, μ, ρ, and $\bar{v}$ fixed. If tests are then repeated for, say, five different lengths,

tube

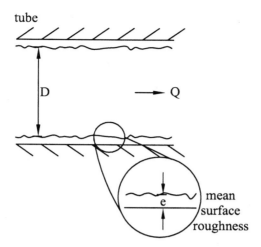

FIGURE 10.13 Magnified view of the inner surface of a tube (or pipe) through which fluid flows. Although we often assume a smooth surface, all surfaces have an inherent roughness; in a blood vessel, this roughness arises from the cobblestone-type arrangement of the endothelial cells. Surface roughness is quantified here via a mean height e.

five different surface roughnesses, five different fluids, and five different flow rates, it is easy to see that the number of necessary experiments can become very large. An obvious question, therefore, is whether we can reduce the number of requisite tests. Fortunately, the answer is yes. Let us now consider the so-called *Buckingham Pi Theorem*, which enables this reduction in experimental complexity.

10.5.2 Recipe

The goal of the Buckingham Pi Theorem is to identify nondimensional groups of parameters from physical parameters in the experiment of interest that allow the experiments to be performed more efficiently. The two key ideas are, first, the difference between fundamental and derived "dimensions" and, second, the importance of "scales" to measure the fundamental dimensions (or units). To appreciate what a fundamental quantity is, note that we typically think of velocity as a *derived* quantity—it is the time rate of change of a position; that is, whereas length and time may be taken as "fundamental," velocity is derived from these. In general, we usually take length, time, mass, and temperature as *fundamental*[2] in continuum mechanics. Herein we shall denote these fundamental dimensions, as L, T, M, and Θ, respectively.

---

[2] These are not unique; many prefer entropy instead of temperature, for example.

The concept of "scales" may require a change in mindset. For example, if asked how we should measure length, time, mass, and temperature, many would respond "in terms of meters, seconds, kilograms, and Kelvins." Here, however, we wish to introduce a different approach. If you look around a classroom, for example, you could measure it in length-dimensions of meters (or feet), but you could also measure it in terms of floor tiles (if present), ceiling tiles (if present), chalk boards (if present), and so forth; that is, a room that is 40 ft wide could also be said to be, for example, 40 floor tiles, 13.3 ceiling tiles, or perhaps 2 chalkboards wide, provided standard sized floor tiles of 1 ft^2, ceiling tiles 3 ft long, and a 20-ft-long chalkboard. In other words, there are multiple, convenient scales, or rulers, available. Indeed, it is interesting to review the historical development of standard weights and measures that led to the introduction of the metric system in 1799 (following the French Revolution, which encouraged change in many areas). For example, older "scales" included the Biblical cubit (distance from the elbow to the tip of the middle finger) and the fathom (distance between outstretched arms), each of which were useful but not precise, for they differ from person-to-person (see Boorstin, 1985, Chapter 51). Alternatively, one could pick a scale that is well defined in a particular problem and precise. For example, a tube could be said to be 1 in. in diameter and 5 in. long or simply 1 diameter in diameter and 5 diameters long. The utility of using such intrinsic scales becomes apparent primarily via illustration in particular examples. To generate such examples, let us now simply list the *five requisite steps* (i.e., a recipe) for employing the Buckingham Pi Theorem:

1. Identify the functional relationship of interest: $x_1 = f(x_2, x_3, \ldots, x_n)$.
2. Select appropriate fundamental dimensions relevant to the problem, like L, T, M, and Θ, and determine the associated dimensions of each variable in step 1: $[x_i] = L^{a_i} T^{b_i} M^{c_i} \Theta^{d_i}$.
3. Pick scales for each fundamental dimension: L_s, T_s, M_s, and Θ_s.
4. Compute the Pi groups: $\pi_i = x_i/(L_s)^{a_i}(T_s)^{b_i}(M_s)^{c_i}(\Theta_s)^{d_i}$.
5. Rewrite the basic equation in terms of the Pi groups: $\pi_1 = g(\pi_2, \pi_3, \ldots, \pi_n)$.

Just as in $C\forall$ analyses, the Buckingham Pi Theorem can be executed without much mathematical difficulty. Rather, *the most challenging aspect is learning how to pick reasonable scales*; this is best accomplished via experience. Hence, let us now consider a few examples to begin to build up the necessary intuition.

Example 10.5 Find a general nondimensional relation for the fluid velocity induced by a moving plate in a Couette flow (cf. Fig. 9.7 of Chapter 9).

Solution: Following the five-step recipe, let us first write a general functional relationship for the velocity in terms of variables that we expect to influence it. For example, the velocity v_x in the fluid may depend on the position y between the plates as well as the velocity U_0 of the moving plate and the gap distance h. Although we would also expect the fluid properties, viscosity and density, to likewise play a role, for simplicity let us consider only the following:

Step 1: $v_x = f(U_0, h, y)$.

Step 2: Next, we must identify the fundamental dimensions, which, in general, are L, T, M, and Θ, but for this isothermal problem, length, time, and mass will suffice. Moreover, we must find the dimensions of each variable, namely[3]

$$[v_x] = \frac{L}{T} = L^1 T^{-1} M^0, \qquad [h] = L = L^1 T^0 M^0,$$

$$[U_0] = \frac{L}{T} = L^1 T^{-1} M^0, \qquad [y] = L = L^1 T^0 M^0.$$

Step 3: This is the most important and indeed the most challenging step: pick appropriate scales. Because each variable in the basic equation depends only on length and time, we only need two scales. Clearly, a convenient and natural intrinsic length scale is the separation distance h between the plates. Picking a timescale is much different. Generally, one tries to find a quantity having dimensions of time based on variables in the list. In this case, we see that dividing a length by a velocity will yield a time. Hence, let

$$L_s = h \quad \text{and} \quad T_s = \left(\frac{h}{U_0}\right).$$

Step 4: Following the formula in the above recipe, we now calculate nondimensional Pi variables for each of our four variables in the original function:

$$\pi_1 = \frac{v_x}{(h)^1 (h/U_0)^{-1} (1)^0 (1)^0} = \frac{v_x}{U_0}, \qquad \pi_3 = \frac{h}{(h)^1 (h/U_0)^0} = \frac{h}{h} = 1,$$

$$\pi_2 = \frac{U_0}{(h)^1 (h/U_0)^{-1}} = \frac{U_0}{U_0} = 1, \qquad \pi_4 = \frac{y}{(h)^1 (h/U_0)^0} = \frac{y}{h}.$$

Step 5: Therefore, the final step is to write our relation of interest,

$$x_1 = f(x_2, x_3, x_4) \Leftrightarrow v_x = f(U_0, h, y)$$

in terms of Pi-groups:

---

[3] Note that we use [x] to denote the dimension of x; this is not to be confused with the use of brackets to denote a matrix.

$$\pi_1 = g(\pi_2, \pi_3, \pi_4) \Leftrightarrow \frac{v_x}{U_0} = g\left(1, 1, \frac{y}{h}\right)$$

or, according to the Buckingham Pi Theorem, we merely need to relate two nondimensional parameters functionally, not four dimensional parameters:

$$\frac{v_x}{U_0} = g\left(\frac{y}{h}\right).$$

Note: The values of unity in the function [e.g., $g(1, 1)$], simply imply constants in the general function and thus do not need to be written explicitly. This clearly reduces the experimental need. Rather than performing experiments wherein we measure velocities for multiple combinations of U_0 and h at multiple values of y, we merely need to relate two nondimensional quantities for any y. This is yet another example where theory tells us what to measure and why (i.e., how to interpret the data). Indeed, if we look back to Example 9.2, we find that a Navier–Stokes solution revealed that $v_x = U_0(y/h)$, which is recovered by the Buckingham Pi result if the function g is simply linear in y/h Clearly, this theorem can aid in the experimental identification of various functional relations of interest.

Example 10.6 Recall from Chapter 8 that the exit velocity from a simple reservoir is $v = \sqrt{2gh}$, which is the same result that one obtains for the velocity of a mass particle that is dropped from a resting position at height h above the surface. If one wishes to determine the latter result (for the first time) experimentally given the general relation

$$v = f(m, g, h)$$

for a particle of mass m, and neglecting air resistance, one might seek to perform multiple experiments for different masses m and heights h. Let us see what Buckingham Pi would suggest, however.

Solution

Step 1:   State the desired functional relationship, $v = f(m, g, h)$.
Step 2a: Select fundamental units such as L, T, and M and determine the unit equations for each variable:

$$[v] = L^1 T^{-1} M^0, \qquad [g] = L^1 T^{-2} M^0,$$
$$[m] = L^0 T^0 M^1, \qquad [h] = L^1 T^0 M^0.$$

Step 3:   Pick scales. Selections of intrinsic scales are obvious for length and mass. For time, however, we could pick a length divided by the impact velocity, or because gravity is key to the problem and it has a time dimension within, we could also consider the square root of h divided by g; that is, we could let

$$L_s = h, \qquad T_s = \sqrt{\frac{h}{g}}, \qquad M_s = m.$$

Step 4: Compute the Pi groups:

$$\pi_1 = \frac{v}{(h)^1 \left(\sqrt{h/g}\right)^{-1} (m)^0} = \frac{v}{\sqrt{gh}}, \qquad \pi_3 = \frac{g}{(h)^1 \left(\sqrt{h/g}\right)^{-2}} \equiv 1,$$

$$\pi_2 = \frac{m}{(m)^1} \equiv 1, \qquad \pi_4 = \frac{h}{(h)^1} \equiv 1.$$

Step 5: Rewriting our original equation, we find

$$\frac{v}{\sqrt{gh}} = f(1, 1, 1) \quad \text{or} \quad \frac{v}{\sqrt{gh}} = c.$$

Thus, Buckingham Pi reveals that $v/\sqrt{gh}$ is a constant and all we need to do is to measure multiple values of v for multiple values of h and find the constant, which, of course, we know is $\sqrt{2}$.

Example 10.7 Recall that we began this section by considering experiments to determine the pressure drop Δp associated with flow through a potentially rough circular pipe, which we expect to depend on several factors. The pressure drop may be affected, for example, by the diameter D of the pipe, the length L of the pipe, the roughness height e of the inner surface of the pipe, the viscosity μ of the fluid, the density ρ of the fluid, and the mean velocity $\bar{v}$ of the fluid. Use the Buckingham Pi Theorem to determine a set of dimensionless groups that can be used to design appropriate experiments and to correlate the associated data.

Solution: Follow the recipe:

Step 1: $\Delta p = f(D, L, e, \mu, \rho, \bar{v})$.
Step 2: The fundamental units that are needed are L, T, and M; thus, for each variable, we have[4]

$$[\Delta p] = \frac{\text{Force}}{\text{Area}} = \frac{ML/T^2}{L^2} = L^{-1}T^{-2}M^1, \qquad [\mu] = \frac{\text{Force/Area}}{1/\text{Time}} = \frac{ML/T^2}{L^2/T}$$

$$= L^{-1}T^{-1}M^1,$$

$$[D] = L^1 T^0 M^0, \qquad\qquad\qquad [\rho] = \frac{\text{Mass}}{\text{Volume}} = L^{-3}T^0 M^1,$$

$$[L] = L^1 T^0 M^0, \qquad\qquad\qquad [\bar{v}] = L^1 T^{-1} M^0.$$

$$[e] = L^1 T^0 M^0,$$

---

[4] In some cases, we may not know the dimensions of a parameter directly, such as the viscosity. In such cases, we recall a simple relation that relates the parameter to those having known dimensions (e.g., $\sigma_{xy} = 2\mu D_{xy}$).

Step 3:   It is reasonable to let the diameter be the length scale and likewise the ratio of diameter to the mean velocity be the timescale. For mass, it is reasonable to take the density times a volume. Although $\pi D^2 L/4$ is the fluid volume over the entire length of the pipe, selecting D^3 as a volume is similarly acceptable. Hence, let

$$L_s = D, \qquad T_s = \left(\frac{\bar{v}}{D}\right)^{-1} = \frac{D}{\bar{v}}, \qquad M_s = \rho D^3.$$

Step 4:   The Pi groups are thus

$$\pi_1 = \frac{\Delta p}{(D)^{-1}(D/\bar{v})^{-2}(\rho D^3)^1} = \frac{\Delta p}{\rho \bar{v}^2},$$

$$\pi_5 = \frac{\mu}{(D)^{-1}(D/\bar{v})^{-1}(\rho D^3)^1} = \frac{\mu}{\rho \bar{v} D} = \frac{1}{\mathrm{Re}},$$

$$\pi_2 = \frac{D}{(D)^1(D/\bar{v})^0(\rho D^3)^0} = \frac{D}{D} = 1,$$

$$\pi_6 = \frac{\rho}{(D)^{-3}(D/\bar{v})^0(\rho D^3)^1} = \frac{\rho}{\rho} = 1,$$

$$\pi_3 = \frac{L}{(D)^1(D/\bar{v})^0(\rho D^3)^0} = \frac{L}{D},$$

$$\pi_7 = \frac{\bar{v}}{(D)^1(D/\bar{v})^{-1}(\rho D^3)^0} = \frac{\bar{v}}{\bar{v}} = 1.$$

$$\pi_4 = \frac{e}{(D)^1(D/\bar{v})^0(\rho D^3)^0} = \frac{e}{D},$$

In particular, note that the combination of $\rho \bar{v} D / \mu$ terms appears so commonly in fluid mechanics that it is given a special symbol Re and is called the *Reynolds' number*. It has been mentioned earlier, but its utility will be seen in more detail in Section 10.6.

Step 5:   Rewriting the governing functional equation

$$x_1 = f(x_2, x_3, x_4, x_5, x_6, x_7) \Leftrightarrow \Delta p = f(D, L, e, \mu, \rho, \bar{v})$$

in terms of Pi-groups, we have

$$\pi_1 = g(\pi_2, \pi_3, \pi_4, \pi_5, \pi_6, \pi_7) \Leftrightarrow \frac{\Delta p}{\rho \bar{v}^2} = g\left(1, \frac{L}{D}, \frac{e}{D}, \frac{1}{\mathrm{Re}}, 1, 1\right).$$

Hence, according to the Buckingham Pi Theorem,

$$\frac{\Delta p}{\rho \bar{v}^2} = g\left(\frac{L}{D}, \frac{e}{D}, \frac{1}{\mathrm{Re}}\right) = \tilde{g}\left(\frac{L}{D}, \frac{e}{D}, \mathrm{Re}\right).$$

Note: If we have an equation $y = ax^2$, then we say $y = f(x)$. Similarly, if we have an equation $y = a/x^2$, then we again say $y = g(x)$. The key here is the functional dependency. As it turns out, extensive experiments have revealed that π_1 depends linearly on L/D and, thus,

$$\frac{\Delta p}{\frac{1}{2}\rho \bar{v}^2} = \frac{L}{D} f\left(\mathrm{Re}, \frac{e}{D}\right),$$

where the function f is called a *friction factor*. (Note: The function $\tilde{g}$ is arbitrary; hence, we can multiply or divide it by a constant such as $\dfrac{1}{2}$, which we do so as to have a kinetic-energy-type term.) This relation will play a key role in Section 6. First, however, let us consider a specific example from the lung mechanics literature.

Example 10.8 As noted earlier in this chapter, the primary function of the lungs is to facilitate gas exchange between the atmosphere and the blood. Toward this end, the capillary system in the lungs is very different than that found elsewhere. Conforming to the alveolar geometry (Fig. 10.3), capillary blood flow in the lungs is better described as a *sheet flow* rather than a tube flow; that is, the blood flows within the thin planar walls of the alveoli, which appear as parallel membranes separated by hexagonally positioned posts. Fung and his colleagues sought to quantify the pressure–flow relation in this sheet flow and began with a nondimensionalization. Here, let us perform a similar procedure and compare to that reported by Fung (1993).

Solution: Fung considered the pressure drop Δp within a pulmonary capillary to depend on the density and viscosity of the blood (ρ, μ), mean velocity U, circular frequency of oscillation ω, sheet thickness h and width w, post diameter ε and separation distance a, angle between the mean flow and post alignment θ, the hematocrit H and red blood cell diameter D_c, the elastic modulus of the red blood cell E_c, and a ratio between the vascular space and tissue volume (VSTR). Consistent with Step 1 in our Buckingham Pi approach,

$$\Delta p = g(\rho, \mu, U, \omega, h, w, \varepsilon, a, \theta, H, D_c, E_c, \text{VSTR}).$$

It is easy to see that appropriate fundamental units are L, T, and M, where (Step 2)

$$[\Delta p] = L^{-1}T^{-2}M^1, \qquad [\omega] = L^0T^{-1}M^0, \qquad [a] = L^1T^0M^0,$$

$$[\rho] = L^{-3}T^0M^1, \qquad [h] = L^1T^0M^0, \qquad [D_c] = L^1T^0M^0,$$

$$[\mu] = L^{-1}T^{-1}M^1, \qquad [w] = L^1T^0M^0, \qquad [E_c] = L^{-1}T^{-2}M^1,$$

$$[U] = L^1T^{-1}M^0, \qquad [\varepsilon] = L^1T^0M^0,$$

and, of course, $[\theta] = [H] = [\text{VSTR}] = 1$. If we select length, time, and mass scales (Step 3) to be

$$L_s = h, \qquad T_s = \frac{h}{U}, \qquad M_s = (\rho hw)h,$$

then (Step 4)

$$\pi_p = \frac{\Delta p}{\rho U^2}\left(\frac{h}{w}\right), \qquad\qquad \pi_w = \frac{w}{h},$$

$$\pi_\rho = \frac{h}{w}, \qquad\qquad \pi_\varepsilon = \frac{\varepsilon}{h},$$

$$\pi_\mu = \frac{\mu}{\rho U w} = \frac{\mu}{\rho U h}\left(\frac{h}{w}\right), \qquad \pi_a = \frac{a}{h}\left(\frac{\varepsilon}{\varepsilon}\right) = \frac{a}{\varepsilon}\left(\frac{\varepsilon}{h}\right),$$

$$\pi_U = 1, \qquad\qquad \pi_{D_c} = \frac{D_c}{h},$$

$$\pi_\omega = \frac{\omega}{U}h, \qquad\qquad \pi_{E_c} = \frac{E_c}{\rho U^2}\left(\frac{h}{\omega}\right),$$

$$\pi_h \equiv 1,$$

and, thus (Step 5),

$$\frac{\Delta p}{\rho U^2}\left(\frac{h}{w}\right) = \tilde{g}\left(\frac{h}{w}, \mathrm{Re}, \frac{w}{h}, \frac{\varepsilon}{h}, \frac{a}{h}, \frac{D_c}{h}, \frac{\omega}{U}h, \frac{E_c}{\rho U^2}\left(\frac{h}{\omega}\right), \theta, H, \mathrm{VSTR}\right),$$

where the Reynolds' number is $\mathrm{Re} = \rho U h / \mu$. Hence, Buckingham Pi reduced the number of independent variables from 13 to 10, a slight improvement. Fung (1993) actually chose a few different, equivalent nondimensional parameters; they are related to the present ones via

$$\frac{\nabla p h^2}{\mu U} = \frac{(\Delta p/h)h^2}{\mu U} = \frac{\Delta p h}{\mu U}\left(\frac{\mu}{\rho U h}\right)\left(\frac{h}{w}\right) = \frac{\Delta p}{\rho U^2}\left(\frac{h}{w}\right),$$

$$\frac{\mu U}{E_c h} = \frac{\mu U}{E_c h}\left(\frac{\rho U^2}{\rho U^2}\right) = \frac{\mu}{\rho U h}\left(\frac{\rho U^2}{E_c}\right),$$

$$\sqrt{\frac{h^2 \omega \rho}{4\mu}} = \frac{1}{2}\sqrt{\left(\frac{\omega}{U}h\right)\left(\frac{\rho U h}{\mu}\right)},$$

which is to say, our current Pi groups differ from Fung's only through the Reynolds' number $\rho U h / \mu$ and the term h/w. It is interesting that experiments revealed that the Reynolds' number Re and Womersley's number

$$\frac{h}{2}\sqrt{\frac{\omega\rho}{\mu}}$$

are both less than unity and thus negligible in this sheet flow. Note, too, that Fung's parameter $\mu U / E_c h$ is essentially the ratio of the shear stress in a Couette flow between parallel plates (cf. Example 9.2 of Chapter 9) to the modulus of the red blood cell (RBC), which was interpreted as a RBC membrane shear strain despite the flow not being Couette.

Experiments suggested further that, with a minus sign accounting for the pressure gradient being opposite the pressure drop,

$$\frac{\nabla p h^2}{\mu U} = -G_1\left(\frac{D_c}{h}, \frac{\mu_0 U}{E_c h}, H\right) G_2\left(\frac{w}{h}\right) f\left(\frac{h}{\varepsilon}, \frac{\varepsilon}{a}, \theta, \text{VSTR}\right),$$

where

$$\mu G_1\left(\frac{D_c}{h}, \frac{\mu_0 U}{E_c h}, H\right) \equiv \mu_a$$

was taken to be the apparent viscosity, with the form

$$\mu_a = \mu\left[1 + c_1\left(\frac{D_c}{h}\right) H + c_2\left(\frac{D_c}{h}\right) H^2\right],$$

the effect of $\mu_0 U/E_c h$ being yet unexplored. The function G_2 was found to be

$$G_2\left(\frac{w}{h}\right) = \frac{12}{1 - 0.63(w/h)} \approx 12,$$

whereas the function f was called a *geometric friction factor*. It was found experimentally to vary nearly linearly with h/ε with values of f from 1.5 to 5 for h/ε from 1 to 5. Values of VSTR ~ 91, $h \sim 7.4\,\mu m$, $\varepsilon \sim 4\,\mu m$, and $a \sim 12\,\mu m$; f would equal 1 in the absence of posts. Hence, the semi-empirical relation reduced to

$$\nabla p \cong -\frac{12\mu_a U}{h^2} f\left(\frac{h}{\varepsilon}, \frac{\varepsilon}{a}, \theta, \text{VSTR}\right).$$

Because shear flow is two-dimensional, in general, Fung and colleagues thus considered

$$\frac{\partial p}{\partial x} = -\frac{12\mu_a U}{h^2} f_x\left(\frac{h}{\varepsilon}, \frac{\varepsilon}{a}, \theta, \text{VSTR}\right) = -\frac{12\mu_a U}{h^2} f_x,$$

$$\frac{\partial p}{\partial y} = -\frac{12\mu_a V}{h^2} f_y\left(\frac{h}{\varepsilon}, \frac{\varepsilon}{a}, \theta, \text{VSTR}\right) = -\frac{12\mu_a V}{h^2} f_y,$$

where U and V are mean velocities in the x and y directions, respectively, and $f_x \sim f_y \sim 2.5$. In general, the mean values of 2-D velocities within the capillaries are

$$U = -\frac{h^2}{12\mu_a f_x}\left(\frac{\partial p}{\partial x}\right), \qquad V = -\frac{h^2}{12\mu_a f_y}\left(\frac{\partial p}{\partial y}\right).$$

Finally, Fung and colleagues suggested that

$$h = h_0 + \alpha \Delta p$$

based on morphometric data, with $h_0 = 4.28\,\mu m$ in cat lung and $3.5\,\mu m$ in human lung, $\alpha = 0.219\,\mu m/cm\,H_2O$ in cat lung for a $\Delta p \sim 10\,cm\,H_2O$, and $\alpha = 0.127\,\mu m/cm\,H_2O$ in human lung for a $\Delta p \sim 10\,cm\,H_2O$. For more details, see Fung (1984; 1993). The take-home message here is simply that Buckingham Pi can often be used advantageously to guide empirical studies, particularly those associated with complex flows as in the pulmonary capillaries.

10.6 Pipe Flow

Recall from our general control volume analysis that the balance of some total quantity of interest N is given by

$$\left.\frac{dN}{dt}\right|_{sys} = \frac{\partial}{\partial t} \int_{C\kern-0.5em\text{V}} \rho\eta d\kern-0.5em\text{V} + \int_{CS} \rho\eta(\boldsymbol{v} \cdot \boldsymbol{n})dA, \tag{10.63}$$

where η is defined as the quantity per unit mass. For the *first law of thermodynamics*, or balance of energy, let $N \equiv E$ and $\eta = e = u + gz + \frac{1}{2}v^2$ where E is the total energy in a system, with u, gz, and $\frac{1}{2}v^2$ the internal, potential, and kinetic energies per unit mass, respectively. Recall, too, that for energy balance in a system [Eq. (10.1)],

$$\left.\frac{dE}{dt}\right|_{sys} = \dot{Q} + \dot{W}, \tag{10.64}$$

where $\dot{Q}$ is the rate at which heat is added *to* the system and $\dot{W}$ is the rate at which work is done *on* the system. The control volume formulation for the first law is thus [Eq. (10.22)]

$$\dot{Q} + \dot{W} = \frac{\partial}{\partial t} \int \rho\left(u + gz + \frac{1}{2}v^2\right)d\kern-0.5em\text{V} + \int \rho\left(u + gz + \frac{1}{2}v^2\right)(\boldsymbol{v} \cdot \boldsymbol{n})dA. \tag{10.65}$$

Mechanical work is a force $\boldsymbol{f}$ multiplied by some displacement $\boldsymbol{u} = \boldsymbol{x} - \boldsymbol{X}$. The time rate of work is thus given by

$$\frac{dW}{dt} = \dot{W} = \frac{d}{dt}(\boldsymbol{f} \cdot \boldsymbol{u}). \tag{10.66}$$

Assuming a constant force $\boldsymbol{f}$ at the point of interest, the rate of work becomes

$$\dot{W} = \boldsymbol{f} \cdot \frac{d\boldsymbol{u}}{dt} = \boldsymbol{f} \cdot \boldsymbol{v}. \tag{10.67}$$

Let us now focus specifically on the balance of energy in a pressurized "pipe system" and, in particular, with inlet and outlet control surfaces 1 and 2 (Fig. 10.14). The rate at which work is done on a differential area of CS_1 is given by

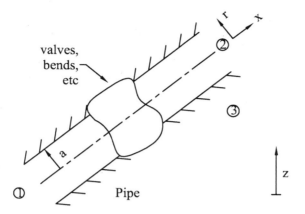

FIGURE 10.14 Flow of fluid through an inclined tube, which may contain intermediate valves, bends, or other geometric discontinuities. This diagram is simply to
illustrate the possible need to include minor losses in a pipe-flow analysis.

$$\frac{dW_1}{dt} = (p_1 dA)\hat{e}_x \cdot v_1\hat{e}_x = p_1 v_1 dA, \tag{10.68}$$

and the rate of work done on a part of CS_2 is similarly given by

$$\frac{dW_2}{dt} = (p_2 dA)(-\hat{e}_x) \cdot v_2\hat{e}_x = -p_2 v_2 dA. \tag{10.69}$$

The rate of work done on a part of CS_3 is given by, at each point,

$$\frac{dW_3}{dt} = [p(x)dA]\hat{e}_r \cdot v + (\sigma_{rx} dA)\hat{e}_x \cdot v. \tag{10.70}$$

If v is assumed to be zero by the no-slip boundary condition, then

$$\frac{dW_3}{dt} = 0. \tag{10.71}$$

Hence, from the first law of thermodynamics, we have for a steady flow,

$$\dot{Q} + \dot{W} = \int \rho e(v \cdot n)dA, \tag{10.72}$$

which can be written as

$$\dot{Q} + \int_{A_1} \frac{p}{\rho} p_1 v_1 dA - \int_{A_2} \frac{p}{\rho} p_2 v_2 dA = \int_{CS} \rho e(v \cdot n)dA, \tag{10.73}$$

wherein

$$\dot{W} = \int_{CS_1} \left(\frac{dW_1}{dt}\right)dA + \int_{CS_2} \left(\frac{dW_2}{dt}\right)dA + \int_{CS_3} \left(\frac{dW_3}{dt}\right)dA, \tag{10.74}$$

and we have multiplied each working term by unity (i.e., ρ/ρ) for reasons to be seen in Eq. 10.78. Assuming that the density is constant at each control surface,

$$\dot{Q} = \rho \int_{CS} \left(u + gz + \frac{1}{2}v^2 \right)(\boldsymbol{v} \cdot \boldsymbol{n})dA - \rho \int_{A_1} \frac{1}{\rho} p_1 v_1 dA + \rho \int_{A_2} \frac{1}{\rho} p_2 v_2 dA, \quad (10.75)$$

or, because $-v_1 = \boldsymbol{v}_1 \cdot -\boldsymbol{n}_1$ and $v_2 = \boldsymbol{v}_2 \cdot \boldsymbol{n}_2$,

$$\dot{Q} = \rho \int_{CS} \left(u + gz + \frac{1}{2}v^2 \right)(\boldsymbol{v} \cdot \boldsymbol{n})dA + \rho \int_{CS} \left(\frac{p}{\rho} \right)(\boldsymbol{v} \cdot \boldsymbol{n})dA, \quad (10.76)$$

or, finally, for our inlet and outlet control surfaces,

$$\dot{Q} = \rho \int_{CS} \left(u + gz + \frac{1}{2}v^2 + \frac{P}{\rho} \right)(\boldsymbol{v} \cdot \boldsymbol{n})dA. \quad (10.77)$$

Thus, assuming that the pressure, gravity, mass density, and internal energy are each well represented by their mean values defined over each cross section,

$$\dot{Q} = \left(\frac{p_2}{\rho_2} - \frac{p_1}{\rho_1} \right)\int \rho|\boldsymbol{v} \cdot \boldsymbol{n}|dA + g(z_2 - z_1)\int \rho|\boldsymbol{v} \cdot \boldsymbol{n}|dA$$

$$+ \int \frac{1}{2}v^2 \rho|\boldsymbol{v} \cdot \boldsymbol{n}|dA + (u_2 - u_1)\int \rho|\boldsymbol{v} \cdot \boldsymbol{n}|dA, \quad (10.78)$$

where the *mass flux* is defined as $\dot{m} = \int \rho|\boldsymbol{v} \cdot \boldsymbol{n}| dA$. Hence, we obtain

$$\dot{Q} = \left(\frac{p_2}{\rho_2} - \frac{p_1}{\rho_1} \right)\dot{m} + g(z_2 - z_1)\dot{m} + \int \frac{1}{2}\rho v^2|\boldsymbol{v} \cdot \boldsymbol{n}|dA + (u_2 - u_1)\dot{m}. \quad (10.79)$$

Here, note that the scalar v is the velocity at each point, which may vary from point to point across the control surface (e.g., parabolically for an incompressible, fully developed, Newtonian flow, as seen in Chapter 9). To address the associated integral, let us introduce a *kinetic energy coefficient* α defined as

$$\alpha = \frac{\int_{CS} \frac{1}{2}v^2 \rho(\boldsymbol{v} \cdot \boldsymbol{n})dA}{\int_{CS} \frac{1}{2}\bar{v}^2 \rho(\boldsymbol{v} \cdot \boldsymbol{n})dA} \rightarrow \alpha \int_{CS} \frac{1}{2}\bar{v}^2 \rho(\boldsymbol{v} \cdot \boldsymbol{n})dA = \int_{CS} \frac{1}{2}v^2 \rho(\boldsymbol{v} \cdot \boldsymbol{n})dA, \quad (10.80)$$

where $\bar{v}$ is the mean value of the velocity (actually speed) v. Because the mean value does not vary with position, note that

$$\int_{CS} \frac{1}{2}v^2 \rho(\boldsymbol{v} \cdot \boldsymbol{n})dA = \left(\frac{1}{2}\alpha_2 \bar{v}_2^2 - \frac{1}{2}\alpha_1 \bar{v}_1^2 \right)\int \rho|\boldsymbol{v} \cdot \boldsymbol{n}|dA, \quad (10.81)$$

or

$$\int_{CS} \frac{1}{2} v^2 \rho(\boldsymbol{v} \cdot \boldsymbol{n}) dA = \left(\frac{1}{2} \alpha_2 \bar{v}_2^2 - \frac{1}{2} \alpha_1 \bar{v}_1^2 \right) \dot{m}. \tag{10.82}$$

Substituting this result into Eq. (10.79), we thus have

$$\dot{Q} = \left(\frac{p_2}{\rho_2} - \frac{p_1}{\rho_1} \right) \dot{m} + g(z_2 - z_1)\dot{m} + \left(\frac{1}{2} \alpha_2 \bar{v}_2^2 - \frac{1}{2} \alpha_1 \bar{v}_1^2 \right) \dot{m} + (u_2 - u_1)\dot{m} \tag{10.83}$$

or, by rearranging,

$$\left(\frac{p_1}{\rho_1} + gz_1 + \frac{1}{2} \alpha_1 \bar{v}_1^2 \right) - \left(\frac{p_2}{\rho_2} + gz_2 + \frac{1}{2} \alpha_2 \bar{v}_2^2 \right) = (u_2 - u_1) - \frac{\dot{Q}}{\dot{m}}, \tag{10.84}$$

where $(u_2 - u_1) - \dot{Q}/\dot{m}$ represents the "unwanted" conversion of mechanical energy to thermal energy (e.g., heat due to frictional effects). Although it can be useful to quantify these quantities in many engineering problems (e.g., in the design of heating devices), here we shall simply consider the thermal terms to represent major losses, h_M, and minor losses, h_m. For pipe flow, therefore, we have, via a control volume analysis,

$$\left(\frac{p_1}{\rho_1} + gz_1 + \frac{1}{2} \alpha_1 \bar{v}_1^2 \right) - \left(\frac{p_2}{\rho_2} + gz_2 + \frac{1}{2} \alpha_2 \bar{v}_2^2 \right) = h_M + h_m. \tag{10.85}$$

Extensive experimentation has revealed that the *major losses* are due to viscous (frictional) losses over significant lengths of the pipe (or tube); conversely, experience has revealed that the *minor losses* arise due to the fluid flowing through complex geometries such as bends, sudden contractions or expansions, valves, and diffusers. Note, too, that we must respect the many assumptions employed when deriving this equation (e.g., steady flow and constant p, ρ, and g over a cross section).

Let us now design an experiment that would allow us to quantify the major loss h_M. Clearly, data reduction would be simplified if there were no minor losses with which to contend. This can be accomplished by simply collecting data within a straight pipe. From Eq. (10.85), it is also clear that our data analysis will be simpler if we eliminate the effect of gravity (e.g., collect data in a horizontal pipe) and if we likewise focus on an incompressible flow (ρ = constant). Given these experimental conditions, Eq. (10.85) reduces to

$$\frac{p_1 - p_2}{\rho} + \frac{1}{2}(\alpha_1 \bar{v}_1^2 - \alpha_2 \bar{v}_2^2) = h_M. \tag{10.86}$$

If we have a constant-diameter pipe, then mass balance requires that $\bar{v}_1 = \bar{v}_2$ (i.e., $Q = \bar{v}_1 A_1 = \bar{v}_2 A_2$); hence, if $\alpha_1 = \alpha_2$, as it would in a fully developed flow, then we merely have

$$\frac{\Delta p}{\rho} = h_M, \tag{10.87}$$

which allows us to determine h_M by knowing how the pressure drop depends on the relevant parameters that define our experiment: pipe diameter and length, fluid viscosity and density, surface roughness, and mean flow. Recall Example 10.7 in Section 10.5 on the Buckingham Pi Theorem in which we did just that; namely from Buckingham Pi, for the proposed experiment,

$$\frac{\Delta p}{\rho \bar{v}^2} = g\left(\frac{L}{D}, \text{Re}, \frac{e}{D}\right) \tag{10.88}$$

or, as we noted earlier based on such experiments,

$$\frac{\Delta p}{\frac{1}{2}\rho \bar{v}^2} = \frac{L}{D} f\left(\text{Re}, \frac{e}{D}\right). \tag{10.89}$$

Hence, from Eqs. (10.87) and (10.89),

$$\frac{\Delta p}{\rho} = f\left(\text{Re}, \frac{e}{D}\right)\left(\frac{L}{D}\right)\frac{\bar{v}^2}{2} = h_M, \tag{10.90}$$

where f is the friction factor, Re is the Reynolds' number, L is the length of the pipe, D is the diameter of the pipe, and $\bar{v}$ is the mean velocity through a cross section. It will prove convenient, therefore, to write the total additive major losses in multiple connected tubes as

$$h_M = \sum f\left(\text{Re}, \frac{e}{D}\right)\left(\frac{L}{D}\right)\frac{\bar{v}^2}{2}. \tag{10.91}$$

Recall that the Reynolds' number is defined as

$$\text{Re} = \frac{\rho \bar{v} D}{\mu}. \tag{10.92}$$

Experiments have revealed that if Re < 2100, then the flow will be laminar (i.e., the fluid will flow as if layers slide one relative to the other) in a pipe. Furthermore, if the flow is laminar, it is easy to show (Exercise 10.35) that the friction factor $f = 64/\text{Re}$. If Re > 2100, however, then the flow is turbulent (i.e., random) and we must use the Moody diagram to find the friction factor f (Fig. 10.15); this diagram represents a wealth of experimental data.

Finally, experience has revealed that it is convenient to write the minor losses as

$$h_m = \sum K\left(\frac{\bar{v}^2}{2}\right), \tag{10.93}$$

where K is a minor loss coefficient. Values of K are listed in Table 10.1, in which it is noted that in some cases, investigators prefer to write K in terms of an "equivalent" length. In other words, a minor loss in a geomet-

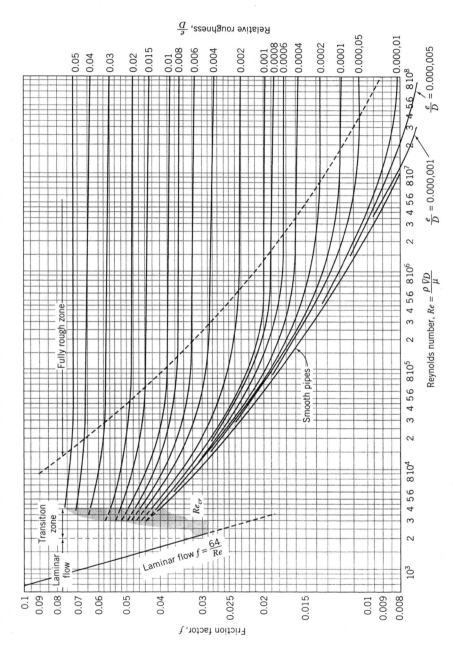

FIGURE 10.15 Moody diagram for the friction factor f as a function of the Reynolds' number Re and the surface roughness e/D. [From Moody (1944), with permission].

TABLE 10.1. Minor loss coefficients for some common geometrical discontinuities in pipe systems. After Fox and McDonald (1992).

Type	Loss
Entrances	

Reentrant $K = 0.78$

Square Edge $K = 0.5$

rounded

r/D	0.02	0.06	≥ 0.15
K	0.28	0.15	0.04

Gradual Contractions

			K for angle θ (deg)			
A_2/A_1	10	15–40	50–60	90	120	180
0.50	0.05	0.05	0.06	0.12	0.18	0.26
0.25	0.05	0.04	0.07	0.17	0.27	0.41
0.10	0.05	0.05	0.08	0.19	0.29	0.43

Valves $K = (f)(L_e/D)$
  Gate valve (open) $L_e/D = 8$
  Globe valve (open) $L_e/D = 340$
  Ball valve (open) $L_e/D = 3$

Standard Elbows
  45° $L_e/D = 16$
  90° $L_e/D = 30$

Return Bend
  180° $L_e/D = 50$

Standard Tee
  Through run $L_e/D = 20$
  Through branch $L_e/D = 60$

ric irregularity could be shown to be equal to a major loss over a particular length. Consequently, some let $K = f(\text{Re}, e/D)(L_e/D)$, whereby the value of K can be computed given the value of the friction factor f and the equivalent length L_e/D.

In summary, the final form of the balance of energy equation for a control volume analysis of flow in a pipe from point 1 to point 2 is

$$\left(\frac{p_1}{\rho_1} + gz_1 + \frac{1}{2}\alpha_1\bar{v}_1^2 \right) - \left(\frac{p_2}{\rho_2} + gz_2 + \frac{1}{2}\alpha_2\bar{v}_2^2 \right) = \sum \left[f\left(\text{Re}, \frac{e}{D}\right)\left(\frac{L}{D}\right) + K \right]\frac{\bar{v}^2}{2}.$$
(10.94)

It is important to note that the derivation of this relation did not specify a particular class of fluids (e.g., inviscid or Newtonian); thus, it is very general, notwithstanding the restrictions needed to evaluate some of the integrals.

Let us now consider the kinetic energy coefficient α for a few special cases. First, if the fluid is ideal (i.e., inviscid and incompressible), then $\mu = 0$ and there are no losses due to internal friction (viscous or geometric) and $\rho_1 = \rho_2$. Hence, the pipe flow equation reduces to

$$\frac{p_1}{\rho} + gz_1 + \frac{1}{2}\alpha_1\bar{v}_1^2 = \frac{p_2}{\rho} + gz_2 + \frac{1}{2}\alpha_1\bar{v}_2^2,$$
(10.95)

which we recognize to be similar to the Bernoulli equation (linear momentum balance for steady flow along a streamline for an ideal fluid [Eq. (8.81)]. Indeed, Bernoulli must hold for the ideal fluid along any streamline s or in an irrotational flow; hence, the kinetic energy coefficient $\alpha_1 = 1 = \alpha_2$ (prove using equation 10.80) for an ideal fluid and the pipe flow (energy) equation recovers the Bernoulli (momentum) equation. For this reason, some refer to Bernoulli as an energy equation, but we do not.

Experiments reveal that the kinetic energy coefficient $\alpha \sim 1.08$ for the case of a turbulent flow; hence, this value should be used whenever Re > 2100. Let us now consider the case of a laminar, fully developed, steady, incompressible flow of a Newtonian fluid. Recall from the Navier–Stokes solution that for a rigid circular tube, the velocity profile is [Eq. (9.45)]

$$v_z(r) = \frac{1}{4\mu}\left(\frac{dp}{dz} \right)(r^2 - a^2).$$
(10.96)

Hence, the mean velocity is given by $\bar{v} = Q/A$, where [Eq. (9.49)]

$$Q = -\frac{\pi a^4}{8\mu}\left(\frac{dp}{dz} \right);$$
(10.97)

that is,

$$\bar{v} = -\frac{a^2}{8\mu}\left(\frac{dp}{dz} \right) \equiv -\frac{a^2}{2}c,$$
(10.98)

where we let $c \equiv (1/4\mu)dp/dz$ for convenience. For fully developed laminar flow of a Newtonian fluid in a circular tube, the kinetic energy coefficient [Eq. (10.80)] thus becomes

$$\alpha = \frac{\int_0^{2\pi}\int_0^a \frac{1}{2}\rho[c(r^2 - a^2)]^2[c(r^2 - a^2)(\hat{e}_z \cdot \hat{e}_z)]rdrd\theta}{\int_0^{2\pi}\int_0^a \frac{1}{2}\rho[(-a^2/2)c]^2[c(r^2 - a^2)(\hat{e}_z \cdot \hat{e}_z)]rdrd\theta}. \tag{10.99}$$

Although the kinetic energy coefficient is introduced because the velocity is difficult to integrate in general, we can evaluate these integrals exactly in this case. The numerator is

$$(2\pi)\left(\frac{1}{2}\rho c^3\right)\int_0^a (r^2 - a^2)^3 rdr = \frac{1}{2}\rho c^3 \int_0^a (r^7 - 3a^2r^5 + 3a^4r^3 - a^6r)dr$$

$$= \pi\rho c^3\left(\frac{1}{8}a^8 - \frac{3}{6}a^8 + \frac{3}{4}a^8 - \frac{1}{2}a^8\right)$$

$$= \pi\rho c^3\left(-\frac{1}{2}a^8\right). \tag{10.100}$$

Integrating the denominator, we obtain

$$(2\pi)\left(\frac{1}{2}\right)\rho\left(-\frac{a^2c}{2}\right)^2 c\int_0^a (r^2 - a^2)rdr = \frac{\pi\rho c^3 a^4}{4}\int_0^a (r^3 - a^2r)dr$$

$$= \frac{\pi\rho c^3 a^4}{4}\left(\frac{1}{4}a^4 - \frac{1}{2}a^4\right)$$

$$= \pi\rho c^3\left(-\frac{1}{16}a^8\right). \tag{10.101}$$

Hence, we have for a fully developed, steady, incompressible, laminar flow of a Newtonian fluid

$$\alpha = \frac{\pi\rho c^3\left(-\frac{1}{8}a^8\right)}{\pi\rho c^3\left(-\frac{1}{16}a^8\right)} = \frac{16}{8} = 2. \tag{10.102}$$

In summary, theory (system → control volume relation) provided a control volume form of the balance of energy equation whereby differences in pressures, gravitational "heads," and velocities are related to unwanted thermal losses. These losses, in turn, can be shown experimentally to be related to major (viscous) losses and minor (geometric) losses in bends, valves, expansions, and so forth; hence, we have a *combined theoretical – experimental relation*. Specific parameters in this so-called pipe flow relation—the kinetic energy coefficient α and friction factor f—can be determined from theory in simple cases (e.g., $\alpha = 1$ and $f = 0$ for an ideal fluid, whereas $\alpha = 2$ and $f = 64/Re$ for a fully developed, steady, incompressible laminar flow of a Newtonian fluid), but they must be determined experimentally in other cases (e.g., $\alpha \sim 1.08$ in a turbulent flow).

Finally, note that in contrast to the viscous losses in a piping system, the presence of a pump actually augments the flow. This can be thought of empirically as the pump contributing a negative loss, say $-h_p$. Hence, a complete pipe flow equation, in the presence of a pump, is

$$\left(\frac{p_1}{\rho_1} + gz_1 + \frac{1}{2}\alpha_1 \bar{v}_1^2\right) - \left(\frac{p_2}{\rho_2} + gz_2 + \frac{1}{2}\alpha_2 \bar{v}_2^2\right)$$

$$= \sum\left[f\left(Re, \frac{e}{D}\right)\left(\frac{L}{D}\right) + K\right]\frac{\bar{v}^2}{2} - h_p. \qquad (10.103)$$

Manufacturers of pumps list values of h_p for their pumps. In the absence of such information, however, it is easy to design experiments to find the value(s) of h_p. For example, consider the next example.

Example 10.9 Design an experiment to determine h_p for a pump.

Solution: Experimentalists seek to design tests that are theoretically motivated while easily performed and interpreted. To determine h_p for a pump, for example, one would seek to eliminate all minor losses as well as the gravitational effects if possible. Consider, for example, Figure 10.16 for which the pipe flow equation is

$$\left(\frac{p_1}{\rho_1} + \frac{1}{2}\alpha_1 \bar{v}_1^2\right) - \left(\frac{p_2}{\rho_2} + \frac{1}{2}\alpha_2 \bar{v}_2^2\right)$$

$$= f\left(Re, \frac{e}{D}\right)\left(\frac{L_1}{D}\right)\frac{\bar{v}_1^2}{2} + f\left(Re, \frac{e}{D}\right)\left(\frac{L_2}{D}\right)\frac{\bar{v}_2^2}{2} - h_p$$

and from which h_p can be solved in terms of measurable quantities. Of course, balance of mass requires that $Q_1 = Q_2$ and thus $\bar{v}_1 = \bar{v}_2$ in a constant cross-section pipe. If the flow is the same at 1 and 2, both turbulent or both laminar, then our data reduction simplifies. Indeed, if we are able to measure the inlet p_3 and outlet p_4 pressures very near the pump, then

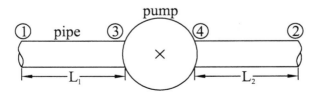

FIGURE 10.16 Simple experimental setup to determine the geometric loss (actually gain) due to a particular pump.

$$\frac{p_3 - p_4}{\rho} = -h_p \rightarrow h_p = \frac{p_4 - p_3}{\rho},$$

where $p_4 > p_3$ due to the pump. Hence, the gain due to a pump is determined primarily by the pressure jump across the pump.

Example 10.10 A basic laboratory setup is shown in Figure 10.17 for the purposes of studying flow in a stenotic artery. Determine the inlet pressure for the test section.

Solution: Here, we must use the pipe-flow equation separately for the upstream and downstream sections relative to the test section. For the upstream section (assuming a steady laminar flow; i.e., Re < 2100), we have

$$\left(\frac{p_1}{\rho} + gh + \frac{1}{2}(2)\bar{v}_1^2\right) - \left(\frac{p_3}{\rho} + \frac{1}{2}(2)\bar{v}_3^2\right)$$

$$= \frac{64}{\text{Re}}\left(\frac{L_1 + L_2}{D}\right)\frac{\bar{v}^2}{2} + \sum(K_{\text{entrance}} + K_{\text{bend}})\frac{\bar{v}^2}{2}.$$

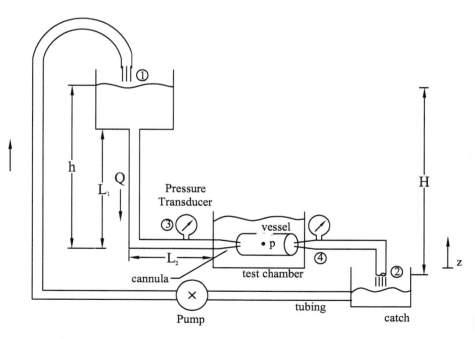

FIGURE 10.17 Possible laboratory setup to examine flow through a stenotic vessel.

If the reservoir is open to atmosphere $p_1 = 0$ (gauge) and if the reservoir is large, then $\bar{v}_1 \ll \bar{v}_3$. Hence, we have

$$p_3 = \rho g h - \frac{1}{2}\rho\bar{v}^2(2 + K_e + K_b) - \frac{32\mu\bar{v}}{D^2}(L_1 + L_2).$$

A similar downstream analysis will provide the value of p_4.

Example 10.11 Just because our focus is biomechanics, this should not mean that we cannot look at everyday problems involving continuum mechanics. Indeed, a student should continually try to understand and explain the wide variety of mechanical phenomena that we experience on a daily basis, for in doing so, one is forced to practice the art of formulating and solving problems. For example, let a hydraulic turbine be supplied with water from a mountain lake through a supply pipe. The pipe diameter is 1 ft and the average roughness e is 0.05 in. Minor losses can be neglected. Flow leaves the pipe at atmospheric pressure at an average velocity $\bar{v} = 27.5$ ft/s. Find the height h if the length L of the pipe is 3000 ft. Let $\mu/\rho = 10.76 \times 10^{-6}$ ft^2/s. Note, too, that the combination μ/ρ is called the *kinematic viscosity* in contradistinction to the *absolute viscosity* μ.

Solution: For pipe flow,

$$\left(\frac{p_1}{\rho_1} + gz_1 + \frac{1}{2}\alpha_1\bar{v}_1^2\right) - \left(\frac{p_2}{\rho_2} + gz_2 + \frac{1}{2}\alpha_2\bar{v}_2^2\right) = h_M + h_m.$$

In the case where we neglect the minor losses, the pipe-flow equation becomes

$$\left(\frac{p_1}{\rho_1} + gz_1 + \frac{1}{2}\alpha_1\bar{v}_1^2\right) - \left(\frac{p_2}{\rho_2} + gz_2 + \frac{1}{2}\alpha_2\bar{v}_2^2\right) = \sum f\left(\mathrm{Re}, \frac{e}{D}\right)\left(\frac{L}{D}\right)\frac{\bar{v}^2}{2}.$$

Applying the given conditions, we have (with $v_1 \ll 1$ in lake)

$$gh - \frac{1}{2}\alpha\bar{v}_2^2 = f\left(\mathrm{Re}, \frac{e}{D}\right)\left(\frac{L}{D}\right)\frac{\bar{v}_2^2}{2}.$$

Before we can calculate the height h, we need to determine if the flow is laminar or turbulent. To do this, we need to calculate the Reynolds' number:

$$\mathrm{Re} = \frac{\rho\bar{v}D}{\mu} = \frac{\bar{v}D}{\mu/\rho} = \frac{(27.5\ \mathrm{ft/s})(1\mathrm{ft})}{10.76 \times 10^{-6}\ \mathrm{ft^2/s}} = 2.556 \times 10^6.$$

Because $2.556 \times 10^6 > 2100$, the flow is turbulent and we must use the Moody diagram to find the value of the friction factor f. Also, $\alpha \sim 1.08$ and $e/D = 0.05$ in./12 in. $= 0.004$. From the Moody diagram (Fig. 10.15), the friction factor $f \sim 0.028$ (which is simply best approximated by eye). Substituting these values into the pipe-flow equation,

$$gh - \frac{1}{2}\alpha\bar{v}_2^2 = f\left(\text{Re}, \frac{e}{D}\right)\left(\frac{L}{D}\right)\left(\frac{\bar{v}^2}{2}\right) \rightarrow h = \frac{1}{g}\left[f\left(\text{Re}, \frac{e}{D}\right)\left(\frac{L}{D}\right)\left(\frac{\bar{v}_2^2}{2}\right) + \frac{\alpha\bar{v}_2^2}{2}\right],$$

we obtain

$$h = \frac{1}{32.2\,\text{ft}/\text{s}^2}\left[0.028\left(\frac{3000\,\text{ft}}{1\,\text{ft}}\right)\left(\frac{(27.5)^2\,\text{ft}^2/\text{s}^2}{2}\right) + \frac{1}{2}(1.08)(27.5)^2\,\frac{\text{ft}^2}{\text{s}^2}\right] = 999\,\text{ft}.$$

Hence, the height h is approximately $1000\,\text{ft}$.

Example 10.12 Find the minor loss due to a medical stopcock.

Solution: Again, we must design an appropriate experiment. Clearly, because the pipe-flow equation is but one scalar equation, we can solve for only one unknown, the K_{sc} (stopcock) of interest. The experiment should thus enable us to measure all other quantities and, in particular, to simplify measurement and interpretation. Recalling the general formula

$$\left(\frac{p_1}{\rho_1} + gz_1 + \frac{1}{2}\alpha_1\bar{v}_1^2\right) - \left(\frac{p_2}{\rho_2} + gz_2 + \frac{1}{2}\alpha_2\bar{v}_2^2\right) = \sum f\left(\text{Re}, \frac{e}{D}\right)\left(\frac{L}{D}\right)\left(\frac{\bar{v}^2}{2}\right) + \sum h_m,$$

let us consider a setup similar to that in Figure 10.16, except with a stop-cock in place of the pump. Hence, ensuring a laminar flow ($\alpha = 2$), constant cross-sectional areas and thus $\bar{v}_1 = \bar{v}_2 = \bar{v}$, and a horizontal system, we are left with

$$\frac{p_1}{\rho} - \frac{p_2}{\rho} = f\left(\text{Re}, \frac{e}{D}\right)\left(\frac{L_1}{D}\right)\left(\frac{\bar{v}^2}{2}\right) + f\left(\text{Re}, \frac{e}{D}\right)\left(\frac{L_2}{D}\right)\left(\frac{\bar{v}^2}{2}\right) + h_m,$$

or

$$h_m = \frac{p_1 - p_2}{\rho} - \frac{64}{\text{Re}}\left(\frac{L_1 + L_2}{D}\right)\left(\frac{\bar{v}^2}{2}\right),$$

with

$$\text{Re} = \frac{\rho\bar{v}D}{\mu}.$$

Hence, to determine the loss coefficient, we must know μ and ρ for our fluid, determine L_1, L_2, and D for the tubing, and measure p_1 and p_2. Most importantly, however, we must ensure that all assumptions are satisfied by the actual flow, as, for example, that it is indeed steady, incompressible, and laminar.

Clearly, like Bernoulli, the pipe-flow equation is algebraic and easily solved. The most important thing, therefore, is to understand how to use these equations, which is to say, to understand the conditions under which they apply.

Observation 10.2 Although we have used nondimensionalization to reduce general functional dependencies in order to design more efficient experiments (via the Buckingham Pi Theorem), nondimensionalization can also be useful in the solution of known governing equations. For example, the governing differential equation for a mass-spring-dashpot system is (recall from physics)

$$m\frac{d^2x}{dt^2} + c\frac{dx}{dt} + kx = f(t),$$

where m is mass, c is a viscous dissipation, k is a spring constant, f is a forcing function, and t is time; x is, of course, a position. Although we do not provide a recipe per se, let us consider the following general approach. First, let us note the units of each variable:

$$[m] = L^0 T^0 M^1,$$

$$[x] = L^1 T^0 M^0,$$

$$[t] = L^0 T^1 M^0,$$

$$[c] = \frac{(\text{Force})(\text{Time})}{\text{Length}} = L^0 T^{-1} M^1,$$

$$[k] = \frac{\text{Force}}{\text{Length}} = L^0 T^{-2} M^1,$$

$$[f] = \text{Force} = L^1 T^{-2} M^1.$$

If we assume that the unperturbed (vertically held) mass-spring-dashpot system has length x_0, then we could pick scales such as

$$L_s = x_0, \qquad T_s = \sqrt{\frac{m}{k}}, \qquad M_s = m.$$

Again, the timescale is the hardest one to pick, but we simply need a characteristic measure having units of time.

Noting that

$$\left[\frac{d^2x}{dt^2}\right] = L^1 T^{-2} M^0, \qquad \left[\frac{dx}{dt}\right] = L^1 T^{-1} M^0,$$

we see that if we let

$$\lambda \equiv \frac{x}{x_0}, \qquad \tau \equiv \frac{t}{\sqrt{m/k}},$$

then

$$\frac{d}{dt} = \frac{d}{d\tau}\frac{d\tau}{dt} = \frac{1}{\sqrt{m/k}}\frac{d}{d\tau},$$

$$\frac{d^2}{dt^2} = \frac{d}{d\tau}\left(\frac{1}{\sqrt{m/k}}\frac{d}{d\tau}\right)\frac{d\tau}{dt} = \frac{k}{m}\frac{d^2}{d\tau^2},$$

and, thus,

$$\frac{dx}{dt} = \frac{x_0}{\sqrt{m/k}}\frac{d\lambda}{d\tau}, \qquad \frac{d^2x}{dt^2} = \frac{kx_0}{m}\frac{d^2\lambda}{d\tau^2}.$$

Consequently, our governing equation can be written as

$$m\left(\frac{kx_0}{m}\frac{d^2\lambda}{d\tau^2}\right) + c\left(\frac{x_0}{\sqrt{m/k}}\frac{d\lambda}{d\tau}\right) + kx_0\lambda = g(\tau),$$

or, by dividing by kx_0, we have

$$\frac{d^2\lambda}{d\tau^2} + \frac{c}{\sqrt{mk}}\frac{d\lambda}{d\tau} + \lambda = \frac{g(\tau)}{kx_0}.$$

If we define a parameter $\delta = c/\sqrt{mk}$ and let $F \equiv g(\tau)/kx_0$ be a nondimensional forcing function, we see that our original dynamic system, which depended on the three parameters m, c, and k, now depends on only one nondimensional parameter δ, namely

$$\frac{d^2\lambda}{d\tau^2} + \delta\frac{d\lambda}{d\tau} + \lambda = F(\tau),$$

which clearly admits a simpler numerical parametric study. Nondimensionalization can thus be very useful in studying the behavior of many differential equations, particularly those that model dynamical systems. This is illustrated further in Chapter 11 in a study of the dynamic stability of aneurysms.

10.7 Conclusion

In summary, we see from the few cases considered in Chapters 8–10 that many different types of flows are of interest to the biomedical engineer. Quantifying in vivo flows in the airways, circulatory system, and urinary system is particularly important for understanding normal physiology and for diagnostic and therapeutic purposes. Quantifying ex vivo flows is likewise important for the purposes of designing basic and applied experiments that range from studying the mechanobiological response of cells to shear stress to determining the minor loss coefficient for a medical stopcock. As we stated in the beginning of Chapter 7, the importance of biofluid mechanics is far reaching; yet, just as we concluded in Chapter 6 for biosolid mechanics, an introductory text is simply that—an introduction. Because of the complex geometries, initial-boundary conditions, and fluid behaviors in biofluid mechanics, interested students are strongly encouraged to pursue advanced courses in fluid mechanics and computational mechanics. We have seen but an introduction as to how equations can be formulated in an Eulerian sense (which gave rise to local and convective parts of acceleration), how differential and control volume equations can each provide the requisite information in particular cases, and how theory and experiment must often be combined to complete an analysis. Much more remains to be learned, however. We have not addressed the important roles of waves (and their reflections) in the vasculature, of diffusion across the vascular wall during flow, of the effect of taper, curves, and bifurcations on the velocity and pressure fields, and so on.

Appendix 10: Thermodynamics

As noted in Section 10.6, the first law of thermodynamics states that the time rate of change of the (total) energy E must balance the rate at which work is done on the body plus the rate at which heat is added to the body. Here, let us derive a differential equation for energy balance (similar to the differential equations for mass balance and linear momentum balance in Sections 8.1 and 8.2 of Chapter 8) for the special case of no mechanical work (e.g., a rigid solid); once done, we then list the more general thermo-mechanical energy equation.

In contrast to Section 10.5, let the energy per unit mass e be given by the internal energy u alone (i.e., no kinetic or potential energy). Moreover, let the heat be added to the body via two means: through the surface, via a heat flux vector q defined per unit area, and volumetrically, via a heat source (scalar) q_s defined per unit mass. Note, too, that convention defines q positive outward, and being an arbitrary vector, we need $q \cdot n$, where n is an outward unit normal vector to the surface through which the flux occurs. Because $q \cdot n$ acts at each point over a differential area and q_s acts at each point over a differential volume and because the first law requires the balance of total energy, we have

$$\frac{d}{dt}\int u\rho dv = -\int q \cdot nda +\int \rho q_s dv. \qquad (A10.1)$$

The minus sign in the flux term accounts for our desire to quantify heat *addition* to the body. Recall from Chapter 7 that the divergence theorem allows us to convert a surface integral to a volume integral, namely

$$-\int q \cdot nda = -\int \nabla \cdot q dv. \qquad (A10.2)$$

Now, if we can exchange the order of the time differentiation and the volume integration in Eq. (A10.1), then we can collect all terms into a single integral and thereby obtain our governing differential equation (cf. Section 8.1). As in Section 8.1, therefore, we seek to relate dv to dV the original differential volume that is independent of time.

Employing the same arguments as in Section 8.1, let the original and current differential volumes both be cuboidal; hence,

$$dv = \frac{\partial x}{\partial X}\frac{\partial y}{\partial Y}\frac{\partial z}{\partial Z} dV \tag{A10.3}$$

and

$$\frac{d}{dt}\int u\rho dv = \int \frac{d}{dt}\left(u\rho \frac{\partial x}{\partial X}\frac{\partial y}{\partial Y}\frac{\partial z}{\partial Z}\right)dV. \tag{A10.4}$$

Using the product rule for the differentiation and exploiting results from Section 8.1, this equation can be written as

$$\int\left[\frac{du}{dt}\rho + u\left(\frac{d\rho}{dt}+\rho\nabla\cdot\boldsymbol{v}\right)\right]\frac{\partial x}{\partial X}\frac{\partial y}{\partial Y}\frac{\partial z}{\partial Z} dV \tag{A10.5}$$

and, consequently, using Eq. (A10.3), Eq. (A10.1) becomes

$$\int\left[\frac{du}{dt}\rho + u\left(\frac{d\rho}{dt}+\rho\nabla\cdot\boldsymbol{v}\right)\right]dv = \int(-\nabla\cdot\boldsymbol{q}+\rho q_s)dv, \tag{A10.6}$$

or

$$\int\left[\frac{du}{dt}\rho + u\left(\frac{d\rho}{dt}+\rho\nabla\cdot\boldsymbol{v}\right)+\nabla\cdot\boldsymbol{q}-\rho q_s\right]dv = 0, \tag{A10.7}$$

which must hold for *all* arbitrary domains (volumes). This can be satisfied if the integrand is always zero; that is,

$$\frac{du}{dt}\rho + u\left(\frac{d\rho}{dt}+\rho\nabla\cdot\boldsymbol{v}\right) = -\nabla\cdot\boldsymbol{q}+\rho q_s \tag{A10.8}$$

where we recall from mass balance (Eq. 8.11) that the second term on the left-hand side must be zero. Thus, our differential equation for energy balance, in the absence of mechanical work terms, is

$$\rho\frac{du}{dt} = -\nabla\cdot\boldsymbol{q}+\rho q_s. \tag{A10.9}$$

This equation is similar to the general equations of linear momentum balance in Section 8.2, in that we have not specified particular material behaviors (i.e., constitutive relations).

The most commonly used constitutive equation for heat flux, however, is "Fourier's law," which states that $\boldsymbol{q}$ is proportional to the temperature gradient, namely

$$q = -k\nabla T, \tag{A10.10}$$

where k is a material constant called the thermal conductivity. Typical values for soft tissues are on the order of $k = 4.76\,\text{mW/cm}^\circ\text{C}$ for normal human aorta and $k = 4.85\,\text{mW/cm}\,^\circ\text{C}$ for a fibrous atherosclerotic plaque, both measured at 35°C. A commonly assumed constitutive relation for the internal energy, in the absence of deformation, is $u = c_v T$, where c_v is the specific heat, at a constant volume, and T is the absolute temperature. Hence, in this special case, Eq. (A10.9) becomes

$$\rho \frac{d}{dt}(c_v T) = -\nabla \cdot (-k\nabla T) + \rho q_s \tag{A10.11}$$

or, for constant c_v and k, the famous heat diffusion equation

$$\rho c_v \frac{dT}{dt} = k\nabla^2 T + \rho q_s \rightarrow \frac{dT}{dt} = \alpha\nabla^2 T + \frac{q_s}{c_v}, \tag{A10.12}$$

where $\alpha = k/\rho c_v$ is the so-called thermal diffusivity (a material property). This equation is widely studied in applied mathematics and allows one to determine temperature fields $T(x, y, z, t)$ in terms of the material parameters (ρ, c_v, k) and the heat supply q_s. Microwave energy is a prime source of q_s.

Finally, note that in the case of mechanical work, Eq. (A10.9) can be shown to become (Humphrey, 2002)

$$\rho \frac{du}{dt} = \sigma_{xx} D_{xx} + \sigma_{yy} D_{yy} + \sigma_{zz} D_{zz} + \sigma_{xy} D_{xy} + \cdots + \sigma_{zx} D_{zx} - \nabla \cdot \boldsymbol{q} + \rho q_s,$$

$$\tag{A10.13}$$

which couples mass, linear momentum, and energy balance equations, thus resulting in a formidable system of four partial differential equations. The solution of such coupled problems is important in biomechanics, but this is the topic of advanced courses. We merely introduce a few coupled problems in Chapter 11.

Exercises

10.1   A water jet "pump" has a jet area of A_0 and jet speed v_0 (Fig. 10.18). The jet is within a secondary stream of water having speed $\frac{1}{10}v_0$. The total area of the duct is $5A_0$. The water is completely mixed and leaves the jet pump in a uniform stream. The pressures of the jet and

duct

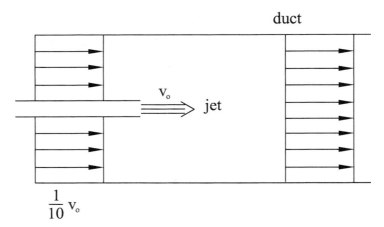

FIGURE 10.18

secondary stream are the same at the pump inlet. Determine the speed at the pump exit and the pressure drop $p_1 - p_2$. Note, of course, that a similar situation could exist within an artery due to a high velocity flow through a needle or catheter.

10.2   Recall from Chapter 9 that the flow within a tube tends to "develop" over a region called the entrance length L_e, beyond which the flow is said to be fully developed. This development of the flow is due to viscous effects between the fluid and the surface of the tube and, in particular, the no-slip condition. This slowing of the flow near the surface is called a *boundary effect*, and the region of the affected flow is called a *boundary layer*. The boundary effect reaches the center-line in a fully developed flow. The easiest way to study boundary layers is to consider air that flows by a rigid, stationary flat plate (Fig. 10.19). The velocity profile in the boundary layer can be shown to be given by

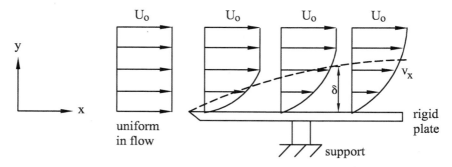

FIGURE 10.19

$$v_x = U\left[\frac{3}{2}\left(\frac{y}{\delta}\right) - \frac{1}{2}\left(\frac{y}{\delta}\right)^3\right],$$

where δ is a measure of the growing boundary layer; it is simply a number at any position x. Find the x-direction reaction force needed to hold the plate in place. Assume the plate is L long and W wide. In $y \in [0, \delta]$, the fluid has a viscosity of μ. Hint: Construct a cuboidal control volume from the leading edge to a downstream location where $y = \delta$.

10.3 If there is some diffusion (radial) through the wall of a circular artery such that the (lost) mass flux is $\dot{m}_3$ and the input flow rate is $Q_1 = \bar{v}_1 A_1$, find the mean velocity $\bar{v}_2$ of the flowing fluid that reaches the end of the vessel.

10.4 In many tests on the biomechanical properties of a soft tissue, one must immerse the specimen in a physiologic salt solution to keep the tissue viable. In some cases, however, measurements must be performed without the tissue immersed [e.g., to find the thermal conductivity using a monochromatic flash system (Davis et al., 2000)]. Hence, we may be very interested in quantifying the rate at which a test chamber is filled (Fig. 10.20). Find dh/dt using a control volume approach.

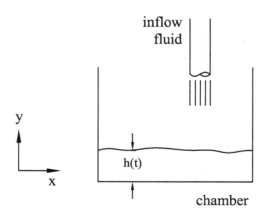

FIGURE 10.20

10.5 If you place your finger in a gentle flow of water from a faucet such that your finger barely touches the water, you will see that the fluid "bends" under your finger rather than deflecting away from it (Fig. 10.21). This is called the *Coanda effect*. Determine if the fluid tends to pull your finger into the flow or if it tends to push your finger away (try it). To formulate and solve this problem, construct a control

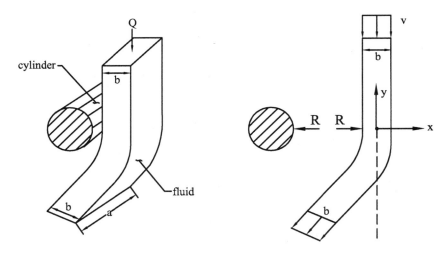

volume for the fluid, and let the force due to the solid cylinder acting on the fluid be R.

10.6   Consider a flow through a sudden expansion (Fig. 10.22). If the flow is incompressible and the effects of shear stress are neglected, show that the pressure drop Δp is given by

$$\frac{\Delta p}{1/2\,\rho\bar{v}_1^2} = 2\left(\frac{d}{D}\right)^2\left[1-\left(\frac{d}{D}\right)^2\right].$$

Why can you *not* use Bernoulli to solve this problem?

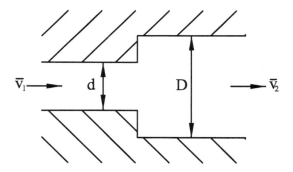

10.7   Water exits a device at velocity U and lifts a flat plate of mass m to height h (Fig. 10.23). Find the requisite value of U. Assume that there is a (negligible) guide wire that keeps the plate centered and that the orifice of the device is d in diameter.

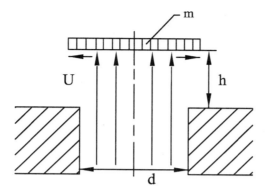

FIGURE 10.23

10.8 Culture media enters a parallel plate device at a uniform velocity U. The flow develops as it progresses through the plates until it becomes fully developed. Find the requisite value of U in terms of the maximum value $v_x)_{max}$ in the fully developed region.

10.9 A conical funnel of half-angle α drains through a small area A at the vertex. The exit velocity of the fluid is $\bar{v} = \sqrt{2gy}$, where y is the distance from the exit to the free surface of the fluid, which changes over time. The funnel is initially filled to height h. Find an expression for the time to drain the funnel in terms of the initial volume V_0 and the initial volumetric flow rate $Q_0 = A\sqrt{2gh}$. [From Fox and McDonald (1992, p. 163).]

10.10 To calibrate an electromagnetic flowmeter, one allows the flowing fluid to drain into a catch basin that is on a digital scale, the output of which can be sampled via a computer. Derive an expression for the measured weight as a function of time and the flow rate Q. Let the basin be $h \times w \times L$ in dimensions.

10.11 A *hydraulic accumulator* is designed to reduce pressure pulsations in a perfusion system (Fig. 10.24). For the instant shown, determine the rate at which the accumulator gains or loses fluid.

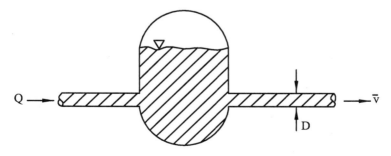

FIGURE 10.24

10.12 An incompressible fluid flows steadily and uniformly into the entrance region of a rigid circular tube of radius a (i.e., $v_{entrance} = U_0 \hat{e}_z$). If the fully developed velocity at a downstream section is

$$v_z = v_z)_{max}\left(1 - \frac{r^2}{a^2}\right),$$

find $v_z)_{max}$ in terms of U_0 and other known quantities.

10.13 Not all of pulmonary inspired/expired air is available for gas exchange. The air contained within the conducting airways accounts for 150 mL of the ~500 mL of air inspired/expired with each breath. If the respiratory rate is 15 breaths per minute, compute the total volumetric flow rate as well as that which is available for gas exchange (so-called alveolar ventilation).

10.14 Fick's principle is used to compute pulmonary blood flow Q_b. This principal states that the consumption of oxygen per minute is equal to the amount of O_2 taken up by the blood in the lungs per minute. If the concentration of O_2 is C_i and C_e on inspiration and expiration, respectively, then O_2 consumption per minute is

$$\dot{V}_{O_2} = Q_b(C_e - C_i).$$

Find the physiologic values of the quantities and compute Q_b.

10.15 According to West (1979), airflow within the complex network of conducting airways is seldom fully developed and laminar, except possibly near the terminal bronchioles, where the Re ~1. Conversely, the flow is unlikely to be fully turbulent either, except possibly in the trachea, especially during exercise. In most cases, therefore, the flow is *transitional*, for which semi-empirical methods are needed. For example, West suggests that in much of the conducting airways,

$$\Delta P = c_1 Q + c_2 Q^2,$$

where c_1 and c_2 are parameters. How does this relation differ from that for a steady laminar flow?

10.16 The Mach number M is defined as the ratio of a characteristic velocity to that of the speed of sound (e.g., V_{sound} ~360 m/s in air under standard conditions). If $M < 0.3$, compressibility effects are negligible and the flow can be treated as incompressible. What is the Mach number in the human trachea in normal and extreme exercise? Can the associated flows be considered incompressible?

10.17 In Section 10.4, we said that the mechanical power per unit volume was determined via the product of stresses times their respective rates of deformation; total power was obtained by integrating over the volume. Show that the result of Section 10.4.1 can be obtained if one computes the total power as the pressure drop Δp times the total volumetric flow rate Q, where $dp/dz \sim -\Delta p/L$. Explain why the total power can be calculated thus.

10.18 In Table A9.4, we see that the trachea has a diameter of ~18 mm and the terminal bronchi (at 11 generations, counting the trachea as the 0th generation) a diameter of ~1.09 mm. Given that bifurcations are reasonably symmetrical in the airways, what would Murray's law suggest for the number of generations? Note, too, that Table A9.4 reports 2050 terminal bronchi. What does this suggest by way of possible shear stress control of airway lumen?

10.19 Given the results for the bifurcation angles

$$\cos\theta_1 = \frac{a_0^4 + a_1^4 - a_2^4}{2a_0^2 a_1^2}, \qquad \cos\theta_2 = \frac{a_0^4 - a_1^4 + a_2^4}{2a_0^2 a_2^2},$$

show that $\theta_1 = \theta_2$ if $a_1 = a_2$. Likewise, show that if $a_1 > a_2$, then $\theta_1 > \theta_2$. Finally, show that if $a_1 \gg a_2$, then $a_2 \sim a_0$ and $\theta_1 \sim \pi/2$.

10.20 Consistent with the previous exercise, show that

$$\cos\theta_1 = \frac{\left(1+\alpha^3\right)^{4/3} + 1 - \alpha^4}{2\left(1+\alpha^3\right)^{2/3}}, \qquad \cos\theta_2 = \frac{\left(1+\alpha^3\right)^{4/3} + \alpha^4 - 1}{2\alpha^2\left(1+\alpha^3\right)^{2/3}}.$$

Hence, if $a_1 = a_2$, $\alpha = 1$ and $\theta_1 = \theta_2 = 37.5°$.

10.21 Repeat Example 10.6 for the relation $v = f(m, g, h)$ using scales $L_s = h$, $T_s = h/v$, and $M_s = m$. Compare to prior results.

10.22 Consider Example 10.7 for the pipe flow. Identify three possible timescales.

10.23 Repeat Example 10.7 for the problem of flow in a pipe using $M_s = D^2 L$.

10.24 In a classic paper, Stokes showed in 1851 that the drag force F_D on a sphere that settles slowly (at velocity $\bar{v}$) in a static fluid of viscosity μ is

$$F_D = 6\pi a \mu \bar{v},$$

where a is the radius of the sphere. Use Buckingham Pi to find appropriate nondimensional groups by which to conduct such an experiment. This solution is valid if $Re = \rho \bar{v} d/\mu < 1$, where $d = 2a$.

10.25 Using Stokes' classic result for the drag force on a freely falling spherical solid in a viscous fluid (previous exercise) and Archimedes' classic result for the buoyant force (Observation 8.2 of Chapter 8), show that one can use such an experiment to measure the viscosity of the fluid, namely

$$\mu = \frac{2(\rho_s - \rho_f)ga^2}{9\bar{v}},$$

where ρ_s and ρ_f are the mass densities of the solid sphere and fluid, respectively, g is the acceleration due to gravity (9.81 m/s^2), and a and

$\bar{v}$ are the radius and the mean velocity of descent of the sphere, respectively. Hint: Recall that Stokes' result was obtained by ignoring the convective acceleration.

10.26 If $\mu = 1.8$ cP, $\rho_s = 1.08\,\text{g/cm}^3$, $\rho_f = 1.01\,\text{g/cm}^3$, and the volume of the sphere is $90\,\mu\text{m}^3$, use the result from Exercise 10.25 to compute $\bar{v}$. This "Stokes' flow" solution is valid of course, only if Re < 1. Is it?

10.27 Fluid having a viscosity μ and a mass density ρ is placed between two rigid concentric cylinders of radius a and b, respectively. Use Buckingham Pi to find a nondimensional relation for the torque T, in terms of the above parameters, that is necessary to rotate both cylinders at a constant angular velocity ω.

10.28 The pressure variation in a partially filled container of liquid rotating with angular velocity ω is a function of the distance r from the axis of rotation, the local depth z from the free surface, the mass density ρ, and g. Using Buckingham Pi, find a nondimensional relation for the pressure p. [From Alexandrou (2001, p. 235).]

10.29 Use Buckingham Pi to find a relation between the meniscus height h in a tube of diameter d for a fluid having specific weight $\gamma = \rho g$ and surface tension σ; that is, study $h = f(d, \gamma\sigma)$. (Fig. 10.25)

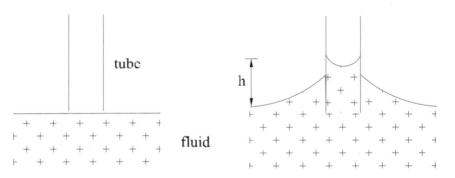

FIGURE 10.25

10.30 We used Buckingham Pi to transform general functional relationships from dimensional to nondimensional form. This is particularly useful in the design of experiments. The same ideas of identifying fundamental units, determining the units of all variables, and picking appropriate scales can similarly be used to nondimensionalize governing equations of motion. Use this method to nondimensionalize the axial Navier–Stokes equation for a steady, laminar, fully developed, incompressible flow in a rigid tube [cf. Eq. (9.37) of Chapter 9]. Hint: The final result should include an appropriate Reynolds' number Re.

10.31 Repeat the nondimensionalization in Observation 10.2 using a timescale of $T_s = m/c$.

10.32 Repeat the nondimensionalization in Observation 10.2 using a timescale of $T_s = c/k$. Although any valid scale is acceptable, experience proves sometimes that particular scales are more natural or useful. The timescale c/k is indeed the one used most often in the dynamics (vibration) of a mass-spring-dashpot system. In practice, therefore, one finds it necessary to compare the use of multiple scales to identify those that are most useful.

10.33 A drug is to be supplied via an intravenous (IV) line. If the solution has a density of $920\,kg/m^3$ and a viscosity of $0.9 \times 10^{-3}\,N\,s/m^2$ and if the 10-mm-diameter syringe is advanced at 0.5 cm/s, find the force needed to move the plunger. Assume a friction factor of 0.0001. (Fig. 10.26)

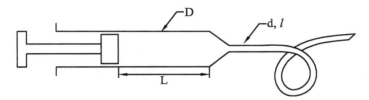

FIGURE 10.26

10.34 An IV drip is set up to deliver medicine at a specified rate. Assume that $\rho = 940\,kg/m^3$ and $\mu = 0.9 \times 10^{-3}\,N\,s/m^2$. Find the number of drops per minute if the drops are spherical with a diameter 12% greater than the inlet tube. Assume a friction factor of 0.0001. (Fig. 10.27)

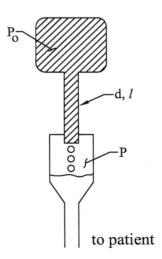

FIGURE 10.27

10.35 Show that for a steady, incompressible, fully developed laminar flow of a Newtonian fluid that the friction factor is

$$f = \frac{64}{\text{Re}}.$$

Hint: Recall from the Navier–Stokes solution that [Eq. (9.49) of Chapter 9]

$$Q = -\frac{\pi a^4}{8\mu}\left(\frac{dp}{dz}\right) \rightarrow Q = \frac{\pi D^2 \Delta P}{16(8)\mu L},$$

where $Q = \bar{v}A$ and from the pipe-flow equation,

$$\frac{\Delta P}{\rho} = (f)\left(\frac{L}{D}\right)\frac{\bar{v}^2}{2}.$$

10.36 Design a simple experiment to determine the value of the minor loss coefficient for a valve. Derive the governing equation to find the value of K.

10.37 Fig. 10.28 is a schema of a standard test to study vascular response to changes in flow and pressure. If the pressure transducers measure upstream and downstream values, use the pipe-flow equation to show how the pressure in the vessel can be determined/estimated.

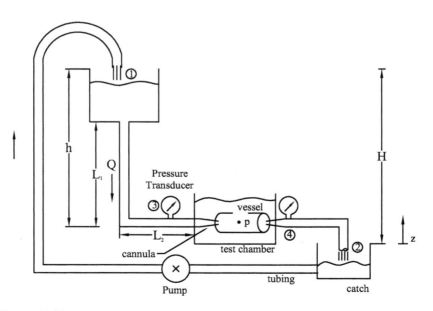

FIGURE 10.28

Part IV
Closure

11
Coupled Solid–Fluid Problems

We considered a variety of problems in Chapters 2–10 that fall within the domain of either biosolid mechanics or biofluid mechanics, each of which is very important in its own right. Whether in the body (in vivo) or in the laboratory (in vitro), however, many "real-life" problems simultaneously involve solid–fluid interactions. For example, although we may seek to determine the stresses in the limbs of a pilot who has ejected from an aircraft, for purposes of identifying safety measures, it is the wind that induces the applied loads of importance; aneurysms may be considered as thin-walled, nearly spherical membranes that exhibit a solidlike character, but the applied loads are due to the internal flowing blood and the surrounding cerebrospinal fluid; mechanotransduction in bone, which exhibits a strong solidlike behavior, appears to be influenced directly by both loads due to weight bearing and those due to the flow of blood and bone fluid within the many different canals within the bone; and an atomic force microscopic examination of the mechanics of a cell may primarily reveal the properties of the cortical membrane and underlying solidlike cytoskeleton, but flow of the cytosol likely plays a key role as well. Hence, from these simple examples and many more like them, we see that solid–fluid interactions are important at the organism, organ, tissue, cellular, and molecular levels. Indeed, although it tends to be convenient to introduce students to a subject by focusing only on that subject, most research and clinical problems require interdisciplinary and multidisciplinary approaches (i.e., analysis and design of *coupled problems*). Such problems are typically complex and require advanced approaches, but here we consider a few introductory examples.

11.1 Vein Mechanobiology

11.1.1 Biological Motivation

The saphenous vein used in a coronary artery bypass surgical procedure (cf. Fig. 3.8 of Chapter 3) will grow and remodel in response to its altered environment, which includes marked changes in both pressure and flow. To understand the associated mechanobiology, we must compute the blood-pressure-induced stresses in the wall (i.e., the solid mechanics problem; Section 3.4 of Chapter 3) as well as the flow-induced wall shear stresses (i.e., the fluid mechanics problem; Section 9.2 of Chapter 9). Here, let us consider a simple problem for which we examine possible morphological changes due to the altered pressure and flow experienced by the vein graft. Consistent with prior analyses, consider as a first approximation the mean circumferential wall stress $\sigma_{\theta\theta}$ due to a quasistatic pressurization and the mean wall shear stress τ_w due to a steady flow.

11.1.2 Theoretical Framework

Let the normal venous pressure and flow be denoted by P (~10 mm Hg) and Q (~20 mL/s, where the mean velocity $\bar{v} \sim 0.01$–0.04 m/s). Moreover, let the arterial values be given by $P_a = \varepsilon_p P$ and $Q_a = \varepsilon_Q Q$, where scaling factors $\varepsilon_p = 120/10 \approx 12$ and $\varepsilon_Q = 40/20 \approx 2$. Hence, in the normal physiologic state, we have [from Eqs. (3.41) and (9.55)]

$$\sigma_{\theta\theta} = \frac{Pa}{b-a}, \qquad \tau_w = \frac{4\mu Q}{\pi a^3}, \tag{11.1}$$

where a is the homeostatic pressurized inner radius and $b - a$ is the associated thickness. In the altered (bypass) state, we have

$$\sigma_{\theta\theta} = \frac{(\varepsilon_p P) r_i}{h}, \qquad \tau_w = \frac{4\mu(\varepsilon_Q Q)}{\pi r_i^3}, \tag{11.2}$$

where r_i and h are the new (altered) inner radius and wall thickness, respectively. If the vein grows and remodels over time such that it restores the wall shear and circumferential stresses to the basal levels, then

$$\frac{4\mu(\varepsilon_Q Q)}{\pi r_i^3} = \frac{4\mu Q}{\pi a^3}, \qquad \frac{(\varepsilon_p P) r_i}{h} = \frac{Pa}{b-a}, \tag{11.3}$$

or

$$r_i = (\varepsilon_Q)^{1/3} a, \qquad h = \varepsilon_p (\varepsilon_Q)^{1/3} (b-a). \tag{11.4}$$

Hence, by comparing morphologically measured values of a versus r_i and $b - a$ versus h in terms of the alterations ε_p and ε_Q, we can test the simple hypotheses that the vascular wall grows and remodels in such a way that it restores both $\sigma_{\theta\theta}$ and τ_w to the basal values. Such a hypothesis is teleologically attractive because we expect endothelial cells to function best at a particular value of τ_w and the medial smooth muscle cells to likewise function best at a particular value of $\sigma_{\theta\theta}$. We also expect, of course, that such growth and remodeling will occur over days to weeks or months. For a more complete understanding, therefore, we need to augment Eqs. (11.1) and (11.2), which come from equilibrium and boundary conditions, with constitutive equations that account for the growth and remodeling. As noted in Humphrey (2002), it appears that growth and remodeling is a result of imbalances in the production and removal of constituents at altered configurations. In the case of the vein graft, its configuration will be altered initially by the increased pressure (due to elastic distension and possibly damage-induced weakening) and the increased flow (due to endothelial production of vasodilators to enlarge the lumen and thereby reduce the wall shear stress). Accounting for the altered production and removal of cells and matrix requires kinetic relations in addition to constitutive relations for stress. As noted in Chapter 12, formulation of appropriate kinetic relations for mechanosensitive changes in tissue constituents is one of the most important needs today in biomechanics. In this simple example, therefore, we see how a biomechanical analysis can be used to investigate the mechanobiology, indeed to identify what needs to be measured and why, and that solid and fluid mechanics must be considered simultaneously in certain problems.

11.2 Diffusion Through a Membrane

11.2.1 Biological Motivation

There are many different types of membranes in the body: plasma membranes which surround/define a cell, the pericardium which surrounds the heart, the pleura which surrounds the lungs, the meninges (i.e., dura mater, pia mater, and arachnoid) which surround the brain, the sheaths that cover tendons, and so on. These membranes serve a variety of biological and mechanical functions. For example, the pericardium appears to restrict gross motions of the heart, which is otherwise suspended freely within the thorax via its connections to major blood vessels at its base. It is also thought that the pericardium, which exhibits a compliant behavior at low strains but a stiff behavior at high strains (cf. Fig. 2.23 of Chapter 2) tends to limit any acute overdistension of the heart (we emphasize acute, for growth and remodeling of the pericardium allows chronic dilatations of the heart as in congestive heart failure). Finally, the pericardium encloses the pericardial space, which is filled with a small amount of lubricating fluid that reduces frictional forces between the beating heart and the protective pericardial sac, and it allows for selective transmural diffusion of molecules, particularly water and water-bound substances. There is often a need, therefore, to understand the *permeability* of biological membranes and how the permeability changes with disease, injury, repair, clinical treatment, functional adaptation, and even normal development. Toward this end, let us design a simple experiment to quantify the permeability of a nonlinear biological membrane.

11.2.2 Theoretical Basis

To induce a fluid to flow across a solid membrane, there must be a "driving force" such as a mechanical or a chemical gradient. For simplicity, let us consider the former. Experience reveals that the net flow of a fluid across a thin permeable membrane depends not only on the pressure gradient across the membrane but also on the properties of the fluid, the properties of the membrane, and the thickness of the membrane. In particular, biological membranes typically consist of a monolayer of cells that is attached to a basement membrane that covers an underlying 2-D plexus of structural proteins, which, in turn, are embedded in a proteoglycan-dominated matrix. Hence, the effective *porosity* of the network of fibers plays a key role in defining the overall permeability. This network may change with finite deformations, of course; hence, we also expect the permeability to vary with strain.

Recalling from Chapter 6 that the deformation gradient $[F]$ is the fundamental measure of motion and thus deformation, we ultimately desire to know how changes in $[F]$ affect the permeability. Because $[F]$ may vary from point to point in general (i.e., it can describe nonhomogeneous deformations) and it is defined by nine independent components, it would be prudent to investigate a possible deformation-dependent permeability first in terms of a simple motion. Consider, therefore, a homogenous deformation of a circular membrane that is defined by the mapping of a generic material particle originally at (R, Θ, Z) to (r, θ, z) in a current configuration whereby

$$r = \beta R, \qquad \theta = \Theta, \qquad z = \lambda Z, \qquad (11.5)$$

and β and λ are stretch ratios (i.e., just numbers for each equilibrium state). This mapping reveals that a material particle will not change its circumferential location; it will merely move along a radial line and move up and down as the stretched or unstretched membrane thins and thickens. From Eq. (6.81) of Chapter 6, therefore, the components of $[F]$ are

$$[F] = \begin{bmatrix} \beta & 0 & 0 \\ 0 & \beta & 0 \\ 0 & 0 & \lambda \end{bmatrix}, \qquad (11.6)$$

which we see are independent of position (R, Θ, Z) as desired; that is, associated measurements of permeability will reveal the influence of a single stretch, not the average effect of different stretches at different points. Note, too, that the deformation is *equibiaxial* in-plane; that is, the stretch β is the same in the radial and circumferential directions in each equilibrium state. Question: How is this the case if particles move in the radial direction, but not in the circumferential direction? The answer, of course, is that as particles move radially outward, the circumference $2\pi r$ increases and $\beta = 2\pi r / 2\pi R = r/R$ (a ratio of lengths) in this homogeneous deformation.

If we further assume that the membrane is mechanically incompressible, then $\det[F] = 1$ and, therefore, $\beta^2\lambda = 1$ or $\beta = 1/\sqrt{\lambda}$. Hence, by simply measuring either the out-of-plane stretch λ or the in-plane equibiaxial stretch β at any point, we can completely quantify the deformation at every point. This clearly simplifies the experimental challenge of quantifying a deformation-dependent permeability and again reveals that theory should guide experiment.

Let us now seek a functional relationship among the net volumetric flow rate Q and the pressure drop Δp, the properties of the fluid (say ρ and μ if Newtonian), the geometry of the membrane (say, initial radius A and thickness H), and the deformation (say, λ); that is, we seek to identify experimentally the specific functional form (recall the acrostic DEICE from Chapter 1) of the relation

$$Q = f(\Delta p, \rho, \mu, A, H, \lambda). \tag{11.7}$$

With six independent variables, such a determination could be very difficult and require many experiments if one tries to hold five of the six variables constant while varying one alone; repeating this process to isolate the effect of each variable could thus result in many, many experiments, each of which would have to be repeated to identify further effects of experimental noise or specimen-to-specimen variation. Thus, let us appeal to the Buckingham Pi (Section 10.5 of Chapter 10) approach to address this.

Recalling the five-step recipe for this approach, note (Step 2) that fundamental units/dimensions for this isothermal problem are length L, time T, and mass M. Moreover, the units of each variable are as follows:

$$[Q] = \frac{\text{Volume}}{\text{Time}} = L^3 T^{-1} M^0,$$

$$[\Delta p] = \frac{\text{Force}}{\text{Area}} = L^{-1} T^{-2} M^1,$$

$$[\rho] = \frac{\text{Mass}}{\text{Volume}} = L^{-3} T^0 M^1,$$

$$[\mu] = \frac{\text{Force}/\text{Area}}{1/\text{Time}} = L^{-1} T^{-1} M^1,$$

$$[A] = \text{Length} = L^1 T^0 M^0,$$

$$[H] = \text{Length} = L^1 T^0 M^0,$$

$$[\lambda] = \text{Non-dimensional} = L^0 T^0 M^0.$$

Next (Step 3), we must pick reasonable scales. As noted in Chapter 10, this is the most important step, and experience often serves one well. Nevertheless, we should find comfort in knowing that such selections are not unique and, consequently, one often tries multiple combinations of scales and evaluates the utility of each. Note, therefore, that possible length scales include

$$A, a, H, h, \tag{11.8}$$

where a and h are the deformed radius and thickness, respectively; that is, $a = \beta A$ and $h = \lambda H$. Possible mass scales include

$$\rho a^2 h, \ \rho A^2 H, \ \rho a^3, \ \rho h^3, \dots \tag{11.9}$$

where the mass density (mass/volume) simply needs to be multiplied by a volume term to yield a mass. As usual, the timescale is often more difficult to select. Possible scales include (verify that each is a unit of time) the following:

$$\sqrt{\frac{h}{g}}, \ \sqrt{\frac{\rho A^2}{\Delta p}}, \ \sqrt{\frac{\rho a^2}{\Delta p}}, \ \frac{a^2 h}{Q}, \ \frac{\mu}{\Delta p}; \tag{11.10}$$

that is, we can exploit any quantity having a unit of time, including velocity, acceleration, volumetric flow rate, force (which is a mass times acceleration, which has a unit of time), pressure, and material properties such as viscosity or even a shear modulus.

Rather than comparing this analysis for multiple sets of scales, as we would do in practice, here we illustrate the procedure using scales that are revealed to be useful in hindsight. Hence, let us select the following scales:

$$L_s = A, \qquad T_s = \frac{\mu}{\Delta p}, \qquad M_s = \rho A^2 H. \tag{11.11}$$

One reason for this selection is that A and H are defined in the undeformed configuration and thus they do not change with the stretch λ or the flow across the membrane. The next step (Step 4) in the Buckingham Pi procedure is to determine the Pi variables. Thus, note that

$$\pi_1 = \frac{Q}{(A)^3 (\mu/\Delta p)^{-1} (\rho A^2 H)^0} = \frac{Q\mu}{\Delta p A^3}, \tag{11.12}$$

$$\pi_2 = \frac{\Delta p}{(A)^{-1} (\mu/\Delta p)^{-2} (\rho A^2 H)^1} = \frac{\mu^2}{\Delta p \rho A H}, \tag{11.13}$$

$$\pi_3 = \frac{\rho}{(A)^{-3} (\mu/\Delta p)^0 (\rho A^2 H)^1} = \frac{A}{H}, \tag{11.14}$$

$$\pi_4 = \frac{\mu}{(A)^{-1} (\mu/\Delta p)^{-1} (\rho A^2 H)^1} = \frac{\mu^2}{\Delta p \rho A H}, \tag{11.15}$$

$$\pi_5 = \frac{A}{(A)^1 (\mu/\Delta p)^0 (\rho A^2 H)^0} = 1, \tag{11.16}$$

$$\pi_6 = \frac{H}{(A)^1 (\mu/\Delta p)^0 (\rho A^2 H)^0} = \frac{H}{A} = \frac{1}{A/H}, \tag{11.17}$$

$$\pi_7 = \lambda. \tag{11.18}$$

Noting that our six original independent variables have been reduced to three independent variables, the final step (Step 5) in the Buckingham Pi procedure yields

$$Q = f(\Delta p, \rho, \mu, A, H, \lambda) \rightarrow \pi_1 = \tilde{f}(\pi_2; \pi_3, \pi_7), \tag{11.19}$$

or

$$\frac{Q\mu}{\Delta p A^3} = \tilde{f}\left(\frac{\mu^2}{\Delta p \rho A H}, \frac{A}{H}, \lambda\right). \tag{11.20}$$

Experimental determination of this relation in terms of three dependent variables certainly has greater promise than that for the original relation, which has six dependent variables. For example, if we are interested primarily in the flow of a particular fluid (e.g., interstitial fluid or pericardial fluid) as a function of the pressure gradient and stretch, then we would simply need to vary π_2 and π_7 for a convenient value of π_3.

11.2.3 Illustration

Studies of diffusion through various materials, including biomembranes, are often based on the so-called *Darcy's law*, which was put forth by H. P. Darcy in 1856 based on the flow of water through soil. Darcy, a French civil engineer, was given the task of providing the city of Dijon with clean water. To do this, he created an elaborate underground aqueduct system; in 1856, he published a paper describing his design process, which included an appendix describing how sand can filter water. Regardless, Darcy's law is often stated in the following form:

$$\frac{Q}{\pi a^2} = \frac{k}{\mu}\left(\frac{\Delta p}{h}\right), \tag{11.21}$$

where πa^2 and h are the current surface area and the thickness, respectively, of the material through which the fluid flows and k is a *permeability coefficient* (having units of length squared). In the special case of no strain of the solid (i.e., $a \equiv A$ and $h \equiv H$), therefore, Darcy's law can be rewritten as

$$\frac{Q\mu}{\Delta p A^3} = K\left(\frac{A}{H}\right) \tag{11.22}$$

where

$$K = \frac{k\pi}{A^2}.$$

Comparison of this relation with Eq. (11.20) reveals that Darcy's law is a special case of our general relation whereby $\tilde{f} = (K)(A/H)$ and K is a nondimensional permeability; that is, Darcy's law requires that $\tilde{f}$ depend linearly on π_3 and be independent of π_2, with $\pi_7 \equiv 1$ for no strain. Use of the more general result from Buckingham Pi is preferred, however, because it does not presuppose a specific functional form.

For a senior project at Texas A&M University, Janna Vaughn (see Vaughn et al., 2002) used a Buckingham-Pi-based approach to guide the design of a device to study the strain-dependent permeability of a representative biomembrane—excised bovine epicardium (which was obtained from a slaughterhouse). Figure 11.1 is an exploded view of the combined stretching–diffusion device. Briefly, a membrane is first mounted in a stress-free configuration between two circular fixtures (Fig. 11.2) and then placed in the device between the upper and lower fluid-filled chambers. A nearly uniform equibiaxial stretch β of the specimen is imposed by pushing the specimen fixture down such that the specimen is pulled over the outer surface of the bottom chamber (this is akin to pulling a rubber membrane over a tube to create a drum); this is accomplished by turning a threaded

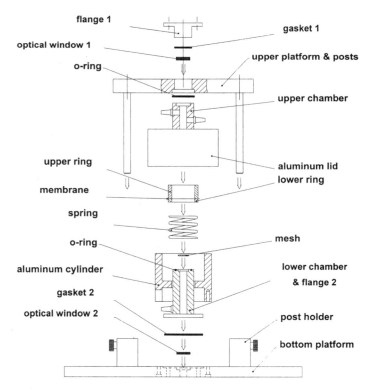

FIGURE 11.1 Exploded view of the combined stretching–diffusion device designed, in part, by Janna Vaughn as part of a senior project at Texas A&M University. From Vaughn et al. (2002), with permission from Elsevier Science.

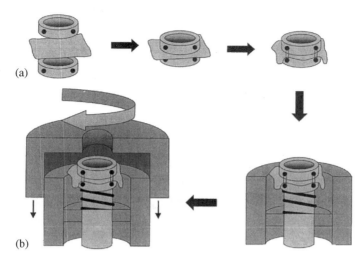

(a)

(b)

Figure 11.2 Expanded view of the specimen fixture from Figure 11.1. Note that the specimen is placed within the fixture in a stress-free configuration. This is facilitated by having a waxed platform inside the restraining rings on which a fluid film can support the specimen prior to mounting. Turning the outer lid pushes down on the membrane holder, which, in turn, imposes a uniform radial stretch of the specimen. With permission from Elsevier Science.

aluminum lid above the upper chamber that pushes the specimen holder down. Also above the upper chamber is an upper platform that applies an overall compressive load to the upper and lower chambers; O-rings ensure a good seal. Fluid flows into the upper chamber, through the membrane, and out of the lower chamber via appropriate fittings. Finally, note that optical windows at the bottom and top allow back-lighting and visualization of markers that are affixed to the surface of the specimen. Tracking the motions of these markers due to stretching of the membrane allows the displacements and their gradients to be computed (Chapter 2).

Figures 11.3 and 11.4 show data from Janna's experiments wherein π_1 is plotted versus π_2 and π_7 ($= \lambda$), for different values of π_7 and π_2, respectively. Over the range of parameters studied, π_1 varied with π_7 much more than it did with π_2, which is not explained by Darcy's law. Indeed, plotting the data as the nondimensional permeability K versus stretch λ (Fig. 11.5) reveals a strong strain dependency. That K increased with increasing stretch may suggest that stretching brings the collagenous fibers closer together, compacting the tissue and thus increasing the resistance to flow. More experiments and analyses are needed to understand this complex solid–fluid coupling, however.

In closing, note that the Δp in this formulation represents the pressure difference across the membrane (Fig. 11.1). Although it is difficult to

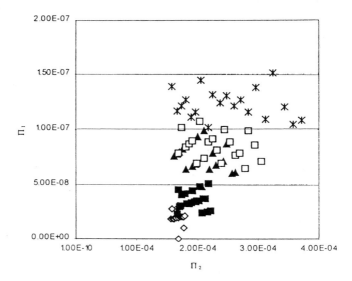

FIGURE 11.3 Data for two of the nondimensional parameters from Eq. (11.20): π_1 versus π_2 for various values of π_7 (denoted by open diamonds, closed squares, closed triangles, open squares, and asterisks for increasing levels of stretch from 1.0 to 1.6, clearly finite deformations). With permission from Elsevier Science.

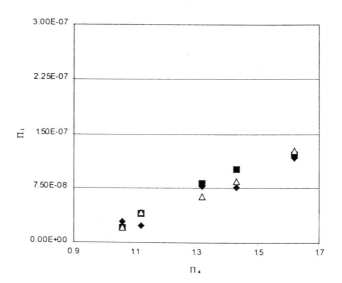

FIGURE 11.4 Similar to Figure 11.3 except for π_1 versus π_7 for various values of π_2 over the range in Figure 11.3. Note that π_7 was denoted as π_4 in the original paper. With permission from Elsevier Science.

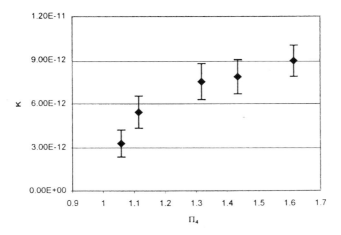

FIGURE 11.5 Similar to Figures 11.3 and 11.4 except for the nondimensional permeability K versus the equibiaxial stretch λ. With permission from Elsevier Science.

measure directly the pressure at these surfaces during flow, the pipe-flow equation [Eq. (10.94) of Chapter 10] can be used to compute p_1 and p_2 based on pressures measured elsewhere in the system (e.g., atmospheric pressure in an open upstream reservoir), differences in height, velocities, and viscous and geometric losses. Indeed, one would use the pipe-flow equation twice: once to get the difference between the pressure at its source and that value at the top surface of the membrane, and once to get the difference between the pressure at the bottom surface of the membrane and that at the ultimate exit point. Hence, as implied in Chapter 10, the pipe-flow equation can be very useful in the design and analysis of important biomechanical experiments. For very slow flows, of course, the system is nearly quasistatic and pressure differences would be due primarily to the gravitational heads $\rho g h$. In general, however, the full pipe-flow equation should be checked to determine the specific values of velocity for which the quasistatic assumption is reasonable.

Observation 11.1 A commonly studied ordinary differential equation in nonlinear dynamics is the so-called *Duffing equation*. It can be written as

$$\eta^2 \ddot{x} + cx + \alpha x^3 + 2\eta\zeta\dot{x} = F(t),$$

where η is the ratio of the forcing and fundamental frequencies, c is a stiffness parameter, α is a nonlinear stiffness parameter, and ζ is the ratio of the actual to the critical damping in the system; x, of course, is the displacement and $F(t)$ is the time-varying forcing function. An overdot implies differentiation with respect to time and a double overdot implies twice differentiation with time. To solve this second-order equation numerically (e.g., via Runge–Kutta), it is useful to rewrite it as a system of first-order equations. Toward this end, consider a change of variables whereby

$$y_0 \equiv x \quad \text{and} \quad y_1 \equiv \dot{x}$$

and, consequently,

$$\dot{y}_0 \equiv \dot{x} \quad \text{and} \quad \dot{y}_1 \equiv \ddot{x}.$$

We see, therefore, that our system of first-order equations is

$$\dot{y}_0 = \dot{x},$$

$$\dot{y}_1 = \ddot{x} \equiv \frac{F(t) - cx - \alpha x^3 - 2\eta \zeta \dot{x}}{\eta^2},$$

or in terms of (y_0, y_1) alone,

$$\dot{y}_0 = y_1,$$

$$\dot{y}_1 = \frac{F(t) - cy_0 - \alpha y_0^3 - 2\eta \zeta y_1}{\eta^2}.$$

Hence, we have differential equations of the form

$$\dot{y}_0 = G(y_0, y_1) \quad \text{and} \quad \dot{y}_1 = H(y_0, y_1)$$

where G and H are known functions. This system of equations is nonlinear in y_0 if $\alpha \neq 0$.

Whereas we consider a few numerical solutions below for the full non-linear system, it is useful to note that qualitative information on the *stability* of the nonlinear system can sometimes be gained by linearizing the system about various equilibria; that is, as in Chapter 5 on the stability of beam columns, we can ask whether the structure would return to its prior equilibrium position if perturbed slightly. It is useful to note, therefore, the following (cf. Figs. 5.21 and 5.22):

If the linearized system is asymptotically stable, then the associated non-linear system is stable about the chosen equilibrium (or fixed) point.

If the linearized system is neutrally stable, then the linearized solution does not provide any useful information about the nonlinear system.

If the linearized system is unstable about a fixed point, then the associated nonlinear system is likewise unstable about that fixed point although it could stabilize about another equilibrium position.

For the Duffing equation, let an equilibrium position (i.e., fixed point) be denoted by a position λ and a zero velocity; that is, $(x, \dot{x}) = (\lambda, 0) = (y_0, y_1)$ at a constant force $F(t) = F_\lambda$. Hence, linearizing the system of first-order equations about the fixed point can be accomplished using a *Taylor series*, namely

$$\dot{y}_0 = G(\lambda, 0) + \frac{\partial G}{\partial y_0}\bigg|_{(\lambda, 0)} (y_0 - \lambda) + \frac{\partial G}{\partial y_1}\bigg|_{(\lambda, 0)} (y_1 - 0) + \text{H.O.T.},$$

$$\dot{y}_1 = H(\lambda, 0) + \frac{\partial H}{\partial y_0}\bigg|_{(\lambda, 0)} (y_0 - \lambda) + \frac{\partial H}{\partial y_1}\bigg|_{(\lambda, 0)} (y_1 - 0) + \text{H.O.T.},$$

where H.O.T. stands for higher-order terms, which are neglected in the process of linearization. For our specific system,

$$\frac{\partial G}{\partial y_0} = 0, \qquad \frac{\partial G}{\partial y_1} = 1,$$

$$\frac{\partial H}{\partial y_0} = \frac{-c - 3\alpha y_0^2}{\eta^2}, \qquad \frac{\partial H}{\partial y_1} = \frac{-2\zeta}{\eta};$$

thus, with $G(\lambda, 0) = 0$ and $H(\lambda, 0) = 0$, the latter because $F_\lambda = c\lambda + \alpha\lambda^3$ at equilibrium, we have

$$\dot{y}_0 = 0(y_0 - \lambda) + 1(y_1 - 0),$$

$$\dot{y}_1 = \frac{-c - 3\alpha\lambda^2}{\eta^2}(y_0 - \lambda) - \frac{2\zeta}{\eta}(y_1 - 0),$$

which can be written in matrix form as

$$\begin{Bmatrix} \dot{y}_0 \\ \dot{y}_1 \end{Bmatrix} = \begin{bmatrix} 0 & 1 \\ \dfrac{-c - 3\alpha\lambda^2}{\eta^2} & \dfrac{-2\zeta}{\eta} \end{bmatrix} \begin{Bmatrix} y_0 - \lambda \\ y_1 \end{Bmatrix}.$$

It can be shown (Strang, 1986) that asymptotic stability in the small requires that tr[ ] < 0 *and* det[ ] > 0; alternatively, neutral stability is given by tr[ ] = 0 and det[ ] > 0 *or* by tr[ ] < 0 and det[ ] = 0; finally, an instability about a fixed point is thus given by tr[ ] > 0 *or* det[ ] < 0. Here, of course, tr[ ] and det[ ] denote the trace and determinant, respectively, of the 2 × 2 matrix [ ] as noted in Appendix 6 of Chapter 6. Clearly, for our system, the trace (i.e., sum of the diagonals) and determinant are

$$\text{tr}[\ ] = \frac{-2\zeta}{\eta}, \qquad \det[\ ] = \frac{c + 3\alpha\lambda^2}{\eta^2} \quad \forall \lambda,$$

respectively. We see, for example, that stability requires that ζ be nonzero and that ζ and η be of the same sign. If, on the other hand, $\zeta = 0$, the system will be only neutrally stable and the linearization will provide no useful information on the nonlinear system. Let us now consider a few numerical examples of the full nonlinear system, which can be interpreted with this backdrop.

Example 11.1 Examine the dynamic behavior of a Duffing system with the following values of the numerical parameters: $\eta = 0.0625$, $c = 1.0$, $\alpha = 0$, and $\zeta = 0.1$, with $F(t) = 1.0 \sin t$. Note that this is a linear system because $\alpha = 0$.

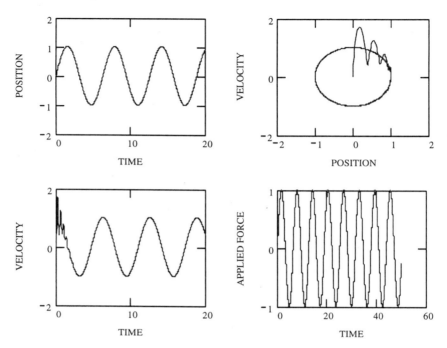

FIGURE 11.6 Results for the Duffing equation ($\alpha = 0$). The upper left panel shows position versus time and the lower left panel shows velocity versus time, both in response to the forcing function shown in the bottom right panel. Finally, the panel in the upper right shows the so-called *phase-plane diagram*, velocity versus position. It is seen that the response, starting at $(x, \dot{x}) = (0, 0.01)$ quickly finds the so-called periodic solution, revealed by the circular path in the phase plane.

Solution: The behavior is best studied graphically in terms of the computed displacement and velocity histories $x(t)$ and $\dot{x}(t)$ as well as the phase-plane plot, $\dot{x}(t)$ versus $x(t)$. This is the so-called geometric method of Poincaré. We use a standard Runge–Kutta numerical method to obtain the solution, which requires that we specify the initial conditions. Here, let the initial conditions be perturbed slightly from equilibrium: $x(0) = 0$ and $\dot{x}(0) = 0.01$. Figure 11.6 reveals the periodic response (i.e., position) given the sinusoidal forcing function. Note, too, that the system recovers quickly from the initial disturbance as expected based on the linearized analysis.

Example 11.2 Repeat the previous example except for a nonlinear case whereby $\alpha = 0.5$. Given that such solutions depend strongly on the initial conditions, also let $x(0) = 1.0$ and $\dot{x}(0) = 1.0$.

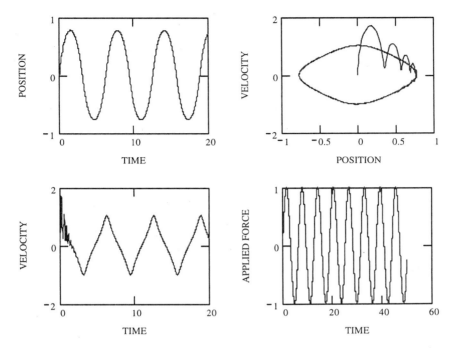

FIGURE 11.7 Similar to Figure 11.6 except that $\alpha = 0.5$, which induces a nonlinearity.

Solution: Figure 11.7 shows the results for $\alpha = 0.5$, $x(0) = 0$, and $\dot{x}(0) = 0.01$. Note the differences with respect to Figure 11.6, but the qualitative similarity. Figure 11.8 shows the dramatic effect of the initial condition, now with $x(0) = 1.0$ and $\dot{x}(0) = 1.0$. In both cases, the response is stable, returning to the periodic solution again consistent with the linearization.

11.3 Dynamics of a Saccular Aneurysm

11.3.1 Biological Motivation

Recall from Section 3.5.1 of Chapter 3 that intracranial saccular aneurysms are thin-walled, balloonlike dilatations of the arterial wall that occur in or near bifurcations in the circle of Willis (Fig. 1.1 of Chapter 1). Based on quasistatic stress analyses, it appears that the intramural wall stress is often on the order of 1 MPa or more, with the rupture strength on the order of 10 MPa. Slight changes in wall thickness or protease degradation of structural proteins within the wall can thereby render these lesions susceptible

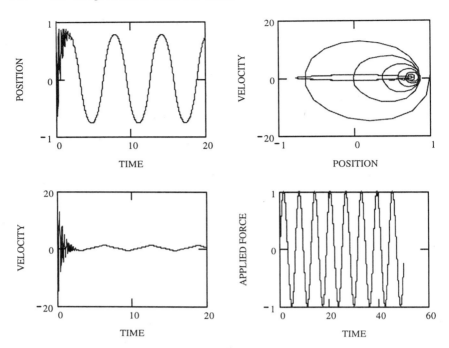

FIGURE 11.8 Similar to Figure 11.6 except that $\alpha = 0.5$ as well as perturbed initial conditions given by $x(0) = 1.0$ and $\dot{x}(0) = 1.0$.

to rupture. There is a need, therefore, to understand better the solid mechanics of the wall of an aneurysm.

It has been suggested that aneurysms may enlarge or rupture due to a dynamic instability called *resonance*. In particular, some investigators have suggested that the pulsatility of the blood pressure may excite a lesion at its natural frequency and thus induce violent vibrations (i.e., resonance). Whereas the heart-induced blood pressure serves as the forcing function for the lesion, the surrounding cerebrospinal fluid (CSF) may affect the dynamic response of the lesion as well; that is, as the aneurysm displaces within the CSF due to the distending blood pressure, by Newton's third law the CSF "pushes back." Toward a better understanding of the solid–fluid coupling related to aneurysm dynamics, let us now consider an idealized case: the pressure-induced distension of a spherical lesion that is surrounded by an external CSF.

11.3.2 Mathematical Framework

It can be shown that the equation of motion (i.e., $\boldsymbol{F} = m\boldsymbol{a}$) for a *spherical membrane* can be written as (Humphrey, 2002)

$$\sigma_{rr}(b) - \sigma_{rr}(a) - 2T\kappa = \rho_s h \frac{d^2 u_r}{dt^2}, \tag{11.23}$$

where $\sigma_{rr}(b)$ and $\sigma_{rr}(a)$ are the stress boundary conditions on the outer and inner surfaces of the membrane, respectively, T is the membrane tension ($T = h\sigma_{\theta\theta} = h\sigma_{\phi\phi}$, where h is the deformed thickness), κ is the curvature ($\kappa = 1/a$, where a is the deformed radius of the sphere), ρ_s is the mass density of the solid (i.e., the membrane), and $u_r(t) = a(t) - A$ is the radial displacement of any point on the aneurysm (with A the undeformed radius). Note that in the absence of dynamic effects, with $\sigma_{rr}(b) = -P_o$ and $\sigma_{rr}(a) = -P_i$ the static pressures, equation 11.23 recovers the equilibrium solution [Eq. (6.62) of Chapter 6], namely

$$2T\kappa = \sigma_{rr}(b) - \sigma_{rr}(a) \to 2h\sigma_{\theta\theta}(1/a) = P_i - P_o, \tag{11.24}$$

or

$$\sigma_{\theta\theta} = \frac{(P_i - P_o)a}{2h} \leftrightarrow T = \frac{Pa}{2}, \tag{11.25}$$

where $P = P_i - P_o$ is the transmural pressure. Note, too, that the wall tension T depends on the deformation of the membrane through its constitutive relation. Here, we let the deformation be tracked via the stretch ratio $\lambda(t) = a(t)/A$, a ratio of deformed to undeformed radii [actually, the ratio of deformed to undeformed circumferences, $2\pi a(t)/2\pi A(t)$ for both the θ and ϕ directions]. If the membrane conserves its volume during deformations, then

$$(4\pi a^2)h = 4\pi A^2 H \to \frac{h}{H} = \frac{1}{(a/A)^2} = \frac{1}{\lambda^2}, \tag{11.26}$$

where H is the undeformed wall thickness. Show that this is consistent with $\det[F] = 1$, where $[F] = \mathrm{diag}[\lambda, \lambda, h/H]$.

We will assume that the boundary condition on stress at the inner wall is given by the time-varying blood pressure, namely

$$\sigma_{rr}(a, t) = -P_i(t), \quad P_i(t) = P_m + \sum_{n=1}^{N} [A_n \cos(n\omega t) + B_n \sin(n\omega t)], \tag{11.27}$$

consistent with the formulation in Section 9.5 of Chapter 9 whereby we recall that P_m is the mean blood pressure, A_n and B_n are Fourier coefficients, and ω is the fundamental circular frequency of the beating heart. To find the outer stress boundary condition, however, we will assume that the CSF is an incompressible, Newtonian fluid. Hence, we need to determine the radial stress in the fluid {recall the Navier–Poisson relation [Eq. (7.65)]},

$$\sigma_{rr}(r, t) = -p(r, t) + 2\mu D_{rr}(r, t), \tag{11.28}$$

which for spherical coordinates is [see Eq. (7.60)]

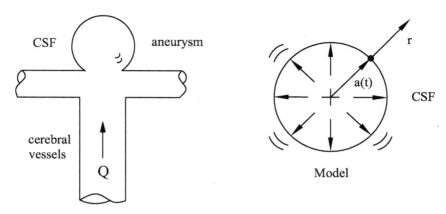

FIGURE 11.9 Simple geometry of a spherical saccular aneurysm surrounded by a fluid that also occupies a spherical domain.

$$\sigma_{rr}(r,t) = -p(r,t) + 2\mu \frac{\partial v_r}{\partial r} \quad r \in [a, \infty), \tag{11.29}$$

where r is a coordinate in the fluid domain (Fig. 11.9; note, too, that because we are modeling the aneurysm as a membrane, we assume that $b = a + h$, where $h \ll 1$, and thus $a \sim b$). By finding the fluid pressure and velocity fields at all r and time t, we can evaluate the stress exerted by the CSF on the membrane and then solve the solid mechanics problem as desired. To solve the CSF problem, therefore, let us assume the following:

1. Incompressible flow ($\nabla \cdot \mathbf{v} = 0$)
2. Newtonian fluid ($\mu = \text{constant}$)
3. Radial unsteady flow only [$v_\theta = 0$, $v_\phi = 0$, $v_r = v_r(r,t)$]
4. Axisymmetric flow ($\partial/\partial\theta = 0$, $\partial/\partial\phi = 0$)
5. Negligible body forces ($\mathbf{g} = \mathbf{0}$)

Mass balance in spherical coordinates is [Eq. (8.16)]

$$\frac{1}{r^2}\frac{\partial(r^2 v_r)}{\partial r} + \frac{1}{r\sin\theta}\frac{\partial(v_\theta \sin\theta)}{\partial\theta} + \frac{1}{r\sin\theta}\frac{\partial v_\phi}{\partial\phi} = 0. \tag{11.30}$$

In spherical coordinates, the appropriate linear momentum balance equation (i.e., the Navier–Stokes equation, $-\nabla p + \mu\nabla^2\mathbf{v} + \rho\mathbf{g} = \rho\mathbf{a}$) requires the following:

$$\hat{e}_r: \quad -\frac{\partial p}{\partial r} + \mu\left(\nabla^2 v_r - \frac{2v_r}{r^2} - \frac{2}{r^2}\frac{\partial v_\theta}{\partial\theta} - \frac{2v_\theta\cot\theta}{r^2} - \frac{2}{r^2\sin\theta}\frac{\partial v_\phi}{\partial\phi}\right) + \rho g_r$$

$$= \rho\left(\frac{\partial v_r}{\partial t} + v_r\frac{\partial v_r}{\partial r} + \frac{v_\theta}{r}\frac{\partial v_r}{\partial\theta} + \frac{v_\phi}{r\sin\theta}\frac{\partial v_r}{\partial\phi} - \frac{v_\theta^2 + v_\phi^2}{r}\right), \tag{11.31}$$

$$\hat{e}_\theta: \quad -\frac{1}{r}\frac{\partial p}{\partial \theta} + \mu\left(\nabla^2 v_\theta + \frac{2}{r^2}\frac{\partial v_r}{\partial \theta} - \frac{v_\theta}{r^2 \sin^2 \theta} - \frac{2\cot\theta}{r^2 \sin\theta}\frac{\partial v_\phi}{\partial \phi}\right) + \rho g_\theta$$

$$= \rho\left(\frac{\partial v_\theta}{\partial t} + v_r\frac{\partial v_\theta}{\partial r} + \frac{v_\theta}{r}\frac{\partial v_\theta}{\partial \theta} + \frac{v_\phi}{r\sin\theta}\frac{\partial v_\theta}{\partial \phi} + \frac{v_r v_\theta}{r} - \frac{v_\phi^2 \cot\theta}{r}\right), \quad (11.32)$$

$$\hat{e}_\phi: \quad -\frac{1}{r\sin\theta}\frac{\partial p}{\partial \phi} + \mu\left(\nabla^2 v_\phi - \frac{v_\phi}{r^2 \sin^2 \theta} + \frac{2}{r^2 \sin^2 \theta}\frac{\partial v_r}{\partial \phi} + \frac{2\cot\theta}{r^2 \sin\theta}\frac{\partial v_\theta}{\partial \phi}\right) + \rho g_\phi$$

$$= \rho\left(\frac{\partial v_\phi}{\partial t} + v_r\frac{\partial v_\phi}{\partial r} + \frac{v_\theta}{r}\frac{\partial v_\phi}{\partial \theta} + \frac{v_\phi}{r\sin\theta}\frac{\partial v_\phi}{\partial \phi} + \frac{v_\phi v_r}{r} + \frac{v_\phi v_\theta}{r}\cot\theta\right), \quad (11.33)$$

where

$$\nabla^2 = \frac{1}{r^2}\frac{\partial}{\partial r}\left(r^2\frac{\partial}{\partial r}\right) + \frac{1}{r^2 \sin\theta}\frac{\partial}{\partial \theta}\left(\sin\theta\frac{\partial}{\partial \theta}\right) + \frac{1}{r^2 \sin^2 \theta}\frac{\partial^2}{\partial \phi^2}. \quad (11.34)$$

After canceling out terms using the above assumptions (do it), we are left with

$$\frac{1}{r^2}\frac{\partial}{\partial r}(r^2 v_r) = 0 \rightarrow v_r(r, t) = \frac{g(t)}{r^2} \quad (11.35)$$

from mass balance, where $g(t)$ is an arbitrary function due to integration (Note: we have an *integration function* rather than an integration constant because the velocity can depend on both position and time) and

$$-\frac{1}{\rho}\frac{\partial p}{\partial r} + \frac{\mu}{\rho}\left(\nabla^2 v_r - \frac{2v_r}{r^2}\right) = \frac{\partial v_r}{\partial t} + v_r\frac{\partial v_r}{\partial r}, \quad (11.36)$$

from linear momentum balance, with

$$\nabla^2 = \frac{1}{r^2}\frac{\partial}{\partial r}\left(r^2\frac{\partial}{\partial r}\right). \quad (11.37)$$

Expanding the linear momentum balance equation, we have

$$-\frac{1}{\rho}\frac{\partial p}{\partial r} + \frac{\mu}{\rho}\left(\frac{\partial^2 v_r}{\partial r^2} + \frac{2}{r}\frac{\partial v_r}{\partial r} - \frac{2v_r}{r^2}\right) = \frac{\partial v_r}{\partial t} + v_r\frac{\partial v_r}{\partial r}. \quad (11.38)$$

If we now substitute the expression for v_r from mass balance into the term in the parentheses, we see that

$$6\frac{g(t)}{r^4} + \frac{2}{r}\left(-\frac{2g(t)}{r^3}\right) - 2\frac{g(t)}{r^4} = 0, \quad (11.39)$$

which is to say, linear momentum balance reduces to

$$-\frac{1}{\rho}\frac{\partial p}{\partial r} = \frac{\partial v_r}{\partial t} + v_r\frac{\partial v_r}{\partial r}. \quad (11.40)$$

It is interesting that this equation is independent of viscosity; indeed, if r is taken to be a streamline direction s, this is the same equation as the s-direction Euler equation (8.58) in the absence of gravity. This aneurysm–CSF problem is thus a special case wherein the same pressure and velocity fields satisfy both the Navier–Stokes and the Euler equations, as discussed in Observation 8.3 of Chapter 8 (page 395). It is very important to recognize, however, that the viscosity of the fluid will still play a role through the stress boundary condition [Eq. (11.29)]. Before exploiting this, let us solve for the pressure field in the CSF. Although the fluid mechanicist would want to know the pressure at all values of r and t, the solid mechanicist only needs to know values at $r = a$, which are felt by the solid (aneurysm). Hence, let us seek p at $r = a$ and thus integrate from a to ∞:

$$-\frac{1}{\rho}\int_a^\infty \frac{\partial p}{\partial r}\,dr = \int_a^\infty \frac{\partial}{\partial t}\left(\frac{g(t)}{r^2}\right)dr + \int_a^\infty v_r \frac{\partial v_r}{\partial r}\,dr, \tag{11.41}$$

or

$$-\frac{1}{\rho}p\Big|_a^\infty = \frac{dg(t)}{dt}\left(-\frac{1}{r}\Big|_a^\infty\right) + \frac{1}{2}v_r^2\Big|_a^\infty, \tag{11.42}$$

which yields

$$-\frac{1}{\rho}(p_\infty - p_a) = \frac{dg}{dt}\left(-\frac{1}{\infty}+\frac{1}{a}\right) + \frac{1}{2}\left(\frac{g^2}{\infty^4}-\frac{g^2}{a^4}\right), \tag{11.43}$$

or

$$p_a = \rho\left(\frac{dg}{dt}\frac{1}{a} - \frac{g^2}{2a^4}\right) + p_\infty. \tag{11.44}$$

Remembering that we are interested in the mechanics of an aneurysm within the head of a human, one might quickly ask the utility of integrating to infinity. In this case, infinity simply means far enough away from the aneurysm, which could be only centimeters (e.g., if $a = 1$ mm and "infinity" was merely taken to be 10 mm, then $1/1-1/10^2 = 0.99$ and $1/1-1/10^4 = 0.9999$, thus revealing that it is reasonable to neglect the terms $1/\infty^n$ in comparison to the $1/a^n$ terms). Note, too, that the deformed radius a and the integration function g are both functions of time; thus, the pressure exerted by the CSF on the aneurysm will likewise vary with time in general. Finally, the radial stress in the fluid is

$$\sigma_{rr} = -p + 2\mu\frac{\partial v_r}{\partial r} = -p + 2\mu\left(-\frac{2g(t)}{r^3}\right) \quad \forall r, t. \tag{11.45}$$

We are interested primarily in the stress at the wall of the aneurysm, where $r = a$. The normal stress in the fluid at this location thus becomes [using Eq. (11.44)]

$$\sigma_{rr}|_{r=a} = -p_a - 4\mu\frac{g(t)}{a^3} = \rho\left(\frac{g^2}{2a^4} - \frac{1}{a}\frac{dg}{dt}\right) - p_\infty - 4\mu\frac{g}{a^3}, \qquad (11.46)$$

whereby we see that the viscosity of the CSF has indeed entered the problem through the stress boundary condition. At this point, note that we still have not found the arbitrary integration function $g(t)$, even though we have used a stress boundary condition at $r = a$ and the condition at infinity in the integration. When similar situations arose in Chapters 3–5 for solids and Chapters 9 and 10 for fluids, we sought additional conditions to generate the requisite number of equations for our unknowns. In particular, we often used kinematic conditions such that the continuity of displacement for two solids (e.g., bone and metal prosthesis in Chapters 3 and 4) or velocities of two materials (e.g., a moving solid plate and underlying fluid in Couette flows in Chapter 9). To find $g(t)$, we can use a similar "matching condition" at $r = a$; that is, by the no-slip condition, we need to match the velocity of the aneurysm and that of the CSF at $r = a$. Given the displacement u_r in the r direction of a material point on the aneurysm, which is the difference between where we are, $a(t)$, and where we were, A, the radial velocity of the aneurysmal wall is

$$\frac{du_r}{dt} = \frac{d}{dt}(a(t) - A) = \frac{da}{dt}, \qquad (11.47)$$

where $a = \lambda A$. Thus, matching this velocity with $v_r(r = a, t)$ in the fluid, we have

$$\frac{du_r}{dt} = A\frac{d\lambda}{dt} = \frac{g(t)}{a^2} \equiv \frac{g(t)}{\lambda^2 A^2} \to g(t) = \lambda^2 A^3 \frac{d\lambda}{dt}. \qquad (11.48)$$

It follows, therefore, that

$$\frac{dg}{dt} = A^3\left[\lambda^2\frac{d^2\lambda}{dt^2} + \frac{d\lambda}{dt}\left(2\lambda\frac{d\lambda}{dt}\right)\right]. \qquad (11.49)$$

Hence, the outer stress boundary condition is

$$\sigma_{rr}|_{r=a} = -\rho\left\{\frac{A^3}{\lambda A}\left[\lambda^2\frac{d^2\lambda}{dt^2} + 2\lambda\left(\frac{d\lambda}{dt}\right)^2\right] - \frac{\lambda^4 A^6}{2\lambda^4 A^4}\left(\frac{d\lambda}{dt}\right)^2\right\} - p_\infty - 4\mu\left[\frac{\lambda^2 A^3}{\lambda^3 A^3}\left(\frac{d\lambda}{dt}\right)\right], \qquad (11.50)$$

which can be simplified to

$$\sigma_{rr}|_{r=a} = -\rho\left[A^2\lambda\frac{d^2\lambda}{dt^2} + 2A^2\left(\frac{d\lambda}{dt}\right)^2 - \frac{1}{2}A^2\left(\frac{d\lambda}{dt}\right)^2\right] - p_\infty - \frac{4\mu}{\lambda}\left(\frac{d\lambda}{dt}\right)$$

$$= -\rho A^2\lambda\frac{d^2\lambda}{dt^2} - \frac{3}{2}\rho A^2\left(\frac{d\lambda}{dt}\right)^2 - \frac{4\mu}{\lambda}\left(\frac{d\lambda}{dt}\right) - p_\infty. \qquad (11.51)$$

The stress acting on the outer surface of the membrane in the positive r direction will be equal and opposite that in the fluid at $r = a$. Hence, the governing differential equation for the solid [Eq. (11.23)] can be written as

$$\left[-p_\infty - \rho A^2 \lambda \frac{d^2\lambda}{dt^2} - \frac{3}{2}\rho A^2 \left(\frac{d\lambda}{dt} \right)^2 - \frac{4\mu}{\lambda}\left(\frac{d\lambda}{dt} \right) \right] - [-P_i(t)] - \frac{2T}{\lambda A} = \frac{\rho_s HA}{\lambda^2}\frac{d^2\lambda}{dt^2},$$
(11.52)

or

$$\left(\frac{\rho_s HA}{\lambda^2} + \rho A^2 \lambda \right)\frac{d^2\lambda}{dt^2} + \frac{3}{2}\rho A^2 \left(\frac{d\lambda}{dt} \right)^2 + \frac{4\mu}{\lambda}\left(\frac{d\lambda}{dt} \right) + \frac{2T}{\lambda A} = P_i(t) - p_\infty,$$
(11.53)

where the time-varying blood pressure $P_i(t)$ is given by Eq. (11.27) and $T = T(\lambda)$ must be given by a constitutive equation for the aneurysm (recall Section 6.4 of Chapter 6). For example, for a Fung-type exponential behavior [cf. Eq. (6.41)], we have

$$T(\lambda) = c\Gamma e^{Q}(\lambda^2 - 1) \quad \text{with} \quad Q = \frac{1}{2}\Gamma(\lambda^2 - 1)^2,$$
(11.54)

where c and Γ are material parameters. Independent of the specific constitutive relation for the solid, however, we see how important the solid–fluid coupling is; the governing equation of motion for the dynamic response of the solid depends on the density ρ, viscosity μ, and far-field pressure p_∞ of the cerebrospinal fluid. This complex, nonlinear ordinary differential equation is solved numerically in Humphrey (2002) using a Runge–Kutta method. Figures 11.10 and 11.11 show one typical result, which suggests that this class of aneurysms is dynamically stable, contrary to the thoughts of many based on simplified linear analyses that neglected the nonlinear stress–stretch relation $T(\lambda)$ and the solid–fluid coupling. Although closed-form solutions are not available for Eq. (11.53), one can obtain analytical solutions for the stability of the aneurysm by linearizing the governing equation about multiple, different equilibrium positions—as noted in Observation 11.1, such linearizations can provide useful information on the nonlinear problem; that is, if one were to inflate the lesion to a particular equilibrium configuration and then perturb it slightly, one could ask whether the perturbed aneurysm (i.e., nonlinear system) would come back to its equilibrium configuration (i.e., be stable) or if it would move away from this configuration (i.e., be unstable). Recall that similar arguments were made in Chapter 5 in the fully linear analysis of the stability of Euler columns. Indeed, by performing a linearized stability analysis (see Humphrey, 2002), it can be shown that it is the fluid viscosity μ alone that renders the aneurysm dynamically stable in the nonlinear case given the assumptions invoked herein.

In summary, based on this analysis, it was concluded that at least one subclass of saccular aneurysms (nearly spherical) is unlikely to be dynamically

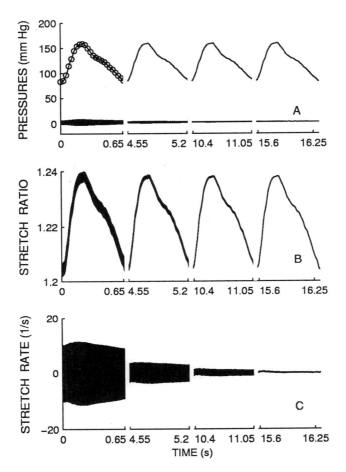

FIGURE 11.10 A representative result for the elastodynamics of an idealized spherical, isotropic saccular aneurysm. Panel A shows the internal forcing function $P_i(t)$, as both original human data (open circles) and their Fourier series representation (solid lines); the bottom curve shows the far-field pressure p_∞. Panels B and C show the associated time-varying stretch and stretch rate given an initial disturbance, which decreases quickly. With permission from Elsevier Science.

unstable as postulated by some and supported by others based on simplified models. This illustrates the importance of modeling well the inherent complexities such as the nonlinear material behavior, the large deformations, and the solid–fluid coupling. Indeed, if one considers a viscoelastic, rather than purely elastic, behavior of the aneurysm, it can be shown that there is further evidence for its dynamic stability. Rather than going into the details of the nonlinear viscoelasticity of solids, which is an important

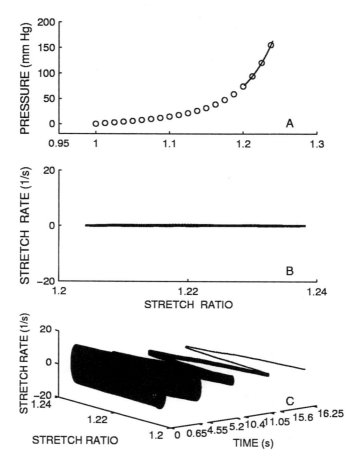

FIGURE 11.11 Further results associated with Figure 11.10. Panel B shows the phase-plane in the case of no initial disturbance, that is, the periodic solution. Panel C shows the phase–time plot in the case of an initial disturbance, the effects of which are seen to decrease quickly such that the periodic solution is recovered (i.e., the periodic solution serves as a strong attractor), thus revealing that the system is dynamically stable. Finally, panel A shows the pressure–stretch behavior both for quasistatic loading (open circles) and for the dynamic case. Clearly, the inertial effects are small and the dynamics can be treated in terms of a series of equilibria. From Shah and Humphrey (1999), with permission from Elsevier Science.

but advanced topic, we consider below an introduction to the linearized theory of viscoelasticity and some associated directions of needed research.

Example 11.3 In Exercise 11.3, we ask that Eq. (11.53) be nondimension-alized to a simpler form. Here, we further neglect the fluid (nondimensional parameters $b = 0$ and $m = 0$) and assume a Fung-type behavior; thus our governing equation reduces to (Humphrey, 2002)

$$\frac{1}{x^2}\ddot{x} + \frac{2}{x}\Gamma e^{Q}(x^2 - 1) = F(\tau),$$

where

$$Q = \frac{1}{2}\Gamma(x^2 - 1)^2.$$

Show that such a dynamic system is dynamically unstable in the small (i.e., in the absence of an external fluid and given small perturbations from an equilibrium position).

Solution: Let the nondimensional distending pressure at equilibrium be F_0 and the associated equilibrium stretch be $x = \alpha$. Moreover, let us consider a change in variables whereby

$$y_0 \equiv x - \alpha \quad \text{and} \quad y_1 \equiv dx/d\tau = \dot{x}.$$

Hence, our single second-order equation can be written in terms of two first-order equations, namely

$$\dot{y}_0 = \dot{x} = y_1,$$

$$\dot{y}_1 = \ddot{x} = \left[F_0 - 2\Gamma e^{Q}\left(x - \frac{1}{x}\right)\right]x^2,$$

or in terms of the fixed point $x = \alpha$,

$$\dot{y}_0 = y_1,$$

$$\dot{y}_1 = \left[F_0 - 2\Gamma e^{Q}\left(y_0 + \alpha - \frac{1}{y_0 + \alpha}\right)\right](y_0 + \alpha)^2,$$

which we can write symbolically as

$$\dot{y}_0 = G(y_0, y_1; \alpha),$$
$$\dot{y}_1 = H(y_0, y_1; \alpha)$$

where G and H are general functions. Expanding in a Taylor series about the fixed point $x = \alpha$, or $y_0 = 0$ and $y_1 = 0$, we have

$$\dot{y}_0 = G(0, 0) + \frac{\partial G}{\partial y_0}\bigg|_{(0,0)}(y_0 - 0) + \frac{\partial G}{\partial y_1}\bigg|_{(0,0)}(y_1 - 0) + \ldots,$$

$$\dot{y}_1 = H(0, 0) + \frac{\partial H}{\partial y_0}\bigg|_{(0,0)}(y_0 - 0) + \frac{\partial H}{\partial y_1}\bigg|_{(0,0)}(y_1 - 0) + \ldots,$$

where

$$G(0,0) = 0, \qquad \frac{\partial G}{\partial y_0} = 0, \qquad \frac{\partial G}{\partial y_1} = 1,$$

$$H(0,0) = 0, \qquad \frac{\partial H}{\partial y_0} \neq 0, \qquad \frac{\partial H}{\partial y_1} = 0.$$

Hence, our resulting system of linearized equations can be written in matrix form as

$$\left\{ \begin{matrix} \dot{y}_0 \\ \dot{y}_1 \end{matrix} \right\} = \begin{bmatrix} 0 & 1 \\ \left. \dfrac{\partial H}{\partial y_0} \right|_{(0,0)} & 0 \end{bmatrix} \left\{ \begin{matrix} y_0 - 0 \\ y_1 - 0 \end{matrix} \right\}.$$

As noted earlier, we know that dynamic stability in the small implies a dynamic stability of the associated nonlinear system. Such (asymptotic) stability requires that the trace of this matrix be negative and its determinant be positive. Note that the trace (i.e., the sum of the diagonal entries) is identically zero, however; thus, an elastic aneurysm cannot be asymptotically stable in the absence of a viscous CSF, which is consistent with results from the numerical solution. In this case, therefore, we see that knowing some results from the theory of systems of first-order, linear, ordinary differential equations provides significant insight without the need to perform complex numerical computations. Indeed, one should always pursue analytical results when possible.

11.4 Viscoelasticity: QLV and Beyond

Whereas the two previous sections address interactions between a "solid" and a "fluid," we now return our attention to the behavior of a single material. Recall from Figure 1.4 of Chapter 1 that it is often convenient in continuum biomechanics to study separately the solidlike (biosolid mechanics) or fluidlike (biofluid mechanics) behavior that is exhibited by a material under conditions of interest; indeed, courses and textbooks are often designed along these separate lines. Reflecting back on Chapters 2–10, however, it should be evident that these divisions often are simply for convenience in particular classes of problems; they are not dictated by the physics per se. For example, our three primary governing differential equations—balance of mass, linear momentum, and energy—can be derived independent of the consideration of a solidlike or a fluidlike behavior. Likewise, our development of constitutive equations can follow a similar procedure (DEICE) regardless of the specific behavior, and it can result in similar relations (e.g., Hooke's law and the Navier–Poisson relation) whereby we relate the concept of stress to displacement gradients (strains) or velocity gradients (shear rates). Clearly, then, it should be very natural mathematically to consider together solidlike and fluidlike behaviors. Indeed, when we recognize that many materials—including glass over long time scales as well as the cytoplasm in a cell, the ligament in a joint, and even bone, to name but a few—simultaneously exhibit both a solidlike and a fluidlike behavior over conditions of interest, we should then pursue a more unified approach.

Traditionally, there have been two primary approaches to considering together solidlike and fluidlike behaviors. One is the so-called *theory of mixtures*, which traces its beginnings to Darcy and Fick in the mid-1800s, but received a more modern and rational treatment by Truesdell in 1957 [see the advanced text by Truesdell and Noll (1965, section 130)]. Briefly, Truesdell postulated that one could model the behaviors of mixtures of multiple constituents, including solids and fluids, by requiring that (1) each constituent obey its own balance and constitutive relations and (2) the overall mixture obey the classical balance relations for mass, momentum, and energy. Such an approach requires that one identify an additional class of constitutive equations, however, one that describes how the constituents exchange mass, momentum, and energy. For example, if a fluid flows through an otherwise stationary porous solid, the flow can induce a motion in the solid. Although we did not do so in Section 11.2, such interactions can be described directly by constitutive relations. V. Mow and colleagues were the first, in 1980, to apply the continuum theory of mixtures to biological tissues. They focused on the mechanical behavior of articular cartilage, which lines the contacting regions of bones in articulating joints and which consists primarily of a type II collagen, extensive proteoglycans, and mobile water. In fact, using the theory of mixtures, they show that the water carries much of the compressive load early in the gait cycle (Mow et al., 1990). Mixture theory continues to enable a much deeper understanding of cartilage mechanics than would have been possible if the solid mechanics had been studied alone. Since 1980, mixture theory has also been used to study the behavior of many different tissues and cells, and it remains as an important area of research, even in the emerging area of modeling growth and remodeling (Humphrey, 2003a). Because of the inherent mathematical complexity, however, the theory of mixtures is beyond the scope of an introductory text.

The second general approach that has been used to address combined solidlike and fluidlike behaviors is the *theory of viscoelasticity*. As the name implies, its goal is to describe behaviors that include both an elastic and a viscous character. Developed in the mid-1800s by savants such as L. Boltzmann (1844–1906), J.C. Maxwell (1831–1879), and Lord Kelvin (1824–1907), this theory developed along two related but separate lines. Boltzmann advocated the use of heredity integrals to describe the history of the mechanical behavior; in contrast, Maxwell and Kelvin advocated the use of differential models to account for the rate effects. Both approaches are useful in biomechanics and both can be developed for linear or nonlinear behaviors. Because the linear relations are much easier to address, however, we shall focus primarily on these.

11.4.1 Linearized Viscoelasticity

Bone and teeth are among the few tissues in the body that exhibit a linear relation between stress and strain over the range of physiologic strains. Although discussed thus far in terms of its solidlike behavior, bone exhibits viscoelastic characteristics. For example, the stress depends on the rate of deformation in addition to the amount of deformation (Fig. 11.12); in general, a viscoelastic response is stiffer at higher rates of deformation. Two other common characteristics that suggest a viscoelastic behavior are *creep* and *stress relaxation*. Creep is a time-dependent deformation under the action of a constant load (Fig. 11.13); stress relaxation is a time-dependent decrease in load at a constant deformation (Fig. 11.14). Toward a quantifi-

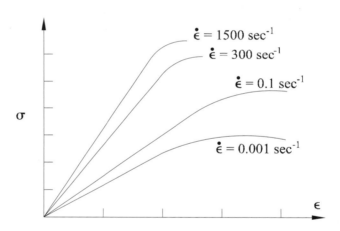

FIGURE 11.12 Strain-rate effects on the mechanical behavior of bone.

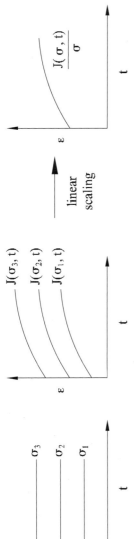

FIGURE 11.13 Characteristic responses to three different constant stresses during a creep test. By definition, *creep* is a continuing deformation (e.g., straining) in the presence of a constant force (or stress). In some cases, normalized creep responses reduce to a single characteristic curve.

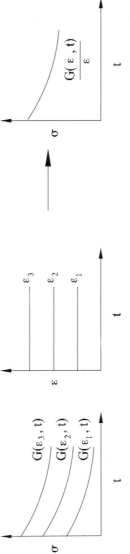

FIGURE 11.14 Similar to Figure 11.13 except for *stress relaxation* (i.e., the continuing decrease in stress in the presence of a constant strain).

cation of such behaviors, let us first consider a linearized, rate-type approach that may be used to describe the viscoelastic behavior of bone and other materials that exhibit a linear material behavior under small strains.

Maxwell Model

Maxwell, Kelvin, and others formulated their rate-type models in terms of 1-D mechanical analog models. Such models are only intended to simulate the macroscopic behavior; they are not designed to provide insight into the underlying molecular basis and they cannot be thought to be based on a general, rigorous mathematical foundation. Rather, analog models simply provide a means to motivate the forms of some constitutive relations.

The Maxwell model consists of a linear spring in series with a linear dashpot (Fig. 11.15). By linear, we mean that the force in the spring is related linearly to its extension and the force in the dashpot is related linearly to its rate of extension (Fig. 11.16). If we denote a scalar uniaxial force by f and the associated extension by δ (which is the current length x minus the original length x_o), this implies that $f_s = k\delta$ and $f_d = c\dot{\delta}$ where the subscripts s and d denote spring and dashpot, respectively, k is the stiffness of the spring, c is the viscosity of the dashpot, and the superimposed dot implies a time derivative. Remember that if a body is in equilibrium, then each of its parts are in equilibrium. Thus, a free-body diagram of the

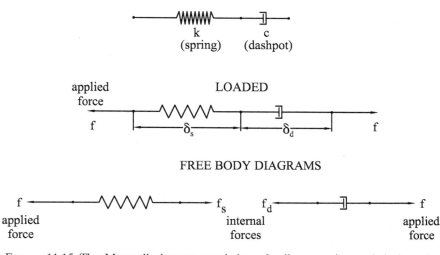

FIGURE 11.15 The Maxwell element consisting of a linear spring and dashpot in series in unloaded and loaded configurations. A free-body diagram reveals that the force (or stress) felt by the spring and the dashpot are the same.

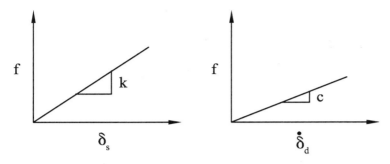

FIGURE 11.16 Mechanical behavior of a linear spring, with modulus k, and a linear dashpot, with viscosity c.

Maxwell element (Fig. 11.15) reveals that f_s and f_d each balance the total applied force f, whereas the total extension δ of the Maxwell element is the sum of the extensions of the spring and the dashpot. Exploiting the latter observation, taking a time derivative, and using the appropriate constitutive relations, we obtain

$$\delta = \delta_s + \sigma_d \rightarrow \dot{\delta} = \dot{\delta}_s + \dot{\delta}_d = \frac{\dot{f}}{k} + \frac{f}{c}. \tag{11.55}$$

Because this is an analog model, it is assumed that an associated 1-D relation in terms of the Cauchy stress σ and linearized strain ε is

$$\dot{\varepsilon} = \frac{\dot{\sigma}}{E} + \frac{\sigma}{\mu}, \tag{11.56}$$

where $\dot{\varepsilon}$ and $\dot{\sigma}$ are the strain rate and stress rate, respectively, E is the Young's modulus, and μ is the viscosity. This is the fundamental constitutive relation for the 1-D Maxwell model of linear viscoelasticity.

Consistent with Figure 11.14, a stress relaxation test can be defined by $\varepsilon = 0$ for $t < 0$ and $\varepsilon = \varepsilon_0$, a constant, for $t \geq 0$. Hence, $\dot{\varepsilon} = 0$ for all $t \geq 0$ and Eq. (11.56) reduces to

$$0 = \frac{\dot{\sigma}}{E} + \frac{\sigma}{\mu} \rightarrow \dot{\sigma} + \frac{E\sigma}{\mu} = 0 \quad \forall t \geq 0. \tag{11.57}$$

This homogeneous, first-order differential equation admits an exponential solution of the form

$$\sigma(t) = c_1 e^{-Et/\mu}, \tag{11.58}$$

which can be verified (do it) by direct substitution. The value of the constant c_1 can be determined by letting $\sigma(t = 0) = \sigma_0$, where the assumption of an instantaneous (elastic) response requires that $\sigma_0 = E\varepsilon_0$. Hence, $c_1 = \sigma_0$ and the stress relaxation of a Maxwell model is given by either

$$\sigma(t) = \sigma_0 e^{-Et/\mu} \quad \text{or} \quad \sigma(t) = \sigma_0 e^{-t/t_R}, \tag{11.59}$$

where $t_R \equiv \mu/E$ is the so-called *relaxation time*. The larger the value of t_R, either via a large μ or a small E, the slower the relaxation. Clearly, at $t = 0$, $\sigma(0) = E\varepsilon_0$, the instantaneous elastic response, whereas for $t \to \infty$, $\sigma(t) \to 0$ with an exponential decay. Because this model relaxes to zero stress, it is sometimes called a Maxwell fluid. It is, nonetheless, a model that accounts for combined solidlike and fluidlike behaviors in general.

In contrast, the creep test is defined by $\sigma = 0$ for $t < 0$ and $\sigma \equiv \sigma_0$ for all $t \geq 0$. Hence, $\dot{\sigma} = 0$ for $t \geq 0$ and $\dot{\varepsilon} = \sigma_0/\mu$, which appears to describe a Newtonian fluidlike behavior consistent with the above. With regard to the associated creep, however, we see that

$$\int \frac{d\varepsilon}{dt} dt = \int \frac{\sigma_0}{\mu} dt \to \varepsilon(t) = \frac{\sigma_0}{\mu} t + c_1, \tag{11.60}$$

where $\varepsilon = \varepsilon_0$ at $t = 0$, the instantaneous response. Hence, with $\varepsilon_0 = \sigma_0/E$, we have $c_1 = \varepsilon_0$ and

$$\varepsilon(t) = \frac{\sigma_0}{\mu} t + \frac{\sigma_0}{E}. \tag{11.61}$$

This suggests that the creep (i.e., lengthening over time) is linear in t. Experimental observations reveal that most "biosolids" that exhibit a viscoelastic character do not relax to zero stress and they do not creep linearly in time. There is clearly a need to consider other models.

Kelvin–Voigt Model

This mechanical analog model is defined by a linearly elastic spring and a dashpot in parallel, not in series (Fig. 11.17). A free-body diagram reveals that $f = f_s + f_d$ and $\delta = \delta_s = \delta_d$. Hence,

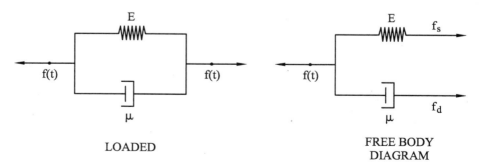

LOADED   FREE BODY
                         DIAGRAM

FIGURE 11.17 The Kelvin–Voigt element consisting of a linear spring and dashpot in parallel.

$$f = k_s \delta + c \dot{\delta}, \tag{11.62}$$

which suggests a 1-D analog model of the form

$$\sigma = E\varepsilon + \mu\dot{\varepsilon} \rightarrow \dot{\varepsilon} + \frac{E}{\mu}\varepsilon = \frac{\sigma}{\mu}. \tag{11.63}$$

In contrast to the Maxwell model, therefore, a stress relaxation test (with $\varepsilon = 0$ for $t < 0$ and $\varepsilon = \varepsilon_0$ for $t > 0$ with $\dot{\varepsilon} = 0$ for $t > 0$) leads to the simple relation that

$$\sigma(t) = E\varepsilon_0 \quad \forall t > 0, \tag{11.64}$$

which states that there is no relaxation for $t > 0$ (see Exercise 11.13 for behavior at $t = 0$). For creep, with $\sigma = \sigma_0$ for $t \geq 0$, we have a simple non-homogenous, first-order differential equation for $\varepsilon(t)$, namely

$$\dot{\varepsilon} + \frac{E}{\mu}\varepsilon = \frac{\sigma_0}{\mu}. \tag{11.65}$$

The homogenous solution is thus exponential:

$$\dot{\varepsilon} + \frac{E}{\mu}\varepsilon = 0 \rightarrow \varepsilon_h(t) = c_1 e^{c_2 t} \tag{11.66}$$

whereby $c_2 = -E/\mu$. The particular solution can be assumed to be constant, $\varepsilon_p(t) = c_3$ whereby $c_3 \equiv \sigma_0/E$. Hence, our solution is

$$\varepsilon(t) = c_1 e^{-Et/\mu} + \frac{\sigma_0}{E}. \tag{11.67}$$

Finally, the condition at $t = 0$ that $\varepsilon(0) = 0$ requires $c_1 = -\sigma_0/E$, thus yielding our creep response:

$$\varepsilon(t) = \frac{\sigma_0}{E}\left(1 - e^{-Et/\mu}\right) = \frac{\sigma_0}{E}\left(1 - e^{-t/t_c}\right), \tag{11.68}$$

where $t_c = \mu/E$ is called the *retardation time*. Note that at $t = 0$, $\varepsilon(0) = 0$, whereas at $t \rightarrow \infty$, $\varepsilon \rightarrow \sigma_0/E$. Hence, the creep is nonlinear, but bounded.

In summary, neither the Maxwell nor the Kelvin–Voigt model reflects commonly observed behavior in soft tissues. In particular, the linear and unbounded creep predicted by the Maxwell model is unrealistic, so, too, is the lack of relaxation for the Kelvin–Voigt model. Given that the instantaneous elasticity and the relaxation of the Maxwell model and the nonlinear, bounded creep of the Kelvin–Voigt model are realistic, one might consider combining these models with each other or perhaps combining them with other spring or dashpot elements. We shall consider such possibilities next. First, however, it proves useful to define two functions: $G(t)$, the *relaxation function* (during a stress relaxation test), and $J(t)$, the *creep function* (during a creep test). In particular, we let

$$G(t) = \frac{\sigma(t)}{\varepsilon_0}, \qquad J(t) = \frac{\varepsilon(t)}{\sigma_0}; \qquad (11.69)$$

hence, for the Maxwell model,

$$G(t) = Ee^{-Et/\mu}, \qquad J(t) = \frac{1}{\mu}t + \frac{1}{E}, \qquad (11.70)$$

whereas for the Kelvin–Voigt model,

$$G(t) = E, \qquad J(t) = \frac{1}{E}(1 - e^{-Et/\mu}). \qquad (11.71)$$

These will prove useful below.

Standard Viscoelastic Solid

It can be shown that combining a linear spring in parallel with a Maxwell element or combining a spring in series with a Kelvin–Voigt element (Fig. 11.18) yields the same differential equation. For example, if E_0 is the stiffness of the spring that is added in parallel to a Maxwell element, the governing equation can be shown to be

$$\dot{\sigma} + \frac{E}{\mu}\sigma = \frac{EE_0}{\mu}\varepsilon + (E + E_0)\dot{\varepsilon}. \qquad (11.72)$$

For stress relaxation, it can be shown that (Wineman and Rajagopal, 2000),

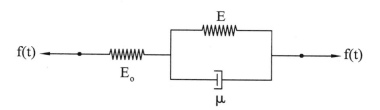

STANDARD MODELS

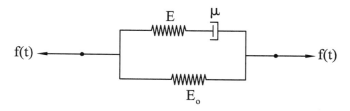

FIGURE 11.18 The standard element: either a spring in series with a Kelvin–Voigt element or a spring in parallel to a Maxwell element.

$$\sigma(t) = \varepsilon_0\left(E_0 + Ee^{-Et/\mu}\right), \tag{11.73}$$

and for creep,

$$\varepsilon(t) = \sigma_0\left[\frac{1}{E_0} + \left(\frac{1}{E_0 + E} - \frac{1}{E_0}\right)\exp\left(\frac{-E_0 E}{\mu(E + E_0)}t\right)\right]. \tag{11.74}$$

Consequently, the relaxation and creep functions can be written as

$$G(t) = E_0 + (E + E_0 - E_0)e^{-t/t_R} \equiv G_\infty + (G_0 - G_\infty)e^{-t/t_R} \tag{11.75}$$

where $G_\infty \equiv E_0$ and $G_0 \equiv E + E_0$, and

$$J(t) = J_\infty + (J_0 - J_\infty)e^{-t/t_c}, \tag{11.76}$$

where $J_0 = 1/G_0$ and $J_\infty = 1/G_\infty$ and the retardation time $t_c = G_0 t_R/G_\infty$. Figures 11.19 and 11.20 compare the associated characteristic responses. The Standard model is thus the simplest mechanical analog model that gives

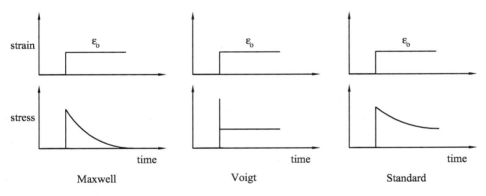

FIGURE 11.19 Characteristic stress relaxation responses of the Maxwell, Kelvin–Voigt, and Standard element consistent with the derivations in the text.

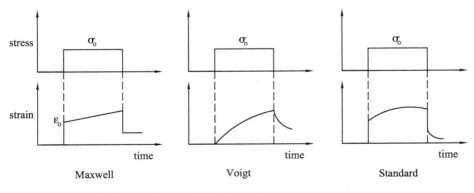

FIGURE 11.20 Similar to Figure 11.19 except for creep.

physically realistic predictions for viscoelastic "solids," including instantaneous elasticity, a nonlinear but bounded creep, and a stress relaxation that tends to a nonzero equilibrium stress.

Example 11.4 Derive the governing equation for a three-element viscoelastic fluid defined by a dashpot (viscosity μ_0) in series with a Kelvin–Voigt element (viscosity μ_v and modulus E_v).

Solution: Let the total stress and strain of the three-element model be σ and ε, respectively. By equilibrium, the stress in the single dashpot σ_0 must balance the stress in the Kelvin–Voigt element σ_v, and both must equal σ. Note, therefore, that

$$\sigma_0 = \mu_0 \dot{\varepsilon}_0, \qquad \sigma_v = E_v \varepsilon_v + \mu_v \dot{\varepsilon}_v.$$

Moreover, the total strain $\varepsilon = \varepsilon_0 + \varepsilon_v$, where the strains in the spring and dashpot of the Kelvin–Voigt element are the same. Hence, with $\sigma = \sigma_v$ and $\varepsilon_v = \varepsilon - \varepsilon_0$, we have

$$\sigma = E_v(\varepsilon - \varepsilon_0) + \mu_v(\dot{\varepsilon} - \dot{\varepsilon}_0),$$

whereby

$$\dot{\sigma} = E_v(\dot{\varepsilon} - \dot{\varepsilon}_0) + \mu_v(\ddot{\varepsilon} - \ddot{\varepsilon}_0).$$

Now, note that $\dot{\varepsilon}_0 = \sigma_0/\mu_0$ and thus $\ddot{\varepsilon}_0 = \dot{\sigma}_0/\mu_0 \equiv \dot{\sigma}/\mu_0$. Hence, we have

$$\dot{\sigma} = E_v\left(\dot{\varepsilon} - \frac{\sigma_0}{\mu_0}\right) + \mu_v\left(\ddot{\varepsilon} - \frac{\dot{\sigma}_0}{\mu_0}\right)$$

or

$$\dot{\sigma}\left(1 + \frac{\mu_v}{\mu_0}\right) + \sigma\left(\frac{E_v}{\mu_0}\right) = E_v\dot{\varepsilon} + \mu_v\ddot{\varepsilon}$$

or

$$\sigma + \dot{\sigma}\left(\frac{\mu_0 + \mu_v}{E_v}\right) = \mu_0\dot{\varepsilon} + \frac{\mu_v\mu_0}{E_v}\ddot{\varepsilon}.$$

Compare this to the three-element (standard) viscoelastic solid.

Boltzmann Model

As noted earlier, Boltzmann advocated a different approach to modeling viscoelastic behavior; he focused on heredity integrals to account for the history of the response, and, in particular, the creep and stress relaxation.

For example, standard linear heredity integrals are

$$\sigma(t) = \int_{-\infty}^{t} G(t-s)\frac{d\varepsilon}{ds}\,ds \qquad (11.77)$$

and

$$\varepsilon(t) = \int_{-\infty}^{t} J(t-s)\frac{d\sigma}{ds}\,ds. \qquad (11.78)$$

It proves useful to evaluate these integrals over two intervals: from $-\infty$ to 0 and from 0 to time t, where

$$\int_{-\infty}^{t}(\ )ds = \int_{-\infty}^{0}(\ )ds + \int_{0}^{t}(\ )ds. \qquad (11.79)$$

Hence, our relations for stress and strain histories can be shown to be

$$\sigma(t) = \varepsilon(0)G(t) + \int_{0}^{t} G(t-s)\frac{d\varepsilon}{ds}\,ds \qquad (11.80)$$

and

$$\varepsilon(t) = \sigma(0)J(t) + \int_{0}^{t} J(t-s)\frac{d\sigma}{ds}\,ds. \qquad (11.81)$$

In many cases, we have zero stress and strain up to and at the time $t = 0$; hence, these equations simplify further. We shall consider such cases below.

In contrast to the aforementioned stress relaxation and creep tests, which are very useful in evaluating viscoelastic responses, let us consider here a class of *sinusoidal straining* tests. Such tests are particularly useful for evaluating "short-time" responses in contrast to the "long-time" responses in creep and relaxation tests. Hence, consider a periodic strain history of the form

$$\varepsilon(t) = \varepsilon_A \sin \omega t \qquad (11.82)$$

where ω is the fundamental (circular) frequency of the test (with $2\pi f = \omega$) and ε_A is the amplitude of the small strain. In order to use Eq. (11.80), with $\varepsilon(0) = 0$, it proves convenient to consider a change of variables: let $t - s = \tau$ whereby $s = t - \tau$ at any fixed time t. Hence,

$$\frac{d}{ds}(\ ) = \frac{d}{d\tau}(\ )\frac{d\tau}{ds} = -\frac{d}{d\tau}(\ ), \qquad ds = -d\tau. \qquad (11.83)$$

Equation (11.80), with $\varepsilon(0) = 0$ and $\tau \in [0,\infty)$, can thus be written as

$$\sigma(t) = \int_{0}^{\infty} G(\tau)\frac{d\varepsilon(t-\tau)}{d\tau}\,d\tau = \int_{0}^{\infty} G(\tau)\dot{\varepsilon}(t-\tau)\,d\tau. \qquad (11.84)$$

Given that $\dot{\varepsilon}(t) = \varepsilon_A \omega \cos \omega t$, we have

$$\sigma(t) = \int_{0}^{\infty} G(\tau)\varepsilon_A \omega \cos \omega(t-\tau)\,d\tau, \qquad (11.85)$$

or by using the standard trigonometric identity that $\cos(\alpha \pm \beta) = \cos \alpha \cos \beta \mp \sin \alpha \sin \beta$, we have

$$\sigma(t) = \int_0^\infty G(\tau)\varepsilon_A \omega(\cos \omega t \cos \omega\tau + \sin \omega t \sin \omega\tau)d\tau, \qquad (11.86)$$

or

$$\sigma(t) = (\varepsilon_A \cos \omega t)\omega \int_0^\infty G(\tau)\cos \omega\tau\, d\tau + (\varepsilon_A \sin \omega t)\omega \int_0^\infty G(\tau)\sin \omega\tau\, d\tau. \quad (11.87)$$

If we now denote

$$G_1(\omega) \equiv \omega \int_0^\infty G(\tau)\sin(\omega\tau)d\tau,$$
$$G_2(\omega) \equiv \omega \int_0^\infty G(\tau)\cos(\omega\tau)d\tau, \qquad (11.88)$$

then

$$\sigma(t) = \varepsilon_A[G_1(\omega)\sin \omega t + G_2(\omega)\cos \omega t], \qquad (11.89)$$

where $G_1(\omega)$ and $G_2(\omega)$ are called the *storage modulus* and the *loss modulus*, respectively.

At this juncture, it is instructive to note that if subjected to a strain history of the form $\varepsilon(t) = \varepsilon_A \sin \omega t$, then we expect a viscoelastic material to respond out of phase, such as $\sigma(t) = \sigma_A \sin(\omega t + \phi)$. Conversely, if the response is purely elastic, we expect $\phi = 0$, whereas if the response is purely viscous, we expect $\phi = \pi/2$. Computation of the value of ϕ in general can thus be revealing. Note, therefore, that given this expression for $\sigma(t)$, we have

$$\sigma(t) = \sigma_A(\sin \omega t \cos \phi + \cos \omega t \sin \phi). \qquad (11.90)$$

Comparison of Eqs. (11.89) and (11.90) thus reveals that

$$\varepsilon_A G_1(\omega) = \sigma_A \cos \phi \rightarrow G_1(\omega) = \frac{\sigma_A}{\varepsilon_A}\cos \phi,$$

$$\varepsilon_A G_2(\omega) = \sigma_A \sin \phi \rightarrow G_2(\omega) = \frac{\sigma_A}{\varepsilon_A}\sin \phi, \qquad (11.91)$$

$$\tan \phi = G_2(\omega)/G_1(\omega).$$

Finally, note that it can be useful to use a complex variable ($z^* = Ax + iy$, with $i = \sqrt{-1}$) representation for these sines and cosines. It can be shown that if $\varepsilon^*(t) = \varepsilon_A e^{i\omega t}$ and $\sigma^*(t) = \sigma_A e^{i(\omega t+\phi)}$, then a *complex modulus*

$$G^* = \frac{\sigma^*}{\varepsilon^*} = G_1 + iG_2. \qquad (11.92)$$

In this case, it is evident that G_1 is the ratio of that part of the stress that is in-phase with the strain to the strain itself, whereas G_2 is the ratio of that part of the stress that is $\pi/2$ out-of-phase with the strain to the strain itself.

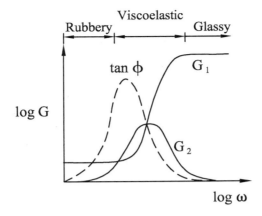

FIGURE 11.21 Characteristic complex moduli for viscoelastic behaviors in rubbery and glassy regimes.

For a purely elastic (i.e., Hookean) response, therefore, $G_1 = 1$ and $G_2 = 0$, with $\phi = 0$, and for a purely viscous (i.e., Newtonian) response, $G_1 = 0$ and $G_2 = 1$, with $\phi = \pi/2$. Computation of these moduli can thereby enable one to assess the "degree" of the viscoelastic response [i.e., its deviation from a purely elastic or a purely viscous response (cf. Fig. 11.21)].

Example 11.5 Demonstrate why G_2 is called the loss modulus.

Solution: Consider the energy dissipated during a simple cycle of loading:

$$\xi = \int \sigma \, d\varepsilon \equiv \int_0^{2\pi/\omega} \sigma \frac{d\varepsilon}{dt} \, dt.$$

If we let, consistent with Exercise 11.16,

$$\varepsilon(t) = \varepsilon_A \sin \omega t, \qquad \sigma(t) = \sigma_A \sin(\omega t + \phi),$$

then

$$\xi = \int_0^{2\pi/\omega} \sigma_A (\sin \omega t \cos \phi + \cos \omega t \sin \phi) \varepsilon_A \omega \cos \omega t \, dt$$

$$= (\sigma_A \cos \phi) \varepsilon_A \omega \int_0^{2\pi/\omega} \sin \omega t \cos \omega t \, dt + (\sigma_A \sin \phi) \varepsilon_A \omega \int_0^{2\pi/\omega} \cos^2 \omega t \, dt,$$

which, from integral tables, yields

$$\xi = \left(\frac{\sigma_A \cos \phi}{\varepsilon_A} \right) \varepsilon_A^2 \omega \left(\frac{1}{2\omega} \sin^2 \omega t \Big|_0^{2\pi/\omega} \right) + \left(\frac{\sigma_A \sin \phi}{\varepsilon_A} \right) \varepsilon_A^2 \omega \left(\frac{1}{2} t + \frac{1}{4\omega} \sin 2\omega t \Big|_0^{2\pi/\omega} \right)$$

or, from Exercise 11.16,

$$\xi = G_1\varepsilon_A^2\left(\frac{1}{2}\right)(0) + G_2\varepsilon_A^2\omega\left[\left(\frac{1}{2}\right)\left(\frac{2\pi}{\omega}\right)\right] = \pi G_2\varepsilon_A^2,$$

thereby revealing that the dissipation (i.e., loss) in this linear model is given entirely by the "loss modulus" G_2.

In summary, this has been but a brief introduction to a few aspects of *linear* viscoelasticity. For more on this topic, the reader is encouraged to consult Ferry (1980) or Wineman and Rajagopal (2000).

11.4.2 Quasilinear Viscoelasticity

Because of the nonlinear material behavior exhibited by most soft tissues, linearized relations for viscoelasticity do not apply. In an attempt to account for the nonlinear "elastic" behavior while preserving the mathematical machinery of linear viscoelasticity, Fung proposed a so-called *quasilinear theory of viscoelasticity*, often referred to as QLV.

Briefly, following Boltzmann's approach, Fung suggested that, in one dimension, one should consider the following relationship for the first Piola–Kirchhoff stress Σ_{11} and stretch λ (Fung, 1990):

$$\Sigma_{11}(t) = \int_{-\infty}^{t} G(t-\tau)\frac{\partial \Sigma_{11}^e}{\partial \lambda}\frac{d\lambda}{d\tau}\,d\tau, \tag{11.93}$$

where G is a reduced relaxation function, with $G(0) = 1$, $\Sigma_{11}^e(\lambda)$ is a nonlinearly elastic response function (in terms of the first Piola–Kirchhoff stress), and λ is an axial stretch ratio. Fung noted that it is common to express the relaxation function in terms of a finite sum of exponential decay functions. Noting problems common to such approaches, including finding $G(\infty)$, Fung further noted that the observed relative insensitivity of the hysteresis during cyclic loading of many soft tissues suggests the need for a continuous relaxation spectrum. The literature reveals many subsequent applications of Fung's QLV theory. We will leave it as an exercise to explore such applications, however.

11.4.3 Need for Nonlinear Theories

This subsection could simply be entitled "Beyond QLV." Despite the success of QLV in fitting data from various experiments, numerous investigators have shown that QLV is not sufficiently general to describe many of the complicated behaviors exhibited by soft tissues, including a strain-dependent relaxation and fundamentally different short-term and long-term viscoelastic responses. Building upon the many fundamental advances in nonlinear viscoelasticity since World War II [by Green, Rivlin, Pipkin, and Bernstein et al., among others; see Ferry (1980)], various approaches have been proposed. These include the single integral finite strain model of Johnson et al. (1996), the combined differential–integral model of Pioletti and Rakotomanana (2000), the generalized elastic-Maxwell model of Holzapfel and Gasser (2001) and Holzapfel et al. (2002), and the modified superposition model of Provenzano et al. (2002), to name but a few. Like mixture theory, however, nonlinear viscoelasticity is not an introductory topic; thus, the reader is simply encouraged to seek advanced courses on this topic. When doing so, remember that science is but relative truth; thus, each theory and approach is limited. Noting any restrictions or limitations, particularly those in small strain theories, is of paramount importance as we seek to use biomedical engineering design and analysis to improve health care.

11.5 Lubrication of Articulating Joints

11.5.1 Biological Motivation

Human diarthroidal joints can function well for seven or more decades despite the relatively high loads and at times low speeds of relative motion that they experience. For example, tissue stresses can reach $18\,\text{MPa}$ in the knee during just normal walking. Recall from Chapter 7 that synovial fluid serves as a tremendous lubricant that reduces wear of cartilage in articulating joints such as the knee and hip; the coefficient of friction in joints is an amazing 0.003–0.03, much less than values attained by man-made lubricants. The mechanics of lubrication is a well-developed area of study in mechanical engineering, called *tribology*, and one may be tempted to apply results from tribology directly to the analysis of knee or hip mechanics. Indeed, as recently as the 1930s, it was suggested that the efficiency of healthy diarthroidal joints could be explained via the theory of hydrodynamic lubrication; that is, it was thought that loads are transferred between articulating surfaces via a thin, pressurized layer of fluid lubricant. Let us briefly consider a simple example of such a theory of lubrication.

11.5.2 Hydrodynamic Lubrication

Consider the "hydrodynamic slider bearing" shown in Figure 11.22. Let us assume that the fluid behavior is Newtonian and that the flow is incompressible, steady, and laminar. It would seem reasonable, therefore, to assume the velocity field of the fluid to be of the form $v = v_x(x, y)\hat{\boldsymbol{i}} + v_y(x, y)\hat{\boldsymbol{j}}$ in the absence of gravity. If we assume further that the Reynolds' number is small (Re < 1), then viscous effects dominate inertial effects. Moreover, if we assume that the gap distance $h(x) \ll L$, then the

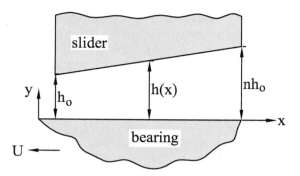

FIGURE 11.22 The slider bearing, which supports a compressive load W due to hydrodynamic lubrication. Let the length of the slider be L.

x-direction momentum equation is most important and changes in v_x with respect to y may be considered to be less than those with respect to x. Hence, the Navier–Stokes equation in x [Eq. (8.34)] reduces to

$$\frac{dp}{dx} = \mu \frac{d^2 v_x}{dy^2} \tag{11.94}$$

and our problem is similar to that in Chapter 9 for flow between parallel flat plates. In particular, our general solution is

$$v_x(y) = \frac{1}{\mu}\left(\frac{dp}{dx}\right)\frac{y^2}{2} + c_1 y + c_2. \tag{11.95}$$

Now, for boundary conditions. Here, let us consider the simplified case of a step-slider (Fig. 11.23). Hence, we have

$$v_x(0) = -U, \qquad v_x(h) = 0, \qquad 0 \le x < L - bL, \tag{11.96}$$

$$v_x(0) = -U, \qquad v_x(nh) = 0, \qquad L - bL < x \le L, \tag{11.97}$$

and Eq. 11.95 becomes

$$v_x(y) = \frac{1}{2\mu}\left(\frac{dp}{dx}\right)(y^2 - nhy) + U\left(\frac{y}{nh} - 1\right) \tag{11.98}$$

for $n = 1$ or n, in general. Hence, the volumetric flow rate is

$$Q = \int_0^w \int_0^h v_x(y)\, dy\, dz = -\frac{n^3 h^3 w}{12\mu}\left(\frac{dp}{dx}\right) - \frac{Unhw}{2} \tag{11.99}$$

for $n = 1$ or n. Because Q must be a constant

$$Q_1 = -\frac{h^3 w}{12\mu}\left(\frac{dp}{dx}\right)_1 - \frac{Uhw}{2} = -\frac{n^3 h^3 w}{12\mu}\left(\frac{dp}{dx}\right)_2 - \frac{Unhw}{2} = Q_2, \tag{11.100}$$

which is to say, (dp/dx) must be a constant over $0 \le x < L - bL$ and also over $L - bL < x \le L$; that is, p must be linear in each of these two domains (Fig. 11.24). We let

$$\left(\frac{dp}{dx}\right)_1 = \frac{p_m - p_o}{L(1-b)}, \qquad \left(\frac{dp}{dx}\right)_2 = \frac{p_o - p_m}{bL} \qquad (11.101)$$

where p_m is the value of the pressure at $x = L(1-b)$ and p_o is the uniform pressure outside the slider on both the right and the left ends. Let us now solve for $p_m - p_o$ using the constraint that $Q_1 = Q_2$. We find that

$$-\frac{h^3 w}{12\mu}\left(\frac{p_m - p_o}{L(1-b)}\right) - \frac{Uhw}{2} = -\frac{n^3 h^3 w}{12\mu}\left(\frac{p_o - p_m}{bL}\right) - \frac{Unhw}{2}, \qquad (11.102)$$

or

$$(p_m - p_o)\left(\frac{h^3 w}{12\mu L}\right)\left(\frac{1}{1-b} + \frac{n^3}{b}\right) = \frac{Uhw}{2}(n-1). \qquad (11.103)$$

Hence,

$$p_m - p_o = \frac{6\mu U L w b}{h^2}\left(\frac{(n-1)(1-b)}{b + n^3(1-b)}\right). \qquad (11.104)$$

The associated total vertical load W that the slider can support is

$$W = p_o(LW) + \frac{1}{2}(L - bL)(p_m - p_o)w + \frac{1}{2}(bL)(p_m - p_o)w$$

$$= p_o(LW) + \frac{1}{2}Lw(p_m - p_o) = W_{static} + W_{dyn}. \qquad (11.105)$$

Defining the dynamic load-carrying capability as W_{dyn}, we can define a nondimensional *load coefficient* C_w as

$$C_w = W_{dyn}\left(\frac{h^2}{U\mu L^2 w^2}\right) = \frac{3b(n-1)(1-b)}{b + n^3(1-b)}. \qquad (11.106)$$

This formula can be used by the design engineer to determine preferred values of b and n to maximize C_w given any other design constraints. Note, however, that if $n = 1$ or $b = 0$, the step disappears and the dynamic load-carrying capacity is lost. This is the case in parallel plates, or the so-called Couette flow. Finally, note that a frictional drag coefficient C_f can also be defined as the total drag force F_D divided by the total vertical load-bearing capacity W. It can be shown that

$$C_f = \frac{F_D}{W} = \frac{h(n-1)}{L} + \frac{h[n(1-b) + b][b + n^3(1-b)]}{3Lbn(n-1)(1-b)}. \qquad (11.107)$$

Obviously, we seek to minimize C_f while maximizing C_w.

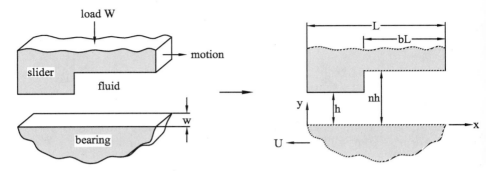

FIGURE 11.23 A step-slider bearing showing important loads and dimensions.

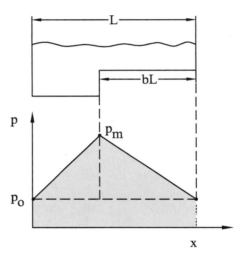

FIGURE 11.24 Computed pressure distribution for the step-slider bearing shown in Figure 11.23.

Whereas a step-slider (Fig. 11.23) allows us to begin to appreciate some aspects of hydrodynamic lubrication, such a sharp geometric discontinuity is clearly unrealistic in the case of an articulating joint, even a prosthetic one. Hence, one would be more likely to consider the *inclined slider* (Fig. 11.22). The solution of this problem is similar but more complex because $h = h(x)$. It can be shown, however, that if $h(x)$ changes linearly, then

$$p(x) - p_o = \frac{\mu U L w}{h_o^2}\left(\frac{6(n-1)(1-x/L)(x/L)}{(n+1)[n+(n-1)(x/L)]^2}\right), \quad (11.108)$$

where $h(x) = h_o[1 + (n-1)(x/L)]$. For $L \gg h_o$, one finds a parabolic distribution of pressure with load and frictional load coefficients

$$C_w = \frac{6}{(n-1)^2}\left(\ln n - \frac{2(n-1)}{n+1}\right), \tag{11.109}$$

$$C_f = \frac{h_o(n-1)}{L}\left(\frac{\ln n}{6\ln n - 12(n-1)/n+1} + \frac{1}{2}\right). \tag{11.110}$$

Despite the simplicity of this and similar analyses, mechanical engineers can describe well and thus design many efficient bearings. Yet, this conventional hydrodynamic lubrication theory reveals that a continuous, high-speed operation (i.e., speed U) is needed between the two opposing solid surfaces to maintain a sufficient pressure and thus thickness of the lubricant. High-speed relative motion is not a characteristic of the articulating joint, however; thus, the mechanics of diarthroidal joints is more complex. We are reminded, therefore, that biomechanics is not just mechanics applied to biology; many times, it must include the development or extension of mechanics to solve a biologically important problem.

In the 1960s, attention in orthopedic biomechanics turned toward coupled theories that included both the flow of the lubricant and the deformation of the solid load-bearing surfaces. Such *elastohydrodynamic* theories were based on the assumption that compression of the cartilage caused it to spread out and thereby increase the surface area over which the load was applied. It was argued that lower stresses would induce less wear. Nevertheless, such theories could not explain many empirical findings and they still suggested the need for relatively high speeds of relative motion between the load-bearing surfaces; there remained a need for more appropriate analyses. One such idea was called the *squeeze-film* theory. Recall from Figure 7.14 that synovial fluid exhibits a non-Newtonian character. Basically, it was suggested that a hydrostatic pressure would be generated in the synovial fluid when the two load-bearing solid surfaces were brought closer together and that this was due in part to the high apparent viscosity of the synovial fluid at low shear rates and at long times of applied stress. Nevertheless, it seemed that an important aspect of the biophysics was still being neglected.

From the 1960s to the 1980s, others began to consider the "porous" nature of the cartilage. One suggestion was that when the cartilage was loaded, interstitial fluid would be exuded and this fluid would aid in the lubrication. Such theories were referred to as *weeping lubrication*, which allowed for a load-dependent, self-pressurizing mechanism. Conversely, others suggested a so-called *boosted lubrication* whereby the water portion of the synovial fluid was forced into the cartilage, thus leaving a higher-viscosity, strongly non-Newtonian solution consisting primarily of hyaluronic acid in the joint space. Consideration of such scenarios (i.e., cases of fluids diffusing within solids) led to the widespread use of the aforementioned theory of mixtures in cartilage mechanics, which we discuss briefly in the following section.

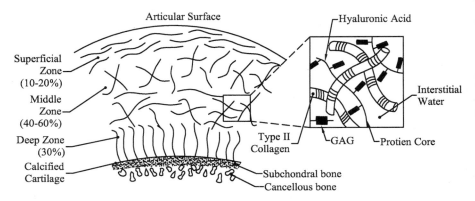

FIGURE 11.25 Schema of articular cartilage showing the strong nonhomogeneity in composition and its solid–fluid mixture constitution, which is dominated by water, proteoglycans, and type II collagen. Recall, for example, that tendons, ligaments, and bones (other structurally significant members of an articulating joint) consist largely of type I collagen, not type II. Thus, cartilage possesses a unique structure consistent with its unique function.

11.5.3 Need for a Mixture Theory

As noted earlier, in 1980, Mow and colleagues proposed the use of the theory of mixtures to account for the combined fluid and solidlike behavior exhibited by articular cartilage. Here is a very brief synopsis, which comes from Mow et al. (1990). Cartilage consists of ~50–73% type II collagen, 15–30% proteoglycan, and ~5% chondrocytes by dry weight; by wet weight, cartilage consists of 58–78% water. These constituents are organized in a highly complex, nonhomogeneous fashion (Fig. 11.25). In particular, the collagen is oriented differently in the superficial, middle, and deep "zones," and the collagen and proteoglycans form a complex composite microstructure. The collagen is packed more densely in the superficial and deep zones; thus, most of the proteoglycans are located in the middle zone. The proteoglycans contribute significantly to the compressive and swelling properties. Recall from Chapter 1 that proteoglycans consist of a protein core on which are bonded various glycosaminoglycans (GAGs); cartilage has abundant chondroitin sulfate and some keratan sulfate. Mow and colleagues idealized this structure as a porous matrix (collagen + proteoglycans, or solid phase) that is swollen with water (a liquid phase), both bound and unbound. In other words, there was no attempt to model directly the complex interactions (physical entanglements, electrostatic bonds, and excluded volume) between proteoglycans and collagen; attention was focused on the volume-averaged mean composite behavior of the solid portion of the tissue. Of particular importance, however, is the ability of applied loads to cause the interstitial fluid to redistribute within the cartilage or to flow out of or into the cartilage. One of the key parameters describing such a behavior is thus the permeability of the tissue, which was measured via confined compression tests wherein a sample was placed above a rigid porous filter and loaded from above. The permeability was found to decrease nonlinearly with applied compressive strain, typical values being on the order of $(0.5–2) \times 10^{-15} \, \text{m}^4/\text{N s}$. Other important material parameters included the moduli for the solid (Young's modulus and Poisson's ratio because of the assumption of a linear elastic behavior by the solid portion) and diffusive drag coefficient. The latter was a measure of the ease that the fluid could diffuse through the solid and thus was inversely related to the permeability. Indeed, rather than modeling the viscosity of the interstitial fluid directly, the diffusive drag coefficient was used to capture this dissipative characteristic. Viscoelastic effects can thus be modeled, in part, through solid–fluid interactions even though more recent studies suggest the need to account for the intrinsic viscoelastic behavior of the "solid." Although the basic formulation is straightforward, solution of even simple initial and boundary value problems is nontrivial; in particular, it is very difficult to identify appropriate boundary conditions between a non-Newtonian (synovial) fluid and a solid–fluid mixture (cartilage). Thus, the reader is referred to advanced texts. The take-home message here is that, again, we see the importance of modeling coupled phenomena in soft tissue biomechanics.

11.6 Thermomechanics, Electromechanics, and Chemomechanics

Continuum mechanics rests upon three fundament balance relations: mass, linear momentum, and energy. With the exception of the development of the pipe-flow equation in Chapter 10, we have focused on the balance of mass and linear momentum. One reason for this is that the energy equation is particularly important in nonisothermal problems, but the body tends to regulate its temperature within a narrow range ($37 \pm 2°C$). Nevertheless, there are many cases wherein *biothermomechanics* is important (Humphrey, 2003b). In particular, advances in laser, microwave, radio-frequency, and other technologies has motivated the widespread use of supraphysiologic temperatures to treat a wide variety of diseases and injuries. Examples include the treatment of joint laxity (e.g., severe sprains), visual problems (e.g., LASIK surgery and secondary cataracts), skin defects (e.g., port wine stains and melanomas), chronic pain, cardiovascular disorders (e.g., atrial fibrillation and obstructive atherosclerotic lesions), gynecological disorders (e.g., endometriosis), prostate problems (e.g., benign prostatic hyperplasia as well as malignancies), and so on. Such treatments denature proteins and kill cells thereby affecting the biomechanical properties of the treated tissues as well as the associated mechanobiology; the latter being particularly important with regard to the post-treatment healing response. For example, if a thermal treatment alters the biomechanical properties, and perhaps the geometry, of a tissue or organ, then different stresses will result due to normal in vivo loads. An altered state of stress, in turn, will alter cellular activity via mechanotransduction mechanisms, including possible changes in cell migration, proliferation, apoptosis, and the production and removal of the matrix. Hence, not only is the response to the initial (thermal) injury important, so too is the biomechanics. There is, therefore, a pressing need to study coupled thermomechanical problems. For example, soft tissues will often change their extensibility and hydration in response to a thermal damage. Hence, there is a need to address the fluidlike, solidlike, and thermal behaviors together. Again, however, biothermomechanics is inherently complex and nonlinear, and detailed design and analysis are beyond the scope of an introductory text.

Nonetheless, to illustrate the importance of combining thermal and mechanical analyses, let us consider a simple example. Recall from Chapters 2–4 that metallic implants are commonly used in joint replacements, particularly in the hip and knee. Moreover, although there are multiple ways of securing such devices within the host tissue, one commonly used method is to "cement" the device in place. Bone cements, such as (poly)methylmethacrylate (or PMMA), are injected into the space between the bone and prosthesis and allowed to polymerize. This process involves an exothermic reaction, which is to say, one that gives off "heat." A concern, therefore, is whether the transfer of heat from the curing cement to the bone might thermally damage the bone cells, which, in turn, would compromise the bone–implant interface. As a first approximation of this problem, one could consider a structure consisting of three concentric, circular layers: the inner metallic prosthesis, the layer of PMMA, and the outer layer of cortical bone. The bioheat transfer problem could be assumed to be axisymmetric; hence, from Eq. (A10.12) of Chapter 10, the basic heat transfer equation could be written in terms of temperature T as

$$\rho c_v \frac{\partial T}{\partial t} = k \left[\frac{1}{r} \frac{\partial}{\partial r} \left(r \frac{\partial T}{\partial r} \right) \right] + \rho q_s \qquad (11.111)$$

for each layer, where c_v and k are the specific heat and thermal conductivity, respectively, of each material. Recall, too, that q_s is a so-called volumetric heat source or sink. One of the important realizations in biomechanics is that flowing blood is capable of convecting away significant amounts of heat. In 1948, H. Pennes suggested that one could model this convective loss through the q_s term (i.e., as a heat sink). Determination of a reasonable constitutive relation for q_s thus requires that we couple the analyses of heat transfer and blood flow. Similarly, recall from Eqs. (2.69) of Chapter 2 that the strains in a solid depend on both the state of stress and the temperature. Hence, changes in temperature within the three materials (prosthesis, cement, and bone) will change their states of stress and strain. Hence, the solid mechanics should be coupled with that of the analysis of the heat transfer. Without going into mathematical detail, therefore, we see that even in a simple example, solid mechanics, fluid mechanics, and heat transfer should be addressed in coupled fashion to design well a common clinical procedure.

With the exception of a brief discussion in Chapter 6, we have not addressed the mechanics of muscle. There are a number of reasons for this, not the least of which is that our understanding of muscle mechanics remains inadequate. Nevertheless, it should be clear that the biomechanics of athletic performance, rehabilitation, cardiac health, the vasculature, and many other areas depends primarily on an understanding of muscle mechanics. In particular, heart disease remains a leading cause of morbidity and mortality, and biomechanics has great potential to impact treatment in many ways—from the design of artificial hearts and assist devices to understanding thermal treatments such as transmyocardial revascularization. Cardiac function (i.e., the filling and ejecting of blood from the heart to the rest of the body) depends on a strong *electromechanical coupling*; that is, the propagation of electrical signals dictates the order of contraction of cardiac muscle fibers, which, in turn, dictates cardiac output. There is, therefore, a pressing need to couple our study of the electrical activity and mechanical performance of the heart. Being an advanced topic, however, we merely refer the reader to the literature, including the excellent books by Glass et al. (1991) and Panfilov and Holden (1997).

Finally, although not emphasized earlier, it should be clear from Chapters 2–10 that the mechanics is coupled strongly with the biochemistry in many cases. For example, with regard to biological growth and remodeling, A. Turing recognized in the 1950s that one must not only quantify the effects of stress and strain on tissue response (e.g., Wolff's law of bone remodeling), one must also account for the rate at which various molecules (e.g., morphogens) are produced by the cells and the diffusion of such molecules throughout the tissue. Cellular production and molecular diffusion are each affected directly by the mechanics (e.g., pressure gradients); thus, such studies must address the *chemomechanical* coupling. Indeed, recent data suggest that even the contractility of smooth muscle cells in early hypertension may be governed, in part, by a pressure-induced increase in the conversion of G-actin to F-actin in the cytoplasm (i.e., a pressure-induced chemical reaction called polymerization).

Similarly, we mentioned earlier that there is a need to couple mechanics and thermodynamics in studies of the use of heat to treat disease and injury. One of the most commonly used equations in biothermomechanics is that of S. Arrhenius, which states that the rate k of a chemical reaction (e.g., denaturation of a protein) depends on the temperature given two material parameters, the activation energy E_a and gas constant R, namely

$$k(T) = A \exp\left(-\frac{E_a}{RT}\right). \tag{11.112}$$

Recent data suggest, however, that the rate of thermal denaturation also depends on the state of stress in the tissue: A higher stress tends to delay the denaturation at a given temperature. It should be noted, therefore, that the parameter A in Arrhenius's relation can be shown to be related to an activation entropy. It is well known that tissue elasticity is primarily entropic (i.e., determined primarily by load-induced changes in the conformations of molecules), not energetic (as in metals). It is clear, therefore, that the chemical reactions responsible for the thermal denaturation of a protein or the thermal death of a cell depend on the thermodynamics and the solid mechanics.

In summary, most real-world problems necessitate that we address coupled problems—solid–fluid, thermomechanical, electromechanical, chemomechanical, and, indeed, thermomechanochemical and so on. Toward this end, we should first learn well the basics in each discipline, knowing that important contributions will come from appropriate synthesis. Interdisciplinary and multidisciplinary research teams thus hold great promise and should be pursued vigorously.

Exercises

11.1   Repeat the nondimensional analysis of Section 11.2 using as length, time, and mass scales:

$$L_s = A, \qquad T_s = \frac{A^2 H}{Q}, \qquad M_s = \rho A^2 H.$$

Compare the results with those in Section 11.2

11.2   Repeat Exercise 11.1 using

$$L_s = A, \qquad T_s = \sqrt{\frac{\rho A^2}{\Delta p}}, \qquad M_s = \rho A^2 H.$$

11.3   Recall from Observation 10.2 (p. 541) that the Buckingham Pi method can be used to nondimensionalize known equations. Show that the governing differential equation of motion for the aneurysm [Eq. (11.53)] can be written in nondimensional form as

$$\left(\frac{1}{x^2} + bx\right)\ddot{x} + \frac{3}{2}b\dot{x}^2 + 4m\frac{\dot{x}}{x} + 2\frac{f(x)}{x} = F(\tau),$$

where

$$x \equiv \lambda, \qquad f = \frac{T}{c},$$

$$b = \frac{\rho A}{\rho_s H}, \qquad F = \frac{PA}{c},$$

$$m = \frac{\mu}{\sqrt{\rho_s c H}}, \qquad \tau = \frac{t\sqrt{c}}{\sqrt{\rho_s A^2 H}},$$

and c is a material parameter having units of force/length.

11.4 If the CSF surrounding a spherical aneurysm is assumed to be ideal, then the governing equation reduces to

$$\left(\frac{1}{x^2} + bx\right)\ddot{x} + \frac{3}{2}b\dot{x}^2 + 2\frac{f(x)}{x} = F(\tau).$$

If $F(\tau) = F_0$, a constant, show that the equation can be integrated once in time to yield

$$\frac{1}{2}\dot{x}^2 + \frac{1}{2}b\dot{x}^2 x^3 + 2\int xf(x)dx - \frac{1}{3}F_0 x^3 = \text{constant}.$$

This form of the equation can be related to the first law of thermodynamics (e.g., $\dot{x}^2/2$ is a nondimensional kinetic energy term, whereas $-F_0 x^3/3$ is a work-type term related to a pressure times volume). Hint: Note that

$$\ddot{x}\dot{x} = \frac{d}{dt}\left(\frac{1}{2}\dot{x}^2\right), \quad x^2\ddot{x} = \frac{d}{dt}\left(\frac{1}{3}x^3\right), \quad dx = \frac{dx}{dt}dt.$$

11.5 If $F(\tau) \equiv F_0$, a constant, if $f(x) = f_0$, a constant surface tension, and if there is no external fluid, then the governing equation of motion of a spherical "soap bubble" is (cf. equation in Exercise 11.3)

$$\frac{1}{x^2}\ddot{x} + \frac{2f_0}{x} = F_0.$$

Solve this equation for $x(\tau)$ and comment on its interpretation.

11.6 In Section 11.3, we saw that the pressure field $p(r, t)$ in a radial flow in a spherical domain for an incompressible, Newtonian fluid is independent of the viscosity; that is, we have found a special flow wherein the same pressure field satisfies both the Navier–Stokes and the Euler equations. Because the flow is radial, we can also define a radial streamline $s \equiv r$, where $ds = dr$. Show, therefore, that an unsteady Bernoulli equation can be written in the form

$$p_A + \rho g z_A + \frac{1}{2}\rho v_A^2 = p_B + \rho g z_B + \frac{1}{2}\rho v_B^2 + \rho\int_A^B \frac{\partial v_s}{\partial t}ds.$$

Hint: Integrate the appropriate Euler equation along a radial streamline.

11.7   Use the unsteady Bernoulli equation in Exercise 11.6, with point A at $r = a$, point B at $r = \infty$, and $v_s \equiv v_r = g(t)/r^2$ from our mass balance relation, to show that one obtains the same pressure field $p(r, t)$ from Bernoulli as obtained from Navier–Stokes.

11.8   Repeat Example 11.2 for $\eta = 0.0625$, $\zeta = -0.1$ $c = 1.0$, $\alpha = 0.5$, $F(t) = 1.0 \sin t$, $x(0) = 0$, and $\dot{x}(0) = 1$. What does the negative value of ζ induce?

11.9   Repeat Exercise 11.8 with $x(0) = 1$ and $\dot{x}(0) = 1$.

11.10 Viscoelastic characteristics include instantaneous elasticity, creep, stress relaxation, instantaneous recovery, delayed elasticity, permanent set, and hysteresis. Define and discuss each characteristic.

11.11 We found in Eq. (11.59) that the stress relaxation $\sigma(t)$ in a Maxwell model is given by an exponential decay. Compute the rate of change of stress as a function of time. Observe that the relaxation is initially very rapid; indeed, show that only 37% of the initial stress remains at time $t = t_R$, the relaxation time.

11.12 Similar to the previous exercise, investigate the creep response of a Kelvin–Voigt model. In particular, compute $\dot{\varepsilon}(t)$ and sketch it versus time. Show, too, that only 37% of the asymptotic strains remains to be realized after $t = t_c$.

11.13 As we saw in Eq. (11.64), the Kelvin–Voigt model does not allow stress relaxation. Intuitively, we realize that a step change in ε from 0 to ε_0 at time $t = 0$ can only be accomplished via an infinite stress (if that were possible) because a viscous dashpot cannot otherwise extend instantaneously. Thereafter, the stress in the viscous element drops to zero, for it requires a strain rate to produce a stress, thus the spring sustains the constant extension with a constant stress. In a similar way, qualitatively discuss the creep response of the Kelvin–Voigt model and, in particular, justify why this behavior is sometimes referred to as a *delayed elasticity* (cf. Findley et al., 1976, p. 56).

11.14 Recall again the stress relaxation of a Maxwell model:

$$\sigma(t) = \sigma_0 e^{-t/t_R},$$

where t_R is the so-called relaxation time. Because the model is linear, superposition holds (cf. Section 5.5 of Chapter 5). Thus, a set of Maxwell models in parallel (Fig. 11.26) and under a constant strain ε_0 will stress relax according to

$$\sigma(t) = \varepsilon_0 \sum_{i=1}^{n} E_i e^{-t/t_R^i},$$

where $t_R^i = \mu_i/E_i$ for all elements $i = 1, 2, \ldots n$. For a continuous distribution of relaxation times, from 0 to ∞, we have

$$\sigma(t) = \varepsilon_0 \int_0^{\infty} R(t_R) e^{-t/t_R} dt_R,$$

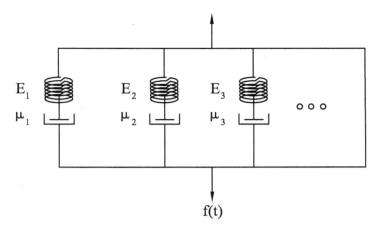

FIGURE 11.26.

where $R(t_R)$ is called the *relaxation spectrum*—a distribution of relaxation times. Formulate a similar analysis of the creep response of a set of Kelvin–Voigt elements in series whereby

$$\varepsilon(t) = \sigma_0 \sum_{i=1}^{n} C_i \left(1 - e^{-t/t_c^i}\right),$$

or

$$\varepsilon(t) = \sigma_0 \int_0^\infty C(t_c)(1 - e^{-t/t_c}) dt_c,$$

where $C(t_c)$ is called the retardation spectrum.

11.15 A Maxwell model in series with a Kelvin–Voigt model is a four-parameter model sometimes called a *Burgers* model. If the spring stiffness and dashpot viscosity are given by (E_1, μ_1) and (E_2, μ_2) for the Maxwell and Kelvin–Voigt components, respectively, show that the creep function (or compliance) for the Burgers model is

$$J(t) = \frac{1}{E_1} + \frac{t}{\mu_1} + \frac{1}{E_2}(1 - e^{-E_2 t/\mu_2}).$$

11.16 Assume that a uniaxial member is subjected to a strain of the form $\varepsilon(t) = \varepsilon_A \sin \omega t$. If the material behaves elastically, one would expect a stress response of the form $\sigma(t) = \sigma_A \sin \omega t$ (i.e., in phase). For a viscoelastic response, however, one would expect the stress response to be out of phase with the strain. Hence, let

$$\sigma(t) = \sigma_A \sin(\omega t + \phi) = \sigma_A (\sin \omega t \cos \phi + \cos \omega t \sin \phi),$$

where ϕ is the phase angle. This form suggests that a complex representation may be useful, namely

$$\varepsilon(t) = \varepsilon_A e^{i\omega t}, \qquad \sigma(t) = \sigma_A e^{i(\omega t + \phi)},$$

where $i = \sqrt{-1}$. Show that

$$\frac{\sigma(t)}{\varepsilon(t)} = \frac{\sigma_A}{\varepsilon_A} e^{i\phi} = G_1 + iG_2,$$

where G_1 and G_2 are called the *storage modulus* and the *loss modulus*, respectively. Show, too, that

$$G_1 = \frac{\sigma_A}{\varepsilon_A}\cos\phi, \qquad G_2 = \frac{\sigma_A}{\varepsilon_A}\sin\phi, \qquad \tan\phi = \frac{G_2}{G_1}.$$

Note that, for example, $G_1 \sim 10^9 \, \text{Pa}$, $G_2 \sim 10^7 \, \text{Pa}$, and $\phi \sim 0.01$ for a typical polymer.

11.17 Following up on Exercise 11.16, note that $G^* = G_1 + iG_2$ is called the *complex modulus*. Show that for the sinusoidal straining in Exercise 11.16, the magnitude of $G^* = \sigma_A/\varepsilon_A$. Note, too, that $\tan\phi$ is often called the *mechanical loss*.

11.18 If a Standard model consists of a spring in series with a Kelvin–Voigt element, and the spring has a stiffness $E = 1\,\text{GPa}$, whereas the Kelvin–Voigt element has stiffness $10\,\text{kPa}$ and viscosity $10^7\,\text{P}$ (poise), plot $\log J_1$, $\log J_2$, and $\log(\tan\phi)$ versus $\log\omega \in [-8, 8]$, where

$$J^* = J_1 - iJ_2 = \frac{\varepsilon_0}{\sigma_0} e^{-i\phi}$$

for creep.

11.19 Show that for a Maxwell model subjected to an oscillatory motion,

$$G^* = \frac{\mu^2\omega^2/E}{1 + \mu^2\omega^2/E^2} + i\left(\frac{\mu\omega}{1 + \mu^2\omega^2/E^2}\right)$$

and

$$\tan\phi = \frac{G_2}{G_1} = \frac{E}{\omega\mu}.$$

11.20 Consistent with the prior exercise, plot $\tan\phi$ and $|G^*|$ versus $\log\omega \in [-4, 4]$ given values of $E = 1\,\text{GPa}$ and $\mu = 5 \times 10^9\,\text{P}$. Repeat for the same E but with $\mu = 5 \times 10^{10}$ and $5 \times 10^8\,\text{P}$. Discuss the behavior in terms of changes in t_R. Note that these values are reasonable for a polymer that may be used in a biomedical device.

11.21 Show that the frictional drag coefficient C_f for the step-slider in Section 11.5.2 is correct as given. Hint: The total drag force F_D can be computed by integrating the shear stresses over the bearing surface; that is,

$$F_D = w\int_0^{L(1-b)} \tau_1 \, dx_1 + \int_{L(1-b)}^{bL} \tau_2 \, dx_2,$$

where, in general,

$$\tau \equiv \mu \frac{\partial v_x}{\partial y}\bigg|_{y=0}.$$

11.22 The so-called *Reynolds' equation* governs general flows in hydrodynamic lubrication theory. In one dimension, show that it can be written as

$$\frac{d}{dx}\left(\frac{h^3(x)w}{\mu}\left(\frac{dp}{dx}\right)\right) + 6U\frac{dh}{dx} = 0.$$

Hint: Use the result for the velocity $v_x(y)$ in Section 11.5.2 and compute the volumetric flow rate

$$Q = w\int_0^{h(x)} v_x(y)dy.$$

Exploit the fact that Q is constant with respect to x even though the gap distance is $h = h(x)$.

12
Epilogue

Biomechanics is often defined as "mechanics applied to biology" (Fung, 1993). Although this is certainly true, it is hoped that the reader now appreciates that biomechanics can and must be much more. Because of the complexity of tissue structure and behavior, there is a need for new, sophisticated theoretical frameworks; because of the continuing lack of data, there is a need for new, clever experiments; because of the geometric complexity of cells, tissues, and organs, there is a need for robust computational methods; and because of the significant morbidity and mortality that results from disease and injury, there is a need for improved modalities for diagnosis and treatment. Clearly then, we must continue to expand the scope of biomechanics, to seek new concepts, postulates, technologies, and techniques upon which a rigorous understanding can be based. Biomechanics is thus better defined as the development, extension, and application of mechanics to answer problems of importance in biology and medicine. Biomechanics is a vibrant field—one with great promise.

This book was designed to be but an introduction to biosolid and biofluid mechanics. There are, of course, many other areas of introductory biomechanics (e.g., studies of whole-body motions, such as gait analysis and athletic performance) that were not covered. Reflecting on what was presented in Chapters 1–11, however, one of the most important things to realize is that *mechanics offers a consistent and rigorous method of approach to study the wide variety of initial and boundary value problems that arise in biology and medicine.* Another important thing to realize is that biomechanics requires advanced study; hence, this book is a beginning, not an end. Finally, it is important to know that there are many ways that a young biomechanicist can contribute to basic science and health care delivery.

12.1 Future Needs in Biomechanics

The first author (JDH) was recently asked to review some of the past achievements of the Continuum Mechanics of Soft Biological Tissues and to suggest areas that are in need of further study (Humphrey, 2003a). Among the many needs and promises, it was suggested that fundamental research is needed in eight particular areas:

Molecular and Cell Biomechanics
Developmental Biomechanics
Biomechanics of Growth and Remodeling
Injury Biomechanics and Rehabilitation
Functional Tissue Engineering
Muscle Mechanics
Solid–Fluid Interactions
Biothermomechanics

Cells are the fundamental units of life; understanding their biomechanical behavior will thus reveal may new insights into the biology and mechanics of health, disease, injury, and clinical treatment. Cell mechanics is essential, for example, for explaining basic processes such as cell adhesion, contraction, division, migration, spreading, and even phagocytosis (i.e., the engulfing and digestion of extracellular material). Likewise, it appears that cellular apoptosis (i.e., programmed cell death), the synthesis and degradation of the matrix, and the production of growth regulatory molecules, cytokines, and cell surface receptors are also influenced greatly by the mechanics. Cells consist of a multitude of different types of molecules, however; thus, to understand the cell, we must ultimately understand the mechanics of the associated proteins, phospholipids, and even nucleotides. Of primary interest, at present, is how the three primary cytoskeletal proteins (actin, the intermediate filaments, and microtubules) change their organization in response to mechanical loads. Inasmuch as the function of these three proteins is controlled in large part by a host of accessory proteins (e.g., α-actinin, myosin, and talin), there is a similar need to understand the contribution of the accessory proteins to the mechanics. Cells interact mechanically, chemically, and electrically with other cells, and they likewise interact with the extracellular matrix. There is a pressing need to understand the mechanics of the molecules, particularly the adhesion molecules, that govern these interactions. For example, extracellular matrix–integrin–cytoskeletal interactions are clearly important to the mechanobiology, but much remains unknown. Finally, we must realize the importance of understanding better how the mechanics directly affects the chemistry. It is generally believed that deformations of the cells can change the conformations of the molecules, which, in turn, can change binding energies. This, too, must be understood better.

It is purported that Aristotle (384–322 B.C.) stated that "Here and elsewhere we shall not obtain the best insights into things until we actually see them growing from the beginning." Developmental biology clearly holds many keys to unlocking secrets of importance to clinical care. As noted by the biologist A. K. Harris (1994), however, "without the aid of mechanicians, and others skilled in simulation and modeling, developmental biology will remain a prisoner of our inadequate and conflicting physical intuitions and metaphors."

For obvious reasons—in particular the smallness of tissues and organs in the embryo and fetus—biological development has attracted less attention in biomechanics than many other areas. Fortunately, however, the desire to understand molecular- and cellular-level phenomena has led to technological advancements (e.g., atomic force microscope) that can also be useful in the study of development. An interesting example of how understanding development may increase our understanding in many other areas, not the least of which is tissue engineering, is a comparison of aortic development versus changes induced in maturity due to hypertension. During development, the blood pressure increases from ~0 before the heart beats to ~120/80 mm Hg in maturity. This increase in pressure is followed by a concomitant increase in wall thickness, which appears to maintain the wall stress at a "preferred" value. In development, this thickening is accomplished by adding more and more layers of elastin–collagen–smooth muscle, each of equal thickness. Conversely, in hypertension in maturity, the wall also thickens in response to an increasing pressure so as to return the wall stress toward its preferred value, but this thickening occurs via the addition of material to extant layers, not by adding new layers. Why? We do not yet know the answer to this simple question, but this example should illustrate that understanding development will likely provide important clues for those who seek to understand many issues in biomechanics.

Murray (1926) suggested that biological "organization and adaptation are observed facts, presumably conforming to definite laws because, statistically at least, there is some sort of uniformity or determinism in their appearances. Let us assume that the best quantitative statement embodying the concept of organization is a principle which states that the cost of operation of physiological systems tends to be a minimum. . . ." Over the years, many investigators have used the concept of optimization to understand and predict various aspects of biological growth and remodeling. One such case was discussed in Chapter 10. The key question, however, is optimization of what? In 1952, Turing showed that we must also consider the production and removal of morphogens as well as their possible diffusion. In 1981, Skalak showed that we must also consider the pointwise kinematics of growth, and, soon thereafter, Fung showed that we must address stress-mediated changes in mass. Biological growth and remodeling is clearly complex, involving changes in morphology, reaction–diffusion chemistry, kinematics, stress, and mass production. We are only beginning to scratch the surface of this important area.

Tissues are susceptible to a variety of injuries: abrasion, crushing, dissection, rupture, and tearing, to name a few. Whereas such injuries are typically thought to be due to accidental trauma, often in athletics, falls, or vehicular crashes, others are purposefully induced clinically. An example of the latter is balloon angioplasty, the procedure wherein a balloon-tipped catheter is inflated within a diseased artery for the purpose of enlarging a lumen that is compromised by an obstructive atherosclerotic plaque. Angioplasty works, in part, by weakening (i.e., damaging) the wall, fracturing the atherosclerotic plaque, and sometimes by creating small dissections between the plaque and wall. Although referred to as a "controlled injury," the actual level of control is poor because we do not understand the details of the injury/damage process. Perhaps a greater understanding could help reduce the 20–30% failure (i.e., restenosis) rate. Understanding damage mechanics likewise holds promise in the area of robotic-assisted surgery. Whereas a robot can perform certain operations much more repeatedly and precisely than a human surgeon, it lacks the tactile feedback and control that is second nature to the skilled surgeon. To prevent robot-induced damage, we must understand the associated strength of the tissues involved.

Related to the general topic of injury biomechanics is the process of healing. For example, whereas it may seem natural to immobilize, and thereby protect or reduce pain in an injured limb, findings over the last 35 years suggest that this may be naive. It appears that immobilized collagenous tissues undergo histological changes that include a loss of material and, thus, strength. Indeed, it appears that the production of new tissue (e.g., wound healing) is hastened by certain levels of mechanical loading. To understand some aspects of healing, therefore, we need to understand better the associated biomechanics and mechanobiology.

According to Butler et al. (2000), "the goal of 'tissue engineering' is to repair or replace tissues and organs by delivering implanted cells, scaffolds, DNA, proteins, and/or protein fragments at surgery." Toward this end, the U.S. National Committee on Biomechanics suggested the following needs (Butler et al., 2000): (1) In vivo stress and/or in vivo strain histories need to be measured in normal tissues for a variety of activities; (2) the mechanical properties of the native tissues must be established for subfailure and failure conditions; (3) a subset of these mechanical properties must be selected and prioritized (i.e., we cannot expect a tissue-engineered material to mimic exactly the native tissue; hence, we must determine which properties are most important with regard to functionality); (4) standards must be set when evaluating the repairs or replacements after surgery so as to determine "how good is good enough?;" (5) we must determine what physical regulation cells experience in vivo as they interact with an extracellular matrix; and (6) we must determine how physical factors influence cellular activity in bioreactors and how cell–matrix implants can be mechanically stimulated before surgery to produce a better outcome. Clearly, continuum biomechanics has a key role to play in achieving most, if not all, of these objectives.

In 1983, Fung noted that without a theory of muscle mechanics, we cannot understand human athletic performance or much of rehabilitation engineering; we cannot develop a theory of the heart or autoregulation of the vasculature; we cannot understand asthma or accommodation of the eye; indeed, we cannot even understand activities of the cell such as migration. Clearly, the mechanics of muscle and motor proteins is fundamental to understanding key activities of life at the organism, organ, tissue, and cellular level. Fortunately, we have learned much about muscle since the 1950s and the early work by Huxley and others. Yet, early ideas that muscle contraction is one dimensional still pervades the literature even though it is now clear that the force generation due to muscle contraction can be multiaxial. A better constitutive equation for muscle is thus imperative.

As noted briefly in Chapter 11, the division of continuum mechanics into disciplines such as "solid mechanics" and "fluid mechanics" is artificial and simply a natural consequence of historical developments. In the body, however, solidlike and fluidlike behavior go hand in hand. Whether it be the removal of wastes by the renal system, the transport of blood by the cardiovascular system, the functioning of an articulating joint, or even the response of an individual cell to an abrupt change in load, solid–fluid interactions are critical. In the future, therefore, there will continue to be a pressing need for research and teaching to address directly such couplings.

Advances in laser, microwave, radio-frequency, and similar technologies continue to encourage the use of thermal energy (heat) to treat disease and injury. Most clinical applications have been motivated primarily by the availability of the technology, however, not a detailed understanding of the associated biothermomechanics. There is a need, therefore, to understand better the effects of heat on cells and tissues and, in particular, to determine optimal dosing protocols in terms of clinically measurable and controllable parameters such as the temperature level, state of stress during heating, and the duration of heating. For example, until recently it was not commonly appreciated that mechanical load can play just as important of a role in the thermal denaturation of proteins as the temperature level; that is, whereas the effect of temperature appears to affect the denaturation through the activation energy via an Arrhenius-type process (Section 11.6), the effect of mechanical load appears to affect the process through the activation entropy. Because tissue elasticity is due more to changes in the configurations of the underlying proteins (i.e., changes in configurational entropy) rather than to changes in bond energies (i.e., energetic elasticity), understanding better the biothermomechanics may yield new insights into tissue mechanics as well as thermal treatments.

12.2 Need for Lifelong Learning

Much has been learned in and through biomechanics, particularly over the last 35 years, and there is now an extensive literature. We must build upon prior understanding and achievements, of course; thus, there is a need to appreciate that which is in the literature. That said, we must also be careful not to be bound by past methods or concepts. New technologies are revealing much more detail about the fundamental building blocks of life—genes, proteins, and cells—and new hypotheses and theories should build upon new observations. The challenges, and likewise the promises, of biomechanics have never been greater.

An introductory course is clearly the beginning, not the end of one's learning. The interested student is thus encouraged to pursue intermediate and advanced courses in biology, mathematics, and mechanics as well as specialized courses in biomechanics. Yet, formal course work, even through the doctoral degree, is not the end of one's learning. Advances are being realized every day; thus, one must continually consult the archival literature to stay abreast of the latest developments. In biomechanics, this means that we should be especially aware of that which is reported in the leading journals: the *Journal of Biomechanics*, which was founded in 1968, the ASME *Journal of Biomechanical Engineering*, founded in 1977, *Computer Methods in Biomechanics and Biomedical Engineering*, founded in 1998, and, most recently, *Biomechanics and Modeling in Mechanobiology*, founded in 2002. These journals and others such as the *Annals of Biomedical Engineering* and the *IEEE Transactions for Biomedical Engineering* continue to promote the growth of biomechanics. Note, too, that new ideas are presented at national and international meetings such as the World Congress of Biomechanics, which began in 1990 via a meeting at San Diego and has been followed by meetings in 1994 at Amsterdam, in 1998 at Sapporo, and in 2002 at Calgary. These focused meetings, as well as symposia at many different technical meetings, promote the exchange of ideas and thus contribute to the rapid growth of continuum biomechanics. Students should try to attend such meetings whenever possible.

We must also remember that biomechanics is part of a larger, multidisciplinary activity whose goal is to understand better the conditions of health as well as those of disease and injury. Consequently, biomechanics has and will continue to benefit greatly from developments in the basic life sciences, medical sciences, mathematics, and materials science. Indeed, it would be hard to find an archival paper on biomechanics that does not refer to research in these allied fields and, conversely, it would be hard to find archival journals in these allied fields (e.g., the *American Journal of Physiology*, the *Biophysical Journal*, the American Heart Association's *Circulation Research*, the ASME *Journal of Applied Mechanics*, *The Journal of Orthopedic Research*, and so on) that do not contain papers on biomechanics. These sources must be consulted as well, and we must seek to work in multidisciplinary teams consisting of experts from the many allied areas of study.

12.3 Conclusion

Biomechanics is intellectually stimulating and challenging. More importantly, however, it is vitally important. Whereas physicians see human pain and suffering on a daily basis, few engineers do. Nonetheless, we must continually remind ourselves that the ultimate goal of biomechanics is to contribute to the improvement of health care delivery, a goal that deserves our very best effort.

References

Alberts B, A Johnson, J Lewis, M Raff, K Roberts, P Walter (2002) Molecular Biology of the Cell, Garland Publishing, New York.

Alexandrou AN (2001) Principles of Fluid Mechanics. Prentice-Hall, Englewood Cliffs, NJ.

Askeland DR (1994) The Science and Engineering of Materials. Third Edition. PWS Kent Publishing, Boston.

Ayad S, R Boot-Handford, MJ Humphries, KE Kadler, A Shuttleworth (1994) The Extracellular Matrix FactsBook. Academic Press, New York.

Bell ET (1986) Men of Mathematics. Simon & Schuster, New York.

Binnig G, CF Quate, C Gerber (1986) Atomic force microscope. Phys Rev Letters 56: 930–933.

Birk DE, JF Southern, EI Zycband, JT Fallon, RL Trelstad (1989) Collagen fiber bundles: A branching assembly unit in tendon morphogenesis. Development 107: 437–443.

Boorstin DJ (1985) The Discoverers. Vintage Books, New York.

Boresi AP, OM Sidebottom, FB Seely, JO Smith (1993) Advanced Mechanics of Materials. John Wiley & Sons, Chichester.

Boyer CB (1949) The History of the Calculus. Dover, New York.

Butler DL, HA Awad (1999) Perspectives on cell and collagen composites for tendon repair. Clin Orthoped Relat Res 367S: S324–S332.

Butler DL, SA Goldstein, F Guilak (2000) Functional tissue engineering: The role of biomechanics. ASME J Biomech Eng 122: 570–575.

Carrel A, CC Guthrie (1906) Results of the biterminal transplantation of veins. Am J Med Sci 132: 415–422.

Carter DR, GS Beaupré (2001) Skeletal Function and Form: Mechanobiology of Skeletal Development, Aging, and Regeneration. Cambridge University Press, Cambridge.

Carver W, ML Nagpal, M Nachtigal, TK Borg, L Terracio (1991) Collagen expression in mechanically stimulated cardiac fibroblasts. Circ Res 69: 116–122.

Comroe JH, RD Dripps (1977) The Top Ten Clinical Advances in Cardio-vascular–Pulomonary Medicine and Surgery 1945–1975 (Vols. I and II). Public Health Service Document 017-043-00084-6. US Government Printing Office, Washington, DC.

Costa KD, FCP Yin (1999) Analysis of indentation: Implications for measuring mechanical properties with atomic force microscopy. ASME J Biomech Eng 121: 462–471.

Cowin SC (2001) Bone Biomechanics Handbook. Second Edition. CRC Press, Boca Raton, FL.

Cui Y, C Bustamante (2000) Pulling a single chromatin fiber reveals the forces that maintain its high order structure. Proc Natl Acad Sci USA 97: 127–132.

Davidson JM, MG Giro (1986) Control of elastin synthesis: Molecular and cellular aspects. In: RP Mecham, ed. Regulation of Matrix Accumulation, Academic Press, New York, pp. 177–217.

Davis SE, DJ Doss, JD Humphrey, NT Wright (2000) Effects of heat-induced damage on the radial component of thermal diffusivity of bovine aorta. ASME J Biomech Eng 122: 283–286.

Delp MD, PN Colleran, MK Wilkerosn, MR McCurdy, J Muller-Delp (2000) Structural and functional remodeling of skeletal muscle microvasculature is induced by simulated microgravity. Am J Physiol 278: H1866–H1873.

Dorland's Illustrated Medical Dictionary (1988) WB Saunders, Philadelphia.

Fawcett DW (1986) Bloom and Fawcett: A Textbook of Histology. Eleventh Edition. WB Saunders, Philadelphia.

Ferry JD (1980) Viscoelastic Properties of Polymers. John Wiley & Sons, Chichester.

Findley WN, JS Lai, K Onaran (1976) Creep and Relaxation of Nonlinear Viscoelastic Materials. Dover, New York.

Fox RW, AT McDonald (1992) Introduction to Fluid Mechanics. Fourth Edition. John Wiley & Sons, New York.

Frangos JA (1993) Physical Forces and the Mammalian Cell. Academic Press, New York.

Fung YC (1984) Biodynamics: Circulation. Springer-Verlag, New York.

Fung YC (1990) Biomechanics: Motion, Flow, Stress, and Growth. Springer-Verlag, New York.

Fung YC (1993) Biomechanics: Mechanical Properties of Living Tissues. Second Edition. Springer-Verlag, New York.

Fung YC, S Liu (1991) Changes of zero-stress state of rat pulmonary arteries in hypoxic hypertension. J Appl Physiol 70: 2455–2470.

Fung YC, S Liu (1992) Strain distribution in small blood vessels with zero-stress state taken into consideration. Am J Physiol 262: H544–H552.

Galbraith CG, R Skalak, S Chien (1998) Shear stress induces spatial reorganization of the endothelial cell cytoskeleton. Cell Motility Cytoskel 40: 317–330.

Gelman RA, DC Poppke, KA Piez (1979) Collagen fibril formation in vitro. The role of the nonhelical terminal regions. J Biol Chem 254: 11,741–11,745.

Glass L, PJ Hunter, AD McCulloch (1991) Theory of Heart. Springer-Verlag, New York.

Gooch KJ, T Blunk, G Vunjak-Novakovic, R Langer, LE Freed (1998) Mechanical forces and growth factors utilized in tissue engineering. In: CW Patrick Jr, AG Mikos, LV McIntire, eds. Frontiers in Tissue Engineering. Pergamon, Oxford.

Han HC, DN Ku (2001) Contractile responses in arteries subjected to hypertensive pressure in seven day organ culture. Ann Biomed Eng 29: 467–475.

Han HC, L Zhao, M Huang, LS Hou, YT Huang (1998) Postsurgical changes of the opening angle of canine autogenous vein graft. ASME J Biomech Engr 120: 211–216.

Harris AK (1994) Multicellular mechanics in the creation of anatomical structures. In: N Akkas, ed. Biomechanics of Active Movement and Division of Cells. Springer-Verlag, New York, pp. 87–129.

Harris H (1999) The Birth of the Cell. Yale University Press, New Haven, CT.

Holzapfel GA, TC Gasser (2001) A viscoelastic model for fiber-reinforced composites at finite strains: Continuum basis, computational aspects, and applications. Comp Meth Appl Mech Eng 190: 4379–4403.

Holzapfel GA, M Stadler, CAJ Schulze-Bauer (2002) A structural model for the viscoelastic behavior of arterial walls: Continuum formulation and finite element analysis. Eur J Mech A: Solids 21: 441–463.

Humphrey JD, PB Canham (2000) Structure, mechanical properties, and mechanics of intracranial saccular aneurysms. J Elast 61: 49–81.

Humphrey JD (2001) Stress, strain and mechanotransduction in cells. ASME J Biomech Eng 123: 638–641.

Humphrey JD (2002) Cardiovascular Solid Mechanics: Cells, Tissues, and Organs. Springer-Verlag, New York.

Humphrey JD (2003a) Continuum biomechanics of soft tissues. Proc R Soc (Lond) 459: 3–46.

Humphrey JD (2003b) Continuum thermomechanics and the clinical treatment of disease and injury. Appl Mech Rev 56: 231–260.

Ingber DE, SR Heidemann, P Lamoureux, RE Buxbaum (2000) Opposing views on tensegrity as a structural framework for understanding cell mechanics. J Appl Physiol 89: 1663–1678.

Johnson AT (1991) Biomechanics and Exercise Physiology. John Wiley & Sons, New York.

Johnson GA, GA Livesay, SLY Woo, KR Rajagopal (1996) A single integral finite strain viscoelastic model of ligaments and tendons. ASME J Biomech Eng 118: 221–226.

Johnson MA, MF Beatty (1995) The Mullins effect in equibiaxial extension and its influence on the inflation of a balloon. Int J Eng Sci 33: 223–245.

Khan AS, S Huang (1995) Continuum Theory of Plasticity. John Wiley & Sons, New York.

Kucharz EJ (1992) The Collagens: Biochemistry and Pathophysiology. Springer-Verlag, Berlin.

Leckband D (2000) Measuring forces that control protein interactions. Annu Rev Biophys Biomol Struct 29: 1–26.

Lefevre M, RB Rucker (1980) Aorta elastin turnover in normal and hypercholesterolemic japanese quail. Biochim Biophys Acta 630: 519–529.

Levesque MJ, RM Nerem (1985) The elongation and orientation of cultured endothelial cells in response to shear stress. J Biomech Eng 107: 341–347.

Lodish H, A Berk, SL Zipurski, P Matsudaire, D Baltimore, J Darnell (2000) Molecular Cell Biology. W.H. Freeman, New York.

McAnulty RJ, GJ Laurent (1987) Collagen synthesis and degradation in vivo. Evidence for rapid rates of collagen turnover with extensive degradation of newly synthesized collagen in tissues of the adult rat. Collagen Rel Res 7: 93–104.

Milnor W (1989) Hemodynamics. Williams and Wilkens, Baltimore, MD.

Moody LF (1944) Friction factors for pipe flow. Trans of the ASME. 66: 671–684.

Mow VC, WC Hayes (1991) Basic Orthopedic Biomechanics. Raven Press, New York.

Mow VC, A Ratcliff, SLY Woo (1990) Biomechanics of Diarthrodial Joints. Volumes 1 and 2. Springer-Verlag, New York.

Mow VC, RM Hochmuth, F Guilak, R Trans-Son-Tay (1994) Cell Mechanics and Cellular Engineering. Springer-Verlag, New York.

Murray CD (1926) The physiological principle of minimum work. I. The vascular system and the cost of blood volume. Proc Natl Acad Sci USA 12: 207–214.

Niedermuller H, M Skalicky, G Hofecker, A Kment (1977) Investigations on the kinetics of collagen-metabolism in young and old rats. Exp Gerontol 12: 159–168.

Nigg BM, W Hertzog (1994) Biomechanics of the Musculoskeletal System. John Wiley & Sons, Chichester.

Nimni ME (1992) Collagen in cardiovascular tissue. In: GW Hastings, ed. Cardiovascular Biomaterials. Springer-Verlag, New York.

Ninomiya Y, BR Olsen, T Ooyama (1998) Extracellular Matrix–Cell Interaction: Molecules to Disease. Karger, Basel.

Oberhauser AF, PE Marszalek, HP Erickson, JM Fernandez (1998) The molecular elasticity of the extracellular matrix protein tenascin. Nature 393: 181–185.

Özkaya N, M Nordin (1999) Fundamentals of Biomechanics: Equilibrium, Motion, and Deformation. Springer-Verlag, New York.

Panfilov AV, AV Holden (1997) Computational Biology of the Heart. John Wiley & Sons, London.

Pioletti DP, LR Rakotomanana (2000) Non-linear viscoelastic laws for soft biological tissues. Eur J Mech A: Solids 19: 749–759.

Popov EP (1999) Engineering Mechanics of Solids. Prentice-Hall, Englewood Cliffs, NJ.

Pries AR, TW Secomb, P Gaehtgens (1995) Design principles of vascular beds. Circ Res 77: 1017–1023.

Provenzano PP, RS Lakes, DT Corr, R Vanderby Jr (2002) Application of non-linear viscoelastic models to describe ligament behavior. Biomechan Model Mechanobiol 1: 45–57.

Radmacher M, RW Tillmann, M Fitz, HE Gaub (1992) From molecules to cells: Imaging soft samples with the atomic force microscope. Science 257: 1900–1905.

Ratner BD (2003) Biomaterials Science: Introduction to Materials in Medicine. Academic Press, San Diego, CA.

Roark RJ, WC Young (1975) Formulas for Stress and Strain. McGraw-Hill, New York.

Robert L, W Hornbeck (1989) Elastin and Elastases, Vol I, CRC Press, Boca Raton FL.

Roesler H (1987) The history of some fundamental concepts in bone biomechanics. J Biomech 20: 1025–1034.

Rosen LE, TM Hollis, MG Sharma (1974) Alterations in bovine endothelial histidine decarboxylase activity following exposure to shearing stresses. Exp Mol Pathol 20: 329–343.

Rubin CT, L Lanyon (1985) Regulation of bone mass by mechanical loading: The effect of peak strain magnitude. Calcif Tissue Int 37: 441–447.

Shah AD, JD Humphrey (1999) Finite strain elastodynamics of saccular aneurysms. J. Biomech 32: 593–599.

Skalak R, S Chien (1987) Handbook of Bioengineering. McGraw-Hill, New York.

Slattery JC (1981) Momentum, Energy, and Mass Transfer in Continua. Krieger, New York.

Stamenovic D, DE Ingber (2002) Models of cytoskeletal mechanics of adherent cells. Biomech Model Mechanobiol 1: 95–108.

Sten-Knudsen O (1953) Torsional elasticity of the isolated cross striated muscle fibre. Acta Physiol Scand 28: 7–240.

Strang G (1986) Introduction to Applied Mathematics. Wellesley–Cambridge Press, Boston.

Tanner RI (1985) Engineering Rheology. Oxford University Press, Oxford.

Ten Cate AR, DA Deporter (1975) The degradative role of the fibroblast in the remodelling and turnover of collagen in soft connective tissue. Anat Rec 182: 1–14.

Timoshenko SP, JN Goodier (1970) Theory of Elasticity. Third Edition. McGraw-Hill, New York.

Truesdell C, W Noll (1965) The Nonlinear Field Theories of Mechanics. In: S Flugge, ed. Handbuch der Physik. Springer-Verlag, Berlin, Vol. III/3.

Tsao Y-MD, E Boyd, DA Wolf, G Spaulding (1994) Fluid dynamics within a rotating bioreactor in space and earth environments. J Spacecraft Rockets 31: 937–943.

Vaughn J, A Czipura, JD Humphrey (2002) Measurement of the finite strain dependent permeability of biomembranes. J Biomech 35: 287–291.

Waldman LK, YC Fung, JW Covell (1985) Transmural myocardial deformation in the canine left ventricle: normal in vivo three-dimensional finite strains. Circ Res 57: 152–163.

Weibel ER (1963) Morphometry of the Human Lung. W.B. Saunders, Philadelphia.

West JB (1979) Repiratory Physiology—The Essentials. Williams and Wilkens, Baltimore, MD.

Wiebers DO et al. (1998) Unruptured intracranial aneurysms—risk of rupture and risks of surgical intervention. International study of unruptured intracranial aneurysms investigators. N Engl J Med 339: 1725–1733.

Wineman AS, KR Rajagopal (2000) Mechanical Properties of Polymers: An Introduction. Cambridge University Press, Cambridge.

Wolf DA, RP Schwarz (1991) Analysis of Gravity-Induced Particle Motion and Fluid Perfusion Flow in the NASA-Designed Rotating Zero-Head-Space Tissue Culture Vessel. NASA Technical Paper 3143. NASA, Washington, DC.

Wolff J (1986) The Law of Bone Remodeling. Springer-Verlag, Berlin.

Zamir M (2000) The Physics of Pulsatile Flow. Springer-Verlag, New York.

Zhu C, G Bao, N Wang (2000) Cell mechanics: Mechanical response, cell adhesion, and molecular deformation. Annu Rev Biomed Eng 2: 189–226.

Zubkov YN, BM Nokiforov, VA Shustin (1984) Balloon catheter technique for dilatation of constricted cerebral arteries after aneurysmal SAH. Acta Neurochir 70: 65–79.

Zwolak RM, MC Adams, AW Clowes (1987) Kinetics of vein graft hyperplasia: Association with tangential stress. J Vasc Surg 5: 126–136.

Index

Acceleration
  Cartesians 336, 338
  Convective 337
  Cylindricals 339
  Local 337
  Sphericals 340
Actin 437
Adventitia 121, 305
Aneurysms 124, 132, 305, 571
Angiogenesis 119
Angiography 119
Angioplasty
  Balloon 119, 150, 444, 615
  Neuroangioplasty 293
Angular velocity 341
Anisotropy 90
Anticlastic bending 228
Apoptosis 16
Arterial grafts 444, 502
Artery 304, 309
Atherosclerosis 305, 444
Atomic force microscope 116, 236, 276

Beams
  Curved 268
  Deflections 227
  Shear/bending moment diagrams 203
  Straight 203–245
  Stresses in 211–218
Bernoulli equation 396, 403

Bessel functions 470
Biaxial experiments 287, 560
Bifurcations 515
Bingham plastics 362
Biomechanics
  Definition 3, 612
  Father of 3
  Scope of 3, 5–8
Bioreactor 130, 453, 484
Blood
  Cells 360
  Clotting 359
  Constituents 359–361
  Constitutive relation 358
  Viscosity 357
Body force 383
Bone
  Cancellous 95
  Cortical 95
  Mechanobiology 154
  Properties 95
  Structure 94, 156
Boundary conditions
  Cantilever support 30, 229
  Free-end 229
  Free-surface 441, 480
  No-slip 352, 431
  Simple-support 229
Boundary layers 546
Buckingham Pi Theorem 519
Bulk modulus 99
Buoyant force 391

Cable 30
Cardiac cycle 466
Cardiopulmonary resuscitation 268
Cartesian coordinates
    Acceleration 338
    Deformation gradient 272
    Equilibrium 106
    Hooke's Law 87
    Navier-Poisson 353
    Navier–Stokes 387
    Vorticity 344
    Shear rates 348
    Strain 70
Cartilage 365, 602
Catheters 293, 483, 486, 509
Cell
    Cytoskeleton 12
    Mechanics 23, 115, 275, 613
    Structure 11, 236
Cell types
    Blood cells 360
    Endotheliel cells 427, 436
    Fibroblasts 22
    Platelets 357, 361
Centroids 142–147
Chordae tendineae 86, 110, 114
Continuum mechanics 8–11, 16, 24,
    26
Collagen 17–19
Columns 202, 245–254
Conservation of
    Mass 379, 400, 498
    Linear momentum 379, 499
    Energy 379, 499
Constitutive relations
    Bone 96
    General formulation 25, 81, 351
    Fung exponential 279, 285, 291
    Hooke's law 87, 91
    Mooney-Rivlin 328
    Navier–Poisson equation 353
    Neo-Hookean 298
    Newtonian 351
    non-Newtonian 473
Coanda effect 548
Control volume 495
Coordinate systems 28, 48, 52, 54, 72
CPR 269
Couette flow 351, 440

Couple 28
Creep 584
Curl 344
Curvatures 124, 297
Cylindrical coordinates 54
    Acceleration 339
    Deformation gradient 310
    Equilibrium 107
    Hooke's law 92
    Navier–Stokes 388
    Vorticity 345
    Shear rates 349
    Strain 72
    Stress components 54–55
Cytochalasin B 437

Darcy's law 563
Deformation gradient 271, 310
DEICE 25, 82, 90, 290, 351
Del operator
    Cartesian 337, 370
    Cylindrical 372
    Spherical 375
Delamination 96
Denude 506
Design
    Optimal biological 513–518
    Transducer 189
Determinant 324
Diagrams
    Bending 203–211
    Control volume 495
    Free-body 29
    Shear force 203–211
Differential equations 419–421
Diffusion 493
Dilatant fluid 356, 358
Displacement 67
Displacement gradient 69–70
Divergence
    Of a vector 370, 375
    Theorem 373–374
Ductile 200, 226
Duffing equation 567

Elastic behavior 82
Elastin 17, 20

Elastohydrodynamics 601
Elastodynamics 571–580
Endothelial Cells 7, 22, 427, 436
Endothelin 427
Energy equation
    Differential equation 545
    Pipe-flow equation 531
Entrance length 433, 546
Equilibrium 29, 105
Eulerian description 338
Euler's equation 392
Experimental
    Designs 188, 239, 287, 460, 518
    Needs 119
    Set-ups 102, 411, 507, 538
Extracellular matrix 16–22
Eye 132, 301–303

Failure (material) 45, 244
Fatigue 244
Fibroblasts 22
Finite elements 166
First law of thermodynamics 379, 499,
    528, 543, 601
Flexure formula 216
Flow
    1-D, 2-D, 3-D 341
    Creeping 389
    Secondary 405
    Separation 405
    uni-directional 341
    uniform 546
Fluid 331
Fourier series 466, 573
Fracture toughness 187
Free-body diagram 29
Free surface 441, 480
Friction
    Belt 34–36
    Coefficient of 34, 597
    Factor (fluids) 532–535
    Static 33
Fully developed flow 433

Gradient
    Del operator 372
    Green strain 70

Growth and remodeling 613–614
Growth factors 16, 22

Haversian canals 156
Heart 120, 170
Heat equation 545
Hematocrit 358
Hemodynamics 359
Hemoglobin 360
Homogeneous behavior 83
Hooke's law 87–92
Hoop stress 121
Hydraulic accumulator 549
Hydrodynamic lubrication 597
Hydrostatic pressure 306, 333, 355, 393
Hypertension 605, 614
Hysteresis 83

Ideal fluid 389, 396, 535
Identity matrix 272
Incompressibility 88, 306
Incompressible flow 346, 382
Induction 25
Initial condition 419
Integrins 15
Intermediate filaments 13
Interpolation 75, 289
Invariants 97, 286
Inverse of a matrix 283, 323
Inviscid fluid 355, 389
Irrotational flow 345, 395
Isotropy 84

Jacobian 323

Kinematic constraint 244
Kinetic energy 528
Kinetic energy coefficient 530
Knee 365, 597

Lagrangian description 336
Lamé constants 99
Laminar flow 433, 532
Laplace's equation 464

Laser tweezers 116
Least squares 292
Length
    Entrance 434, 448
    Scale 520
Lifelong learning 617
Linear momentum balance 379
Losses
    Major (viscous) 531
    Minor (geometric) 532
    Tabulated values 534
Lubrication theory 597–602
Lungs
    Basic anatomy 492–495
    Pulmonary dimensions 478
    Sheet flow in 525

Mach number 550
Macrophages 22
Manometer 412–414
Mass
    Balance 379, 400
    Density 334, 380
    Scales 520
Matrix operations 321–324
Mechanobiology 4, 557
Mechanotransduction 22–24, 81, 604
Media 121, 305
Membranes
    Biaxial stretching 287
    Definition 287
    Constitutive behavior 291
    Inflation 297
    Instability of 295–301
Metals 83, 97
Method of pins 37
Microgravity 453
Microscopy
    Atomic force 116, 236–239, 276
    Electron 170, 463
    Light 235, 309
Mixture theory 583, 602
Modulus
    Loss 594, 610
    Shear 88
    Storage 594, 610
    Young's 87
Mohr's circle 67

Molecular mechanics 116–118
Moment-curvature relation 215
Moment due to force 27
Moments of area
    First 142–147
    Second 194–197
    Polar 175, 196
Mullin's effect 294
Muscle 14
Myocardium 84, 190

Navier–Poisson relations 353
Navier–Space equation 109
Navier–Stokes equation
    Cartesians 387, 417
    Cylindricals 388, 418
    Sphericals 574
    Worksheets 417–418
Neuroangioplasty 293
Neutral axis 216
Newtonian fluid 353
Nitric oxide 427
Nondimensionalization 541
Non-Newtonian fluid 356, 473
No-slip condition 352, 431, 455, 462, 489

Optical tweezers 116
Orthostatic intolerance 453
Orthotropy 91
Osteoporosis 116

Papillary muscle 189–193
Parallel axis theorem 254–260
Parallel-plate flow 428
Pascal 424, 438, 461, 478
Pathline 392
Pericardium 202, 559
Permeability 559, 567, 603
Phase plane 570
Pitot tube 413
Plasma 357–359
Plasticity 85
Platelets 357, 361
Poise 358, 461, 478
Poiseuille flow 431, 450

Poisson's ratio 87, 89
(poly)methylmethacrylate 171, 187, 604
Power 513
Power-law fluid 473
Preconditioning 279
Pressure
  Absolute 390, 412
  Drop 485
  Dynamic 451
  Gauge 390, 443
  Hydrostatic 306, 333
  Stagnation 413
Principal values
  Strain 76
  Stress 62–66
Prosthesis (hip) 165, 184
Proteoglycans 17, 21
Pseudoelasticity 83, 278
Pseudoplastic 356
Pulley 33
Pulsatile flow 465–473

Rate-of-deformation 346–349
Receptors (cell surface) 12
Relaxation
  Function 589
  Spectrum 609
  Stress 585
Residual stress 303–315
Resultants 209
Retardation spectrum 609
Reynolds' number 436, 452, 524
Rheopectic 359
Rheology 361
Right-hand rule 172
Rigid-body motion 74
Rod 202
Rotations 73–74, 274, 341–345
Rouleaux 357
Rubber 83, 299–301

Serum 359
Scales for nondimensionalization 520
Shaft 202
Shear modulus 88, 99
Shear-rates 348–350

Sheet flow in lungs 525
Sickle cell 363
Sign convention 49, 55, 204, 435, 447
Skin friction 451
Smooth muscle cells 316
  Constitutive behavior 317
Specific gravity 402, 422
Specific weight 402
Spherical coordinates
  Acceleration 340
  Divergence 375
  Equilibrium 107
  Navier-Stokes 575
  Shear-rates 349
  Strain 72
  Stress components 54–55
  Vorticity 345
Stability
  Beams 248–254
  Dynamic 568
  General 246–248
  Inflated membranes 293–301, 571–582
Statically indeterminate 167, 184, 242
Statics 29–39
Stenosis 339
Stent 502
Stiffness 279
Streamline 392
Steady flow 337
Strain
  Gauges 78–80
  Green 70
  Infinitesimal (or small) 71, 74
  Microstrain 88, 198
  Plane 93
  Principal 76–78
  Shear 74
  Transformations 76–77
St. Venant's principle 111
Strain energy 100, 285
Stress
  Cauchy 47–50, 115, 148
  Concentration 227
  First Piola–Kirchhoff 115, 148, 283
  Hydrostatic 66
  Plane 93
  Principal 62–66

Relaxation 585
Residual 121, 307
Second Piola-Kirchhoff 285
States of 54
Transformations 56–60
Yield 201
Superposition 239
Surface force 283
Surface tension 552, 607
Synovial fluid 359, 363
System 495

Taylor series 104
Temperature
First law of thermodynamics
543–545
Viscosity dependence on 478
Tension 291
Theory 80–81, 399, 411
Thixotropic 359
Time, scales 520
Tissue engineering 121, 129, 444
Torsion 169–194
Transducer design
AFM 235
Load cells 262
Torque cells 189
Trace 569
Transformations (see strain and stress)
Transpose 322
Transverse isotropy 91
Tribology 582
Trigonometric identies 58
Truss 36–37
Turbulent flow 404, 433

Uniform flow 546
Uniform stress 112
Universal solutions 114, 118, 123, 128,
134

Vasospasm 293
Vectors
Curl 369
Divergence 369

Scalar product 368
Vector product 369
Vein grafts 119, 445, 557
Velocity 335
Velocity gradients 349
Ventricular-assist device 405
Viscoelasticity
Boltzmann model 592
Burger model 609
Characteristic behaviors 583
Kelvin–Voigt model 588–590
Maxwell model 586–588
Quasilinear model 596
Standard model 590–592
Vinculin 437
Viscometer
Cone-and-plate 366
Concentric cylinder 458–462
Descent of a sphere 458
Parallel plate 366, 368
Viscosity
Absolute 11, 352, 539
Apparent 362, 473
Kinematic 539
of blood 357
of plasma 359
of water 359
Vorticity
Cartesian 344
Cylindrical 345
Spherical 345
Volume 89
Volumetric flow 400

Wall shear stress 367, 435, 451, 460
Work 528–531
Working 530
Worksheets 417–418
Womersley's number 472

Yield criterion 201
Yield stress 85, 201
Young's modulus 87

Zero-stress state 307

About the Authors

Jay D. Humphrey is a Professor of Biomedical Engineering and a Fellow of the M.E. DeBakey Institute at Texas A&M University. He received the B.S. degree from Virginia Tech and the M.S. and Ph.D. degrees from The Georgia Institute of Technology, all in Engineering Science and Mechanics, and he completed a postdoctoral fellowship in Cardiovascular Research at Johns Hopkins University. Professor Humphrey has authored a graduate textbook entitled *Cardiovascular Solid Mechanics: Cells, Tissues, and Organs*, coedited a book entitled *Cardiovascular Soft Tissue Biomechanics*, and published over 100 journal articles and book chapters. He is Co-Editor of the international journal *Biomechanics and Modeling in Mechanobiology* and Fellow of the American Institute for Medical and Biological Engineering.

Sherry L. Delange is a recent graduate of Texas A&M University. An honors student, she received her B.S. degree in 2000 and her M.S. degree in 2002, both in Biomedical Engineering. Her M.S. thesis focused on the biomechanics of the lens capsule, the membrane that invests the lens of the eye and is vital to accommodation. She was the first to measure the multi-axial mechanical properties of the lens capsule and the first to show regional differences in vitro in the multiaxial finite strains.